Direct Microbial Conversion of Biomass to Advanced Biofuels

Direct Microbial Conversion of Biomass to Advanced Biofuels

Edited by

Michael E. Himmel

Biosciences Center, National Renewable Energy Laboratory (NREL), Golden, CO, USA

ELSEVIER

AMSTERDAM • BOSTON • HEIDELBERG • LONDON • NEW YORK • OXFORD
PARIS • SAN DIEGO • SAN FRANCISCO • SINGAPORE • SYDNEY • TOKYO

Elsevier
Radarweg 29, PO Box 211, 1000 AE Amsterdam, Netherlands
The Boulevard, Langford Lane, Kidlington, Oxford OX5 1GB, UK
225 Wyman Street, Waltham, MA 02451, USA

Notices
Knowledge and best practice in this field are constantly changing. As new research and experience broaden our
understanding, changes in research methods, professional practices, or medical treatment may become necessary.

Practitioners and researchers must always rely on their own experience and knowledge in evaluating and using any
information, methods, compounds, or experiments described herein. In using such information or methods they
should be mindful of their own safety and the safety of others, including parties for whom they have a professional
responsibility.

To the fullest extent of the law, neither the Publisher nor the authors, contributors, or editors, assume any liability
for any injury and/or damage to persons or property as a matter of products liability, negligence or otherwise, or
from any use or operation of any methods, products, instructions, or ideas contained in the material herein.

ISBN: 978-0-444-59592-8

British Library Cataloguing in Publication Data
A catalogue record for this book is available from the British Library

Library of Congress Cataloging-in-Publication Data
A catalog record for this book is available from the Library of Congress

For information on all Elsevier publications
visit our website at http://store.elsevier.com/

Working together
to grow libraries in
developing countries

ELSEVIER   Book Aid
             International

www.elsevier.com • www.bookaid.org

Transferred to Digital Printing in 2015

Contents

Contributors

Hal S. Alper   Department of Chemical Engineering, The University of Texas at Austin, Austin, Texas, USA; Institute for Cellular and Molecular Biology, The University of Texas at Austin, Austin, Texas, USA

Edward A. Bayer   Department of Biological Chemistry, The Weizmann Institute of Science, Rehovot, Israel

Mary J. Biddy   National Bioenergy Center, National Renewable Energy Laboratory (NREL), Golden, CO, USA

Yannick J. Bomble   Biosciences Center, National Renewable Energy Laboratory (NREL), Golden, CO, USA

Steven D. Brown   Biosciences Division, Oak Ridge National Laboratory, Oak Ridge, TN, USA; BioEnergy Science Center, Oak Ridge National Laboratory, Oak Ridge, TN, USA; Bredesen Center for Interdisciplinary Research and Graduate Education, University of Tennessee, Knoxville, TN, USA

Roman Brunecky   Biosciences Center, National Renewable Energy Laboratory (NREL), Golden, CO, USA

Nicholas C. Carpita   Purdue University, Botany and Plant Pathology, West Lafayette, IN, USA

Ryan E. Davis   National Bioenergy Center, National Renewable Energy Laboratory (NREL), Golden, CO, USA

R. den Haan   Department of Biotechnology, University of the Western Cape, Bellville, South Africa

Bryon S. Donohoe   Biosciences Center, National Renewable Energy Laboratory (NREL), Golden, CO, USA

Christopher K. Dugard   Purdue University, Botany and Plant Pathology, West Lafayette, IN, USA

Muhammad Ehsaan   Clostridia Research Group, Centre for Biomolecular Sciences, BBSRC Sustainable Bioenergy Centre, Nottingham Digestive Diseases Centre NIHR Biomedical Research Unit, School of Life Sciences, University of Nottingham, University Park, Nottingham, UK

Adam M. Guss   Biosciences Division, Oak Ridge National Laboratory, Oak Ridge, TN, USA; BioEnergy Science Center, Oak Ridge National Laboratory, Oak Ridge, TN, USA; Bredesen Center for Interdisciplinary Research and Graduate Education, University of Tennessee, Knoxville, TN, USA

Michael E. Himmel   Biosciences Center, National Renewable Energy Laboratory (NREL), Golden, CO, USA

Sarah E. Hobdey   Biosciences Center, National Renewable Energy Laboratory (NREL), Golden, CO, USA

Rumana Islam   Department of Biosystems Engineering, University of Manitoba, Winnipeg MB, Canada

David K. Johnson Biosciences Center, National Renewable Energy Laboratory (NREL), Golden, CO, USA

Katalin Kovács Clostridia Research Group, Centre for Biomolecular Sciences, BBSRC Sustainable Bioenergy Centre, Nottingham Digestive Diseases Centre NIHR Biomedical Research Unit, School of Life Sciences, University of Nottingham, University Park, Nottingham, UK

Nathan Kruer-Zerhusen Department of Molecular Biology and Genetics, Cornell University, Ithaca, NY, USA

Wouter Kuit Clostridia Research Group, Centre for Biomolecular Sciences, BBSRC Sustainable Bioenergy Centre, Nottingham Digestive Diseases Centre NIHR Biomedical Research Unit, School of Life Sciences, University of Nottingham, University Park, Nottingham, UK

D.C. la Grange Department of Biochemistry, Microbiology and Biotechnology, University of Limpopo, Sovenga, South Africa

Sadhana Lal Department of Biosystems Engineering, University of Manitoba, Winnipeg MB, Canada

Sun-Mi Lee Department of Chemical Engineering, The University of Texas at Austin, Austin, Texas, USA; Clean Energy Research Center, Korea Institute of Science and Technology, Seongbuk-gu, Korea

David B. Levin Department of Biosystems Engineering, University of Manitoba, Winnipeg MB, Canada

James C. Liao Department of Chemical and Biomolecular Engineering, University of California, Los Angeles, California, USA

Xiaoxia N. Lin Department of Chemical Engineering, University of Michigan, Ann Arbor, MI, USA

Maureen C. McCann Purdue University, Biological Sciences, West Lafayette, IN, USA

Nigel P. Minton Clostridia Research Group, Centre for Biomolecular Sciences, BBSRC Sustainable Bioenergy Centre, Nottingham Digestive Diseases Centre NIHR Biomedical Research Unit, School of Life Sciences, University of Nottingham, University Park, Nottingham, UK

Jeremy J. Minty Department of Chemical Engineering, University of Michigan, Ann Arbor, MI, USA

Ashutosh Mittal Biosciences Center, National Renewable Energy Laboratory (NREL), Golden, CO, USA

Luc Moens National Bioenergy Center, National Renewable Energy Laboratory (NREL), Golden, CO, USA

Sarah Moraïs Department of Biological Chemistry, The Weizmann Institute of Science, Rehovot, Israel

Riffat Munir Department of Biosystems Engineering, University of Manitoba, Winnipeg MB, Canada

Jessica Olstad National Bioenergy Center, National Renewable Energy Laboratory (NREL), Golden, CO, USA

Bryan W. Penning Purdue University, Biological Sciences, West Lafayette, IN, USA

Heidi Pilath National Bioenergy Center, National Renewable Energy Laboratory (NREL), Golden, CO, USA

Umesh Ramachandran Department of Biosystems Engineering, University of Manitoba, Winnipeg MB, Canada

S.H. Rose Department of Microbiology, University of Stellenbosch, Stellenbosch, South Africa

Kyle B. Sander Biosciences Division, Oak Ridge National Laboratory, Oak Ridge, TN, USA; BioEnergy Science Center, Oak Ridge National Laboratory, Oak Ridge, TN, USA; Bredesen Center for Interdisciplinary Research and Graduate Education, University of Tennessee, Knoxville, TN, USA

Christopher J. Scarlata National Bioenergy Center, National Renewable Energy Laboratory (NREL), Golden, CO, USA

John Schellenberg Department of Microbiology, University of Manitoba, Winnipeg MB, Canada

Katrin Schwarz Clostridia Research Group, Centre for Biomolecular Sciences, BBSRC Sustainable Bioenergy Centre, Nottingham Digestive Diseases Centre NIHR Biomedical Research Unit, School of Life Sciences, University of Nottingham, University Park, Nottingham, UK

Arjun Singh National Bioenergy Center, National Renewable Energy Laboratory (NREL), Golden, CO, USA

Richard Sparling Department of Microbiology, University of Manitoba, Winnipeg MB, Canada

Jennifer L. Takasumi Department of Chemical and Biomolecular Engineering, University of California, Los Angeles, California, USA

Eric C.D. Tan National Bioenergy Center, National Renewable Energy Laboratory (NREL), Golden, CO, USA

Ling Tao National Bioenergy Center, National Renewable Energy Laboratory (NREL), Golden, CO, USA

Melvin P. Tucker National Bioenergy Center, National Renewable Energy Laboratory (NREL), Golden, CO, USA

W.H. van Zyl Department of Microbiology, University of Stellenbosch, Stellenbosch, South Africa

Tobin J. Verbeke Department of Microbiology, University of Manitoba, Winnipeg MB, Canada

Todd B. Vinzant Biosciences Center, National Renewable Energy Laboratory (NREL), Golden, CO, USA

Lawrence P. Wackett Department of Biochemistry, Molecular Biology and Biophysics, University of Minnesota, Minneapolis, MN, USA

Wei Wang Biosciences Center, National Renewable Energy Laboratory (NREL), Golden, CO, USA

Hui Wei Biosciences Center, National Renewable Energy Laboratory (NREL), Golden, CO, USA

Carrie M. Wilmot Department of Biochemistry, Molecular Biology and Biophysics, University of Minnesota, Minneapolis, MN, USA

David B. Wilson Department of Molecular Biology and Genetics, Cornell University, Ithaca, NY, USA

Klaus Winzer Clostridia Research Group, Centre for Biomolecular Sciences, BBSRC Sustainable Bioenergy Centre, Nottingham Digestive Diseases Centre NIHR Biomedical Research Unit, School of Life Sciences, University of Nottingham, University Park, Nottingham, UK

Edward J. Wolfrum National Bioenergy Center, National Renewable Energy Laboratory (NREL), Golden, CO, USA

Chia-Wei Wu Biosciences Division, Oak Ridge National Laboratory, Oak Ridge, TN, USA

Qi Xu Biosciences Center, National Renewable Energy Laboratory (NREL), Golden, CO, USA

Shihui Yang   National Bioenergy Center, National Renewable Energy Laboratory (NREL), Golden, CO, USA

John M. Yarbrough   Biosciences Center, National Renewable Energy Laboratory (NREL), Golden, CO, USA

Eric M. Young   Department of Chemical Engineering, The University of Texas at Austin, Austin, Texas, USA

Min Zhang   National Bioenergy Center, National Renewable Energy Laboratory (NREL), Golden, CO, USA

Ying Zhang   Clostridia Research Group, Centre for Biomolecular Sciences, BBSRC Sustainable Bioenergy Centre, Nottingham Digestive Diseases Centre NIHR Biomedical Research Unit, School of Life Sciences, University of Nottingham, University Park, Nottingham, UK

Foreword

The outlook for affordable next-generation or advanced fuels from biomass is as complex and multidimensional as are the possible routes for its attainment. Advanced biofuel is defined as follows by the Cornell University Law School's Legal Information Institute: "The term 'advanced biofuel' means fuel derived from renewable biomass other than corn kernel starch." Direct microbial conversion (DMC), also referred to as consolidated bioprocessing, is a promising biomass processing strategy originally introduced by Professor L. Lynd (Dartmouth College) because it reduces process complexity and energy input requirements compared with classical simultaneous saccharification and fermentation. Although a powerful strategy, DMC is currently limited by three key process engineering and scientific considerations: (1) the titers of biomass-degrading enzymes produced by aerobic, non-naturally cellulolytic DMC microorganisms have not yet reached that of dedicated enzyme production hosts; (2) aerobic DMC production pathways, required primarily for high-yield biosynthesis of hydrocarbons and lipids, may be difficult to economically scale up to large volumes (e.g., >1 ML aerobic bioreactors) because of challenging gas–liquid mass transfer requirements at such scales; and (3) metabolic pathways in anaerobic, naturally cellulolytic DMC microbes are not optimized for fuel production. For example, it is not known if for DMC processes cellulase titers must rival the ultra-high levels obtainable by dedicated enzyme production strains. One envisioned process scheme for a DMC-capable fungus or yeast would be a high volumetric inoculation into an aerobic culture containing biomass slurry to produce sufficient hydrolytic enzymes, followed by forced anaerobiosis for fermentation of the biomass sugars to ethanol or related products. Another DMC process scheme envisions large-scale anaerobic cultures of primarily cellulolytic thermophiles engineered to produce ethanol or butanol at economically relevant titers. Moreover, to date, highly reduced, deoxygenated products requiring cellular respiration (Krebs cycle) for their production have been produced only in modest titers and scales; challenges also remain in economically recovering such products.

To outline the current state of the art, this book is composed of three sections. First, overviews of the benefits from consolidated fermentations are discussed in the context of green processes (Wei et al., Chapter 1), followed by a detailed review of microbial hydrocarbon production (Wackett et al., Chapter 2) and then by a technoeconomic analysis of advanced biofuels production (Scarlata et al., Chapter 3). In the second section, **Biomass Structure**

and Recalcitrance, the outlook for engineering biomass for advanced fuels production is presented (McCann et al., Chapter 4), which sets the stage for four reviews highlighting recent advances in understanding and engineering improved cellulase enzymes (Wilson, Kruer-Zerhusen, and Wilson; Hobdey et al., and Morais et al., Chapters 5–8, respectively). In the third section, **Fuels from Fungi and Yeast**, van Zyl et al. review advances in engineering cellulase production in yeast (Chapter 9), followed by Chapters by Yang et al. and Xu et al., Chapters 10 and 11, respectively, discussing the suitability of the cellulolytic fungus, *Trichoderma reesei*, as a DMC host. In Chapter 12 by Lee et al., metabolic engineering challenges for yeasts are presented. In the final section, **Fuels from Bacteria**, Schwarz et al. (Chapter 13) discuss in considerable detail the recent development and outlook for new tools for genetic manipulation of Clostridia. On a related topic, Takasumi and Liao present the outlook specifically for butanol production from cellulolytic Clostridia in Chapter 14. Yarbrough et al. (Chapter 15) next report the effects of particle size on DMC of poplar wood by *Clostridium thermocellum*, followed by a review from Brown et al. regarding the metabolic pathway engineering required for DMC to ethanol by *C. thermocellum* (Chapter 16). The next two chapters describe the challenges and potential solutions for advanced fuels production using co-cultures, first from an "omics" perspective (Levin et al., Chapter 17) and then from a pathway engineering and fermentation model point of view (Minty et al., Chapter 18). A novel route to hydrocarbons via chemical synthesis from microbially produced polyhydroxybutyrate closes the book (Chapter 19).

The editor thanks the authors for their contributions, as well as the Department of Energy (DOE) Bioenergy Technologies Office and the DOE Office of Biological and Environmental Research through the BioEnergy Science Center for research support. I also acknowledge Peter Ciesielski (National Renewable Energy Laboratory) for the original art used for the cover of this book.

Michael E. Himmel
Biosciences Center, National Renewable Energy
Laboratory (NREL), Golden, CO, USA

PART 1

Direct Microbial Conversion of Biomass to Advanced Biofuels

Feedstock Engineering and Biomass Pretreatments: New Views for a Greener Biofuels Process

Hui Wei[1], Wei Wang[1], Melvin P. Tucker[2], Michael E. Himmel[1], Roman Brunecky[1]
[1]*Biosciences Center, National Renewable Energy Laboratory (NREL), Golden, CO, USA;*
[2]*National Bioenergy Center, National Renewable Energy Laboratory (NREL), Golden, CO, USA*

Feedstock Engineering Aiming to Provide More Pretreatable and Digestable Biomass

Throughout this book, authors discuss microbial processing technologies for biomass conversion to fuels and chemicals as well as challenges and opportunities for biomass production. Concepts regarding green technologies have not been emphasized. In general, a green process is defined as a production process with the lowest consumption of resources while also avoiding or minimizing the use and generation of chemicals hazardous to the environment.[1,2]

Note that the use of colors to describe this terminology is not restricted to the case of biomass conversion. In fact, because of today's communication needs with the media, government, and the public, color-coding has been used to distinguish different topics in biotechnology. Whereas green biotechnology refers to technologies applied to Agri-food processes, other colors are used to describe the application to marine and aquatic processes (blue biotechnology), medical processes (red biotechnology), and general industrial processes (white biotechnology), respectively.[3–5] Of note, the principles of the above-mentioned green process and green chemistry are important for effective, "sound-bite" style communication to the public and the government necessary for raising their awareness and rallying their support.

The aforementioned definition of green processes prompts us to propose that the green biofuels process should include at least the following components:

1. "Green production" of feedstocks, with low consumption of water, fertilizer, and energy and yet generating biomass with maximal pretreatability and digestibility.
2. "Green pretreatment" of biomass, with low energy consumption, little or no use of hazardous chemicals, and no generation of hazardous waste.

Direct Microbial Conversion of Biomass to Advanced Biofuels. http://dx.doi.org/10.1016/B978-0-444-59592-8.00001-4

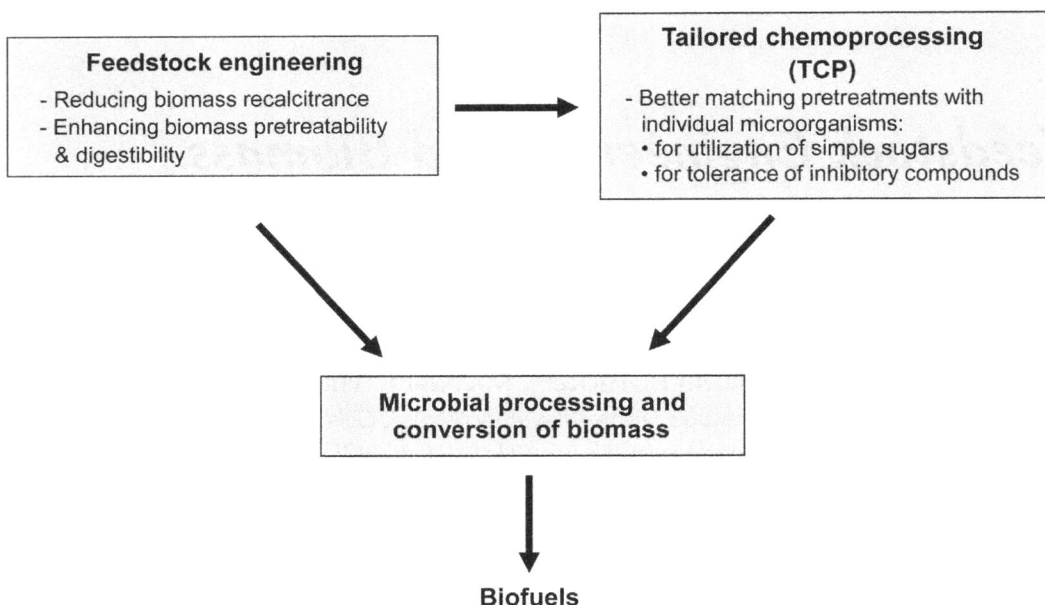

Figure 1
Scheme of project flow for integrating feedstock engineering and chemical/microbial processes that enable more efficient biomass conversion to biofuels.

Feedstock engineering is an integral part of the green biofuels process. The goal of feedstock engineering is to generate novel bioenergy crops to produce biomass with traits designed for easier downstream processing of biomass during pretreatment and/or digestion steps in a bioconversion process (Figure 1). Traditional methods for introducing exogenous enzymes to biomass particles, as used today in simultaneous saccharification and fermentation (SSF) processes are limited by multilength scale diffusion barriers, from the level of the biomass chip to the cellular structure of the plant cell wall. There are multiple macroscale and microscale factors believed to contribute to the recalcitrance of lignocellulosic feedstocks to thermochemical pretreatment and subsequent enzymatic saccharification. On the gross anatomical level, macroscale factors include plant structural effects, such as the epidermal tissue protecting the plant stem, the arrangement and density of the vascular bundles in cell walls, and the relative amount of sclerenchymatous (thick wall) plant tissues. On a microscale, important factors include the degree of lignification as well as the structural heterogeneity and complexity of cell-wall constituents, such as cellulose microfibrils and matrixing polymers, including the hemicelluloses and pectins. In the context of the biorefinery, these chemical and structural features of biomass affect liquid penetration and/or enzyme accessibility and activity and ultimately, conversion costs.[6]

The current state of the art for the biofuels industry relies on multiple processing steps to achieve the conversion of lignocellulosic biomass to a liquid fuel, such as but not limited to ethanol. Each processing step also has a cost-per-gallon fuel cost associated with it. The two primary processing steps under consideration today—thermochemical pretreatment and the enzymatic conversion—are significant contributors to the overall minimum selling price of biofuels and can be affected by reduced recalcitrance plant technology.[7–9]

These two bottlenecks for cost-effective conversion are detailed in a 2009 report that states, "In order to further reduce costs, process improvements must be made in several areas, including pretreatment, enzymatic hydrolysis, and fermentation."[7] Furthermore, in their very recent article, Klein-Marcuschamer et al. stated, "Analysis shows that, in general, the vast majority of the literature to date has significantly underestimated the contribution of enzyme costs to biofuel production."[9] This highlights the sensitivity of the current industry to enzyme and pretreatment costs, and although there have been significant advances in enzyme technology over the years, the cost of enzymes remains a key issue and the cost of pretreatment remains high.

In Planta Engineering for Reduced Recalcitrance Traits

To date, no recombinant plant technology with reduced recalcitrance traits is utilized by the biofuels industry on a commercial scale. However, there have been early attempts to reduce the enzyme costs associated with a biorefinery. The primary idea has been to express large amounts of glycoside hydrolase (GH) enzymes in planta, thus shifting the cost of enzyme production from expensive fungal sources, which is the current method of production, to a cheaper, plant biofactory model in which the enzymes necessary for enzymatic deconstruction are expressed within the plants themselves.[10–12] Although an interesting approach thus far, it has not met with commercial success and remains problematic in several senses. For example, the large amounts of enzymes required place metabolic burdens on the plant and require additional inputs of nitrogenous fertilizers. Furthermore, it is logical that expression of large amounts of highly active plant deconstructing enzymes is deleterious to plant health and is a critical consideration when expressing active GHs in planta. In addition, from a process prospective, this approach incurs additional capital and operating costs by adding the very expensive extra processing steps of having to first extract the GH enzymes from the "plant factories" intact and active and then add them back in at a later step.

A recent workaround to the problems of expressing large amounts of GH enzymes is to use GH enzymes that are expressed in planta in an inactive form using intein self-excising elements in an attempt to avoid negative effects from the heterologous GH enzymes. In this case, enzymes are not extracted from the plant tissue, as in the previous approach. Rather, after the plant is fully grown and senesced, a low-temperature and low-pressure pretreatment from which the enzymes can survive is used. Pretreatment then "activates" the enzymes utilizing various "intein-trigger" mechanisms, which then deconstruct the plant.[13]

Another promising example of reduced recalcitrance technology is to utilize plants that express low levels of GH enzymes that are expressed and targeted to cell walls during plant growth and development. Brunecky et al. demonstrated a 15% reduction in the recalcitrance of corn stover and tobacco plants by expressing the thermostable endoglucanse E1 in planta at very low enzyme titers (ng cellulase/mg tissue) in the wall.[14] This approach provided significant improvements in the digestibility of the engineered plants, and by utilizing very low levels of expressed cellulase, no negative phenotypes were observed in the growing plants.

Another recent direction in feedstock research is the demonstration that modifying or reducing the level of lignin in the plant also reduces pretreatment requirements. By altering the lignin content and composition by independently targeting multiple steps of the lignin pathway, it was shown that lignin content in alfalfa is inversely related to recalcitrance.[15] By targeting caffeic acid *O*-methyltransferase (COMT), it was possible to improve cell-wall enzymatic saccharification efficiency without a reduction in postharvest biomass yield in switchgrass,[16] and this technology has been extended to target transcriptional regulators of the lignin pathway (PvMYB4, *Panicum virgatum* MYB4) with even greater reductions in recalcitrance.[17] The COMT lines are already in commercial field trials, and both the COMT and MYB4 lines have been shown to possess reduced recalcitrance and support enhanced ethanol yields at reduced enzyme loadings. Note that in general, 'classical' MYB (myeloblastosis) transcription factors are involved in the control of the cell cycle in higher eukaryotes, whereas plant MYB transcription factors can be involved in many aspects of plant secondary metabolism, as well as cell morphogenesis or cell fate.

Chen et al. have reported that various modifications to the lignin biosynthesis pathway in the model crop alfalfa yielded on the order of 20–40% improvements in enzymatic hydrolysis efficiency after a mild acid pretreatment, and perhaps a 5% improvement in sugar release improvement using only a mild acid pretreatment.[15] Some of these lignin modifications have also been reported in bioenergy crops; in their recent paper, Fu et al. showed that compared with control plants, transgenic switchgrass with COMT downregulation showed significant increases in saccharification efficiency with or without mild acid pretreatment.[16] They showed that the transgenic plants had a 16.5–21.5% increase in saccharification efficiency with mild pretreatment and a 29.2–38.3% increase without any pretreatment.

Furthermore, in a traditional SSF fermentation scheme, these plants yielded 30–38% more ethanol by SSF compared with control plants. Overexpression of the MYB4 lignin repressor in switchgrass gives even greater reductions in recalcitrance.[17] Given that the approaches utilize unique and distinct mechanisms of action, we believe that in the future expressing GHs in a COMT-switchgrass-deficient plant should result in a significant improvement in saccharification efficiency, even when assuming that these effects are nonsynergistic. However, it is well known that delignification of biomass is highly synergistic with our proposed combination of GH expression and lignin modification; thus, the actual improvement may be higher.[18]

Therefore, we support the notion that the introduction of reduced recalcitrance feedstocks would have a transformational, not incremental, effect on reducing the overall costs of the bioconversion of lignocellulosic feedstocks. A key advantage of this solution is that it is a downstream drop-in fit for current and proposed future bio-based fuel production processes, requiring only minimal changes to plant design and operating procedures.

Mild and Green Pretreatments of Biomass for Lower Toxicity in Lignocellulosic Hydrolysates and Solid Residues

Current pretreatment technologies are designed to achieve the highest yield of fermentable substrates (including simple sugars) from biomass. To achieve this goal, the so-called pretreatment severity factor, based on treatment pH, temperature, and reaction time, is tailored to the process objectives.[19,20] For example, most pretreatment schemes strive to improve substrate "accessibility," now known to be a key factor affecting substrate–enzyme interactions.[21–26] Going forward, another goal of pretreatment technology will be to improve substrate fractionation, depending upon the needs of new process designs. In addition to producing fermentable substrates, most pretreatment processes today also generate compounds inhibitory or toxic to fermentative organisms. These compounds are generally the products of sugar and lignin degradation that include furfural, hydroxymethyl furfural, soluble phenolics, and a host of sugar-lignin condensation products that are not fully characterized.[27]

Another dilemma is that the higher severity of pretreatments will also result in a higher extent of degradation of sugars that inhibit the downstream enzymatic hydrolysis and fermentation steps. Therefore, there is a delicate balance in controlling the pretreatment conditions between maximizing the sugar yield and minimizing the sugar degradation and inhibitory compound formation. Therefore, from the standpoint of toxicity mitigation and cost, a mild and environmentally benign pretreatment process would be a promising future improvement.

One approach to developing a mild, yet effective pretreatment technology is by designing the optimized pretreatment reactor vessels for this goal. From recent work, it is clear that more insights into the effects of reactor design related to biomass digestibility are needed.[28] For example, we reported that corn stover, acid-pretreated under the same severity but in three different types of reactors (i.e., ZipperClave, Steam gun, and Horizontal reactor), exhibited different enzymatic digestibility. The corn stover pretreated in the Horizontal reactor and Steam gun achieved much higher enzymatic digestions, 95% and 88% cellulose converted to glucose, respectively, after 96 h, compared with 69% for the ZipperClave pretreated sample. Among the chemical and physical characteristics examined, particle size varied the most among the three treated samples. The Horizontal reactor treated sample produced the smallest particle size distribution, which is directly related to cellulose conversion. Microscopic analysis showed a more delaminated and defibrillated structure for pretreated samples from the Horizontal reactor, which was likely due to the shearing effect of the reactor's internal screws. This study indicates that reactor designs that augment the thermal and chemical energy applied to the pretreatment of biomass with mechanical energy can substantially aid in overcoming the recalcitrance of biomass through the breakdown of the physical structure of the plant cell wall.

These results also support results from previous research showing that increasing substrate accessibility is critical to increasing the efficiency of enzymatic hydrolysis.

Given the aforementioned results, we conclude that it is possible to produce a highly digestible substrate at moderate chemical pretreatment severities by utilizing appropriate reactor design. For example, recalcitrance can be reduced by integrating an accessory mechanical stage into the reactor, which will increase cell wall delamination and defibrillation. An example of this approach is the integrated mild alkali pretreatment and mechanical refining process proposed recently by National Renewable Energy Laboratory researchers.[29] The digestibility of the feedstock generated using this process was comparable to that generated by a dilute-acid pretreatment process at higher severity.

In addition to the favorable digestibility of the feedstock, a great advantage of a green pretreatment process is that it is environmentally friendly and microorganism-benign. The use of high concentrations of chemicals in pretreatment processes leads to the corrosion of the reactors, lines, and/or pumps and the additional economic burdens of chemical recycling or disposal. Usage of high concentrations of acids and bases also leads to high process water salinity, increasing the overall plant water requirement. Milder pretreatments may also somewhat mitigate the generation of toxic compounds from biomass.

A New Concept of Tailored Chemoprocessing for Individual Microorganisms

It is known that different pretreatment processes generate various toxic compounds and that biofuel-producing microorganisms have various levels of tolerance to these inhibitory compounds. Matching the reaction design with the appropriate conversion microorganism is an obvious opportunity that is made more attractive by modern microbial genetic engineering. Here, we propose the concept of tailored chemoprocessing (TCP) using platform industrial microorganisms.

There are two layers of compatibility between pretreatments and individual microorganisms:

1. The first layer is the carbohydrate compatibility between the array of simple sugars that pretreatments generate and the set of sugars that microorganisms can utilize. This should be a straightforward analysis to perform using modern computational methods.
2. The second layer of compatibility between pretreatments and individual microorganisms is the toxicity aspect. A simplified example is that if a strain is not sensitive to phenolic compounds, then an alkaline pretreatment process could be a good viable match. However, to better match toxicity parameters between pretreatments and microorganisms, a more rational strategy is required.

One particular approach we emphasize is to replace the existing chemical(s) used in pretreatment with chemicals that have similar effectiveness during pretreatment but display lower inhibition in downstream enzymatic hydrolysis or lower toxicity to microbial growth. The assumption is that different chemicals within the same category may have similar function in targeting plant cell-wall components and cleaving certain chemical linkage bonds, but they vary in the level of toxicity to downstream microbial fermentation. For example, a recent review compiles a list of halotolerant cellulases produced by *Bacillus* sp. and *Martelella mediter-ranea* that are enhanced by some metal ions, such as Fe^{2+} and Cu^{2+}, but inhibited by other metal ions, such as Cd^{2+} and Co^{2+}.[30] These results suggest that compared with Cd^{2+} and Co^{2+}, Fe^{2+} and Cu^{2+} are more suitable for use in dilute-acid/metal co-catalyzed biomass pretreatments,[31,32] which will better match the downstream conversion of biomass to fuels using microorganisms that produce enzymes with similar metal ion sensitivity profiles.

Building Unified Chemobiomass Databases and Libraries of Chemicals

There are currently multiple public databases available related to plant cell wall-related genes and proteins; biomass chemical compositions; and the chemical reactions, interactions, and processes of general chemicals (Table 1). Among them, the databases available for plant cell wall proteins and biomass-degrading enzymes—mainly Carbohydrate-Active enZymes (CAZy)—are useful sources for identifying candidate genes and proteins for the aforementioned feedstock engineering research aimed at providing more pretreatable and digestible biomass feeding into greener conversion processes.

Other than the two "generic" databases listed in Table 1 as well as the Chemical Thesaurus reaction chemistry database and the Ionic Liquids database (ILThermo), to the best of our knowledge, there are no chemical databases today designed specifically to provide information about chemical catalysts acting on biomass. To facilitate the development of the aforementioned TCP, here we propose building a unified Chemobiomass Database that focuses on an efficient collection and management of information related to the action of chemicals on biomass. To construct this database, one needs to systematically explore candidate chemical compounds and to identify all possible chemicals that can efficiently depolymerize cellulose, hemicellulose, and lignin. Especially important to target are the covalent linkages between cell wall polymers, such as the polysaccharide–polysaccharide glycosidic bonds (branching) and lignin-carbohydrate ester bonds.

In addition to the proposed construction of a Chemobiomass Database, this approach would require a library of chemicals and a library of biomass model substrates and/or derivatives. Together, they will provide a comprehensive chemical and molecular informatics framework for optimizing the chemicals used in pretreatments, which will lead to the generation of lower toxicity hydrolysates and residues for the downstream fermentation of individual biofuel-producing microorganisms.

Overall, the conversion of biomass to biofuels is an integrated, systematic process that starts with the feedstock engineering and optimal pretreatment technologies, which then lay the foundation for development of novel microbial technologies for conversion of sugars to fuels

Table 1: List of databases for plant cell wall-related genes and proteins, biomass-degrading enzymes, biomass chemical characterization, and the chemicals with potentials for biomass pretreatments.

Database	Note	Website
Plant Cell Wall-Related Genes and Proteins		
Cell wall genomics	A resource for genetic analyses of cell wall-related genes in *Arabidopsis* and maize	http://cellwall.genomics.purdue.edu/
Cell wall navigator	Protein families involved in plant cell wall metabolism	http://cellwall.ucr.edu/Cellwall/
WallProtDB	A collection of cell wall proteomic experimental data	http://www.polebio.scsv.ups-tlse.fr/WallProtDB/
Biomass-Degrading Enzymes		
Carbohydrate-active enzymes (CAZy)	Classification and associated information for the enzymes that assemble, modify, and break down oligo- and poly-saccharides	http://www.cazy.org
Cazymes analysis toolkit (CAT)	Tools for analyzing and annotating CAZYmes	http://mothra.ornl.gov/cgi-bin/cat/cat.cgi
Biomass Chemical Characterization		
Biomass feedstock composition and property database	Chemical, thermal, and mechanical properties of various biomass feedstock materials	http://www.afdc.energy.gov/biomass/progs/search1.cgi
Plant ionomics database	Mineral nutrient and trace element composition for *Arabidopsis*, maize, rice, and soybean, etc.	http://www.ionomicshub.org/home/PiiMS
General Chemical Databases with Potentials for Biomass Pretreatments		
Chemical thesaurus reaction chemistry database	Information about chemical entities, chemical reactions, interactions, and processes	http://www.chemthes.com
Ionic Liquids database (ILThermo)	Chemical and physical properties of ionic liquids related to various industrial applications	http://ilthermo.boulder.nist.gov/

production, as illustrated in the following chapters in this book. It is noteworthy that this biomass conversion process, at a high level, is a fully integrated, circular system in which the progress in microbial technology development will affect the direction taken by feedstock engineering, followed again by new rounds of microbial technology development until process economic targets are met.

Conclusions

The chemical and structural complexity of plant biomass contributes to the high cost of lignocellulosic biofuels produced by biochemical processes. Feedstock engineering to generate biomass with reduced recalcitrance can provide a raw materials foundation for better pretreatment and microbial conversion technologies. In addition, new green pretreatment technologies and the TCP concept, enabled by the proposed construction of new Chemobiomass Databases and physical libraries of biomass model substrates and derivatives, will enable detailed profile matching of individual biofuel-producing microorganisms and pretreatments for cost-effective advanced biofuels.

Acknowledgments

The review of in planta expression was funded by the Laboratory Directed Research and Development (LDRD) program at the National Renewable Energy Laboratory. The remainder of the work was funded by the Center for Direct Catalytic Conversion of Biomass to Biofuels (C3Bio), an Energy Frontier Research Center funded by the US Department of Energy, Office of Science, Office of Basic Energy Sciences under Award Number DE–SC0000997.

References

1. Diwekar U. Green process design, industrial ecology, and sustainability: a systems analysis perspective. *Resour Conserv Recycl* 2005;**44**:215–35.
2. Horvath IT, Anastas PT. Innovations and green chemistry. *Chem Rev* 2007;**107**:2169–73.
3. Lorenz P, Zinke H. White biotechnology: differences in US and EU approaches? *Trends Biotechnol* 2005;**23**:570–4.
4. Hartmann EM, Durighello E, Pible O, Nogales B, Beltrametti F, Bosch R, et al. Proteomics meets blue biotechnology: a wealth of novelties and opportunities. *Mar Genomics* 2014. [Epub ahead of print].
5. Bauer MW. Distinguishing red and green biotechnology: cultivation effects of the elite press. *Int J Publ Opin Res* 2005;**17**:63–89.
6. Himmel ME, Ding SY, Johnson DK, Adney WS, Nimlos MR, Brady JW, et al. Biomass recalcitrance: engineering plants and enzymes for biofuels production. *Science* 2007;**315**:804–7.
7. Aden A, Foust T. Technoeconomic analysis of the dilute sulfuric acid and enzymatic hydrolysis process for the conversion of corn stover to ethanol. *Cellulose* 2009;**16**:535–45.
8. Kabir Kazi F, Fortman J, Anex R, Kothandaraman G, Hsu D, Aden A, Dutta A. *Techno-economic analysis of biochemical scenarios for production of cellulosic ethanol* 2010. NREL Tech Rep NREL/TP-6A2-46588.
9. Klein-Marcuschamer D, Oleskowicz-Popiel P, Simmons BA, Blanch HW. The challenge of enzyme cost in the production of lignocellulosic biofuels. *Biotechnol Bioeng* 2012;**109**:1083–7.
10. Hood EE, Love R, Lane J, Bray J, Clough R, Pappu K, et al. Subcellular targeting is a key condition for high-level accumulation of cellulase protein in transgenic maize seed. *Plant Biotechnol J* 2007;**5**:709–19.
11. Ransom C, Balan V, Biswas G, Dale B, Crockett E, Sticklen M. Heterologous *Acidothermus cellulolyticus* (1,4)-β-endoglucanase E1 produced within the corn biomass converts corn stover into glucose. *Appl Biochem Biotechnol* 2007:207–19.
12. Taylor LE, Dai Z. Decker SR, Brunecky R, Adney WS, Ding SY, et al. Heterologous expression of glycosyl hydrolases in planta: a new departure for biofuels. *Trends Biotechnol* 2008;**26**:413–24.
13. Shen B, Sun X, Zuo X, Shilling T, Apgar J, Ross M, et al. Engineering a thermoregulated intein-modified xylanase into maize for consolidated lignocellulosic biomass processing. *Nat Biotechnol* 2012;**30**:1131–6.
14. Brunecky R, Selig M, Vinzant T, Himmel M, Lee D, Blaylock M, et al. In planta expression of *A. cellulolyticus* Cel5A endocellulase reduces cell wall recalcitrance in tobacco and maize. *Biotechnol Biofuels* 2011;**4**:1.
15. Chen F, Dixon RA. Lignin modification improves fermentable sugar yields for biofuel production. *Nat Biotechnol* 2007;**25**:759–61.
16. Fu C, Mielenz JR, Xiao X, Ge Y, Hamilton CY, Rodriguez M, et al. Genetic manipulation of lignin reduces recalcitrance and improves ethanol production from switchgrass. *Proc Natl Acad Sci USA* 2011;**108**:3803–8.

17. Shen H, He X, Poovaiah CR, Wuddineh WA, Ma J, Mann DGJ, et al. Functional characterization of the switchgrass (*Panicum virgatum*) R2R3-MYB transcription factor PvMYB4 for improvement of lignocellulosic feedstocks. *New Phytol* 2012;**193**:121–36.

18. Selig M, Vinzant T, Himmel M, Decker S. The effect of lignin removal by alkaline peroxide pretreatment on the susceptibility of corn stover to purified cellulolytic and xylanolytic enzymes. *Appl Biochem Biotechnol* 2009;**155**:94–103.

19. Wyman CE, Dale BE, Elander RT, Holtzapple M, Ladisch MR, Lee Y. Coordinated development of leading biomass pretreatment technologies. *Bioresour Technol* 2005;**96**:1959–66.

20. Wyman CE, Dale BE, Balan V, Elander RT, Holtzapple MT, Ramirez RS, et al. *Comparative performance of leading pretreatment technologies for biological conversion of corn stover, poplar wood, and switchgrass to sugars. Aqueous pretreatment of plant biomass for biological and chemical conversion to fuels and chemicals* 2013. p. 239–59.

21. Lee D, Yu AHC, Wong KKY, Saddler JN. Evaluation of the enzymatic susceptibility of cellulosic substrates using specific hydrolysis rates and enzyme adsorption. *Appl Biochem Biotechnol* 1994;**45–6**:407–15.

22. Valdeir A, Jack S. Access to cellulose limits the efficiency of enzymatic hydrolysis: the role of amorphogenesis. *Biotechnol Biofuels* 2010;**3**:4–14.

23. Laureano-Perez L, Teymouri F, Alizadeh H, Dale BE. Understanding factors that limit enzymatic hydrolysis of biomass. *Appl Biochem Biotechnol* 2005;**121**:1081–99.

24. Zhang YHP, Lynd LR. Toward an aggregated understanding of enzymatic hydrolysis of cellulose: noncomplexed cellulase systems. *Biotechnol Bioeng* 2004;**88**:797–824.

25. Chandra R, Bura R, Mabee W, Berlin A, Pan X, Saddler J. Substrate pretreatment: the key to effective enzymatic hydrolysis of lignocellulosics? *Biofuels* 2007:67–93.

26. Jeoh T, Ishizawa CI, Davis MF, Himmel ME, Adney WS, Johnson DK. Cellulase digestibility of pretreated biomass is limited by cellulose accessibility. *Biotechnol Bioeng* 2007;**98**:112–22.

27. Klinke HB, Thomsen A, Ahring BK. Inhibition of ethanol-producing yeast and bacteria by degradation products produced during pre-treatment of biomass. *Appl Microbiol Biotechnol* 2004;**66**:10–26.

28. Wang W, Chen X, Donohoe BS, Ciesielski PN, Katahira R, Kuhn EM, et al. Effect of mechanical disruption on the effectiveness of three reactors used for dilute acid pretreatment of corn stover Part 1: chemical and physical substrate analysis. *Biotechnol Biofuels* 2014;**7**:1–13.

29. Chen X, Shekiro J, Pschorn T, Sabourin M, Tao L, Elander R, et al. A highly efficient dilute alkali deacetylation and mechanical (disc) refining process for the conversion of renewable biomass to lower cost sugars. *Biotechnol Biofuels* 2014;**7**:98.

30. Brunecky R, Donohoe BS, Selig MJ, Wei H, Resch M, Himmel ME. In: Goldman SL, Kole C, editors. *Compendium of bioenergy plants: corn.* CRC Press; 2014. p. 33–77.

31. Nguyen Q, Tucker M. *Dilute acid/metal salt hydrolysis of lignocellulosics.* US Patent 6423145, USA; 2002.

32. Wei H, Donohoe BS, Vinzant TB, Ciesielski PN, Wang W, Gedvilas LM, et al. Elucidating the role of ferrous ion cocatalyst in enhancing dilute acid pretreatment of lignocellulosic biomass. *Biotechnol Biofuels* 2011;**4**:48.

CHAPTER 2

Hydrocarbon Biosynthesis in Microorganisms

Lawrence P. Wackett, Carrie M. Wilmot

Department of Biochemistry, Molecular Biology and Biophysics, University of Minnesota, Minneapolis, MN, USA

Introduction

Fossil fuels are society's major energy source, accounting for more than 80% of energy needs.[1] In the transportation sector, liquid petroleum-based fuels account for 95% of the market. Even with the advent of hydraulic fracturing, which has ushered in a boom of new exploration, US petroleum production is not yet close to meeting demand. While wind or solar energy can potentially be used for meeting electrical needs, there is still a pressing need for energy-dense, liquid fuels in the transportation and manufacturing sectors. Electric cars may benefit from improvements in energy storage density for batteries, but jets have an absolute requirement for liquid fuel to achieve the needed power and engine reliability. Likewise, ship transport will likely remain dependent on liquid hydrocarbon fuels for the foreseeable future.

There has been a dichotomy within the liquid fuel industries in recent years: biofuels consisting largely of ethanol plus some fatty acid esters, and petroleum-based fuels that are principally hydrocarbons. In 1925, Henry Ford called ethanol "the fuel of the future."[2] Since that time, ethanol has persisted as an alternative energy source, but it has never become the dominant fuel that Ford envisioned. Although ethanol from biomass is renewable and petroleum is not, the latter has significant advantages. Petroleum hydrocarbons pack more energy per unit mass, are not hygroscopic like ethanol, and provide a much higher energy return on energy invested for recovery and transport. Thus, renewable hydrocarbons represent a way to harness the best traits of both fuel classes.

Humans around the world have produced ethanol biologically for thousands of years, but there has been limited exploration, production, and repurposing of microbial hydrocarbons as renewable fuel sources. There have been numerous reports of microbial hydrocarbon production in soil and water environments over the past 70 years, and this has laid the groundwork for more recent gene discovery and metabolic engineering. The early studies largely consisted of identifying structures of hydrocarbons that partitioned with neutral lipids in solvent

Direct Microbial Conversion of Biomass to Advanced Biofuels. http://dx.doi.org/10.1016/B978-0-444-59592-8.00002-6

extractions.[3] The demonstration of different structural types presaged the existence of disparate biochemical mechanisms for the biosynthesis of hydrocarbons.

Small and large companies have recognized the potential for novel processes and intellectual property derived from a new class of biofuels that more closely resemble today's prevalent petroleum-based gasoline and diesel fuels. Different companies have developed distinct platforms seeking to produce hydrocarbons alternatively from sugars, cellulosic biomass, or from photosynthesis using carbon dioxide. The underlying microbiology and enzymology of microbial hydrocarbon biosynthesis is discussed in more detail below.

Finally, it should be noted that biogas (methane from methanogenic bacteria) has been used for decades. In addition, there are historic reports that in some parts of China, people used bamboo pipes to transport natural gas to use for cooking, although that gas was likely of thermogenic origin.[4]

Microbiology and Hydrocarbon Products

Microorganisms biosynthesize different types of hydrocarbons: alkanes, alkenes, arenes, and isoprenoid compounds. These compounds are found within phylogenetically diverse microbes, suggesting that the mechanisms are ancient and widespread. In many cases, however, it is not known why specific microbes biosynthesize hydrocarbons. In plants and animals, the biological function of biogenic hydrocarbons is more apparent.[5] For example, plants make waxy hydrocarbons to coat leaf surfaces and protect against desiccation. Some insects protect their eggs with a coating of solid hydrocarbon to prevent both desiccation and predation. Many isoprenoid hydrocarbons offer protection against ultraviolet radiation, and their biosynthetic pathways have been relatively well studied. However, the function and biosynthetic mechanisms for non-isoprenoid, non-gaseous hydrocarbons are largely unknown. Further work is needed for both fundamental understanding and commercial applications.

The biosynthesis of gaseous methane by methanogenic bacteria has been relatively well studied.[6] Methanogens use carbon dioxide and hydrogen or acetate to generate methane in overall energy yielding metabolism. These bacteria often live in close association with other bacteria that degrade more complex organic matter in anaerobic ecosystems. By contrast, much less is known about the microbial production of higher molecular weight gaseous alkanes. Ethane and propane have been detected in copious amounts within deeply buried ocean sediments, and several years ago a biological origin was proposed, although no biochemical mechanisms are available.[7] More recently, other groups have demonstrated enzymatic formation of ethane, propane, and other hydrocarbons. Surprisingly, the enzyme nitrogenase, which functions biologically to reduce dinitrogen gas to ammonia, can react with carbon monoxide to produce gaseous alkanes. Specifically, the vanadium-containing nitrogenase variant has been shown to catalyze ligation and reduction of carbon monoxide into hydrocarbon chains.[8,9] Although a molybdenum-containing nitrogenase does not do this naturally, laboratory-constructed variants of the enzyme will produce alkanes.[10] These reactions are now being examined in more detail as examples of a biological Fischer–Tropsch reaction. A branched alkene, isobutene, is biosynthesized by the yeast *Rhodotorula minuta* and is a gas at room temperature. The carbon atoms are derived from branched chain amino acids via the intermediate isovalerate that undergoes decarboxylation to yield isobutene.[11]

Isoprene is gaseous at 37 °C, a temperature commonly used to grow *Escherichia coli* and other bacteria, making it an interesting target for engineered microbial synthesis. Natural isoprene biosynthesis is widespread. Terrestrial plants, marine plankton, and bacteria produce these compounds in enormous quantities, about 500 million tons annually.[12] Many bacteria produce isoprene, but of strains analyzed in the laboratory, *Bacillus* strains have been observed to produce the highest levels. The methylerythritol phosphate pathway is implicated in mediating isoprene production in *Bacillus subtilis*.[13] Overall, the isoprenoid class of compounds is enormous and range in size from C_{10} to C_{110}.[14] More than 50,000 isoprenoid compounds are known, with many being produced by microorganisms.

Longer-chain alkanes are produced in low amounts by eukaryotic and prokaryotic microbes, many of which are found in marine environments.[15] It is reported that brown algae produce *n*-pentadecane and red algae produce *n*-heptadecane. Another alga, *Dunaliella salina,* produces 6-methyl hexadecane and 4-methyl octadecane. A more recent report indicated *Vibrio furnissii* M1 produced copious quantities of C_{16}-C_{28} alkanes,[16] but those findings turned out to not be reproducible.[17]

Long-chain alkenes have been identified in high G + C gram positive bacteria *Micrococcus* species,[15] *Stenotrophomonas maltophilia*,[15] and *Arthrobacter* species.[18] The alkene chains are in the range of C_{23}-C_{31}, and the structures are inconsistent with derivation from an isoprenoid biosynthetic pathway. The biological function of these long-chain alkenes is currently unknown. However, *Micrococcus* and related bacteria are common human skin inhabitants, and it is interesting to speculate that the alkenes might enhance their survival on exposed skin.

Cyclic hydrocarbons such as alkylbenzenes have been identified in Archae from the genera *Thermoplasma* and *Sulfolobus*. Members of the genus *Alicyclobacillus* produce novel cycloheptane ring structures, presumably to increase the stability of their membranes.[19] Another distinct class of hydrocarbons is the hopanoids that consist of multiple, nonaromatic rings.[20] These compounds structurally resemble cholesterol and may serve a similar function of modulating cytoplasmic membrane fluidity. Hopanoids are found in cyanobacteria, *Streptomyces* sp., and *Zymomonas mobilis*. Sterane, consisting of four fused alicyclic rings, was once thought to be a biomarker for eukaryotes, but it is now known to be biosynthesized by some methanotrophic proteobacteria.

Enzymes and Mechanisms of Hydrocarbon Biosynthesis
Ole-Catalyzed Synthesis of Long-Chain Olefins

Microbial long-chain olefin biosynthesis was studied by Albro and Dittmer in 1969, but without molecular genetic techniques, their studies were limited to crude cell extract experiments.[21–24] However, those early studies were able to establish that fatty acyl groups were undergoing condensation at the carboxyl ends of the chains. Thus, the biosynthesis was denoted as a "head-to-head" condensation reaction with the loss of one carbon atom as carbon dioxide.

Research at LS9, Inc. elucidated the genes involved, which were subsequently confirmed by a group working at the Joint Bioenergy Institute.[25] These were denoted as *ole*, or olefin synthesizing genes. The genes encode four proteins in most microorganisms, although in several the *oleBC* genes are fused. The four proteins belong to different protein superfamilies (Table 1).[26] OleA is homologous to members of the thiolase superfamily, also known as the condensing enzyme superfamily.[27,28] OleB is a member of the α/β hydrolase superfamily. OleC is a member of the AMP-dependent ligase/synthase superfamily, also known as the acetyl-CoA synthetase-like superfamily. OleD is a member of the short chain dehydrogenase/reductase superfamily. Sequence identities of the Ole proteins to the closest corresponding superfamily proteins with different physiological functions are generally low, on the order of 20–30%.[27,28]

A survey of 3558 bacterial genomes demonstrated clear evidence of the *ole* genes in 1.9% of the genomes.[26] The functionality of these operons was experimentally tested using a selection of 14 microorganisms from across multiple phyla, and all were demonstrated to produce long-chain olefins (Table 2). The majority, such as *Shewanella oneidensis* MR-1, produced a single C_{31} polyolefinic hydrocarbon, 3,6,9,12,15,19,22,25,28-hentriacontanonaene (Figure 1). In *S. oneidensis* MR-1, this polyolefin was found to localize to the membrane[29] and was produced at higher levels in cells grown at a lower temperature.[30] This led to the proposal that the polyolefin content maintains constant membrane fluidity at different temperatures. In further work, the C_{29} olefins from four *Arthrobacter* strains, along with the previously studied *Stenotrophomonas maltophilia* and *Micrococcus luteus*, were rigorously identified through comparison to synthesized standards and observed to correspond to the branched fatty acids found in those organisms (Table 2, Figure 2).[18] *Xanthomonas campestris* is interesting in that it produces a diverse range of long-chain olefins: at least 15 in vivo with chain lengths varying from C_{28} to C_{31}, although the predominant product is a C_{29} olefin.[26] The *ole* gene products from this organism appear to have a promiscuity and plasticity in substrate utilization that makes them attractive targets for bioengineering for the production of desirable commodity products. In vitro, it has been demonstrated that the first enzyme in the pathway, *X. campestris* OleA, can handle considerably shorter substrates than its in vivo profile, to give hydrocarbon products down to C_{15} in length.[31]

In *S. oneidensis* MR-1, the polyolefin product of the Ole proteins localizes to the membrane.[29] The *ole* operon sequences give no indication that any of the proteins are targeted or anchored to the membrane, which suggests that the enzymes exist in the cytosol, although it is possible that they associate with an anchored or integral membrane protein. The substrates are hydrophobic long-chain hydrocarbons that have limited solubility in aqueous solution, thus making

Table 1: Homology of the four Ole proteins to known protein superfamilies

Ole protein	Superfamily name (alternative name(s))	Known biological functions within the superfamily
OleA	Thiolase (Condensing enzymes)	Acyl-ACP synthase, thiolase (degradative and biosynthetic), 3-hydroxyl-3-methylglutaryl-CoA synthase, fatty acid elongase, stage V sporulation protein, 6-methylsalicylate synthase, *Rhizobium* nodulation protein NodE, chalcone synthase, stilbene synthase, naringenin synthase, β-ketosynthase domains of polyketide synthase
OleB	α/β-Hydrolase	Esterase, haloalkane dehalogenase, protease, lipase, haloperoxidase, lyase. Epoxide hydrolase, enoyl-CoA hydratase/isomerase, MhpC C–C hydrolase (carbon–carbon bond cleavage)
OleC	AMP-dependent ligase/synthetase (LuxE; acyl-adenylate/thioester Forming, acetyl-CoA synthetase-like)	Firefly luciferase, nonribosomal peptide synthase, acyl-CoA synthase (AMP forming), 4-chlorobenzoate:CoA ligase, acetyl-CoA synthetase, O-succinylbenzoic acid-CoA ligase, fatty acyl ligase, 2-acyl-glycerophospho-ethanolamine acyl transferase, enterobactin synthase, amino acid adenylation domain, dicarboxylate-CoA ligase, crotonbetaine/carnitine-CoA ligase
OleD	Short-chain dehydrogenase/reductase	Nucleoside-diphosphate sugar epimerase/dehydratase/reductase, aromatic diol dehydrogenase, steroid dehydrogenase/isomerase, sugar dehydrogenase, acetoacetyl-CoA reductase, 3-oxoacyl-ACP reductase, alcohol dehydrogenase, carbonyl reductase, 4-α-carboxysterol-C3-dehydrogenase/C4-decarboylase, flaonol reductase, cinnamoyl CoA reductase, NAD(P)-dependent cholesterol dehydrogenase

Adapted from Sukovich et al. (2010) Ref. 26.

Table 2: Head-to-head olefins produced by different bacteria

Microorganism	Carbon chain length	Predominant hydrocarbon
Chloroflexus aurantiacus J-10-fl[a]	C_{31}	$C_{31}H_{58}$
Kocuria rhizophilia DC2201[a]	C_{24}-C_{29}	$C_{27}H_{54}$
Brachybacterium faecium ATCC 15993[a]	C_{27}-C_{29}	$C_{29}H_{58}$
Xanthomonas campestris pv. *campestris*[a]	C_{28}-C_{31}	$C_{31}H_{46}$
Shewanella oneidensis MR-1[a]	C_{31}	$C_{29}H_{54}$
Shewanella putrefaciens CN-32[a]	C_{31}	$C_{31}H_{46}$
Shewanella baltica PS185[a]	C_{31}	$C_{31}H_{46}$
Shewanella frigidimarina NCIMB 400[a]	C_{31}	$C_{31}H_{46}$
Shewanella amazonensis SB2B[a]	C_{31}	$C_{31}H_{46}$
Shewanella denitrificans OS217[a]	C_{31}	$C_{31}H_{46}$
Colwellia psychreryhtraea 34H[a]	C_{31}	$C_{31}H_{46}$
Geobacter bemidjiensis Bem[a]	C_{31}	$C_{31}H_{46}$
Opitutaceae TAV2[a]	C_{31}	$C_{31}H_{46}$
Planctomyces maris DSM8797[a]	C_{31}	$C_{31}H_{46}$
Stenotrophomonas maltophilia ATCC 17674[b]	C_{27}-C_{31}	$C_{30}H_{60}$
Micrococcus luteus ISU[b,c]	C_{23}-C_{29}	$C_{29}H_{58}$
Arthrobacter aurescens TC1[b]	C_{29}-C_{31}	$C_{29}H_{58}$
Arthrobacter chlorophenolicus A6[b]	C_{27}-C_{31}	$C_{29}H_{58}$
Arthrobacter crystallopoietes ATCC 15481[b]	C_{27}-C_{19}	$C_{29}H_{58}$
Arthrobacter oxydans ATCC 14358[b]	C_{29}	$C_{29}H_{58}$

[a]Data from Sukovich et al., 2010[26].
[b]Data from Frias et al., 2009[18].
[c]Data from Albro & Dittmer, 1969[24].

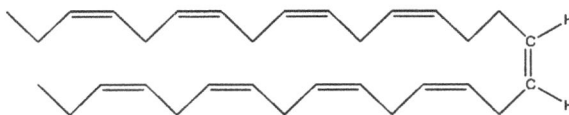

Figure 1
The polyolefin produced by *Shewanella oneidensis* MR-1,
3,6,9,12,15,19,22,25,28-hentriacontanonaene.

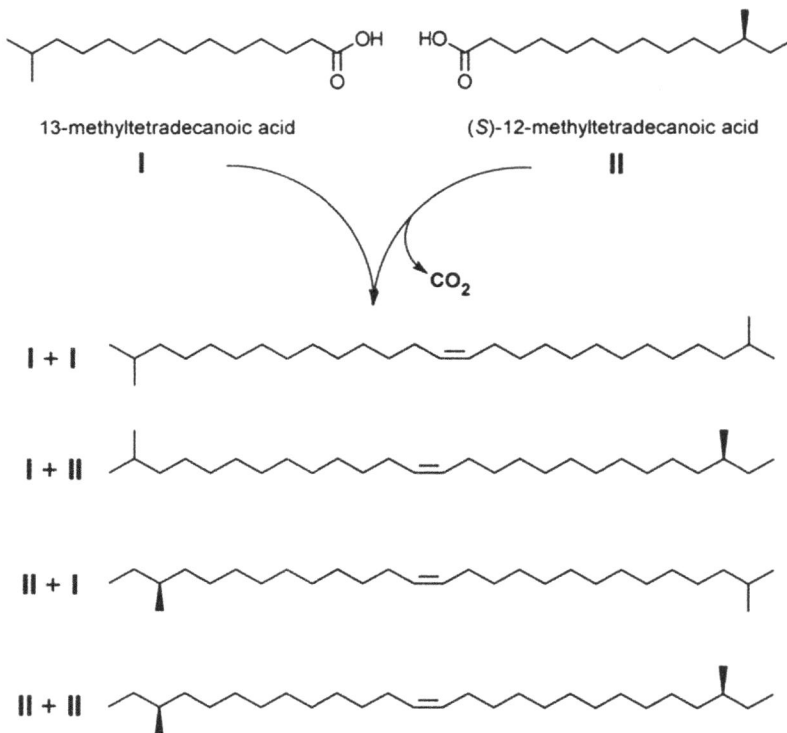

13-methyltetradecanoic acid

I

(*S*)-12-methyltetradecanoic acid

II

CO_2

I + I

I + II

II + I

II + II

Figure 2
The C_{29} olefins produced by *Arthrobacter* strains form branched chain fatty acids.

it unlikely they would be released into the cytosol between Ole enzymes. Therefore, it is likely that the Ole proteins form a catalytic complex that traffics the product of one enzyme into the active site of the next.

The Chemistry of the Ole Gene Products

To date, there have been several published studies on purified Ole proteins that have begun to establish structure/function paradigms for this class of proteins. The OleA protein catalyzes a non-decarboxylative Claisen condensation to produce a β-keto acid intermediate.[31] The OleD

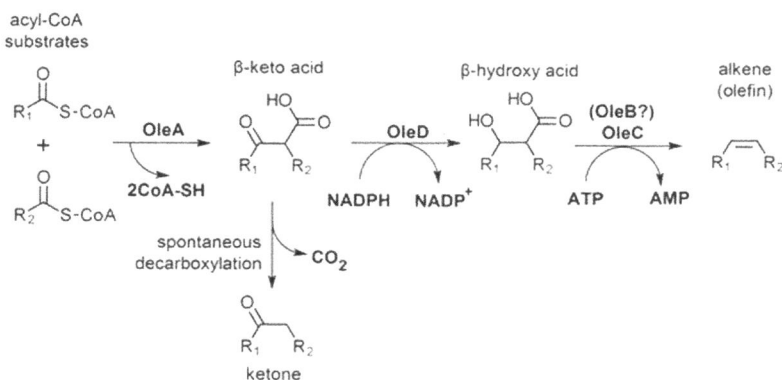

Figure 3
Proposed scheme of microbial olefin biosynthesis by the Ole proteins.

protein has been shown to catalyze an NADPH-dependent reduction of the β-keto acid produced by OleA[32]. OleC is obligately required to produce the final olefin, but OleB is not.[31] In vitro incubations with OleA, OleC, and OleD produce an olefinic hydrocarbon, but product levels do not increase when OleB is added to incubation mixtures. The current overall reaction scheme is shown in Figure 3.

OleA

OleA catalyzes the initial reaction on the pathway to hydrocarbons: a condensation reaction that has been demonstrated using acyl-CoA substrates in vitro.[25,31] Fatty acids are typically catabolized while tethered to CoA, but fatty acid elongations and polyketide biosynthesis occur via iterative reactions while attached to acyl carrier protein (ACP). The final step in fatty acid biosynthesis is the release of ACP, and so both acyl-CoA and acyl-ACP are potential substrates of OleA. In fact, a patent application described the OleA from *S. maltophilia* as condensing acyl groups attached to either CoA or ACP carriers.[27] However, all the current OleA mechanistic work has exclusively used acyl-CoA substrates. Figure 4 shows the OleA reaction mechanism, during which both CoA moieties are displaced. The second CoA moiety is displaced during the central condensation step, leading to a β-keto acid product that is no longer tethered to a carrier, and becomes the substrate of OleD. Thus, it is clear that the overall OleABCD reaction pathway to hydrocarbon proceeds with untethered intermediates in contrast to polyketide biosynthetic pathways.[33] In the Ole biosynthetic pathway, it is thought that the β-keto acid remains sequestered in the Ole complex, and diffuses to the next enzyme in the pathway, OleD, for subsequent reduction to the corresponding β-hydroxy acid.[32] It is important to note that the β-keto acid produced by OleA is inherently unstable, even at neutral pH and ambient temperature. Thus, some decarboxylated products, ketones, are observed in all in vitro studies conducted to date with reconstituted enzyme systems (Figure 3). However, the ketones are rarely observed in vivo with native bacteria containing *ole* genes, suggestive of better coupling in natural systems. The decarboxylation reaction to yield the ketone has been demonstrated in enzyme reaction mixtures, and with synthetic β-keto acid subjected to the same conditions. The β-keto acid is sufficiently unstable such that ketone decarboxylation products are observed within minutes at 25 °C and at neutral pH (Figure 3).

Interestingly, olefin can be generated using OleBCD proteins from one species coupled to OleA from a different species, suggesting that interaction between OleA and OleD is not very specific. This has been demonstrated for *S. maltophilia* OleA with *S. oneidensis* MR-1 OleBCD[26,30] (sequence identity between species' OleA is 35% and OleD 49%) and *Kineococcus radiotolerans* OleA with *S. maltophilia* OleBCD (sequence identity between species' OleA is 49% and OleD 40%).[32] The OleA is the primary determinant of the product formed, with OleBCD from other microorganisms being able to handle β-keto acid OleA products that they do not encounter in vivo.

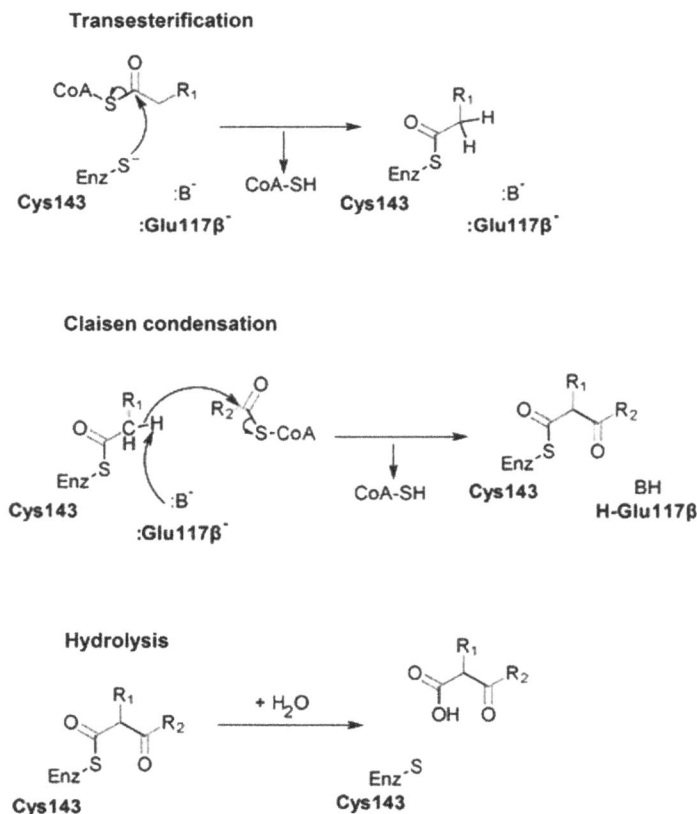

Figure 4

Catalytic steps in *Xanthomonas campestris* OleA-catalyzed condensation of CoA-charged substrates. B represents the proposed catalytic base, E117β. R_1 and R_2 are C_8 to C_{16} CoA-charged fatty acids.

The only Ole protein structure available at this time is the crystal structure of OleA from *X. campestris*, which has been solved to 1.85 Å and shows the typical thiolase homodimer (Figure 5(a)).[34,35] The active site contains the strictly conserved Cys143 that forms a covalent acyl intermediate with the first acyl-CoA substrate via a thioester bond (Figure 5(b)). The active site base required to activate the covalent intermediate for condensation with the second acyl-CoA substrate is likely a glutamate (Glu117β) from the other monomer of the dimer: the first observation within the thiolase superfamily of both monomers contributing active site residues. Oxyanion holes that lower the activation energy of transient tetrahedral species are also conserved in the active site.

OleA is unusual in that it requires three long, independent substrate channels.[35] All thiolase substrates are CoA or ACP charged, and so all possess a pantetheinate channel to accommodate the CoA or ACP tether, and OleA is no exception. Unlike 3-hydroxy-3-methylglutaryl-CoA (HMG-CoA) synthase the substrates of which do not have long acyl chains, OleA must bind the Cys143 covalently tethered alkyl chain from the first substrate and the second acyl-CoA substrate simultaneously for Claisen condensation to occur (Figure 6). In the fatty acid biosynthesis (Fab) enzymes only one of the substrates has a long acyl chain, as these enzymes elongate long-chain hydrocarbons by only two carbons at a time, and so require only two channels: a pantetheinate and an alkyl channel. An alkyl chain channel (designated B), lying orthogonal to the pantetheinate channel, has been well described in crystal structures of fatty acid biosynthesis (Fab) enzymes; the complex of C112A FabH with lauroyl-CoA[36] (equivalent to C143A OleA), FabH bound to decane-1-thiol,[37] and FabB bound to the C_{11} irreversible covalent inhibitor

(a)

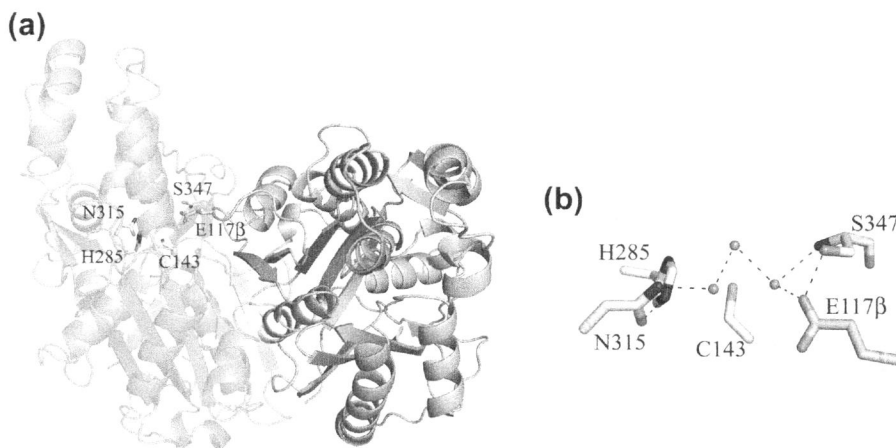

Figure 5

OleA dimer and active site. (a) The physiological dimer of OleA. The two monomers are drawn in gray and tan (dark gray in print versions) cartoon. Each monomer contains one active site. The gray cartoon active site residues are drawn in stick. Note that E117β derives from the neighboring monomer. (b) The OleA active site. Ordered solvent molecules are represented by red (dark gray in print versions) spheres. This figure was produced using PyMOL (http://www.pymol.org/). *Reproduced from Goblirsch et al. (2012) Ref. 35.*

HMG-CoA

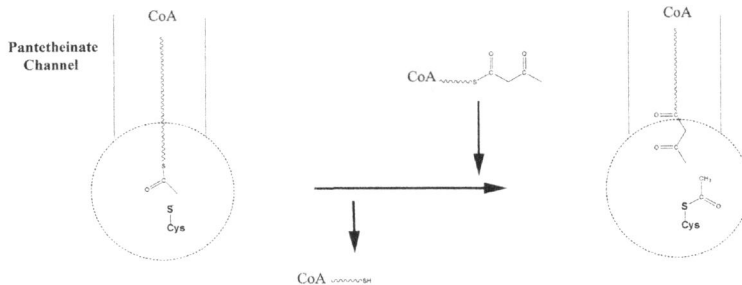

FabH

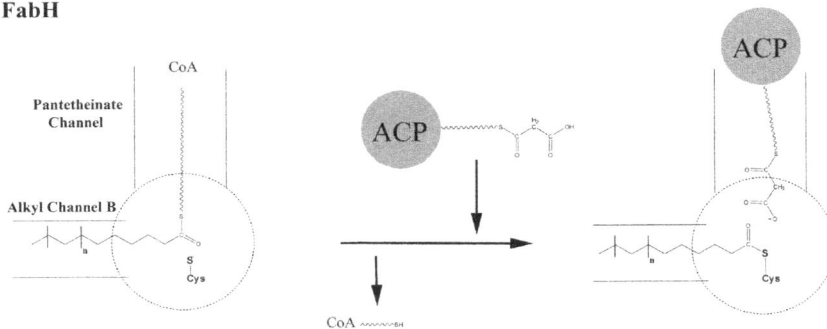

OleA

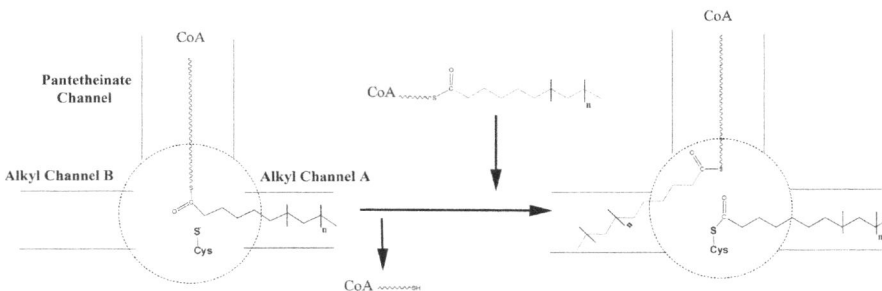

Figure 6

Substrate binding channels in HMG-CoA synthase, FabH, and OleA. Each enzyme follows a ping-pong reaction to complete turnover. All three enzymes require a pantetheinate channel for binding CoA or ACP thioester charged substrates. The CoA or ACP moiety is represented as a yellow or blue sphere, respectively. In addition, FabH requires an alkyl channel (alkyl channel B) in substrate binding while OleA requires two alkyl channels (alkyl channels A and B). The solid black lines perpendicular to the propagating alkyl chain illustrate how FabH and OleA can use fatty-acyl-CoA substrates of different alkyl chain lengths (n = number of carbon atoms). For OleA, the substrates are colored as in Figure 4. *This figure was produced using PyMOL (*http://www.pymol.org/*).*

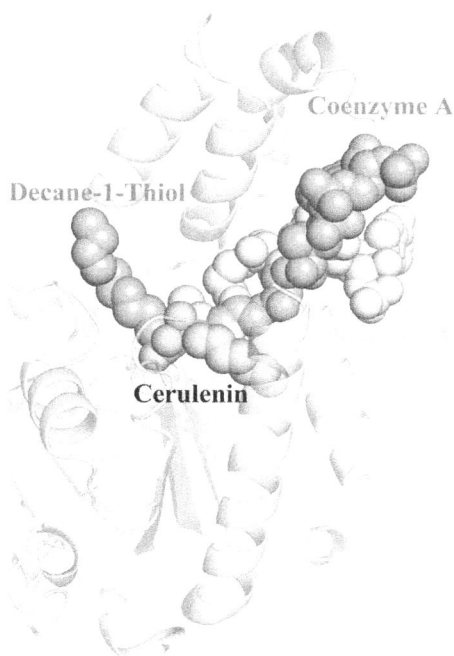

Figure 7

OleA bound with cerulenin overlaid with FabH (bound with decane-1-thiol and coenzyme A, PDB ID 2QX1) and HMG-CoA synthase (bound with coenzyme A, PDB ID 1TXT). OleA is drawn in gray (lighter gray in print versions) cartoon and cerulenin as a green (gray in print versions) space-filling model. For clarity, the monomers of FabH and HMG-CoA synthase are omitted and their bound ligands are drawn in space-filling model (pink (dark gray in print versions) and yellow (light gray in print versions), respectively). Decane-1-thiol occupies the alkyl channel B in FabH. Coenzyme A occupies the pentetheinate channel in both FabH and HMG-CoA synthase. *This figure was produced using PyMOL* (http://www.pymol.org/).

cerulenin (Figure 6).[38] Cerulenin is a natural product that inhibits fatty acid biosynthesis by forming a covalent adduct with the active site Cys.[39,40] Crystal structures of *X. campestris* OleA in complex with cerulenin, whose alkyl chain length matches that of a bona fide substrate, or xenon, a hydrophobic gas used to delineate accessible hydrophobic spaces within proteins, identifies a second channel (designated A) (Figure 7). This channel is orthogonal to the pantetheinate channel and diametrically opposed to channel B. The characteristics of channel B in the Fab enzymes are mirrored in OleA. One side of channel B is formed by a mobile loop that in different FabH crystal structures is either ordered or disordered and has been proposed to act as a gate that enables easy access to the channel, and a more energy efficient exit for product then through the pantetheinate channel (Figure 8).[37] Similarly in OleA, the crystallographically independent loop structures show varying degrees of order.[35] Channel A is composed of two well-ordered pieces of secondary structure. Thus, it seems likely that the OleA product will exit

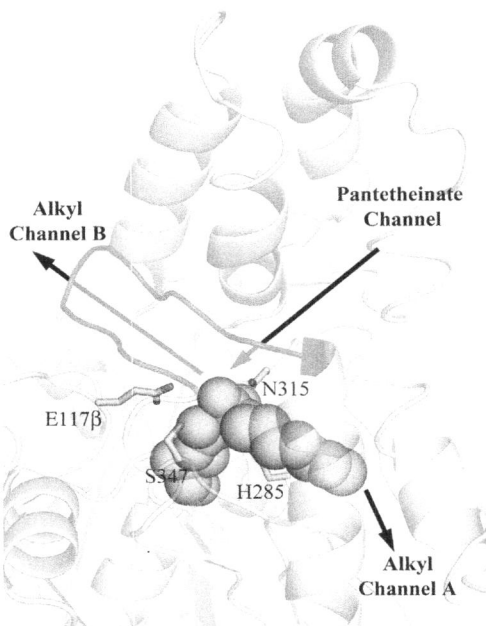

Figure 8

Putative assignment of OleA binding channels. Active site residues are drawn in stick (carbon green (gray in print versions)) and bound cerulenin drawn in green (gray in print versions) space-filling. Labeled arrows indicate the position of the alkyl and pantetheinate channels in the OleA monomer (gray cartoon (lighter gray in print versions)). An 11 residue (239–249) β-hairpin that could only be modeled in one monomer of the $P2_12_12_1$ crystal form is drawn as red (light gray in print versions) cartoon. *This figure was produced using PyMOL* (http://www.pymol.org/). *Reproduced from Goblirsch et al. (2012) Ref. 35.*

from channel B, pulling the product out of channel A and through the active site. This also suggests that interaction with the next enzyme in olefin biosynthesis, OleD, likely occurs at the OleA face containing the channel B loop. The channel B loops from each monomer lie close to each other in a saddle on the dimer surface, and therefore the β-keto acid products from each OleA active site would exit next to each other on the OleA surface. Interestingly, cerulenin binds in channel A in OleA, but channel B in FabB.[38] This suggests that channel A may house the acyl chain of the first substrate, with the second binding in channel B. This likely explains the observed propensity for the OleA covalent acyl intermediate to undergo futile hydrolysis in vitro that releases free acid, as when channel B is empty the loop will be mobile exposing the active site to solvent. Ordering of the channel B loop on binding of the second acyl-CoA substrate would protect the active site chemistry and promote Claisen condensation.

OleD

The substrate of OleD is the product of OleA, and it is an obligate enzyme for olefin biosynthesis. It converts the β-keto acid product of OleA to the β-hydroxy acid. Only the OleD from *Stenotrophomonas maltophilia* has been characterized biochemically.[32] The *S. maltophilia* OleA has been shown to selectively use mono- and di-unsaturated and saturated acyl thioesters as substrates (it is unknown if acyl-CoA or acyl-ACP is the physiological substrate). In vivo the olefin product profile is complex, containing chain lengths between C_{27} and C_{32}, with C_{29} and C_{30} being the most prevalent, and branching at or near the termini (e.g., Figure 2).[18,26,41,42] As such, *S. maltophilia* OleD has a potentially broad range of branched substrates.

In the in vitro study of *S. maltophilia* OleD, Bonnett et al. synthesized a range of substrates of differing chain lengths and stereochemistry.[32] The enzyme was demonstrated to reversibly catalyze the stereospecific reduction of 2-alkyl-3-ketoalkanoic acids in an NADPH-dependent reaction. The α-carbon of the OleA β-keto acid product is chiral, and the *S. maltophilia* OleD was enantioselective for the (*2R, 3S*) isomer, strongly suggesting that this is the chirality of the products of *S. maltophilia* OleA. Of the substrates that have been tested, *syn*-2-decyl-3-hydroxytetradecanoic acid was the most catalytically efficient, with a five- to eight-fold higher k_{cat}/K_m than the shorter chain substrates. As might be expected this manifested itself primarily in K_m, and matches the longer chain substrates that dominate in vivo.

Like *X. campestris* OleA, *S. maltophilia* OleD is a homodimer.[32,35] This suggests that the proposed OleABCD complex may contain two copies of each polypeptide. The overexpression and purification of the *X. campestris* OleD has also been reported and shown to have catalytic activity, but no detailed biochemical studies have been undertaken.[31]

OleC

OleC is a member of the LuxE AMP-dependent acyl-protein synthetase superfamily.[26] It is the third obligate Ole protein required for olefin biosynthesis, and is presumed to convert the β-hydroxy acid product of OleD to olefin. Although no extensive biochemical studies have been conducted on OleC, the enzyme from *S. maltophilia* has been overexpressed and purified.[43] It was demonstrated to be catalytically active, as expected requiring ATP and $MgCl_2$ for activity.[31] Interestingly, the production of olefin was demonstrated using purified *X. campestris* OleA and OleD with myristoyl-CoA substrate. Preliminary X-ray diffraction data from crystals of *S. maltophilia* OleC have been reported, but no structure is currently available.[43]

OleB

OleB is not required for olefin biosynthesis, nor does its presence affect the levels of olefin product, so its role remains enigmatic.[31] However, it is always present in the *ole* operon, and so presumably plays an important role. In some organisms, the *oleB* and *oleC* genes are fused, suggesting a linkage to the activity of OleC.[25,26] OleB belongs to the α/β hydrolase superfamily, suggesting an enzymatic activity, but it is also possible that it plays a scaffolding or regulatory role within the proposed Ole complex.

Aldehyde Deformylating Oxygenase (Formerly Decarbonylase)

There have been many reports of microbial biosynthesis of diesel-length alkanes[3,44]; however, some observations have proven irreproducible.[17] Investigations into the mechanism of this biosynthesis have taken on a new imperative with the push for renewable hydrocarbons. These hydrocarbons are considered to be "drop-in" fuels that require no modification, because straight-chain alkanes in the range of C_{13}-C_{17} are a perfect diesel fuel.

A key insight into the mechanism of cyanobacterial alkane biosynthesis was made by a group of scientists at LS9, Inc. who used comparative genomics with the genomes of cyanobacteria that naturally produce or do not produce the alkanes.[45] This analysis identified two genes that were subsequently shown to lead to hydrocarbon production when heterologously expressed in *E. coli*. The two genes encoded a fatty acyl-CoA reductase and an enzyme denoted at the time as fatty aldehyde deformylase. The enzymes encoded by those genes catalyze the reduction of fatty acyl-CoA or –ACP to an aldehyde and transform the aldehyde to an alkane and a C_1 product, respectively. The C_1 product was initially presumed to be carbon monoxide,[45] but a later study showed it to be formic acid.[46] That study demonstrated that formate was produced in a 1:1 stoichiometry with alkane, and furthermore, the aldehyde hydrogen atom was retained in the formate. Although initial experiments suggested an anaerobic reaction that was formally hydrolytic,[47,48] it is now widely accepted that molecular oxygen is a substrate.[49–52] Studies using $^{18}O_2$ demonstrated that one of the oxygen atoms in the formate product contained ^{18}O, and thus the deformylating enzyme is an oxygenase.[51,52] It is an atypical oxygenase reaction, because the reaction of an aldehyde yielding an alkane and a carboxylic acid is overall redox neutral. In total, these studies have used orthologues from *Prochlorococcus marinus*, *Nostoc punctiformes*, *Synechococcus* sp. RS9917, and *Synechocystis* sp. PCC6803, which were overexpressed in *E. coli* and shown to have catalytic activity. The name aldehyde deformylating oxygenase (ADO) has been proposed for the enzyme.

The X-ray crystal structure of the *Prochlorococcus marinus* ADO had been determined in a structural genomics study (Protein Data Bank ID: 2OC5). The structure reveals interesting similarities to soluble methane monooxygenase (sMMO). The protein has a dinuclear metal cluster bridged by a glutamic acid carboxylate and an oxygen ligand of unknown derivation (Figure 9). A fatty acid, presumably derived from the *E. coli* overexpression vehicle and modeled as C_{18}, is tightly bound in the active site, presumably mimicking binding of the normal aldehyde substrate. The overall ligand type and spatial arrangement is very similar to that of the di-iron cluster of sMMO, and that is consistent with the current model of the reaction as an oxygenative cleavage of the aldehyde C–C bond. As in sMMO, the metal cluster is most likely di-iron.

A significant impediment to studies on ADO is the low turnover observed by all of the groups that have published on the system. It is possible that the isolated enzyme is largely inactivated or that the bound fatty acid acts as an inhibitor. Alternatively, it might be an extremely low turnover enzyme as the level of alkanes in native cyanobacteria is low. This is a key issue for both mechanistic studies and efforts to overproduce alkanes in engineered cells for biotechnological purposes.

Alpha Olefins via Cytochrome P450

In a reaction somewhat analogous to ADO, some bacteria have been reported to oxidatively decarboxylate fatty acids to the corresponding one carbon shorter alkane.[53] There had been reports that some bacteria produce intermediate chain length alkenes, and the demonstration that they were terminal, or alpha, olefins suggested a unique mechanism. An enzyme activity from a *Jeotgalicoccus* species was purified and determined to be a cytochrome P450 monooxygenase based on sequence. The gene was identified by genome sequencing of the organism. The gene was cloned and expressed in *E. coli* and the recombinant strain produced α-olefins.

A mechanism was proposed in which the β-carbon is oxidized, with subsequent elimination of carbon dioxide producing the α-olefin. There is precedent in the literature for similar reactions catalyzed by cytochrome P450 monooxygenases. In 1996, Davis et al. demonstrated the conversion of aldehydes to the corresponding carboxylic acids and α-olefins by the cytochrome P-450$_{BM-3}$ (CYP102A1).[54] They proposed a different mechanism with initial attack at the oxygenated carbon. In a more analogous reaction, the carboxylic acid drug diclofenac was observed to catalyze an oxidative decarboxylation at a position β to an aromatic ring carbon. The authors proposed initial oxidation at the carboxylic acid carbon prior to decarboxylation. In another precedent, a cytochrome P450 from a *Rhodotorula* yeast

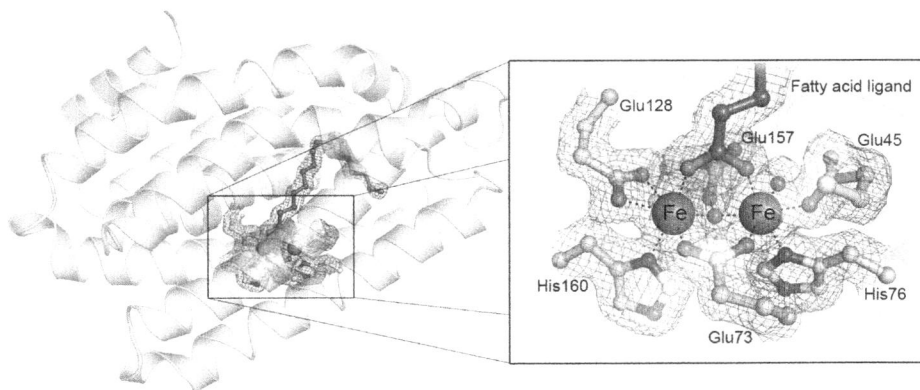

Figure 9

Structure of *Prochlorococcus marinus* MIT9313 aldehyde deformylating oxygenase (ADO) with bound fatty acid. ADO is drawn in cartoon, with active site residues drawn in stick (carbons green (light gray in print versions)). Irons are represented by gold (black in print versions) spheres, and waters by red (dark gray in print versions) spheres. The fatty acid is drawn in stick (carbons dark grey (darker gray in print versions)). 2Fo-Fc electron density (blue (lighter gray in print versions) mesh) is contoured at 1.0 σ. *This figure was produced using PyMOL (http://www.pymol.org/).*

catalyzed a decarboxylation reaction to produce an olefinic hydrocarbon, isobutene. The decarboxylation reaction is catalyzed by a specific cytochrome P450 monooxygenase that has been purified to homogeneity and the gene sequence has been determined.[55] Further studies on the *Jeotgalicoccus* enzyme with isotopic labeled substrates should help resolve the differences in the proposed decarboxylation mechanisms.

Alpha Olefins via a Polyketide-Type Pathway

Cyanobacteria were described that produce C_{19} α-olefins.[56] These α-olefins are longer than the C_{16} and C_{18} fatty acids observed in the cells, which suggested they did not arise from a direct decarboxylation of cellular fatty acids. A plausible mechanism would be a two-carbon elongation of a C_{18} fatty acid followed by a decarboxylation. It was recognized that a polyketide pathway produces a natural product, curacin A, with a terminal carbon–carbon double bond and that a similar mechanism could be operative in cyanobacteria.[57] The genome sequence of *Synechococcus* sp PCC 7002 was scanned for a polyketide-like gene. Specifically, a *curM* gene homolog was found. CurM is involved in the biosynthesis of curacin A. The role of the polyketide gene was first indicated by genetic knockout experiments that led to the elimination of α-olefin biosynthesis. A second biological approach was to replace the natural promoter region with a stronger promoter. That led to an increase in α-olefin biosynthesis, further corroborating the involvement of the polyketide gene, which was subsequently denoted as the *ols* gene.

Based on homology arguments and feeding experiments, a mechanism for α-olefin biosynthesis was proposed that was clearly distinct from previously-known pathways involving a cyto-chrome P450 monooxygenase reaction.[57] The *ols* gene product acts on C_{18} fatty acids, catalyz-ing a two-carbon elongation reaction via a standard decarboxylative Claisen condensation reaction with malonyl-CoA. There is a ketoreductase domain in Ols, suggesting reduction of the β-keto acid to a β-hydroxy acid. Additionally, there is a sulfotransferase domain. This is reminiscent of curacin A biosynthesis in which sulfation of the alcohol intermediate helps activate the β-carbon for an elimination reaction. This would be required for an eliminative decarboxylation reaction proposed to be catalyzed by the C-terminal thioesterase domain.

Conclusions

The presence of hydrocarbons in microorganisms has been known for decades. However, the biochemical basis of hydrocarbon biosynthesis has, in most instances, been revealed only within the last several years. The driving force for many of those biochemical investigations has been the need for developing renewable alternatives to hydrocarbon fuels and feedstocks.

In nature, hydrocarbon biosynthesis is likely an ancient function. One suggestion of this is the observation that the *ole* gene cluster is found throughout the prokaryotic tree of life, in deep branching lineages. The large diversity of mechanisms also suggests a widespread and lengthy occurrence of various hydrocarbon biosynthetic genes.

Clearly, studies on these systems will continue and new mechanisms will likely be discovered. Microbial hydrocarbon biosynthesis will remain a rich area of study for many years.

References

1. Yergin D. *The quest: energy, security and the remaking of the modern world.* New York: Penguin Group; 2011.
2. Kovarik B. Henry Ford, Charles F. Kettering and the fuel of the future. *Automot Hist Rev* 1998;**32**:7–27.
3. Ladygina N, Dedyukhina EG, Vainshtein MB. A review on microbial synthesis of hydrocarbons. *Process Biochem* 2006;**41**:1001–14.
4. Natural gas: history. http://naturalgas.org/overview/history/ September 20, 2014.
5. Eisner T, Eisner M, Siegler M. *Secret weapons: defenses of insects, spiders, scorpions, and other many-legged creatures.* Harvard University Press; 2005.
6. Thauer RK. Biochemistry of methanogenesis: a tribute to Marjory Stephenson. *Microbiol-UK* 1998;**144**:2377–406.
7. Hinrichs KU, Hayes JM, Bach W, Spivack AJ, Hmelo LR, Holm NG, et al. Biological formation of ethane and propane in the deep marine subsurface. *Proc Natl Acad Sci USA* 2006;**103**:14684–9.
8. Hu YL, Lee CC, Ribbe MW. Extending the carbon chain: hydrocarbon formation catalyzed by vanadium/molybdenum nitrogenases. *Science* 2011;**333**:753–5.
9. Lee CC, Hu YL, Ribbe MW. Vanadium nitrogenase reduces CO. *Science* 2010;**329**:642.
10. Yang ZY, Dean DR, Seefeldt LC. Molybdenum nitrogenase catalyzes the reduction and coupling of CO to form hydrocarbons. *J Biol Chem* 2011;**286**:19417–21.
11. Fujii T, Ogawa T, Fukuda H. Preparation of a cell-free, isobutene-forming system from *Rhodotorula-minuta.* *Appl Envir Microbiol* 1988;**54**:583–4.
12. Fall R, Copley SD. Bacterial sources and sinks of isoprene, a reactive atmospheric hydrocarbon. *Envir Microbiol* 2000;**2**:123–30.
13. Julsing MK, Rijpkema M, Woerdenbag HJ, Quax WJ, Kayser O. Functional analysis of genes involved in the biosynthesis of isoprene in *Bacillus subtilis.* *Appl Microbiol Biotechnol* 2007;**75**:1377–84.
14. Walsh CT. Revealing coupling patterns in isoprenoid alkylation biocatalysis. *ACS Chem Biol* 2007;**2**:296–8.
15. Tornabene TG, Holzer G, Peterson SL. Lipid profile of the halophilic alga, *Dunaliella-Salina.* *Biochem Biophys Res Comm* 1980;**96**:1349–56.
16. Park MO, Heguri K, Hirata K, Miyamoto K. Production of alternatives to fuel oil from organic waste by the alkane-producing bacterium, *Vibrio furnissii* M1. *J Appl Microbiol* 2005;**98**:324–31.
17. Wackett LP, Frias JA, Seffernick JL, Sukovich DJ, Cameron SM. Genomic and biochemical studies demonstrating the absence of an alkane-producing phenotype in *Vibrio furnissii* M1. *Appl Envir Microbiol* 2007;**73**:7192–8.
18. Frias JA, Richman JE, Wackett LP. C29 olefinic hydrocarbons biosynthesized by *Arthrobacter species. Appl Envir Microbiol* 2009;**75**:1774–7.
19. Rawlings BJ. Biosynthesis of fatty acids and related metabolites. *Nat Prod Rep* 1998;**15**:275–308.
20. Madigan M, Martinko J. *Brock biology of microorganisms.* 11th ed. Prentice Hall; 2005.
21. Albro PW, Dittmer JC. The biochemistry of long-chain, nonisoprenoid hydrocarbons. IV. Characteristics of synthesis by a cell-free preparation of *Sarcina lutea. Biochemistry* 1969;**8**:3317–24.
22. Albro PW, Dittmer JC. The biochemistry of long-chain, nonisoprenoid hydrocarbons. III. The metabolic relationship of long-chain fatty acids and hydrocarbons and other aspects of hydrocarbon metabolism in *Sarcina lutea. Biochemistry* 1969;**8**:1913–8.
23. Albro PW, Dittmer JC. The biochemistry of long-chain, nonisoprenoid hydrocarbonss. II. The incorporation of acetate and the aliphatic chains of isoleucine and valine into fatty acids and hydrocarbon by *Sarcina lutea* in vivo. *Biochemistry* 1969;**8**:953–9.

24. Albro PW, Dittmer JC. The biochemistry of long-chain, nonisoprenoid hydrocarbons. I. Characterization of the hydrocarbons of *Sarcina lutea* and the isolation of possible intermediates of biosynthesis. *Biochemistry* 1969;**8**:394–404.

25. Beller HR, Goh EB, Keasling JD. Genes involved in long-chain alkene biosynthesis in *Micrococcus luteus*. *Appl Envir Microbiol* 2010;**76**:1212–23.

26. Sukovich DJ, Seffernick JL, Richman JE, Gralnick JA, Wackett LP. Widespread head-to-head hydrocarbon biosynthesis in bacteria and role of OleA. *Appl Envir Microbiol* 2010;**76**:3850–62.

27. Friedman L, Da Costa B. Hydrocarbon producing genes and methods of their use. *World Patent* 2008;WO/2008/147781.

28. Friedman L, Rude M. Process for producing low molecular weight hydrocarbons from renewable resources. *World Patent* 2008;WO/2008/113014.

29. Pinzon NM, Aukema KG, Gralnick JA, Wackett LP. Nile red detection of bacterial hydrocarbons and ketones in a high-throughput format. *MBio* 2011;**2**:e00109–00111.

30. Sukovich DJ, Seffernick JL, Richman JE, Hunt KA, Gralnick JA, Wackett LP. Structure, function, and insights into the biosynthesis of a head-to-head hydrocarbon in *Shewanella oneidensis* strain MR-1. *Appl Envir Microbiol* 2010;**76**:3842–9.

31. Frias JA, Richman JE, Erickson JS, Wackett LP. Purification and characterization of OleA from *Xanthomonas campestris* and demonstration of a non-decarboxylative Claisen condensation reaction. *J Biol Chem* 2011;**286**:10930–8.

32. Bonnett SA, Papireddy K, Higgins S, del Cardayre S, Reynolds KA. Functional characterization of an NADPH dependent 2-alkyl-3-ketoalkanoic acid reductase involved in olefin biosynthesis in *Stenotrophomonas maltophilia*. *Biochemistry* 2011;**50**:9633–40.

33. Koglin A, Walsh CT. Structural insights into nonribosomal peptide enzymatic assembly lines. *Nat Prod Rep* 2009;**26**:987–1000.

34. Haapalainen AM, Merilainen G, Wierenga RK. The thiolase superfamily: condensing enzymes with diverse reaction specificities. *Trends Biochem Sci* 2006;**31**:64–71.

35. Goblirsch BR, Frias JA, Wackett LP, Wilmot CM. Crystal structures of *Xanthomonas campestris* OleA reveal features that promote head-to-head condensation of two long-chain fatty acids. *Biochemistry* 2012;**51**:4138–46.

36. Musayev F, Sachdeva S, Scarsdale JN, Reynolds KA, Wright HT. Crystal structure of a substrate complex of *Mycobacterium tuberculosis* beta-ketoacyl-acyl carrier protein synthase III (FabH) with lauroyl-coenzyme A. *J Mol Biol* 2005;**346**:1313–21.

37. Sachdeva S, Musayev FN, Alhamadsheh MM, Scarsdale JN, Wright HT, Reynolds KA. Separate entrance and exit portals for ligand traffic in *Mycobacterium tuberculosis* FabH. *Chem Biol* 2008;**15**:402–12.

38. Price AC, Choi KH, Heath RJ, Li Z, White SW, Rock CO. Inhibition of beta-ketoacyl-acyl carrier protein synthases by thiolactomycin and cerulenin. Structure and mechanism. *J Biol Chem* 2001;**276**:6551–9.

39. Omura S. The antibiotic cerulenin, a novel tool for biochemistry as an inhibitor of fatty acid synthesis. *Bacteriol Rev* 1976;**40**:681–97.

40. Ronnett GV, Kleman AM, Kim EK, Landree LE, Tu Y. Fatty acid metabolism, the central nervous system, and feeding. *Obes (Silver Spring)* 2006;**14**(Suppl. 5):201S–7S.

41. Suen Y, Holzer GU, Hubbard JS, Tornabene TG. Biosynthesis of acyclic methyl branched poly-unsaturated hydrocarbons in *Pseudomonas-maltophilia*. *J Ind Microbiol* 1988;**2**:337–48.

42. Tornabene TG, Peterson SL. *Pseudomonas-maltophilia*—identification of hydrocarbons, glycerides, and glycolipoproteins of cellular lipids. *Can J Microbiol* 1978;**24**:525–82.

43. Frias JA, Goblirsch BR, Wackett LP, Wilmot CM. Cloning, purification, crystallization and preliminary X-ray diffraction of the OleC protein from *Stenotrophomonas maltophilia* involved in head-to-head hydrocarbon biosynthesis. *Acta Crystallogr Sect F Struct Biol Cryst Commun* 2010;**66**:1108–10.

44. Blumer M, Guillard RRL, Chase T. Hydrocarbons of marine phytoplankton. *Mar Biol* 1971;**8**:183–9.

45. Schirmer A, Rude MA, Li XZ, Popova E, del Cardayre SB. Microbial biosynthesis of alkanes. *Science* 2010;**329**:559–62.

46. Warui DM, Li N, Norgaard H, Krebs C, Bollinger JM, Booker SJ. Detection of formate, rather than carbon monoxide, as the stoichiometric coproduct in conversion of fatty aldehydes to alkanes by a cyanobacterial aldehyde decarbonylase. *J Am Chem Soc* 2011;**133**:3316–9.

47. Das D, Eser BE, Han J, Sciore A, Marsh ENG. Oxygen-independent decarbonylation of aldehydes by cyanobacterial aldehyde decarbonylase: a new reaction of diiron enzymes. *Angew Chem Int Ed* 2011;**50**:7148–52.

48. Eser BE, Das D, Han J, Jones PR, Marsh ENG. Oxygen-independent alkane formation by non-heme iron-dependent cyanobacterial aldehyde decarbonylase: investigation of kinetics and requirement for an external electron donor. *Biochemistry* 2011;**50**:10743–50.

49. Das D, Eser BE, Han J, Sciore A, Marsh ENG. Oxygen-independent decarbonylation of aldehydes by cyanobacterial aldehyde decarbonylase: a new reaction of diiron enzymes. *Angew Chem Int Ed* 2012;**51**:7881.

50. Eser BE, Das D, Han J, Jones PR, Marsh ENG. Oxygen-independent alkane formation by non-heme iron-dependent cyanobacterial aldehyde decarbonylase: investigation of kinetics and requirement for an external electron donor. *Biochemistry* 2012;**51**:5703.

51. Li N, Chang WC, Warui DM, Booker SJ, Krebs C, Bollinger JM. Evidence for only oxygenative cleavage of aldehydes to alk(a/e)nes and formate by cyanobacterial aldehyde decarbonylases. *Biochemistry* 2012;**51**:7908–16.

52. Li N, Norgaard H, Warui DM, Booker SJ, Krebs C, Bollinger JM. Conversion of fatty aldehydes to alka(e)nes and formate by a cyanobacterial aldehyde decarbonylase: cryptic redox by an unusual dimetal oxygenase. *J Am Chem Soc* 2011;**133**:6158–61.

53. Rude MA, Baron TS, Brubaker S, Alibhai M, Del Cardayre SB, Schirmer A. Terminal olefin (1-alkene) biosynthesis by a novel P450 fatty acid decarboxylase from *Jeotgalicoccus species*. *Appl Environ Microbiol* 2011;**77**:1718–27.

54. Davis SC, Sui ZH, Peterson JA, de Montellano PRO. Oxidation of omega-oxo fatty acids by cytochrome P450(BM-3) (CYP102). *Arch Biochem Biophys* 1996;**328**:35–42.

55. Fujii T, Ogawa T, Fukuda H. Characterization of cytochrome P450rm, benzoate 4-hydroxylase, from *Rhodotorula minuta*. *FASEB J* 1997;**11**:A813.

56. Winters K, Parker PL, Van Baalan C. Hydrocarbons of blue-green algae: geochemical significance. *Science* 1969;**163**:467–8.

57. Mendez-Perez D, Begemann MB, Pfleger BF. Modular synthase-encoding gene involved in alpha-olefin biosynthesis in *Synechococcus* sp strain PCC 7002. *Appl Environ Microbiol* 2011;**77**:4264–7.

CHAPTER 3

Perspectives on Process Analysis for Advanced Biofuel Production

Christopher J. Scarlata, Ryan E. Davis, Ling Tao, Eric C.D. Tan, Mary J. Biddy

National Bioenergy Center, National Renewable Energy Laboratory (NREL), Golden, CO, USA

Introduction

The purpose of process analysis is to provide information to make judgments about the challenges and relative merits of a conversion process. Process analysis draws information from one or more of three complementary parts, the process model, the economic model, and the life-cycle model (Figure 1). In brief, the process model describes the plant size, unit operations, process conditions, product yields, flow rates, and associated energy and material balance information for a given modeled technology. The economic model translates the results of the process model into financial results, such as the minimum fuel selling price (MFSP). Similarly, the life-cycle model uses the results of the process model to generate sustainability metrics such as carbon efficiency and greenhouse gas profiles.

The three models are often generated and refined in an iterative manner. The level of rigor in analysis depends on the stage of development and the analysis objectives. Data from research and development (R&D) is used to develop and refine the process model. The output from the process model informs the generation of the economic and life-cycle models that may lead to the subsequent revision of the process model, and so on. Results from concurrent economic and life-cycle analyses can highlight areas of the process model that need further development and identify gaps in science and engineering knowledge that illustrate opportunities for additional research.

The discussion below is intended to present a high-level, qualitative overview of process analysis for three biological approaches for producing hydrocarbons from biomass; aerobic respiration, anaerobic fermentation, and consolidated bioprocessing (CBP), by describing conceptual process designs and their implications on integrated commercial models. The aerobic respiration process is presented in detail as a base-case, and the other processes are contrasted to it.

Direct Microbial Conversion of Biomass to Advanced Biofuels. http://dx.doi.org/10.1016/B978-0-444-59592-8.00003-8

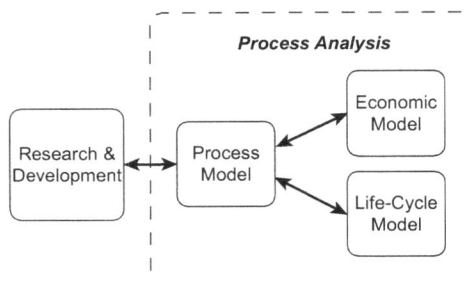

Figure 1
Process analysis is the evaluation of the implications from the results of process, economic, and life-cycle models. There is an iterative give and take of information in process analysis. The results of the process model inform the economic and life-cycle models and vice-versa. Data from research and development informs the process model, and implications from process analysis can be used to show where the R&D opportunities lie to improve process performance, economics, and sustainability.

Overview of Process Analysis

Table 1 lists examples of the diverse assumptions, inputs, and outputs required for a typical biorefinery process analysis using a biochemical/biological-based technology pathway as a guide. Information for the process, economic, and life-cycle models have been included to illustrate the connections between those areas. There are assumptions, such as governing theoretical yields that are provided by literature values and R&D operations. Other process assumptions, such as annual operating factor (i.e., hours of operation per year), may be generated from comparisons with other industry values or from subject matter experts. Financial assumptions may come from public policy (e.g., tax and depreciation rates) or, again, from comparisons with other industry values and subject matter experts.

Inputs to the process model may be provided by experimental data such as enzyme loading, chemical demands, or reactor operating conditions. Some inputs, such as labor costs, may be tailored to specific markets depending on the targeted plant location. Similarly, model outputs can be process-focused, financial-focused, or sustainability-focused. Calculated plant size and flow rates are used to generate capital costs and help determine the variable operating costs for raw materials and energy. Some fixed operating costs can be based on simplifying assumptions (e.g., using a multiplier of the installed capital cost[12]). Summary values from a process model include the bottom line MFSP that is often used as a basis for comparing competing process configurations or to quantify the cost implications associated with measured R&D progress.

Process model

Process analysis begins with the creation of a process model. The model incorporates information from R&D, engineering studies, and published literature. That information is used to generate process flow diagrams, reactor and unit operation simulations, and rigorous material and energy balances typically using chemical process modeling software like Aspen Plus or

Table 1: Examples of the diverse assumptions, inputs, and outputs that are required to complete a biochemical conversion process analysis

	Assumptions	Inputs	Outputs
Process	Governing theoretical yields Operating factor/on-stream time Equipment overdesign Feedstock rate Feedstock composition	Enzyme loading Metabolic yield Microbial productivity Operating conditions Chemical requirements Unit operation performance Reaction conversions Product losses	Plant size Product yield Water consumption Facility heat and power usage Flow rates for all streams Co-product yields Input/output inventories
Economic	Tax rate Discount rate Internal rate of return Depreciation method Construction financing Working capital Indirect capital factors Plant life Loan terms Debt-to-equity ratio Construction schedule Startup time	Process outputs (sets capital and operating costs) Feedstock price Labor and overhead costs Periodic replacement of consumables Equipment design and cost Chemical/material prices Maintenance, tax, and insurance costs	Product selling price or MFSP Annual cash flows Net present value Payback period Cost by production area Co-product revenue Capital vs operating cost allocations
Life-Cycle	Global-warming potential time horizon System boundary Environmental burden allocation LCA types (attributional/consequential)	Life-cycle inventories (process outputs) Natural resource consumption Nonrenewable energy consumption Water consumption Waste water production GHG emissions Criteria air pollutants Solid waste Product and byproduct generation	Global-warming potential Consumptive water use Energy return on investment Net energy value Carbon efficiency

LCA = life cycle analysis, GHG = greenhouse gas.

other modeling packages. The resulting outputs from process modeling can be used to estimate unit-level mass and energy flow rates and biorefinery plant size.

A rigorous biorefinery model in Aspen Plus could have: more than 100 unit operations; more than 500 streams for material, heat, and work; scores of control blocks; and dozens of components. Component properties may come from physical property databases or from experimental data generated by R&D functions and play a key role in thermodynamic property interactions. This level of rigor enables the development of robust models that account for critical process elements such as vapor–liquid equilibrium calculations, heat balances around reactors or distillation columns, product stream characteristics, and other factors that weigh

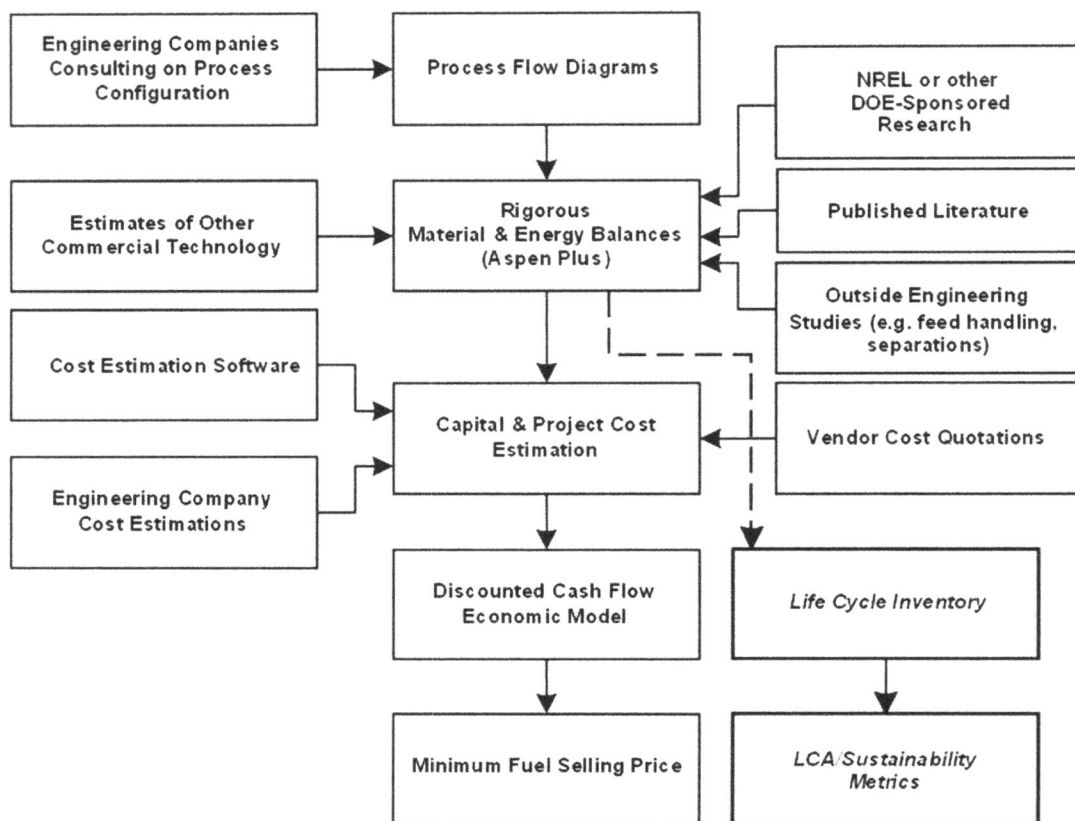

Figure 2
A flow chart of NREL's approach to process design, economics, and life cycle analysis.[13,14]

on overall model integration and resulting techno-economic analysis (TEA) and subsequent life-cycle analysis (LCA) outputs. Figure 2 is a more detailed view of NREL's approach to biorefinery process modeling. The software, in this case Aspen Plus, is at the center of the methodology, and the inputs and assumptions described above form the basis of the models. The figure also illustrates the connection to the economic and life-cycle models.

The process models described in the following sections have unit operations that are common among biochemical conversion technologies. This includes pretreatment reactors, on-site enzyme production, enzymatic hydrolysis and biological conversion reactors, product recovery and finishing operations, wastewater treatment systems, and systems for heat and power integration. The specific size and configuration of these unit operations can be quite different among aerobic, anaerobic, and CBP technologies. Some of these systems may be combined or significantly modified between these technologies, with important economic and sustainability implications.

Techno-economic analysis

TEA has been an indispensable tool over the last several decades to determine biofuel production costs for economic feasibility assessments.[14–25] Many of the approaches to converting biomass into advanced fuels are in ongoing development, and require guidance from process analyses to help prioritize research directions. TEA provides information needed to make informed judgments about the viability of a given conversion process; it is particularly useful to identify technical barriers and measure progress toward overcoming those barriers.[26]

In TEA, plant size and flow rates from the process model are translated into financial metrics such as capital and operating costs. Vendor quotations or other cost-estimating tools are used to generate equipment costs for a given process configuration. Operating costs from the consumption of raw materials, heat and power, and labor can be estimated from the process model. These values can be incorporated into a discounted cash flow model for financial analysis and summarized as the minimum fuel selling price or MFSP. The MFSP is the fuel price in which the net present value, the time-adjusted value of incoming and outgoing cash flows, of a project is zero. Today, there is a lack of statistical data on discount rates used by industry, but Short et al. have suggested that a 10% rate has been appropriate for the evaluation of renewable energy technology.[27]

To avoid artificial penalties on biorefinery economics attributed to "pioneer-plant" or first-of-a-kind facilities with early entry into the market, TEA typically uses design, cost, and performance assumptions intended to be reasonable for an "nth" plant facility—a facility constructed after a sufficient number of earlier facilities using similar technology have been built and operated such that the technology has progressed beyond a learning curve intrinsic to any new commercial-scale process. This allows for TEA to focus on the technology itself and not on economic and process scale-up uncertainty, such as higher-risk financing, delayed start-ups, equipment overdesign, reduced on-stream factors, and other risk premiums associated with pioneer plants that may increase modeled production costs.

Life-cycle analysis

The success of the biofuels industry depends not only on economic viability but also on environmental sustainability. A biorefinery process that is economically viable (e.g., has been optimized for cost) but suffers from a key sustainability drawback is not likely to be a long-term solution to replace fossil-derived fuels. Therefore, an important aspect of evaluating biomass-derived fuel processes is the assessment of resource consumption and environmental emissions, broadly termed life-cycle analysis. LCA provides a framework from which the environmental sustainability of a given process may be measured. Life-cycle inventory (LCI) data are generated from process model input and output flows, associated both with the biorefinery process itself for biofuel production ("direct emissions") and with material inputs and outputs to and from the biorefinery ("indirect emissions"). LCI data are used to quantify the consumption of natural resources, including water, energy, and raw materials, as well as emissions to air, land, and water associated with the production of biofuels. LCI data are then used to evaluate sustainability metrics like consumptive water use, carbon efficiency, or greenhouse gas and fossil energy profiles.

To quantify sustainability impacts, material and energy balances from process modeling are used to develop life-cycle metrics for a modeled biorefinery.[28–30] Software packages such as SimaPro are used to quantify life-cycle impacts, making use of databases such as Ecoinvent and the US Life Cycle Inventory (US LCI). The Ecoinvent parameters may be modified to reflect US or other country-specific conditions (e.g., replacing the default European electricity mix with the US electricity mix), and the US LCI processes are adapted to account for embodied greenhouse gas (GHG) emissions and fossil energy usage associated with the production process for a particular material. The LCI of the bioconversion step captures the impact of input raw materials and outputs, such as emissions and waste, and may be provided by Aspen Plus process modeling outputs. In addition to SimaPro, other common LCA software tools include GREET (US Department of Energy), TRACI (US Environmental Protection Agency), EIO-LCA (Carnegie-Mellon University), Gabi4 (University of Stuttgart), and Spine (Chalmers University).

Implications for the biorefinery business model

Process models are based on four broad considerations; conversion technology, type of feedstock, plant size, and range of products.[13] There are many choices for process technology and feedstock type.[1–11] The size of the plant is often a tradeoff between feedstock availability, economies of scale, and capital costs. However, the choice of products may be an open question. Biorefineries will need to balance the production of commodity fuels, higher value co-products, and heat and power with implications therein on TEA, LCA, and co-product market volume considerations. This is particularly challenging in the context of the revenue model for petroleum refiners.

Petroleum refinery margins are small compared to the margins for the cost of crude oil. Historical data from the Energy Information Agency suggests that refiners earned an average 13% margin on gasoline and 15% on diesel, as a fraction of the retail price at the pump, over the last 10 years.[31] Refining margins have had a high degree of risk as well. Gasoline margins have fluctuated ±50%, and diesel margins have only been slightly more stable at ±35% (Figure 3). In contrast, an average 61% and 57% of the respective cost at the pump is from the margin on crude oil. Most petroleum refiners also have a stake in crude oil production, which is a significant benefit to their bottom line. Biorefineries must develop and maintain business models that enable them to operate profitability. This could take the form of lowering the cost of production, producing higher value co-products, and converting all fractions of the biomass feedstock into useful products. Process analysis can be used to quantify and illuminate the most viable routes to a profitable, sustainable biorefinery business model.

With this overview in mind on the purpose and rationale for process analysis, the following discussion provides a number of examples into the thought process and methodologies behind process analysis for biochemical conversion pathways. The focus is primarily on process modeling with brief discussion on economic implications. LCA is an important part of a well-rounded process analysis; however, it is beyond the scope of this discussion.

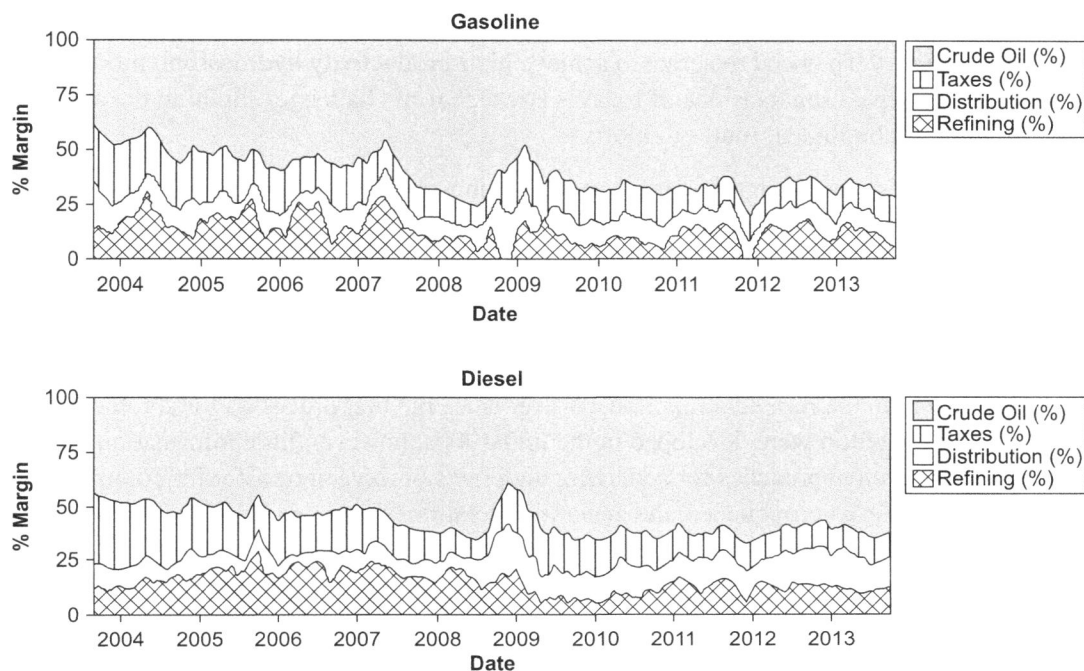

Figure 3
The historical revenue margins for refining, distribution (including marketing), taxes, and crude oil as a percentage of retail gasoline prices (top) and diesel prices (bottom) from 9/2003–9/2103.[31]

The discussion begins with a description of the aerobic production of hydrocarbons (e.g., renewable diesel blendstock or RDB). Hydrocarbon biofuels are advantageous because they would be compatible (i.e., fungible) with existing infrastructure and may be used as a replacement for fossil-derived fuels or as supplemental blendstocks (as in the case of RDB here). This aerobic process is presented in detail, as previously described in Ref. 32 and serves as the basis of comparison for two other processes, anaerobic fermentation and consolidated bioprocessing of biomass to RDB. The discussion focuses on qualitative aspects of process analysis with a particular focus on the contrasts between conceptual process models for each of the technologies. Readers who are interested in detailed techno-economic analysis are encouraged to view the publically available design reports cited in the references and available online (see: http://www.nrel.gov/publications/).

Aerobic Bioprocess
Process Design Details

Initial designs for the biological production of RDB from biomass have been based on microbes that require oxygen to function.[14] Therefore, aeration plays an essential role in biological sugars-to-hydrocarbon production, yet it is costly to implement. Oxygen has a low solubility in aqueous solutions and must be supplied continuously to a submerged, aerobic microbial culture.[33] The need to aerate to achieve high-productivity hydrocarbon production from biomass-derived sugars is one of today's key technical challenges, limiting the ability to scale-up low-cost biological routes to biofuels.

Numerous reviews of oxygen transfer for aerating submerged cultures are available, and most are for stirred tank or airlift/bubble column bioreactor systems.[33–37] Recent literature on oxygen transfer has focused primarily on understanding scale-up of mammalian and plant cell cultures in which bioreactor sizes of 20,000 L are considered "large scale." The scarcity of new information on aeration systems for extremely large-volume applications like commodity biofuels is attributed to two factors. First, aeration of submerged microbiological cultures has been a focus within the fermentation industry ever since the first processes for aerobic production of penicillin were developed in the mid-1900s; however, little information has become available on approaches for achieving high rates of oxygen transfer for commodity scale operations. As a consequence, the general correlations for oxygen transfer in a hypothetical RBD process remain similar to those developed many decades ago.[38] Second, while much of the pioneering work on aerobic submerged production of penicillin was carried out in the public domain at US national laboratories (e.g., US Department of Agriculture's laboratories in Peoria, IL.), more recent developments on scaling up aerobic submerged cultivation processes have been carried out in the private sector, where information is often closely held. The physical engineering challenges of achieving effective gas–liquid mass transfer and the basic approaches for maximizing oxygen transfer at scale remain unresolved in the public domain for low-cost commodity biofuel applications.

The aerobic biofuel production pathway described below follows similar steps as from NREL's 2011 ethanol design report that describes cellulosic sugar conversion.[16] Downstream operations are modified, primarily around sugar conversion and product recovery, which carry important implications for process analysis. A high-level overview of the process is shown as a block diagram in Figure 4.

The conceptual process begins with feed handling (not shown). Feedstock handling operations may consist of loading/unloading equipment, storage, conveyors, preprocessing such as grinding or other operations to achieve particle-size targets, drying, and other steps required to store, prepare, and deliver the feedstock to the throat of the pretreatment reactor. Costs associated with these operations may be considered inside or outside the scope of a process and economic analysis; in NREL's analyses, typically all costs associated with feed-handling operations are included in a given delivered feedstock price, tied to other analyses on feedstock logistics operations. From there, the biomass is conveyed to the pretreatment reactor.[10,39]

Pretreatment and conditioning: In this section, the biomass is treated with dilute sulfuric acid at a moderately high temperature for a short time to liberate hemicellulose sugars and make the biomass susceptible to enzymatic hydrolysis. The pretreated slurry is adjusted to a pH

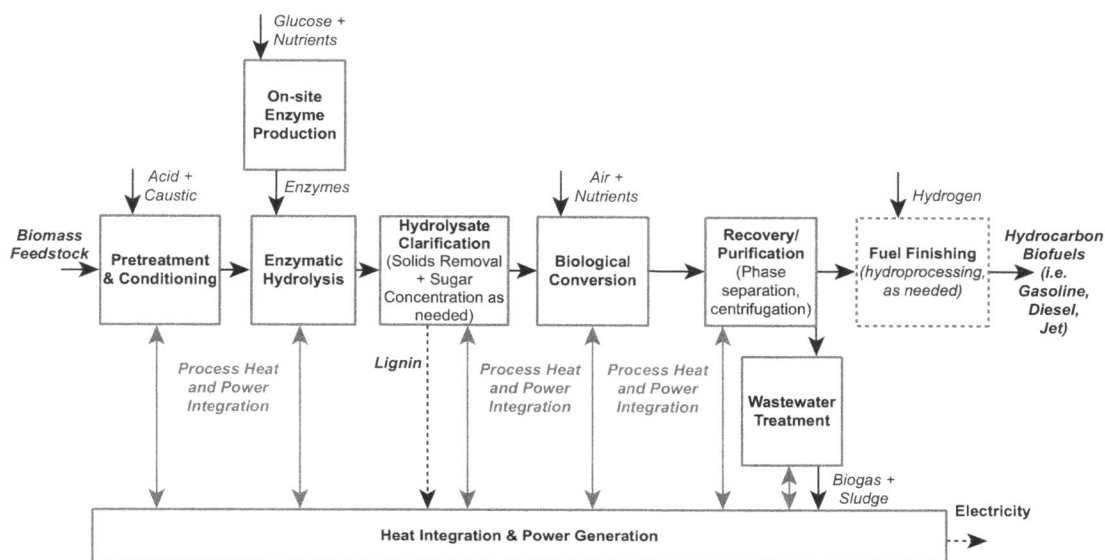

Figure 4
Example process schematic for hydrocarbon production via aerobic bioconversion.[32]

near five for subsequent enzymatic hydrolysis. Pretreatment is operated at 30 wt% total solids, consistent with NREL's previously published design cases.[14,16] Note that other pretreatment technologies could be used in place of dilute sulfuric acid.

During the 2012 state of technology pilot-scale demonstration efforts at National Renewable Energy Laboratory (NREL), strategies focused on improving process integration were important to reducing conversion costs. One modification to the 2011 NREL biochemical ethanol design case[16] explored the addition of a deacetylation preprocessing step, by means of which feedstock is first soaked in dilute sodium hydroxide and then drained to remove a significant portion of acetate prior to dilute-acid pretreatment. Acetate is a known inhibitor for both enzymatic hydrolysis and ethanol fermentation. Adding this upfront preprocessing step not only improved sugar yields from enzymatic hydrolysis and ethanol yields from fermentation during pilot-scale operational trials, it also reduced modeled processing costs by removing a portion of the unfermentable fraction of the biomass, thus reducing overall volumetric throughputs through all downstream operations (resulting in lower capital costs) and lowering the pretreatment severity requirements (also resulting in lower operating costs).[40] This is a prime example of the benefit of integrated process analysis alongside experimental research, in that beyond the observed benefit of reduced acetate inhibition and increased yields, a secondary but equally important benefit of reduced costs could be realized and quantified.

The benefit of a deacetylation preprocessing step in RDB production pathway is less clear, because hydrocarbon-producing organisms may tolerate acetate.[41] Process and cost analysis may be leveraged to serve as an effective mechanism for cost–benefit evaluation between acetate use (yield improvement) and lower biomass throughput (cost improvement) in terms of overall modeled product-selling prices. A continued research emphasis on understanding the process integration of a conceptual biorefinery will be critical to improving the overall economics and maximizing carbon yields.

Optimizing upstream processes, including preprocessing and pretreatment, will be important to produce a hydrolysate stream with the qualities and composition best suited for hydrocarbon production. Further expanding opportunities for producing value-added co-products from currently underused fractions of the biomass, including the lignin and acetate fractions, will also be a key driver toward improving the economic viability of this conversion pathway.

Enzymatic hydrolysis: Enzymatic hydrolysis is carried out in a high-solids continuous reactor (24-h residence time) using a cellulase enzyme that may be prepared on-site. The partially hydrolyzed slurry is transferred to one of several parallel reactors. Hydrolysis is completed in the batch reactors, which are modeled in NREL's processes as one million gallon tanks operating at a 60-h batch time.[14,16] While these hydrolysis conditions may remain constant in transitioning toward hydrocarbon production, enzyme loading is a well-documented primary cost driver in the biochemical process design and remains an important means toward achieving further cost reductions.

In the 2012 state of technology pilot-scale demonstration runs at NREL, overall glucan-to-glucose conversions of 75–90% were observed using an enzyme loading of 20–30 mg enzyme protein/g cellulose.[40,42,43] These results suggest that favorable glucose yield and low enzyme loading are achievable. Further room for improvement exists to maintain high glucose yields while continuing to reduce enzyme dosage as improved enzymes are developed.

Hydrolysate clarification: After enzymatic hydrolysis is completed, the slurry may be clarified using a filter press to remove insoluble solids, primarily lignin-rich residues. This is a shift from anaerobic ethanol processing in which lignin-rich solids are not removed until after fermentation is complete.

In an aerobic bioprocess, the presence of solids is anticipated to interfere with gas–liquid oxygen mass transfer and limits oxygen uptake rates.[44,45] Current literature studies for microbial production of hydrocarbon biofuels tend to focus on using clean, insoluble-solids-free sources of commodity sugars (e.g., glucose, sugarcane juice, corn syrup). It remains unclear to what extent the removal of residual insoluble solids is necessary for aerobic hydrocarbon production pathways, a point which requires experimental validation to resolve. A filter press could be used to clarify the hydrolysate liquor but may result in sugar loss, because some of the soluble sugars will remain in the solids stream. Performing the solids removal process downstream of sugar conversion would be preferable to maximize yield, although it could present new challenges for product purification. The lignin-rich residues removed from the process could be a potential source of fuels and co-products through development of new conversion methods.

The clarified sugar stream may then be sent directly to the biological conversion step or may be further processed to concentrate the sugars by methods such as evaporation, reverse osmosis, or nanofiltration. While using dilute (100–150 g/L) sugars is the approach taken in biochemical cellulosic ethanol production,[16] concentrated commodity sugars (≥500 g/L) have been used in literature reports on hydrocarbon biofuels production.[46] Different processing schemes for the biological conversion step also require different optimum sugar concentrations, with higher sugar concentrations being more conducive to fed-batch operation. Although overall bioreactor volumes can be similar in either case if volumetric productivities (i.e., g/L/h) are similar, there are additional impacts to downstream unit operations within the context of an integrated process model if concentrated sugars are used.

To reach product cost targets, it will be important to understand and quantify the tradeoffs between biological conversion of dilute sugars versus the conversion of more concentrated sugars that require additional costs for sugar concentration but potentially result in higher product titers and/or more efficient water management and bioreactor use, again an example of tradeoffs that may be quantified through process analysis.

Biological conversion: It is convenient to contrast the aerobic bioconversion of sugars to RDB to the well-established cellulosic ethanol processes. Biological RDB conversion processes are modified considerably from the cellulosic ethanol design cases previously published.[16,47] These modifications include the addition of air compressors, use of smaller bioreactor vessels, and powerful agitation systems to achieve targeted levels of oxygen gas–liquid mass transfer. One of the largest bioreactors yet reported publicly for biological upgrading of sugars to hydrocarbons is 130,000 gallons,[48] contrasted with NREL's 2011 ethanol design case using one million gallon anaerobic vessels. It is expected that with further optimization it will be possible to increase maximum aerobic vessel size beyond this initial value.

Smaller bioreactor volumes translate to economy of scale penalties for aerobic processes relative to the one million gallon fermentors modeled for lignocellulosic ethanol production. Additionally, aeration is costly to implement. Air compressors and powerful motors are needed to supply the large quantities of air and vigorous levels of agitation necessary to maintain adequate oxygen transfer rates. A preliminary assessment of an industrial process for producing ethanol, yeast, and lignin products found the cost of aeration for yeast cultivation to be roughly equivalent to the cost of enzymes for cellulose hydrolysis.[49]

Although batch sugar fermentation is stipulated in the NREL biochemical ethanol design model,[16] running the process in fed-batch mode could potentially reduce processing and economic challenges to achieving increased hydrocarbon yields and titers.[14] A fed-batch mode is more conducive to operations for aerobic systems because it allows for controlling the feed rate of substrates and overall reaction conditions to maximize carbon efficiency to fuel product and minimize parasitic energy losses. In light of these considerations, to achieve economic viability it will be important to evaluate the engineering design of microbial fuel production in detail to establish a realistic design at which aerobic conversion may proceed at an acceptably large scale and cost. Further development of microbial catalysts will also be vital to the development of economically viable biomass to RDB processes.

The best cases for minimizing downstream processing requirements are for the RDB product to be secreted from the cell and to separate into a product-rich liquid phase. Some microorganisms (such as heterotrophic algae) may accumulate fuel precursor molecules intracellularly and will require dedicated extraction steps to recover the fuel product. Preliminary scenario modeling indicates that this may incur substantial cost and energy penalties, a point consistent with other analyses that suggested a threefold higher energy efficiency ratio demonstrated today for hydrocarbon product secretion pathways relative to intracellular storage and extraction.[50] Product secretion will therefore remain a preferred route toward achieving economic targets, which, combined with the need to obtain high product titers, also dictates the need for the microorganism to be resistant to toxicity effects of the secreted products.

Product recovery/processing: For pathways in which the product is secreted from the cell, the bioreactor broth primarily contains the RDB product and water. Because most of the insoluble solids are removed prior to this step, only a small amount of insoluble solids such as microbial cell mass are present in the broth. A distinct advantage of diesel-range hydrocarbon products over short-chain alcohols is their low solubility in water, which may be exploited to allow for product separation and recovery via simple phase-separation methods rather than more costly and energy-intensive distillation.

The lighter phase containing the long-chain hydrocarbon product may first be concentrated in a standard decanter vessel. The resulting hydrocarbon-rich phase may then be centrifuged to recover the desired product at high purity. Because the aqueous phase exiting product separation contains high levels of inorganic salts such as ammonium sulfate, excess nutrients, and soluble inorganic compounds, the aqueous stream is directed to wastewater treatment for cleanup and chemical and energy recovery. The current product recovery scheme, which resembles those reported in the literature,[2] may result in considerable savings in energy and capital costs as compared with energy-driven separation processes (e.g., azeotropic distillation) or mass separation (e.g., solvent extraction or absorbent-based) schemes. As noted above, some microbial hydrocarbon production and recovery pathways may require additional extraction or cleanup steps to be incorporated, incurring additional costs. The exact recovery yields and product losses in these designs needs further quantification and potentially further optimization. For example, product recovery yields may be reduced in a secreted product scenario due to emulsification with extracellular material or adsorption to cells or other surfaces.

Some biological products may already be a "final product" blendstock (e.g., paraffins such as pentadecane); most classes of molecules, including fatty acids, fatty alcohols, or isoprenoids, will require an upgrading step such as hydrotreating to saturate the molecule(s) and remove oxygen to produce the final fuel or blendstock product.[2] Even so, hydrotreating operations for such products are likely to be milder and less costly than severe hydroprocessing operations such as that required for pyrolysis oil upgrading.[51] Reaction pathways for oxygen removal may proceed by rejection of CO_2 (decarboxylation) or H_2O (hydrodeoxygenation), most likely with both pathways participating with a favorable preference for one dictated by catalyst and reaction conditions. Each pathway for oxygen rejection incurs tradeoffs, with decarboxylation resulting in lower carbon efficiency (net fuel yields) due to CO_2 formation, but hydrodeoxygenation requiring more hydrogen to reject oxygen as H_2O.

Process analysis allows for an informed cost–benefit evaluation to be conducted optimizing reaction pathways and operating conditions, based on the specific biological intermediate product being considered. Further tradeoffs also exist between intermediate molecules such as fatty acids and isoprenoids, in which the latter is already devoid of oxygen, but instead requires hydrogen for saturation. The resulting products from hydrotreating are largely straight and/or branched saturated paraffins, high in cetane value for use as diesel blendstock materials.

Cellulase enzyme production: An on-site enzyme production section is included in this example with conceptual design and cost assumptions documented in NREL's biochemical design reports.[14,16] Corn syrup is the primary carbon source for enzyme production. Media preparation involves a step in which a portion of the glucose is converted to sophorose to induce cellulase production. The enzyme-producing fungus, modeled after *Trichoderma reesei*, is grown aerobically in fed-batch bioreactors. The entire enzyme production broth, containing the secreted enzyme, is fed to the enzymatic hydrolysis reactor. Other opportunities to further reduce enzyme production cost include optimizing the production process to use lower-cost biomass-derived sugars rather than corn syrup as the primary carbon source. Alternative models for enzyme sourcing exist and may be more realistic in the near-term, namely purchasing commercial enzyme cocktails from third-party vendors.

Wastewater treatment: Wastewater streams are treated by anaerobic and aerobic digestion. The methane-rich biogas from anaerobic digestion is sent to the combustor, where sludge from the digesters is also burned. The treated water is suitable for recycling and is returned to the process. Outside battery limit (OSBL) processes such as wastewater treatment, solids combustion/steam generation, utilities, etc. can add up to contribute substantial costs to overall integrated process economics and should not be neglected as "minor" secondary operations.

Heat and Power Integration: The insoluble lignin-rich residues from the solids separation step, the solids from wastewater treatment, and the biogas from anaerobic digestion are combusted to produce high-pressure steam for electricity production and process heat in one possible design configuration. Any excess steam may be converted to electricity for use in the plant and sold to the grid as a co-product. Opportunities exist to improve process economics by developing higher-value uses for one or more of these residual solids streams.

Aerobic Bioprocess Discussion

Among options discussed here for biochemical hydrocarbon production including aerobic and anaerobic conversion of sugars or more direct consolidated bioprocessing, aerobic conversion is the most complex in terms of number of steps and system design requirements, which carry cost implications. Such issues include the likely need for solids removal prior to conversion (requiring higher cost separation equipment and incurring sugar losses), sugar concentration, fed-batch operation for bioconversion using smaller, more costly bioreactors, and increased power demands for agitation and aeration balanced by achievable oxygen transfer rates. Such considerations call for detailed process analysis informed by experimental data at scale to understand and optimize an appropriate system design.

Aside from engineering and design challenges, from a process standpoint the primary cost drivers for aerobic bioconversion are product yield (g product/g sugar substrate) and volumetric productivity (g/L/h of product being produced), both of which require manipulation of the microorganism and optimization of the production process as key research and development strategies to ultimately achieve viability. Recent literature suggests that the current state of technology for microbial conversion to hydrocarbon biofuels includes product titers ranging from 0.1 to 24 g/L of long-chain hydrocarbons,[2,46,52–54] with times for batch or fed-batch production ranging from 2 to 7 days.[55]

There remains considerable room for improvement in the efficiency of sugar conversion to fuel, particularly for utilization of pentose sugars, across a variety of hydrocarbon-producing microorganisms.[41,56,57] Further improving specific productivity rates (g product/g cell/h) by targeting a lower diversion of sugar to microbial cell growth and/or engineering ways to recover and reuse microbial cells (e.g., cell retention or cell recycle bioreactor configurations) as well as mitigating potential hydrocarbon product toxicity effects will increase overall process yields and improve economic viability.[50]

Another issue frequently overlooked in process and economic analysis is contamination with implications on yield or operational downtime penalties. Contamination poses an important concern for anaerobic fermentation pathways such as ethanol production. It poses an even

Concentration, Reported Yield, Volumetric Productivity by Product

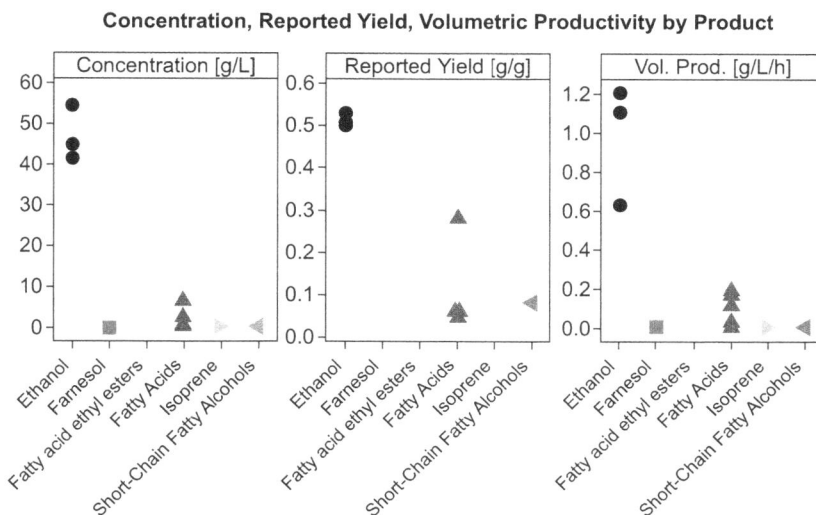

Figure 5

Comparison of titer, sugar utilization, and volumetric productivity for ethanol versus selected hydrocarbon products produced by *Escherichia coli*, excerpted from Ref. 14 and as originally reported in Ref. 58.

greater challenge for aerobic bioprocesses and can quickly spoil yields in aerobic systems by diverting carbon away from the desired fuel product if sterility is not maintained. This calls for the need to develop robust organisms and strict system control and longer vessel turn-around time between batches for cleaning and sterilization.

As an example for illustrative purposes of microorganism development needs for hydro-carbon pathways relative to ethanol benchmarks, Figure 5 shows a subset of published performance metrics for concentration, yield, and volumetric productivity for *E. coli* The figure was excerpted from the 2013 NREL biochemical hydrocarbon design report[14] and based on literature values presented.[58] While it is not intended to be exhaustive, this figure shows that aerobic hydrocarbon pathways require considerable improvements to reach performance on-par with ethanol in terms of yield (directly impacts economics), volumetric productivity (impacts bioconversion capital costs by way of more bioreactor vessels at lower achievable productivities), and product concentration (impacts costs for product purification/separation). Furthermore, important differences exist between specific product classes within the aerobic bioconversion pathway, with respect to ultimate carbon and energy yield potential that a given organism/product combination may realistically be able to achieve. These differences are dictated by theoretical metabolic yields, with a selection of such yield limits for a variety of product classes shown in Table 2, as presented in literature; the resulting process and economic implications of these governing yield limits is included further in Figure 7.

Table 2: Theoretical metabolic yields for various product pathway classes via aerobic bioconversion[2,50,59–61]

	Mass Yield	Carbon Yield	Energy Yield (HHV Basis)
Pentadecane	29%	62%	88%
Farnesene (DXP pathway)	29%	64%	85%
Farnesene (MVA pathway)	25%	56%	74%
Fatty acid (palmitic acid)	36%	67%	89%
Fatty ester (ethyl palmitate)	35%	67%	90%

Anaerobic Bioprocess
Process Design Details

Anaerobic conversion of sugars to hydrocarbons uses a similar process as that discussed above for aerobic conversion, with a number of notable exceptions that carry important implications for process and economic analysis. Anaerobic pathways may allow for a hydrolysis and conversion step similar to that used for ethanol production, namely conducting enzymatic hydrolysis and bioconversion in the same physical vessel without the need for intermediate hydrolysate conditioning (primarily solids removal), or equipment required for high aeration and agitation capabilities. In effect, such a process would be identical to NREL's 2011 biochemical ethanol design report[16] from a unit operation sequencing standpoint except for the product purification and upgrading requirements. A high-level overview of the process is shown in Figure 6. Key process design differences for an anaerobic bioconversion process to hydrocarbons relative to the aerobic process are listed below. All other steps would be largely consistent with the above discussion and are not repeated here.

Hydrolysate clarification: While the aerobic pathway is envisioned to require dedicated hydrolysate clarification steps between enzymatic hydrolysis and bioconversion, namely solids separation and sugar concentration, such steps are primarily attributed to integration needs for submerged aerobic cultivation. Thus, similar to integrated process modeling for anaerobic ethanol production, these steps may not be necessary for anaerobic RDB production pathways.

In this conceptual process, these steps may be removed such that hydrolysis and bioconversion sequentially proceed in the same physical vessel, each at a volume of one million gallons. In this case, lignin and other insoluble solids would be removed and sent to the boiler using a lower-cost lignin press downstream of bioconversion during the product recovery step. This mitigates issues with soluble sugar losses into the solids phase as in the aerobic pathway. However, a tradeoff may be incurred in that the presence of high amounts of insoluble solids (in addition to enzymes and cell biomass) may complicate the subsequent product

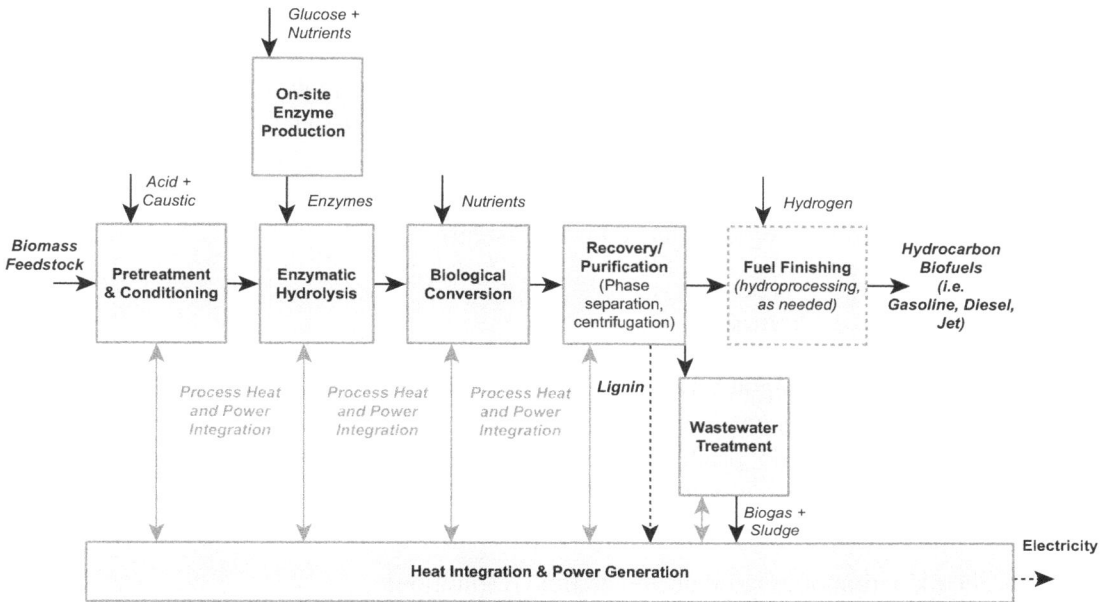

Figure 6

Example process schematic for hydrocarbon production via anaerobic bioconversion.

purification operations based on phase separation principles. This could be particularly true in more complex liquid phase behavior such as the formation of emulsions, potentially requiring more costly purification steps than decantation and centrifugation.

Biological conversion: From a system design standpoint, the bioconversion step would be similar to that described,[16] namely a single vessel by which process conditions are tailored to first allow for hydrolysis to occur at an optimum temperature of 48 °C and then anaerobic fermentation to proceed at lower temperatures of 32–38 °C (separate hydrolysis and fermentation or SHF). Because oxygen mass transfer and associated stringent system design and control is not relevant in this case, a number of preferential design implications would be realized, including large one million gallon vessels (maximizing economy of scale benefits), substantially lower agitation demands merely required for maintaining adequate mixing of the reactor broth, and elimination of aeration compressors and associated power demands.

Microbial pathways to hydrocarbons could be produced through anaerobic fermentation. As proof of concept, NREL researchers have demonstrated using anaerobic fermentation in the laboratory to produce small quantities of hydrocarbon fuel using engineered *Zymomonas mobilis*.[62] Additionally, many anaerobic conversion pathways to produce intracellular and extracellular RDB products and intermediates exist in various bacteria and yeast microorganisms.[63] Thus, while the ability to produce hydrocarbons anaerobically at high yields remains to be demonstrated, potential pathways are available to produce hydrocarbon biofuels by microbial conversion without the need to incorporate more complex and costly aeration capabilities into bioreactor systems. However, it is likely that the development and readiness of such pathways for near-term deployment will lag that of already developed and partially demonstrated aerobic biological pathways to hydrocarbon-like products such as fatty acids, triglycerides, isoprenoids, and paraffins.[2,50,59,60] Compared to the inherent process design challenges for aerobic pathways, which carry considerable expense, anaerobic options hold important potential as a means to achieve lower-cost targets.

Anaerobic Bioprocess Discussion

Anaerobic fermentation generates less metabolic energy (i.e., ATP) than aerobic respiration. This can have profound implications for product yields from a biofuel process. Microbial cell growth is proportional to ATP production and this suggests that more carbon will be diverted to cell growth during aerobic production of biofuels, decreasing product yields when compared to an anaerobic process that produces less ATP.

Consider the oxidation of glucose during aerobic respiration;

$$\text{Respiration: } C_6H_{12}O_6 + 6O_2 \rightarrow 6CO_2 + 6H_2O \quad \Delta G = -2870 \text{ kJ/mol}$$

The reaction generates 2870 kJ of energy per mole of glucose converted and produces 30–32 mol of ATP/mol.[64] Previous research has estimated that 10.5 g of cell mass are produced per mol of ATP when microorganisms are grown on glucose, after adjusting for energy consumed by cell maintenance.[65,66]

In contrast, the energy generated by the anaerobic fermentation of glucose to ethanol is much lower and it produces less ATP per mole of substrate;

$$\text{Fermentation: } C_6H_{12}O_6 \rightarrow 2C_5H_5OH + 6CO_2 \quad \Delta G = -218 \text{ kJ/mol}$$

This reaction generates about 8% of the energy of the aerobic process and only produces 1–2 mol ATP/mol glucose. The impact of decreased ATP production and its effect on increasing product yields has been described in the classic studies on the metabolism of *Z. mobilis* versus *Saccharomyces cerevisiae*. Ethanol fermentation in *Z. mobilis* generates one ATP per glucose via the Entner-Doudoroff pathway. *Saccharomyces cerevisiae* generates two ATP per glucose via Embden–Meyerhof–Parnas pathway. In batch fermentations, *Z. mobilis* was reported to produce a higher yield of ethanol and lower cell mass production than *S. cerevisiae*.[67–70] A more recent paper suggested that the impact could be substantial for the aerobic production of a palmitate ethyl ester, an RDB precursor, where a third or more of the carbohydrate supplied to a biocatalyst may be used for cell growth.[50] Further work is needed to fully understand the impact of bioenergetics and yield for RDB production as there is no commercially applicable strain with a well described pathway and ATP balance in the public literature.

As general proof of concept of the potential cost advantage of anaerobic RDB production, Figure 7 provides a high-level comparison of process economic potential for a number of possible metabolic pathways to RDB products. This includes fatty acids, fatty alcohols, fatty acid methyl/ethyl esters, paraffins, and isoprenoids, with several potential anaerobic pathways, based on applying a consistent set of assumptions regarding sugar utilization efficiencies as a percent of each pathway's theoretical metabolic limit.[14] Presented in terms of fractional cost differences, this demonstrates that the two theoretical anaerobic pathways (to fatty alcohols and to farnesene) have the greatest economic potential to reduce production costs, primarily by way of lower capital and operating costs enabled by a simpler engineering design alluded to above. Additionally, the figure shows important cost differences within the aerobic subset of pathways, driven by differences in theoretical metabolic yield to each product normalized by energy content. This demonstrates the fundamental role that process analysis plays in identifying economic tradeoffs between costs and yields, which only become apparent on establishing process models and conducting a holistic evaluation of integrated conceptual processes.

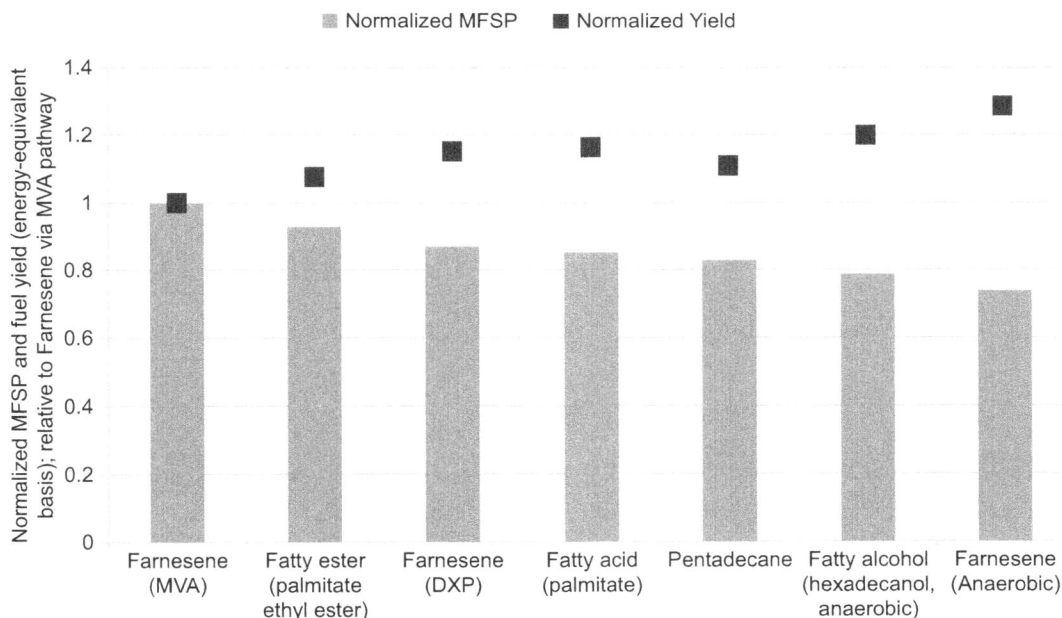

Figure 7

Conceptual comparison of relative economic potential versus fuel yield for various aerobic and anaerobic metabolic pathways to hydrocarbons (MFSP = minimum fuel selling price, adjusted by energy content to $/GGE; fuel yield adjusted by energy content to GGE/ton biomass feedstock; all values are shown as fractional results normalized to the MVA farnesene bioconversion pathway).[14]

Consolidated Bioprocessing
Process Design Details

CBP is an approach to biomass conversion with the potential to combine and simplify multiple processing steps, cellulase production, enzymatic hydrolysis, and bioconversion, in one operation.[71] Many process designs involve four biologically mediated transformations to break down lignocellulose: the production of saccharolytic enzymes; the hydrolysis of polysaccharides in pretreated biomass to monomeric sugars; the fermentation of hexose sugars; and the fermentation of pentose sugars. These four transformations have been designed to occur in a single step in CBP.[71] The ideal CBP system produces saccharolytic enzymes that hydrolyze structural carbohydrates (cellulose and hemicellulose) to oligomers.[72] Then these oligomers are further hydrolyzed to monomers and dimers. Finally these five- and six-carbon sugars are fermented to RDB or other products.[71] By taking place in a single unit operation and avoiding costs for external carbon sourcing for enzymes (depicted in Figure 8), CBP may reduce overall process complexity and, ideally, cost.

Combined CBP Operation: In Figure 8 (a simplified CBP block diagram), seed propagation would not only be used for preparation of an inoculum, but also could be responsible for a significant fraction of overall microbial cell production.[73] It is envisioned that CBP could even be implemented without a separate seed train. Enzyme production and seed propagations could be combined into one area (i.e., on-site enzyme and seed propagation

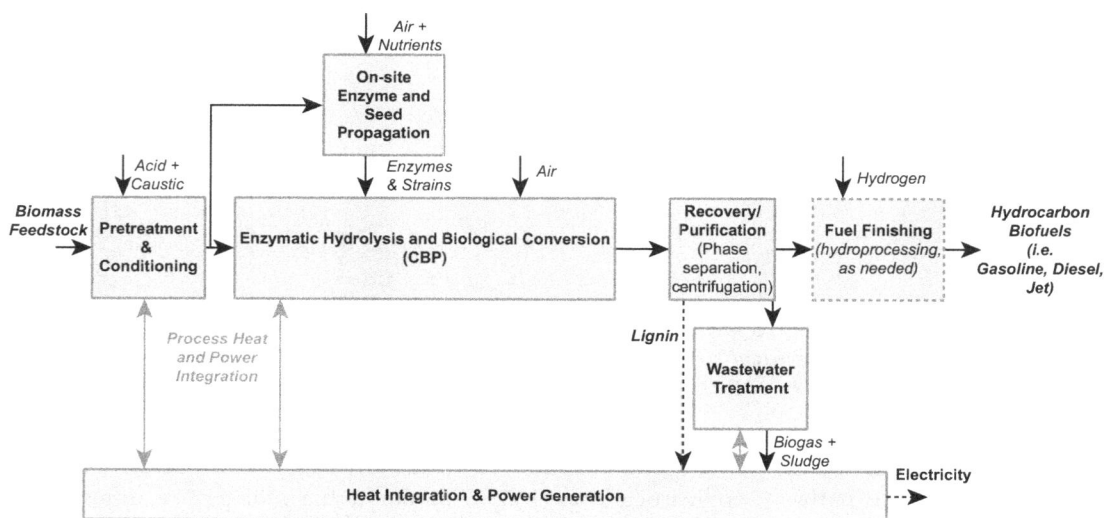

Figure 8
Process schematic for hydrocarbon production via consolidated bioprocessing.

in Figure 8). Solid and liquid phases from the chemical pretreatment step may be separated with 10–30% of the pretreated liquor sent to the enzyme production and seed propagation step to serve as the carbon source. Xylose in the hydrolysate liquor may be used to produce enzyme and cell biomass, instead of requiring an external supply of purchased glucose (e.g., from corn syrup). The remaining hydrolysate liquor and solid streams are sent to the CBP reactors. The enzyme loading could be potentially reduced to as low as 10 mg protein/g cellulose by combining seed propagation and enzyme production. There is no solid–liquid separation step between enzymatic hydrolysis and biological conversion as in the aerobic case (Figure 4), which provides savings for capital expenses and potentially limits yield losses. Sugar from enzymatic hydrolysis is used directly with no concentration step needed. The total bioconversion time can be five days or less, while dedicated enzymatic hydrolysis time is "zero" (i.e., continuous during CBP). Lignin separated by a pressure filter after the biological conversion operation is sent to the combustor, similar to the aerobic and anaerobic conversion cases.

CBP Discussion

CBP offers several opportunities for cost reduction versus other technologies. Relative to aerobic or anaerobic bioconversion processes, capital costs may potentially be reduced by combining enzyme production and seed propagation steps, overall volumetric tank size reduction for enzymatic hydrolysis and fermentation occurring in a single vessel, and eliminating the use of sugar conditioning or concentration operations prior to microbial conversion. Figure 9 presents a simplified tornado chart from NREL's 2013 biological hydrocarbon design case (i.e., the aerobic bioconversion case qualitatively discussed above), formatted to reduce the number of variables down to those which may benefit from CBP concepts.[14] The plot demonstrates cost sensitivities around a single baseline value—in this case, a minimum fuel selling price of $5.10/gallon gasoline equivalent (GGE). Cost sensitivities were established based on individual changes to base-case process or cost input parameters from the design case to quantify their relative impact on overall process economics.

A theoretical 25% total capital expenditure savings, relative to the base aerobic bioconversion process, has the greatest cost impact of the parameters shown, indicating a potential advantage that CBP may have due to its operational simplicity. Additional potential cost savings measures shown in Figure 9 include enzyme loading (implicitly also including associated purchased glucose costs in the baseline aerobic pathway model), elimination of solid–liquid separation operations, and dedicated enzyme production equipment (included in "enzyme production capital"). Taken together, the cost savings potential for these individual cost and process drivers may carry important economic implications for CBP's role in biological hydrocarbon fuel production. There are other opportunities for cost reduction beyond those captured in Figure 9. For example, combining enzyme and seed production may result in increased sugar available for

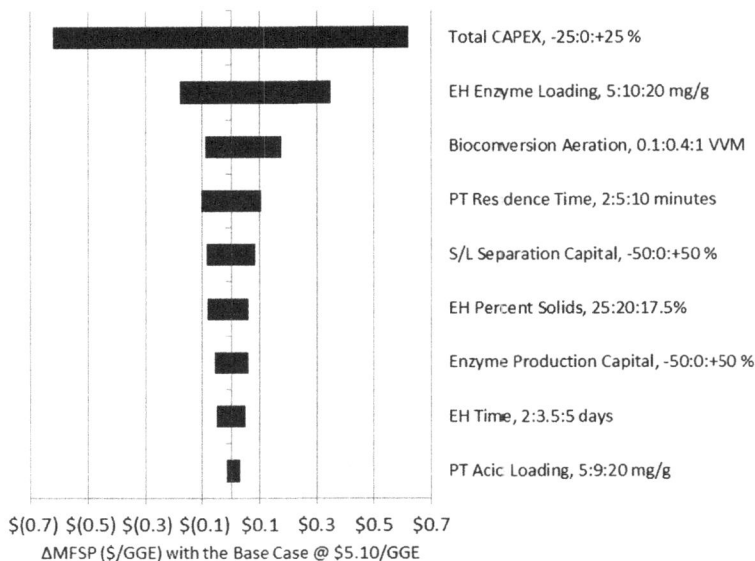

Total CAPEX, -25:0:+25 %

EH Enzyme Loading, 5:10:20 mg/g

Bioconversion Aeration, 0.1:0.4:1 VVM

PT Residence Time, 2:5:10 minutes

S/L Separation Capital, -50:0:+50 %

EH Percent Solids, 25:20:17.5%

Enzyme Production Capital, -50:0:+50 %

EH Time, 2:3.5:5 days

PT Acid Loading, 5:9:20 mg/g

$(0.7) $(0.5) $(0.3) $(0.1) $0.1 $0.3 $0.5 $0.7
ΔMFSP ($/GGE) with the Base Case @ $5.10/GGE

Figure 9

A tornado chart illustrating potential cost impacts of a CBP process relative to an aerobic bioconversion base-case pathway. Excerpted from Ref. 14. CAPEX is facility capital expenditures; EH is enzymatic hydrolysis; PT is pretreatment; S/L is solid/liquid; VVM is volume per volume per minute (volume of air per volume of fermentation broth per minute).

fermentation. This is equivalent to cost savings not only from reduced enzyme loading, but also from increased hydrocarbon yields.

Changes in modeled power consumption are less clear for CBP, but could be in between those for the anaerobic and aerobic pathways. One reference suggests that CBP may allow for up to 80% reduction in process power relative to base-case bioconversion technology.[74] The limited oxygen demands for the CBP concept (oxygen dedicated to cellulase production in Figure 8) may increase power demand compared with anaerobic pathways, but could still potentially present power savings relative to aerobic hydrocarbon pathways. Air demand for biological conversion (i.e., volume of air per volume of fermentation broth per minute or VVM) is one of the indicators for power demand's impact on cost (Figure 9). Power demand for aeration is highly dependent on the type of microorganisms used for CBP. For instance, *E. coli*, *Z. mobilis*, *S. cerevisiae*, and *Pichia stipitis* are the most relevant microorganisms in the context of lignocellulosic ethanol bioprocesses.[75] Some of these microorganisms can produce ethanol (or even hydrocarbons) anaerobically, such as *Z. mobilis*, or with minimal oxygen demand, such as *S. cerevisiae*. Genetically modified *Z. mobilis* or *S. cerevisiae* tailored for CBP purposes could potentially allow for increased energy efficiency compared to standard bioconversion pathways,[50] when combined with the cellulase production function.

Data Gaps, Uncertainties, and Research Needs

The potential pathways for biological conversion of sugars to hydrocarbons allow the opportunity to leverage experience in biochemical processing, specifically cellulose and hemicellulose deconstruction to monomeric sugars as currently used in anaerobic fermentation to ethanol. To ultimately reach economic viability, process analysis allows stakeholders to identify key bottlenecks, uncertainties, and areas for further development as the technology progresses.[32] In the context of biochemical technology pathways to hydrocarbon fuel production, examples of such drivers and areas for further development include:

Investigate synergistic opportunities for sugar/intermediate production and process integration. The sugar production metrics tied to pretreatment and enzymatic hydrolysis will continue to be important areas for further R&D improvement through the use of alternative or milder pretreatment options and/or improved enzyme performance (higher conversion yields and/or lower enzyme doses or cost), including incorporation of new enzyme classes and enzymatic hydrolysis mechanisms. Additionally, components that previously inhibited ethanol fermentation and could be removed early in pretreatment (such as acetic acid) may not pose such inhibitory effects. Developing methods to use biomass-derived intermediates beyond monomeric sugars will also help to improve overall carbon conversion efficiencies in the process. Tailoring the hydrolysate stream to the microorganism tolerance will be essential for improving overall yields and lowering production costs. As previously summarized, consolidated bioprocessing offers another potential pathway to optimize process integration and reduce costs, by which enzymatic hydrolysis and fuel production occur in a single step without the need for external enzymes. Given the relatively high cost of enzyme addition, this approach also warrants continued consideration and research.

Develop separation and conditioning requirements for hydrolysate. A better understanding is needed on the tolerance of hydrocarbon-producing microbes to soluble lignin and other impurities, including organic acids, salts, and other potential inhibitors. The efficacy of insoluble lignin removal following enzymatic hydrolysis may be more challenging and expensive than currently anticipated. Losses of sugar in this removal step will lower yields and increase costs and thus must be minimized. The performance and cost tradeoffs between sugar stream concentration, purity, and microbial hydrocarbon production must be quantified to be able to develop optimal process designs. These tradeoffs are complex and will differ between microbes, fuel products, and process configurations.

Optimize design and scale for aerobic fuel production. The optimal engineering design and operating parameters for the aerobic microbial RDB production process must be identified. This includes determining the most economical bioreactor design that will allow the combination of maximum vessel size, process productivity, and yield while minimizing the requirement and cost for aeration and temperature control. Increased cell growth during aerobic bioprocesses tends to decrease biofuel yields and increase costs. Cell recycling and multistep fermentations strategies can help to minimize cell growth and improve the economics of aerobic bioprocesses.[50] Producing hydrocarbon biofuels by anaerobic microbial conversion and reducing the need to incorporate more complex and costly aeration capabilities into bioreactor systems should also be considered in parallel to improve process economics. The development and demonstration of microbes that can produce hydrocarbons at high rates and yields via anaerobic pathways would be a breakthrough for this field.

Maximize sugar (and/or carbon) utilization and microbe metabolic performance. Better understanding is needed regarding the productivity of the microorganism and the potential of genetic engineering to significantly increase metabolic production rates and yields, minimize side-product formation, and mitigate substrate and/or product toxicity effects. There is currently a scarcity of literature and high quality data in the public domain on sugar conversion and microbial productivity, particularly with respect to production using cellulosic feedstock-derived substrates containing pentose sugars.

Define product separation and final polishing/upgrading requirements. The recovery of products that are secreted directly into the aqueous broth presents challenges such as lowered yields due to products being retained on or within the cell mass, difficulties in breaking emulsions, and incomplete phase separation. Beyond these separation issues, additional finishing steps may be needed to improve product quality to meet fuel specifications. The product recovery and final upgrading operations need to be defined to quantify process costs and equipment requirements.

Evaluate co-product opportunities. The requirement to reach a production cost target on par with gasoline or diesel will likely require simultaneous cost reductions in multiple areas (e.g., by way of engineering improvements and lower chemical/enzyme demands) and higher total product yields. To achieve the latter, carbon efficiency and total yields may need to improve beyond theoretical limits imposed by the metabolic conversion of sugar stream components alone (i.e., sugars derived from cellulose and hemicellulose). It will be important to develop cost-effective technologies to also convert non-sugar components (e.g., lignin, acetate) into value-added co-products, fuels, or fuel precursors. A life-cycle assessment evaluating the tradeoffs associated with diverting some process streams and residues to additional co-products will be critical for developing an economic and sustainable hydrocarbon fuels biorefinery facility.

Conclusion

There is a myriad of biomass conversion processes undergoing various stages of research, development, deployment, and scale-up. The diversity of processes and potential products creates a challenge for policy makers, research organizations, and investors to understand the broader implications of each technology in terms of relative risks, benefits, research needs, and ultimate potential. Process analysis is one tool that can be used to elucidate and quantify such details for biomass-to-hydrocarbon fuel technologies. A number of examples of process analysis concepts were provided in this discussion with a focus on biochemical production of hydrocarbon biofuels via aerobic, anaerobic, or consolidated bioprocessing pathways, but the general methods and concepts are applicable to other technologies as well.

Process analysis takes a holistic view of a conceptual process and extrapolates it to a commercial scale model. It also allows for optimizing the overall process by identifying synergistic opportunities in process integration. The outputs from the model, along with other information, can be used to generate financial (TEA) and life-cycle analysis (LCA). Those metrics allow for the classification of technology choices and may be used to prioritize future research directions, or to assess implications of a specific research function on the overall integrated biorefinery in terms of economics or sustainability. Process analysis can be highly iterative, involving feedback to and from R&D functions and the associated TEA/LCA models.

Process analysis requires a solid technical understanding of underlying scientific principles by which a technology pathway operates. Any process model and subsequent analysis only as good as the inputs that go into it. By applying scientific and engineering expertise coupled with powerful analysis software packages for process simulation and associated economic/sustainability assessment, process analysis allows for a high degree of rigor in translating research concepts to quantified process-level metrics. As such, process analysis is a key tool for stakeholders to use in "de-risking" technologies in any stage of precommercial development, from early exploratory research to scale-up toward demonstration or commercial scales.

Acknowledgment

This work was supported by the US DOE EERE Bioenergy Technologies Office (BETO).

References

1. Kumar A, Jones DD, Hanna MA. Thermochemical biomass gasification: a review of the current status of the technology. *Energies* 2009;**2**:556–81.
2. Rude MA, Schirmer A. New microbial fuels: a biotech perspective. *Curr Opin Microbiol* 2009;**12**:274–81.
3. Brennan L, Owende P. Biofuels from microalgae—A review of technologies for production, processing, and extractions of biofuels and co-products. *Renew Sustain Energy Rev* 2010;**14**:557–77.
4. Balat M. Production of bioethanol from lignocellulosic materials via the biochemical pathway: a review. *Energy Convers Manage* 2011;**52**:858–75.
5. Bridgwater AV. Review of fast pyrolysis of biomass and product upgrading. *Biomass Bioenergy* 2012;**38**:68–94.
6. Jahirul MI, Rasul MG, Chowdhury AA, Ashwath N. Biofuels production through biomass pyrolysis—a technological review. *Energies* 2012;**5**:4952–5001.
7. Viikari L, Vehmaanpera J, Koivula A. Lignocellulosic ethanol: from science to industry. *Biomass Bioenergy* 2012;**46**:13–24.
8. Liu G, Yan B, Chen G. Technical review on jet fuel production. *Renew Sustain Energy Rev* 2013;**25**:59–70.

9. Laskar DD, Yang B, Wang H, Lee JC. Pathways for biomass-derived lignin to hydrocarbon fuels. *Biofuels Bioprod Biorefin-Biofpr* 2013;**7**:602–26.

10. Mood SH, Golfeshan AH, Tabatabaei M, Jouzani GS, Najafi GH, Gholami M, et al. Lignocellulosic biomass to bioethanol, a comprehensive review with a focus on pretreatment. *Renew Sustain Energy Rev* 2013;**27**:77–93.

11. Kumar A, Sharma R. *Principles of business management.* Atlantic Publishers & Distributors Limited; 2000. ISBN: 9788171567157.

12. Peters MS, Timmerhaus KD, West RE. *Plant design and economics for chemical engineers.* New York: McGraw-Hill; 2003.

13. Wooley R, Ruth M, Glassner D, Sheehan J. Process design and costing of bioethanol technology: a tool for determining the status and direction of research and development. *Biotechnol Progr* 1999;**15**:794–803.

14. Davis R, Tao L, Tan E, Biddy M, Beckham G, Scarlata C, et al. Golden (CO): National Renewable Energy Laboratory; 2013. NREL/TP-5100–60223.

15. Dutta A, Talmadge M, Hensley J, Worley M, Dudgeon D, Barton D, et al. Golden (CO): National Renewable Energy Laboratory; 2011. NREL/TP-5100–51400.

16. Humbird D, Davis R, Tao L, Kinchin C, Hsu D, Aden A, et al. Golden (CO): National Renewable Energy Laboratory; 2011. NREL/TP-510–47764.

17. Ringer M, Putsche V, Scahill J. Golden (CO): National Renewable Energy Laboratory; 2006. NREL/TP-510–37779.

18. Wooley RJ, Putsche V. *NREL technical memo.* Golden (CO): National Renewable Energy Laboratory; 1996. NREL/TP-425–20685.

19. Schell DJ, Torget R, Power A, Walter PJ, Grohmann K, Hinman ND. A technical and economic-analysis of acid-catalyzed steam explosion and dilute sulfuric-acid pretreatments using wheat straw or aspen wood chips. *Appl Biochem Biotechnol* 1991;**28**(9):87–97.

20. Eggeman T, Elander RT. Process and economic analysis of pretreatment technologies. *Bioresour Technol* 2005;**96**:2019–25.

21. Tao L, Aden A, Elander RT, Pallapolu VR, Lee YY, Garlock RJ, et al. Process and technoeconomic analysis of leading pretreatment technologies for lignocellulosic ethanol production using switchgrass. *Bioresour Technol* 2011;**102**:11105–14.

22. Galbe M, Sassner P, Wingren A, Zacchi G. Process engineering economics of bioethanol production. *Biofuels* 2007;**108**:303–27.

23. Wright MM, Brown RC. Comparative economics of biorefineries based on the biochemical and thermochemical platforms. *Biofuels Bioprod Biorefin-Biofpr* 2007;**1**:49–56.

24. Dutta A, Dowe N, Ibsen KN, Schell DJ, Aden A. An economic comparison of different fermentation configurations to convert corn stover to ethanol using *Z. mobilis* and *Saccharomyces. Biotechnol Progr* 2010;**26**:64–72.

25. Jones SB, Valkenburg C, Walton CW, Elliott DC, Holladay JE, Stevens DJ, et al. Richland (WA): Pacific Northwest National Laboratory; 2009. PNNL-18284.

26. Aden A, Foust T. Technoeconomic analysis of the dilute sulfuric acid and enzymatic hydrolysis process for the conversion of corn stover to ethanol. *Cellulose* 2009;**16**:535–45.

27. Short W, Packey DJ, Holt T. Golden (CO): National Renewable Energy Laboratory; 1995. NREL/TP-462–5173.

28. Larson ED. A review of life-cycle analysis studies on liquid biofuel systems for the transport sector. *Energy Sustain Dev* 2006;**10**:109–26.

29. Luo L, Voet E, Huppes G, Udo de Haes H. Allocation issues in LCA methodology: a case study of corn stover-based fuel ethanol. *Int J Life Cycle Assess* 2009;**14**:529–39.

30. Cherubini F, Strømman AH. Life cycle assessment of bioenergy systems: state of the art and future challenges. *Bioresour Technol* 2011;**102**:437–51.

31. EIA.gov. In: *Administration, U.S.E.I,* vol. 2013. 2013. http://www.eia.gov/petroleum/gasdiesel/pump_method ology.cfm#3.

32. Davis R, Biddy MJ, Tan EC, Tao L, Jones SB. Golden (CO): National Renewable Energy Laboratory; 2013. NREL/TP-5100–58054, PNNL-22318.

33. Garcia-Ochoa F, Gomez E. Bioreactor scale-up and oxygen transfer rate in microbial processes: an overview. *Biotechnol Adv* 2009;**27**:153–76.

34. Fuchs R, Ryu DDY, Humphrey AE. Effect of surface aeration on scale-up procedures for fermentation processes. *Ind Eng Chem Proc Dd* 1971;**10**:190.

35. van't Riet K. Mass transfer in fermentation. *Trends Biotechnol* 1983;**1**:113–9.

36. Manfredini R, Cavallera V, Marini L, Donati G. Mixing and oxygen-transfer in conventional stirred fermenters. *Biotechnol Bioeng* 1983;**25**:3115–31.

37. Oosterhuis NMG, Kossen NWF. Dissolved-oxygen concentration profiles in a production-scale bioreactor. *Biotechnol Bioeng* 1984;**26**:546–50.

38. Humphrey AE. In: 1st ed. Moo-Young M, editor. *Comprehensive biotechnology: the principles, applications, and regulations of biotechnology in industry, agriculture, and medicine*, vol. 2. Oxford, Oxfordshire; New York: Pergamon Press; 1985.

39. McMillan JD. Pretreatment of lignocellulosic biomass. *Enzym Convers Biomass Fuels Prod* 1994;**566**:292–324.

40. Tao L, Chen X, Aden A, Kuhn E, Himmel ME, Tucker M, et al. Improved ethanol yield and reduced minimum ethanol selling price (MESP) by modifying low severity dilute acid pretreatment with deacetylation and mechanical refining: 2) techno-economic analysis. *Biotechnol Biofuels* 2012;**5**(1):69.

41. *National Advanced Biofuels Consortium (NABC)*. http://www.nabcprojects.org/pdfs/amyris_makes_biofene_from_cellulosic_hydrolysate.pdf; 2012.

42. Chen X, Tao L, Shekiro J, Mohaghaghi A, Decker S, Wang W, et al. Improved ethanol yield and reduced minimum ethanol selling price (MESP) by modifying low severity dilute acid pretreatment with deacetylation and mechanical refining: 1) experimental. *Biotechnol Biofuels* 2012;**5**:60–9.

43. Schell DJ. *2013 DOE Bioenergy Technologies Office (BETO) project peer review*2013. [Alexandria, VA].

44. Chisti Y. *Airlift bioreactors*. London; New York: Elsevier Applied Science; 1989.

45. Chisti Y, Jauregui-Haza UJ. Oxygen transfer and mixing in mechanically agitated airlift bioreactors. *Biochem Eng J* 2002;**10**:143–53.

46. Renninger NS, Mcphee DJ. United States Patent No. 7,846,222; 2008.

47. Aden A, Ruth M, Ibsen K, Jechura J, Neeves K, Sheehan J, et al. Golden (CO): National Renewable Energy Laboratory; 2008. NREL/TP-510–32438.

48. *Solazyme. South San Francisco (CA)*. http://investors.solazyme.com/releasedetail.cfm?releaseid=726878; 2013.

49. Kollaras A, Koutouridis P, Biddy M, McMillan J. *Ethanol producer magazine*. BBI International; 2012. http://www.ethanolproducer.com/articles/8926/multiple-coproducts-needed-to-establish-cellulosic-ethanol-industry.

50. Huang WD, Zhang YHP. Analysis of biofuels production from sugar based on three criteria: thermodynamics, bioenergetics, and product separation. *Energy Envir Sci* 2011;**4**:784–92.

51. Marker TL, Petri J, Kalnes T, McCall M, Mackowaik D, Jerosky B, et al. In: *Energy, U. S. D. o.* UOP LLC; 2005.

52. Steen EJ, Kang YS, Bokinsky G, Hu ZH, Schirmer A, McClure A, et al. Microbial production of fatty-acid-derived fuels and chemicals from plant biomass. *Nature* 2010;**463**:559. U182.

53. Dellomonaco C, Clomburg JM, Miller EN, Gonzalez R. Engineered reversal of the beta-oxidation cycle for the synthesis of fuels and chemicals. *Nature* 2011;**476**:355–9.

54. Peralta-Yahya P, Ouellet M, Chan R, Mukhopadhyay A, Keasling JD, Lee TS. Identification and microbial production of a terpene-based advanced biofuel. *Nat Commun* 2011;**2**:483.

55. Zhang FZ, Rodriguez S, Keasling JD. Metabolic engineering of microbial pathways for advanced biofuels production. *Curr Opin Biotechnol* 2011;**22**:775–83.

56. Hawkins RL. Utilization of xylose for growth by the eukaryotic alga, *Chlorella. Curr Microbiol* 1999;**38**:360–3.

57. Zhou H, Cheng JS, Wang BL, Fink GR, Stephanopoulos G. Xylose isomerase overexpression along with engineering of the pentose phosphate pathway and evolutionary engineering enable rapid xylose utilization and ethanol production by *Saccharomyces cerevisiae*. *Metab Eng* 2012;**14**:611–22.
58. Huffer S, Roche CM, Blanch HW, Clark DS. *Escherichia coli* for biofuel production: bridging the gap from promise to practice. *Trends Biotechnol* 2012;**30**:538–45.
59. DOE. In: Energy, D.O., editor. *Conversion technologies for advanced biofuels (CTAB) roadmapping workshop*. Arlington (VA): Office of the Biomass Program in the Office of Energy Efficiency and Renewable Energy; 2011.
60. Liu TG, Khosla C. Genetic engineering of *Escherichia coli* for biofuel production. *Annu Rev Genet* 2010;**44**:53–69.
61. Dugar D, Stephanopoulos G. Relative potential of biosynthetic pathways for biofuels and bio-based products. *Nat Biotechnol* 2011;**29**:1074–8.
62. Zhang M. Presented at the 2012 SIMB annual meeting; 2012.
63. Ladygina N, Dedyukhina EG, Vainshtein MB. A review on microbial synthesis of hydrocarbons. *Process Biochem* 2006;**41**:1001–14.
64. Lehninger AL, Nelson DL, Cox MM. *Lehninger principles of biochemistry*. 5th ed. New York: W.H. Freeman; 2008.
65. Stouthamer AH. In: Norris JR, Ribbons DW, editors. *Methods in microbiology*, vol. 1. Academic Press; 1969. p. 629–63.
66. Payne WJ. Energy yields and growth of heterotrophs. *Annu Rev Microbiol* 1970;**24**:17–52.
67. Kitani O, Hall CW. Gordon and Breach Science Publishers; 1989. p. 264.
68. Stewart GG, Panchal CJ, Russell I, Sills AM. *International symposium on ethanol from biomass*. The Royal Society of Canada; 1982. pp. 4–57.
69. Lawford GR, Lavers BH, Good D, Charley R, Fein J, Lawford HG. *International symposium on ethanol from biomass*. Winnipeg: Royal Soc Canada; 1982. pp. 482–506.
70. Rogers PL, Lee K, Tribe DE. Kinetics of alcohol production by *Zymomonas mobilis* at high sugar concentrations. *Biotechnol Lett* 1979;**1**:165–70.
71. Lynd LR, van Zyl WH, McBride JE, Laser M. Consolidated bioprocessing of cellulosic biomass: an update. *Curr Opin Biotechnol* 2005;**16**:577–83.
72. Sendich E, Laser M, Kim S, Alizadeh H, Laureano-Perez L, Dale B, et al. Recent process improvements for the ammonia fiber expansion (AFEX) process and resulting reductions in minimum ethanol selling price. *Bioresour Technol* 2008;**99**:8429–35.
73. Lynd LR. Overview and evaluation of fuel ethanol from cellulosic biomass: technology, economics, the environment, and policy. *Annu Rev Energy Envir* 1996;**21**:403–65.
74. Laser M, Jin HM, Jayawardhana K, Lynd LR. Coproduction of ethanol and power from switchgrass. *Biofuels Bioprod Biorefin-Biofpr* 2009;**3**:195–218.
75. Gírio FM, Fonseca C, Carvalheiro F, Duarte LC, Marques S, Bogel-Lukasik R. Hemicelluloses for fuel ethanol: a review. *Bioresour Technol* 2010;**101**:4775–800.

PART 2

Biomass Structure and Recalcitrance

Tailoring Plant Cell Wall Composition and Architecture for Conversion to Liquid Hydrocarbon Biofuels

Maureen C. McCann[1], Bryan W. Penning[1], Christopher K. Dugard[2], Nicholas C. Carpita[2]

[1]*Purdue University, Biological Sciences, West Lafayette, IN, USA;* [2]*Purdue University, Botany and Plant Pathology, West Lafayette, IN, USA*

In the twentieth century, the availability of inexpensive oil enabled the rapid scale-up of the transportation sector and remains a key driver of the global economy. Oil refining and petrochemical production co-evolved with the automotive industry to deliver both affordable vehicles and an efficient fueling infrastructure and achieved versatility in production of chemicals and thousands of derived products. To conserve and extend fossil fuel reserves, mitigate greenhouse gas emissions, and for nations to achieve energy security, the production of biofuels and biomass-derived chemicals needs to be cost-competitive with oil across the full life cycle of production and delivery to the consumer.

While policy and regulation have significant roles to play, the economics of displacing fossil carbon with renewable carbon will become more feasible if key technical challenges are addressed that reduce the costs of feedstock, pretreatment, conversion, separation, and distribution. First, lignocellulosic biomass as the source of renewable carbon creates challenges of added costs of harvest, delivery, storage, and processing to market. Biomass has low energy density, about one-third that of oil, is usually widely distributed, and has relatively low economic value.[1] Second, current conversion technologies to biofuels suffer from low yield and rate of production because of the chemical and structural complexity of biomass feedstocks.

The lignocellulosic biofuels industry faces a classic catch-22 problem. Farmers will not grow energy crops unless there are biofuel refineries within easy shipping distance to buy their crops and unless the prices that they receive are high or higher than those from growing conventional crops like corn or soybean. Industry will not build refineries at scale until there is a steady supply of feedstock.[2] In this chapter, we discuss the potential for synergistic improvement in yield and quality of bioenergy feedstocks that is driven by mechanistic understanding of the downstream conversion process, whether involving biochemical,

Direct Microbial Conversion of Biomass to Advanced Biofuels. http://dx.doi.org/10.1016/B978-0-444-59592-8.00004-X

chemical, or pyrolytic transformations. New capabilities to improve the carbon and energy efficiencies of conversion processes, and the range and versatility of biofuels and bioproducts, by using biomass that is optimized or "tailored" for its end-use, may break this gridlock and enable the next-generation bioeconomy.

Biomass Feedstocks are Already an Abundant Resource

Regional availability of sustainable, affordable, commercial-scale biomass feedstocks is the first prerequisite of a bioenergy and bioproducts supply chain. The US Department of Energy's Billion Ton Study[3] and the 2011 update[4] demonstrate the potential to deliver substantial quantities of feedstock on an annual basis, equivalent to displacing about 3 billion barrels of oil.[5] Annual US oil consumption is seven billion barrels,[5] and so this represents a significant fraction of potential supply from existing resources. In contrast to grain yield, biomass yield has not been a target of selection and plant breeding. There is therefore tremendous potential to maximize production on current acreage using scientific and technological advances and best agricultural practices. As an example of prior success, US corn production increased eight-fold on the same land acreage using fertilizer, herbicides and pesticides, modern tillage methods, and plant breeding to select for traits such as leaves that pointed upright, making them efficient solar energy collectors and decreasing the spacing needed between individual plants. Simply by substituting sweet sorghum or tropical maize varieties on the acreage currently used for corn ethanol could double yield.[2] Exploiting the full range of natural genetic diversity in crops such as maize and sorghum and the use of molecular-assisted plant breeding and tools of genetic modification can intensify agricultural production without increasing land area.

Despite the ample availability of biomass, the magnitude of the feedstock supply creates significant logistical challenges for entraining these feedstocks into the biofuels supply chain. A billion tons of biomass represent the output of a new agricultural system that is larger than the current 800 million tons of all annual agricultural products, including hay and pasture.[4] The intermediate infrastructure between feedstock production and conversion processing is an essential element of the value chain that needs to generate value for both the upstream feedstock element and the downstream conversion sector. The commoditization of biomass feedstocks to define standards for parameters such as water and soil contents[6] and processes for energy-efficient comminution and packing density are needed, adding value on a farm by improved harvesting, drying and baling practices, and delivering biomass in a more uniform format at the biorefinery.

Chemical Structure and Physical Properties of Lignocellulosic Biomass

Lignocellulosic biomass comprises plant cell walls. The fundamental principles of cell-wall composition and architecture are common to all plant species, but the kinds and proportions of the structural components can vary between different cell wall types and in different species.[7,8] All cell walls contain cellulose microfibrils as the main scaffolding components of the wall, varying between 30% and 90% of the dry mass of different cell types within the plant. The cellulose microfibrils are linked together to form a network with cross-linking glycans, which is embedded in a matrix of acidic polysaccharides.

Plants make two distinct kinds of primary walls, those characteristic of grasses and those of dicotyledonous and non-grass-like monocot species, including woody crops. Bioenergy grasses and crop residues including all cereal crops (wheat, rice, maize) and energy crops (switchgrass, Miscanthus, sugarcane, tropical maize, and sorghum), comprise mixtures of cells with primary walls and those with thick secondary (lignified) walls. A distinctive characteristic of grasses compared to dicotyledonous species is that their primary walls are cross-linked with a hydroxycinnamic acid-rich phenylpropanoid network.[9] In the primary walls of grasses, glucuronoarabinoxylans (GAXs) bind to themselves and around cellulose microfibrils. Hydroxycinnamic acids, such as ferulic acid and *p*-coumaric acid, are ester-linked to the arabinosyl units of GAX, where they can serve as initiation sites for network polymerization.

The differences between grass and non-grass primary walls are not as apparent in the secondary walls, but there are three distinct types of secondary walls to consider with respect to bioenergy crops. Angiosperm tree crops, such as poplar, willow, and eucalyptus, produce "wood," in which each heavily lignified secondary cell wall is a multilayered composite of cellulose microfibrils, coated mostly with GAX and some glucomannans, comprising most of the polysaccharide of the wall.[10] The lignin heteropolymer is produced via the oxidative coupling of mostly *p*-coumaryl alcohol, coniferyl alcohol, and sinapyl alcohol subunits (collectively termed monolignols). The polymerization of these subunits leads to the formation of *p*-hydroxyphenyl (H), guaiacyl (G), and syringyl (S) lignin, respectively.[11–13] The H subunits are usually minor components, and the degree to which S and G units are incorporated into the polymer (commonly denoted as the S:G ratio) varies widely among species, tissue types, and even within an individual cell wall. Grasses have secondary walls similar to those of woody angiosperms, but with a lignin rich in hydroxycinnamic acids and higher proportions of H-lignins than found in dicots.[9] Gymnosperm species are differentiated from angiosperms by the high proportions of mannans and glucomannans rather than GAX, as well as a lignin composed of mostly G-lignin.[10]

Although H, G, and S units are widely regarded as the only monomers found in lignin, more sophisticated methods of lignin analysis applied to a broader range of plant species, mutants, and transgenic lines have revealed that other subunits, including aldehydes, side-chain-reduced monolignols, and phenylpropanoid esters and amides, are bona fide lignin components. The diversity of monolignols and the ether and carbon–carbon bonds that link them together in the polymer impart remarkable complexity of structure. In general, the higher the number of methyl ethers, the more limited the points of attachment in the aromatic ring. Other ether– and phenyl–phenyl bonds can tightly link lignin to cell wall polysaccharides.

Differences among species, genotypes, and environmental conditions during growth, harvest, and storage will result in feedstocks with variability in cellulose, hemicelluloses, and lignin compositions and architectures, all of which will impact net energy efficiency and yields of products from conversion.[14] However, the range of polysaccharide and lignin structures is constrained, meaning that fractionation and catalyses of individual wall components is achievable in pathways that are independent of the inherent variability in potential feedstocks. For example, a combined Zn/Pd/C catalyst effectively cleaves the lignin β-O-4 linkage and subsequently hydrodeoxygenates the aromatic fragments, without loss of aromatic functional groups from intact biomass of both tree and grass species, although with different yields and selectivities.[15] Similarly, maleic acid hydrolyzes xylans to xylose at 160 °C with transformation of xylose to furfural at 200 °C in a two-step, one-pot reaction, whether the biomass is switchgrass, corn stover, or milled poplar samples, with a conversion efficiency of between 50% and 70% of initial xylan content.[16] In both catalytic pathways, the extraction of one component increases the efficiency of conversion of residual material, as determined by increased rates of glucose release in saccharification assays.[15,16]

Biochemical, Chemical and Pyrolytic Conversion Pathways Provide Alternative Routes to Fuels

Biochemical routes of conversion depend on hydrolysis of cellulose, xylans, mannans, and other non-cellulosic polysaccharides into their respective monosaccharides for use as carbon sources for fermentative micro-organisms.[17] Desirable reaction pathways for the production of ethanol, butanol, and other fuels can be genetically engineered in bacteria, algae, and yeast. However, the quantity and quality of lignin in biomass crops interferes with the access of hydrolytic enzymes to the polysaccharide components of the plant cell wall, thereby inhibiting their conversion to fermentable monosaccharides.[18] Microorganisms required to ferment sugars to biofuels metabolize some sugars in their own growth, losing up to and even more than one-half of the carbons as carbon dioxide. Thus, only one-third of carbon atoms in the biomass are captured into fuel molecules with today's second-generation technology. Doubling this value would halve the land, water, fertilizer, and energy requirements for growing bioenergy feedstocks.

Because of the high oxygen content of carbohydrates, the energy contents of lignocellulosic biomass from switchgrass and poplar trees are estimated to be only 485 MJ/kmol of carbon (17 MJ/kg of biomass) and 455 MJ/kmol of carbon (19.6 MJ/kg of wood), respectively, in comparison to an energy density of gasoline of 604 MJ/kmol of carbon (32.4 MJ/l).[19] Conversion of lignocellulosic biomass to high energy-density liquid fuels by any conversion process that uses biomass as a sole feedstock is bound to release nearly one-quarter to one-third of carbon as CO_2. This loss of carbon is not due to any process inefficiencies, but simply to conservation of mass.[19] Assuming an energy conversion efficiency of about 75%, biomass carbon release as CO_2 increases to about 40%–50%. For any given unit of biomass, a key goal is to minimize this co-release of CO_2, with the implication that no form of carbon in biomass, including lignin, need be lost from conversion to useful product molecules. Although catalytic transformations that co-produce hydrogen can be envisioned,[20] a source of exogenous hydrogen is required for recovery of all carbon atoms in the biomass. The recent increases in natural gas production from previously inaccessible shales using hydraulic fracturing might provide an opportunity for co-utilization of biomass and methane.

In contrast to biochemical conversion pathways, the power of chemical catalysis to transform biomass components directly to liquid hydrocarbons and aromatic co-products (third-generation biofuels) is an underexplored area of science that has tremendous potential impact. The challenge in using total biomass as fuel and feedstock is deoxygenation of sugar polyols and lignins. Only a few homogeneous catalysts have been investigated for polyol transformations.[21–23] These systems currently rely on precious metals such as Pt, Pd, Rh, and Ru and operate under forcing conditions (>100 °C and >700 psi H_2). Hydrogenation and hydrogenolysis catalysts are needed that are inexpensive and nontoxic earth-abundant metals or precious metal catalysts that are highly robust and efficient under mild conditions and low energy consumption and yielding no waste. Beyond fuels, the success of the petrochemical industry in the 20th century bloomed largely on catalytic chemistry, in which hydrocarbon (from oil cracking) is functionalized with oxygen. For example, polyethylene terephthalate, which accounts for 18% of the world polymer production, is manufactured from ethylene glycol produced from the reaction of ethylene with dioxygen (O_2) and a silver catalyst.[24] Terephthalic acid is produced from the reaction of *p*-xylene and air catalyzed by a Mn^{2+}/Co^{2+} homogeneous catalyst.[25] Analogous catalytic transformations of biomass and its individual polymeric components will play an equally important role in biorefinery development.

Thermochemical conversion pathways are being explored to produce a bio-oil directly from biomass, with subsequent fractionation and catalytic upgrading. Although use of fast-pyrolysis to depolymerize biomass is a well-known process, it results in unusable products.[26] The bio-oils contain hundreds of compounds, have high oxygen content (typically 35–40 wt%), degrade over time, and have a heating value less than half that of gasoline.[27] In contrast, hydropyrolysis of biomass in a fixed-bed mode, in the presence of either an iron-based or a Co/Mo catalyst, has been shown to yield increased amounts of liquid fuels.[28–30] Thermal treatment of biomass results in the formation of vapors, permanent gases, and char, with the relative quantities being temperature- and time-dependent. Much of the available knowledge relates to pyrolysis, which occurs at temperatures in the 400–600 °C range and on the seconds-to-minutes timescale.[31] Because it takes place under milder conditions than gasification, one of the major advantages of this process is that pyrolysis tends to preserve the complex chemical bonds found in biomass. The most efficient processes can convert up to 75% of the starting biomass weight into condensable vapors, which are then cooled and collected.[32] To transform this oil to a useful fuel, it is necessary to remove the oxygen, ideally to less than 1 wt%. A catalytic route is desirable to achieve this goal, but this task is complicated by the presence of such a large number of functionally different and unstable compounds.[33,34]

All three pathways now offer routes to liquid hydrocarbon fuels rather than ethanol, overcoming issues of fungibility with existing transportation fuels and compatible performance characteristics in engines. The one trillion dollar investment that has been made in existing fueling stations and distribution, engines, and pipelines provides a serious challenge to the integration of any new fuel that is not compatible with existing infrastructure.[2]

Tailoring Biomass for Downstream Conversion Processes

Lignocellulosic biorefineries require substantially higher capital expenditure per-gallon capacity than starch/sugar ethanol plants or biodiesel plants because biomass processing is more complex and entails a greater number of unit operations than conventional biofuel facilities. The typical solution to high capital cost is to increase scale by building larger facilities. In the case of biomass processing plants and biorefineries, the costs of transporting biomass greater distances rises rapidly and can offset, or more, any savings from lower per-gallon capital expenditures. It is therefore imperative to maximize the carbon and energy efficiencies of conversion processes such that the yield of fuel and chemicals per unit of biomass is maximized.

An inherent difficulty is the complex nature of biomass feedstocks at multiple length scales, from the heterogeneous structures of cell wall polysaccharides, to the nanoscale structures of cellulose microfibrils and macromolecular lignin, to mesoscale domains of wall architecture, such as cell junctions that have distinct compositions. Specific architectural features that impact the access of hydrolytic enzymes, chemical catalysts, or product extractability include the extent of the cross-links between different polysaccharides, interactions between lignin and carbohydrates, protein cross-linking, cellulose crystallinity and microfibril size, the distribution of lignified tissues in the material, and the ratio of primary to secondary walls.

Plant cell walls have evolved to resist breakdown from microbial and mechanical forces. If cell walls are to be efficiently deconstructed, then we need to take advantage of genetically defined variation in cell wall architecture to simplify sources of heterogeneity and enable carbon-efficient fractionation and catalytic transformations. This knowledge base will provide a basis for the design and optimization of processing methods or for the selection of desired traits by plant breeding or genetic engineering. Feedstocks, such as switchgrass and maize, have cell walls that are characteristic of all grass species, and poplar is an excellent genetic model for fast-growing dicotyledonous trees. All are amenable to genetic modification to generate future biomass crops tailored to catalytic conversion. Below, we discuss genetic strategies for modifying lignin composition and architecture, cellulose and hemicellulose biosyntheses, polysaccharide-lignin interactions, and the regulatory control of secondary wall formation in specific cell types.

Adding Value to Plant Biomass Through Modification of Lignin

A critical need is to add value to the lignin moiety of the biomass that is currently used only for co-firing to produce heat, a loss of one-third of biomass carbon to CO_2. The interest in using abundant and inexpensive lignin as a renewable carbon source for bio-based chemicals and fuels has increased significantly in recent years.[35,36] However, methods for the selective conversion of lignin's complex and heterogeneous phenylpropanoid structure into discrete, low molecular weight aromatic compounds need to be developed. Processes developed to isolate lignin from its native sources[37–39] in various biorefinery scenarios can dramatically alter the initial distribution of substructures. For example, β-ether groups that make up as much as 50% of the interunit linkages in native lignin can be cleaved and nearly eliminated, while the proportion of free phenolic hydroxy groups markedly increases as a result of these isolation processes.[40,41] These features represent a significant opportunity to enhance the ability of the biorefinery to use all components of biomass.

Reduction or modification of lignin composition has been a key focus for overcoming the recalcitrance of the plant cell wall to enzymatic digestion of its carbohydrate constituents to monosaccharide substrates for fermentation.[42–44] However, lowering lignin content, either in mutant or transgenic lines, can compromise biomass yields.[45–48] Further, this strategy is irrelevant to pyrolytic or chemical catalytic conversion processes that use lignin for hydrocarbon fuels and aromatic co-products.[26]

In biochemical conversion pathways, decreasing lignin content may increase release of sugars from cell wall polysaccharides with enzyme treatments.[49–51] In contrast, plants that have been engineered to accumulate chemically altered or higher levels of lignin can provide improved feedstock for direct catalytic conversion.[52] The phenotypic analyses of plants in which phenylpropanoid metabolism is perturbed, as a result of mutation or RNAi inhibition of genes encoding pathway enzymes, have revealed that phenylpropanoid flux can be rerouted from one branch of the pathway to another. For example, by blocking the pathway at ferulate 5-hydroxylase (F5H), aromatics normally converted to 4-hydroxy-3,5-dimethoxy-substituted syringyl (S) lignin monomers in a wild-type plant are diverted to 4-hydroxy-3-methoxy-substituted guaiacyl (G) lignin monomers[53]; by downregulating COMT, the aromatics are blocked at 4,5-dihydroxy-3-methoxy-substituted monomers.[54,55] Defects at other biosynthetic steps lead to pleiotropic phenotypes, such as dwarfing and sterility. Plants carrying mutations in genes encoding cinnamate 4-hydroxylase (C4H) or *p*-coumaroyl shikimate 3′-hydroxylase (C3′H) exhibit dwarfism.[56–58] However, the stunted phenotype of a lignin-deficient Arabidopsis mutant is rescued in a genotypic background mutated in subunits of the mediator complex, indicating that the yield penalty is not a direct consequence of reduced lignin content.[59]

In a recent review, Vanholme et al.[50] describe the lignin pathway and current knowledge of phenolic metabolism and how this knowledge can be used to modify various monomers to make lignin less recalcitrant to commonly used pretreatments. Lignin monomers are transported to the cell wall, where they are polymerized in a combinatorial fashion by free radical coupling mechanisms in a reaction mediated by peroxidases and/or laccases.[13] These include monomers that directly produce a readily cleavable functionality in the polymer (ferulic acid), hydrophilic monomers (guaiacylglycerol, feruloyl quinate, feruloyl glucose, isoconiferin), difunctional monomers and monomer cojugates linked via a readily cleavable functionality (coniferyl ferulate, 3-methoxytyramine ferulate, disinapoyl glucose, diferuloyl sucrose), monomers that minimize lignin-polysaccharide cross-linking (caffeyl alcohol, 5-hydroxyconiferyl alcohol, epicatechin, epigallocatchin), and monomers that give rise to shorter lignin polymers (dihydroconiferyl alcohol, benzenoids).[50] A comprehensive study of the effects of mutations in genes involved in phenylpropanoid biosynthesis via a combination of transcriptomics, metabolomics, and mass spectrometry revealed branched pathways and potential alternative pathways as well as the various mechanisms of compensation.[50] While most of these mutations did not cause visible phenotypes, large changes at the metabolic and transcriptomic levels were detected. The fact that many of these changes can be made without compromising growth indicates that plant developmental plasticity may accommodate large changes to wall composition.

There have been many examples of attempts to modify lignin composition by genetic modification of genes encoding lignin-biosynthetic enzymes. Cinnamyl alcohol dehydrogenase (CAD) catalyzes the last step in monolignol biosynthesis. Downregulation of this gene through RNAi-mediated silencing in switchgrass significantly reduced CAD activity, resulting in significantly less lignin and cutin compared to wild type and higher incorporation of hydroxycinnamyl aldehydes in lignin.[60] Downregulation of switchgrass caffeic acid O-methyltransferase (COMT) also resulted in reduced lignin content, as well as an reduced syringyl:guaiacyl lignin monomer ratio.[61] In a study comparing various lines of transgenic poplar, overexpression of ferulate 5-hydroxylase (F5H) shifted the ratio of lignin monomers, favoring syringyl lignin without impacting total lignin content. Suppression of p-coumarate 3′-hydroxylase (C3′H) significantly reduced total lignin content. After subjecting the two lines to various pretreatments, the C3′H mutant proved less recalcitrant, suggesting that it is total lignin content rather than the monomer ratio that impacts pretreatment.[62]

Many modifications of lignin biosynthesis confer favorable bioprocessing characteristics.[63] Modifying lignin content and S:G ratio have been demonstrated to reduce energy inputs for pulp and paper processing and the recalcitrance of lignocellulosic biomass to enzymatic saccharification.[46,64] In transgenic poplar with almost entirely syringyl (S) lignin,[65] the increased efficiency of kraft pulping[66] and effectiveness of hot water pretreatment in glucose yield after enzymatic digestion[67] were attributed to the linear structures of S lignin and their low degree of polymerization. Differences in processing efficiencies are also observed among natural genetic variants in lignin content in poplar.[68] Simultaneous overexpression of F5H and downregulation of caffeic acid *O*-methyl transferase results in a unique form of lignin with more than 90% benzodioxane units.[63,69] *Mu*-insertional mutant populations can be screened for lignin-biosynthetic genes, such as maize CCR1 showing modified lignin and enhanced digestibility, but normal growth and development.[70] In switchgrass, RNAi-mediated gene silencing of two homologs encoding 4-coumarate:coenzyme A ligase (4CL) reduces G lignin and total lignin content and enhances digestibility.[71] In sorghum, downregulation of one 4CL leads to increased expression of other enzymes to compensate for this change.[72]

Because of our limited understanding of the networks of genes and their feedback interactions, forward genetics remains a powerful tool in biofuels research. Because feedstock behaviors in conversion pathways are complex traits involving multiple genes, several studies focus on the use of populations of recombinant inbred lines to identify cell wall[73] and biofuel production-related quantitative trait loci[74,75] or a meta-analysis of digestibility-related traits as a proxy for saccharification,[76] or naturally occurring poplar populations[77,78] or maize landraces.[74] The extensive numbers of lines in recombinant inbred or association mappling populations imply the need for novel and rapid phenotyping capabilities that are designed to screen for desirable biomass traits. High-throughput pyrolysis molecular-beam mass spectrometry and mid-Infrared (IR) spectroscopy,[79] and Fourier-Transform Near-Infrared (FT-NIR) spectroscopy[80–82] have all been used in various high-throughput screens with glycomics profiling.[83] At lower throughput, solution-state two-dimensional nuclear magnetic resonance (NMR) provides compositional information without the need to deconstruct or fractionate its components.[84]

Studer et al.[68] noted that sugar release was significantly higher in certain poplar genotypes with normal lignin content, inferring that cell wall characteristics other than lignin influence recalcitrance. Thermochemical pretreatments of biomass by steam explosion reach temperatures above the range for lignin phase transition cause lignin to coalesce into molten bodies that redistribute within the biomass.[85] After high-temperature steam pretreatment of cell walls isolated from members of maize recombinant inbred lines,[86] saccharification yield of glucose and xylose is uncorrelated with lignin abundance and quantitative trait loci for lignin abundance and those for saccharification yield of glucose or xylose do not overlap.[87] Genome-wide association studies (GWAS), which take advantage of linkage disequilibrium generated by ancestral recombinations in populations with wide variation, generate a refined list of genes highly likely to contribute to the trait investigated.[88–90] When GWAS was applied to saccharification yield and lignin abundance genotype-structured association panel of 282 maize lines covering 80% of the total genetic diversity,[88] exceptionally strong candidate genes include those that encode several other transcription factors associated with vascularization and fiber formation and components of cellular signaling pathways in addition to those expected to function in cell-wall metabolism.[87] These results provide new insights and strategies beyond modification of lignin composition to enhance yields of biofuels and bioproducts from genetically tailored biomass.

Redesigning Cellulose Microfibrils for Ease of Disassembly

Lowering the high crystallinity of cellulose is a key target for enhancing efficiency of disassembly for conversion.[43] We have gained much knowledge recently on the structure of a bacterial synthase that has inspired more cogent comparisons with plant synthases.[91] An understanding of the active site construction and mechanism of synthesis with knowledge of the unique plant synthase features required for assembly into large complexes enables strategies to redesign the cellulose microfibril.

In plants, the $(1{\to}4)$-β-D-glucan chains of cellulose are synthesized at the plasma membrane by large membrane complexes termed "particle rosettes."[92,93] Particle rosettes are six-membered hexagonal arrays of an estimated six cellulose synthases (CesAs) each. CesAs are intrinsic membrane proteins of about 110 kDa, with channels for extrusion of β-glucan chains comprising eight membrane-panning domains, subtended by their catalytic domains of about 60 kDa extending into the cytoplasm.[94] The Zn-binding "RING" finger domain near the *N*-terminus is thought to couple the CesAs into rosette particles.[95]

The two to three dozen β-glucan chains generated by the rosette complex give an estimated microfibril diameter of around 3.6–3.8 nm for para-crystalline structures of cellulose.[96] However, solid-state ^{13}C-NMR spectroscopy and neutron and X-ray diffraction indicate smaller microfibril diameters of about 2.7–3.0 nm, corresponding to crystalline portions of the microfibril of only 18–24 glucans.[97,98] Because amorphous domains are invisible to X-ray diffraction, up to18 non-crystalline β-glucan chains surface might coat the microfibril in a way that promotes tight interaction with non-cellulosic glycans, such as xyloglucans, glucomannans, and xylans. Higher order of complexity leading to recalcitrance to disassembly can occur by bundling of microfibrils.[99] In summary, altering the microfibril crystallinity by introduction of alternative sugars or linkages into the microfibril, altering the number of β-glucan chains per microfibril to reduce the proportion of crystalline domains in favor of amorphous ones, and interference with bundling with altered non-cellulosic glycans without changing the functional architecture of the cell wall are all targets for improving cellulose availability by easing disassembly.

The three-dimensional crystal structure of the bacterial cellulose synthase A (BcsA) synthase gives a significant conformation of the amino acids that function in uracil-diphosphate glucose (UDP-Glc) binding, chain termination positioning and catalysis of glycosyl transfer.[91] However, structure modeling of a plant CesA is compromised by the addition of two plant-specific sequences, the plant-conserved region (P-CR) and class-specific region (CSR), within the catalytic domain that have no bacterial cognates.[94,100] When the P-CR and CSR are excluded from the catalytic domains, good conservation of the active site defined by the four catalytic motifs and other amino acids with the BcsA is found.[101] Olek et al.[102] used several threading and structure prediction models to show the catalytic motifs give similar active site conservation. Further, the catalytic domain shares substantial structural homology with other related glycosyl transferases to BcsA to classify it as a new nucleotide-binding fold.[102] Although protein engineering can be explored as a means to alter nucleotide-sugar specificity of the active site, the uncertainty about the structural and functional relationships of the P-CR and CSR domains subtending the active site must be resolved before meaningful strategies to modify the cellulose synthase complex can be designed.

Modification of Accessory Proteins for Altering Cellulose Microfibril Structure

Amor et al.[103] first proposed that a sucrose synthase (SuSy) associated with the plasma membrane constituted a "metabolic channel" to funnel a pool of UDP-Glc into the cellulose synthase complex. However, several lines of data have cast doubt on a requirement for SuSy in cellulose synthesis, although it may function facilitatively in vivo. A quadruple mutant that eliminates all detectable SuSy has no effect on rates of cellulose synthesis in Arabidopsis.[104] However, overexpression of SuSy in transgenic poplar results in significant increases in cellulose content.[105] Thus, enhancing amounts of SuSy in the absence of overexpression of the components of the cellulose synthase complex is a reasonable strategy to enhance yields of cellulose.

Several interactions of CesAs with other proteins have been inferred from mutants whose phenotypes include disruption of cellulose synthesis. KORRIGAN, a transmembrane-containing endo-$(1{\rightarrow}4)$-β-D-glucanase, co-localizes with CesAs at the plasma membrane and is required for cell growth and cellulose deposition, but a specific function has never been elucidated.[106] Members of COBRA, a large gene family encoding glycosyl phosphatidylinositol-anchored membrane associated protein, also co-localize to the plasma membrane with CesAs. They do not appear to function directly in synthesis of the glucan chains, but might play a role in the orientation and patterning of both cellulose and lignin during wall deposition.[106] Mutants showed disruption of growth phenotypes traced to defects in wall apposition, with different isoforms associated with normal primary- or secondary-wall cellulose biosynthesis in grasses.[107–109] Their absence gives rise to stem "brittleness" without change in tensile strength in stress-strain experiments.[110] Because some studies indicate that COBRAs might function in crystallization, they are also potential targets to alter recalcitrance properties of cellulose. Modulation of expression of secondary wall COBRAs might have a utility in wall densification or fragmentation during processing.

Modifying Xylan Composition and Architecture in the Interstitial Space

Xylan is the major hemicellulose in cereals and hardwood. The backbone of β-1,4-linked D-xylose backbone is decorated with L-arabinose (arabinoxylans) in grasses, but D-glucuronic acid (glucuronoxylans) in hardwoods. Galactomannan is the major hemicellulose in gymnosperm cell walls (12–15%) with a backbone of β-1,4-linked D-mannose residues.

Secondary wall glucuronoxylan (GX) synthesis in dicots appears to involve a tetrasaccharide initiation sequence, β-D-Xyl-(1→3)-α-L-Rha-(1→2)-α-D-GalA-(1→4)-D-Xyl, located at the reducing end of the xylan polymer.[111] Peña and colleagues found that two mutants deficient in xylan, *irx8* and *fra8*, are essentially devoid of this sequence resulting in a broad distribution of polymer lengths. In contrast, xylans of *irx9* are very short, and nearly all of them contain the tetrasaccharide. York and O'Neill[112] suggest a two-step model in which the tetrasaccharide is a primer for synthesis of short chains of (1→4)-β-D-xylan, which are then spliced together after cleavage of the primer sequence. At least three families of glycosyl transferases are involved in synthesis of the xylan backbone. The GT8 family of retaining-type glycosyl transferase genes encodes IRX8 (GAUT12) and PARVUS (GATL1), and the GT47 family of inverting-type glycosyl transferase genes encode IRX7/FRA8, which are involved in synthesis of the primer.[111,113] The GT43 family of inverting transferase genes encode IRX9 and IRX14, are the apparent synthases of the (1→4)-β-D-xylan oligomeric backbones. Double mutants of IRX10 and IRX10-L from GT47 have greatly reduced GlcA substitutions.[114] Double mutants of *irx15* and *irx15L* have markedly decreased amounts of xylans, and methyl derivatives of GlcA replace the acid form as the xylan side-group.[115,116]

Additional glycosyl transferases add side-group sugars of the GX and GAX polymers.[117] Double mutants with two defective glucuronosyl (GlcA) transferases, *gux1* and *gux2*, produce xylans with reduced GlcA and 4-*O*-methyl GlcA substitutions, resulting in normal plants with more easily extractable xylans.[118,119] GUX1 and GUX2 are not functionally redundant. GUX1 decorates xylan with [Me]GlcA at evenly spaced intervals of mostly eight or 10 residues. In contrast, GUX2 produces more tightly clustered uronosyl residues with more frequent spacing of five to seven xylosyl residues.[119] Further, xylan substitution is not homogeneous, and the existence of differently decorated [Me]GlcA domains might produce xylans with different functional properties, specialized for interaction with cellulose or lignin.[119] The frequency and spacing of [Me]GlcA residues along the xylan chain are expected to have significant structural and functional implications. Xylan chains are generally thought to form three-fold left-handed helical screw axis.[120] Bromley et al.[119] propose that an even [Me]GlcA spacing pattern favors a flat two-fold screw ribbon structure capable of interacting with cellulose microfibrils, placing all GlcA residues on one side of the chain away from the microfibril. This adds surface charge to the microfibril that could impact interaction with other molecules, such as lignin. Conversely, irregularly spaced [Me]GlcA would prevent adoption of a flat ribbon conformation[121,122] and, consequently, block the ability of the xylan to bind to microfibrils.

Proteomic analyses of isolated GAX synthase complexes from wheat membranes implicate a close interaction of GT43 and GT47 family members with a GT75 UDP-Ara mutase, an enzyme that interconverts the UDP-arabinopyranose and UDP-arabinofuranose required for incorporation into the polysaccharide.[123] Xylans and several other polysaccharides can be heavily acetylated, and acetylation can impact wall susceptibility to enzymatic degradation.

Notable is the discovery of a small four-member gene family in Arabidopsis homologous to bacterial *O*-acetyl transferases responsible for reduced wall acetylation (RWA).[124,125] These and other biosynthetic genes represent key targets to alter the fine structures of hemicelluloses, while preserving their functions in modulating the hydrophilicity of microfibril surfaces.

Modulating Gene Expression Networks to Alter Lignin and Carbohydrate Composition and Architecture

Regulatory genes control the transcription of cell wall-related gene networks in specific cell types and at different developmental stages. Control over regulatory gene expression may therefore allow more global modulation of wall composition, particularly because not all components of biosynthetic pathways are known. *Corngrass1* encodes a microRNA that promotes juvenile cell wall morphologies. Its overexpression in switchgrass resulted in up to 250% more starch, releasing more glucose in saccharification assays.[126] Flowering was also inhibited, pointing to potential domestication and limitation of transgene flow into native plants.

Efforts to understand how vascular and fiber cell fate is determined is beginning to yield a wealth of information about regulatory cascades.[127–129] For vascularization of the stem, the breakthrough came a few years ago with the discovery of a class of Arabidopsis NAC domain transcription factors (VND6, VND7, SND1, and NST1) that regulate several downstream transcription factors to specify tracheid and fiber cell fates.[130–132] MYB46 and MYB83 are the direct targets of the NACs,[133,134] which function together to activate a cascade of downstream NAC and MYB domain proteins that directly regulate secondary cell wall programs.[135,136] Orthologous NAC and MYB transcription factors have been identified in woody species[128,129] and grasses.[137,138] Ectopic lignification without polysaccharide deposition occurred in many different cell types as a result of overexpression of MYB58 or MYB63.[139] SND1 and MYB46 regulate expression of MYB58 and MYB63, which in turn regulate genes of the monolignol synthesis pathway. Heterologous expression of SbbHLH1 upregulated transcription factors MYB83, MYB46, and MYB63 and downregulated lignin synthesis genes 4CL1, HCT, COMT, PAL1, and CCR1.[140] SbbHLH1 may be involved in a feedback mechanism that acts on the MYB transcription factors. Aromatic and carbohydrate pathways of secondary wall formation can be independently regulated.

While promoters of most genes of the monolignol biosynthetic pathway possess target AC-response elements for regulation by MYB58 and MYB63, the odd gene out is *F5H*, which lacks the AC-element and is regulated directly by NST1/3.[139,141] Mutations in *NST1* result in impaired lignin monomer synthesis and subsequent reduction in lignin content (Zhao et al., 2010b), but F5H expression is reduced 25-fold, compared to only two-fold reductions in other genes of the biosynthetic pathway.[142] MYB genes also serve as repressors of secondary wall synthesis, and their downregulation is coordinated with the upregulation of the NST genes.[129] Mutations in a WRKY transcription factor that represses secondary wall development in pith cells, is associated with ectopic deposition of cellulose and xylan in these cells, increasing biomass density by almost 50%.[143] A rice homolog of Arabidopsis shine/wax inducer (SHN/WIN), a member of the AP2/ERF family of transcription factors previously shown to be involved in wax/cutin lipid regulation and drought tolerance, was shown to up regulate cellulose synthesis and down-regulate lignin biosynthesis.[144] Discovery of these types of activators and repressors of secondary wall formation may be used to enhance biomass accumulation in multiple cell types.

Conclusions

In this chapter, we envision that revolutionary advances in the production efficiencies by which biomass carbon is converted into biofuels and other energy-rich molecules underpin a bioeconomy derived from a billion tons of biomass. Both the yield and the quality of biomass can be improved using the tools of plant molecular biology and genetic engineering synergistically with improvements in the selectivity and yield of desired products with innovations in catalyst and engineering design. A critical need is a molecular-level understanding of catalyst–biomass interactions and determination of the physical descriptors that control reactivity and selectivity. This new fundamental knowledge will form the basis for the next-generation catalysts and reaction chemistry for conversion of biomass to liquid fuels and high-value chemicals.

References

1. Richard TL. Challenges in scaling up biofuels infrastructure. *Science* 2010;**329**:793–6.
2. Dweikat I, Weil C, Moose S, Kochian L, Mosier N, Ileleji K, et al. Envisioning the transition to a next-generation biofuels industry in the US Midwest. *BioFPR* 2012;**6**:376–86.
3. Perlack RD, Wright LL, Turhollow AF, Graham RL, Stokes BJ, Erbach DC. *Biomass as feedstock for a bioenergy and bioproducts industry: the technical feasibility of a billion-ton annual supply.* DOE GO-102005–2135 Oak Ridge National Laboratory; 2005. 78 pp. http://feedstockreview.ornl.gov/pdf/billion_ton_vision.pdf.
4. U.S. Department of Energy. In: Perlack RD, Stokes BJ, editors. *US billion-ton update: biomass supply for a bioenergy and bioproducts industry.* Oak Ridge (TN): Oak Ridge National Laboratory; 2011. ORNL/TM-2011/224.
5. National Research Council. *Liquid transportation fuels from coal and biomass: technological status, costs, and environmental impacts.* Washington (DC): The National Academies Press; 2009. http://www.nap.edu/openbook.php?record_id=12620≥.
6. Jacobson J, Searcy E, Muth D, Wilkerson E, Sokansanj S, Jenkins B, et al. *Sustainable biomass supply systems.* INL/CON-09–15568. Idaho National Laboratory (INL); 2009.
7. McCann MC, Roberts K. Architecture of the primary cell wall. In: Lloyd CW, editor. *Cytoskeletal basis of plant growth and form.* London: Academic Press; 1991. pp. 109–29.
8. Carpita NC, Gibeaut DM. Structural models of primary cell walls in flowering plants: consistency of molecular structure with the physical properties of the walls during growth. *Plant J* 1993;**3**:1–30.
9. Carpita NC. Structure and biogenesis of the cell walls of grasses. *Annu Rev Plant Physiol Plant Mol Biol* 1996;**47**:445–76.
10. Sarkar P, Bosneaga E, Auer M. Plant cell walls throughout evolution: towards a molecular understanding of their design principles. *J Exp Bot* 2009;**60**:3615–35.
11. Lu FC, Ralph J. Detection and determination of *p*-coumorylated units in lignins. *J Agr Food Chem* 1999;**47**:1988–92.

12. Lu FC, Ralph J. Preliminary evidence for sinapyl acetate as a lignin monomer in kenaf. *Chem Comm* 2002;**1**:90–1.

13. Boerjan W, Ralph J, Baucher M. Lignin biosynthesis. *Annu Rev Plant Biol* 2003;**54**:519–46.

14. US Department of Energy. *Conversion technologies for advanced biofuels: preliminary roadmap and workshop report. December 6–8, 2013*; 2013.

15. Parsell TH, Owen BC, Klein I, Jarrell TM, Marcum CL, Haupert LJ, et al. Cleavage and hydrodeoxygenation (HDO) of C–O bonds relevant to lignin conversion using Pd/Zn synergistic catalysis. *Chem Sci* 2013;**4**:806–13.

16. Kim E, Liu S, Abu-Omar MM, Mosier NS. Selective conversion of biomass hemicellulose to furfural using maleic acid with microwave heating. *Energy Fuels* 2012;**26**:1298–304.

17. US Department of Energy. *Breaking the biological barriers to cellulosic ethanol: a joint research agenda.* Office of Science and Office of Energy Efficiency and Renewable Energy; 2006. http://www.doegenomestolife.org/biofuels/.

18. Xu Q, Adney WS, Ding SY, Himmel ME. Cellulases for biomass conversion. In: Polaina J, MacCabe AP, editors. *Industrial enzymes: structure, function and applications.* London: Springer-Verlag; 2007. pp. 35–50.

19. Agrawal R, Singh NR. Synergistic routes to liquid fuel for a petroleum deprived future. *Amer Inst Chem Eng J* 2008;**55**:1898–905.

20. Cortright RD, Davda RR, Dumesic JA. Hydrogen from catalytic reforming of biomass-derived hydrocarbons in liquid water. *Nature* 2002;**418**:964–7.

21. Drent E, and Jager WW. Hydrogenolysis of glycerol. 2000; US Patent 6 080 898.

22. Schlaf M, Gosh P, Fagan PJ, Hauptman E, Bullock RM. Metal-catalyzed selective deoxygenation of diols to alcohols. *Angew Chem Int Ed* 2001;**40**:3887–90.

23. Schlaf M. Selective deoxygenation of sugar polyols to α, ω-diols and other oxygen content reduced materials- a new challenge for homogeneous ionic hydrogenation and hydrogenolysis catalysis. *Dalton T* 2006;**39**:4645–53.

24. McClellan PP. Manufacture and uses of ethylene oxide and ethylene glycol. *Ind Eng Chem* 1950;**42**:2402–7.

25. Partenheimer W. Methodology and scope of metal/bromide autooxidation of hydrocarbons. *Catal Today* 1995;**23**:69–158.

26. Venkatakrishnan VK, Degenstein J, Delgass WN, Agrawal R, Ribeiro FH. High pressure fast-pyrolysis, fast-hydropyrolysis and catalytic hydro-deoxygenation of cellulose: production of liquid fuel from biomass. *Green Chem* 2014;**16**:792–802.

27. Huber GW, Iborra S, Corma A. Synthesis of transportation fuels from biomass: chemistry, catalysts, and engineering. *Chem Rev* 2006;**106**:4044–98.

28. Pütün AE, Gercel HF, Kockar OM, Ege O, Snape CE, Pütün E. Oil production from an arid-land plant: fixed-bed pyrolysis and hydropyrolysis of *Euphorbia rigida. Fuel* 1996;**75**:1307–12.

29. Pindoria RV, Megaritis A, Herod AA, Kandiyoti R. A two-stage fixed-bed reactor for direct hydrotreatment of volatiles from the hydropyrolysis of biomass: effect of catalyst temperature, pressure and catalyst ageing time on product characteristics. *Fuel* 1998;**77**:1715–26.

30. Rocha JD, Luengo CA, Snape CE. The scope for generating bio-oils with relatively low oxygen contents via hydropyrolysis. *Org Geochem* 1999;**30**:1527–34.

31. Mohan D, Pittman CU, Steele PH. Pyrolysis of wood/biomass for bio-oil: a critical review. *Energy Fuels* 2006;**20**:848–89.

32. Bridgwater AV, Meier D, Radlein D. An overview of fast pyrolysis of biomass. *Org Geochem* 1999;**30**:1479–93.

33. Furimsky E. Catalytic hydrodeoxygenation. *Appl Catal A* 2000;**199**:147–90.

34. Elliott DC. Historical developments in hydroprocessing bio-oils. *Energy Fuels* 2007;**21**:1792–815.

35. Bozell JJ, Tice NC. Organometallic carbohydrate chemistry: metal nucleophiles in the Ferrier reaction. In: *Abstr Pap Am Chem Soc*, Vol. 234. 2007. Abstr. 19-IEC. Boston, MA.

36. Zakzeski J, Jongerius AL, Weckhuysen BM. Transition metal catalyzed oxidation of Alcell lignin, soda lignin, and lignin model compounds in ionic liquids. *Green Chem* 2010;**12**:1225–36.

37. Mosier N, Wyman C, Dale B, Elander R, Lee YY, Holtz apple M, et al. Features of promising technologies for pretreatment of lignocellulosic biomass. *Bioresour Technol* 2005;**96**:673–86.

38. Bozell JJ, Black SK, Myers M, Cahill D, Miller WP, Park S. Solvent fractionation of renewable woody feedstocks: organosolv generation of biorefinery process streams for the production of biobased chemicals. *Biomass Bioenerg* 2011;**35**:4197–208.

39. Bozell JJ, O'Lenick CJ, Warwick S. Biomass fractionation for the biorefinery: heteronuclear multiple quantum coherence-nuclear magnetic resonance investigation of lignin isolated from solvent fractionation of switchgrass. *J Agric Food Chem* 2011;**59**:9232–42.

40. Li JB, Gellerstedt G, Toven K. Steam explosion lignins; their extraction, structure and potential as feedstock for biodiesel and chemicals. *Bioresour Technol* 2009;**100**:2556–61.

41. Samuel R, Pu YQ, Raman B, Ragauskas AJ. Structural characterization and comparison of switchgrass ball-milled lignin before and after dilute acid pretreatment. *Appl Biochem Biotechnol* 2010;**162**:62–74.

42. Chen F, Dixon RA. Lignin modification improves fermentable sugar yields for biofuel production. *Nat Biotech* 2007;**25**:759–61.

43. Himmel ME, Ding S-Y, Johnson DK, Adney WS, Nimlos MR, Brady JW, et al. Biomass recalcitrance: engineering plants and enzymes for biofuels production. *Science* 2007;**315**:804–7.

44. Novaes E, Kirst M, Chiang V, Winter-Sederoff H, Sederoff R. Lignin and biomass: a negative correlation for wood formation and lignin content in trees. *Plant Physiol* 2010;**154**:555–61.

45. Li X, Chapple C. Understanding lignification: challenges beyond monolignol synthesis. *Plant Physiol* 2010;**154**:449–52.

46. Simmons BA, Loqué D, Ralph J. Advances in modifying lignin for enhanced biofuel production. *Curr Opin Plant Biol* 2010;**13**:313–20.

47. Fu C, Mielenz JR, Xiao XR, Ge YX, Hamilton CY, Rodriguez M, et al. Genetic manipulation of lignin reduces recalcitrance and improves ethanol production from switchgrass. *Proc Natl Acad Sci USA* 2011;**108**:3803–8.

48. Fu C, Xiao X, Xi YJ, Ge YX, Chen F, Bouton J, et al. Downregulation of cinnamyl alcohol dehydrogenase (CAD) leads to improved saccharification efficiency in switchgrass. *Bioenerg Res* 2011;**4**:153–64.

49. Boudet AM, Kajita S, Grima-Pettenati J, Goffner D. Lignins and lignocellulosics: a better control of synthesis for new and improved uses. *Trends Plant Sci* 2003;**8**:576–81.

50. Vanholme R, Morreel K, Ralph J, Boerjan W. Lignin engineering. *Curr Opin Plant Biol* 2008;**11**:278–85.

51. Weng J-K, Li X, Bonawitz ND, Chapple C. Emerging strategies of lignin engineering and degradation for cellulosic biofuel production. *Curr Opin Biotech* 2008;**19**:166–72.

52. McLaughlin SB, Samson R, Bransby D, Weislogel A. Evaluating physical, chemical, and energetic properties of perennial grasses as biofuels. In: *Bioenergy '96, proc. seventh national bioenergy conference: partnerships to develop and apply biomass technologies, pp. 1–8. Nashville, TN*. 1996.

53. Chapple CCS, Vogt T, Ellis BE, Somerville CR. An *Arabidopsis* mutant defective in the general phenylpropanoid pathway. *Plant Cell* 1992;**4**:1413–24.

54. Atanassova R, Favet N, Martz F, Chabbert B, Tollier M-T, Monties B, et al. Altered lignin composition in transgenic tobacco expressing *O*-methyltransferase sequences in sense and antisense orientation. *Plant J* 1995;**8**:465–77.

55. Van Doorsselaere J, Baucher M, Chognot E, Chabbert B, Tollier MT, Petit-Conil M, et al. A novel lignin in poplar trees with a reduced caffeic acid 5-hydroxyferulic acid *O*-methyltransferase activity. *Plant J* 1995;**8**:855–64.

56. Franke R, Hemm MR, Denault JW, Ruegger MO, Humphreys JM, Chapple C. Changes in secondary metabolism and deposition of an unusual lignin in the *ref8* mutant of Arabidopsis. *Plant J* 2002;**30**:47–59.

57. Abdulrazzak N, Pollet B, Ehlting J, Larsen K, Asnaghi C, Ronseau S, et al. A coumaroyl-ester-3-hydroxylase insertion mutant reveals the existence of nonredundant *meta*-hydroxylation pathways and essential roles for phenolic precursors in cell expansion and plant growth. *Plant Physiol* 2006;**140**:30–48.

58. Schilmiller AL, Stout J, Weng JK, Humphreys J, Ruegger MO, Chapple C. Mutations in the cinnamate 4-hydroxylase gene impact metabolism, growth and development in Arabidopsis. *Plant J* 2009;**60**:771–82.

59. Bonawitz ND, Kim JI, Tobimatsu Y, Ciesielski P, Anderson NA, Ximenes E, et al. Disruption of mediator rescues the stunted growth of a lignin-deficient *Arabidopsis* mutant. *Nature* 2014;**509**:376–80.

60. Saathoff AJ, Sarath G, Chow EK, Dien BS, Tobias CM. Downregulation of cinnamyl-alcohol dehydrogenase in switchgrass by RNA silencing results in enhanced glucose release after cellulase treatment. *PLoS One* 2011;**6**:e16416.

61. Tschaplinski TJ, Standaert RF, Engle NL, Martin MZ, Sangha AK, Parks JM, et al. Down-regulation of the caffeic acid *O*-methyltransferase gene in switchgrass reveals a novel monolignol analog. *Biotechnol Biofuels* 2012;**5**. Art No. 71.

62. Mansfield SD, Kang KY, Chapple C. Designed for deconstruction—poplar trees altered in cell wall lignification improve the efficacy of bioethanol production. *New Phytol* 2012;**194**:91–101.

63. Bonawitz ND, Chapple C. The genetics of lignin biosynthesis: connecting genotype to phenotype. *Annu Rev Genet* 2010;**44**:337–63.

64. Li X, Chapple C. Understanding lignification: challenges beyond monolignol biosynthesis. *Plant Physiol* 2010;**154**:449–52.

65. Stewart JJ, Akiyama T, Chapple C, Ralph J, Mansfield SD. The effects on lignin structure of overexpression of ferulate 5-hydroxylase in hybrid poplar. *Plant Physiol* 2009;**150**:621–35.

66. Huntley SK, Ellis D, Gilbert M, Chapple C, Mansfield SD. Significant increases in pulping efficiency in C4H-F5H-transformed poplars: improved chemical savings and reduced environmental toxins. *J Agric Food Chem* 2003;**51**:6178–83.

67. Li X, Ximenes E, Kim Y, Slininger M, Meilan R, Ladisch M, et al. Lignin monomer composition affects *Arabidopsis* cell-wall degradability after liquid hot water pretreatment. *Biotechnol Biofuels* 2010;**3**:27–33.

68. Studer MH, DeMartini JD, Davis MF, Sykes RW, Davison B, Keller M, et al. Lignin content in natural *Populus* variants affects sugar release. *Proc Natl Acad Sci USA* 2011;**108**:6300–5.

69. Weng J-K, Mo H, Chapple C. Over-expression of F5H in COMT-deficient *Arabidopsis* leads to enrichment of an unusual lignin and disruption of pollen wall formation. *Plant J* 2010;**64**:898–911.

70. Tamasloukht B, Lam MSJWQ, Martinez Y, Tozo K, Barbier O, Jourda C, et al. Characterization of a cinnamoyl-CoA reductase 1 (CCR1) mutant in maize: effects on lignification, fibre development, and global gene expression. *J Exp Bot* 2011;**62**:3837–48.

71. Xu B, Escamilla-Trevino LL, Sathitsuksanoh N, Shen ZX, Shen H, Zhang YHP, et al. Silencing of 4-coumarate:coenzyme A ligase in switchgrass leads to reduced lignin content and improved fermentable sugar yields for biofuel production. *New Phytol* 2011;**192**:611–25.

72. Saballos A, Sattler SE, Sanchez E, Foster TP, Xin ZG, Kang C, et al. *Brown midrib2 (Bmr2)* encodes the major 4-coumarate: coenzyme A ligase involved in lignin biosynthesis in sorghum (*Sorghum bicolor* L. Moench). *Plant J* 2012;**70**:818–30.

73. Hazen SP, Hawley RM, Davis GL, Henrissat B, Walton JD. Quantitative trait loci and comparative genomics of cereal cell wall composition. *Plant Physiol* 2003;**132**:263–71.

74. Barrière Y, Mechin V, Denoue D, Bauland C, Laborde J. QTL for yield, earliness, and cell wall quality traits in topcross experiments of the F838 × F286 early maize RIL progeny. *Crop Sci* 2010;**50**:1761–72.

75. Lorenzana RE, Lewis MF, Jung H-JG, Bernardo R. Quantitative trait loci and trait correlations for maize stover cell wall composition and glucose release for cellulosic ethanol. *Crop Sci* 2010;**50**:541–55.

76. Truntzler M, Barrière Y, Sawkins MC, Lespinasse D, Betran J, Charcosset A, et al. Meta-analysis of QTL involved in silage quality of maize and comparison with the position of candidate genes. *Theor Appl Genet* 2010;**121**:1465–82.

77. Rae AM, Pinel MPC, Bastien C, Sabatti M, Street NR, Tucker J, et al. QTL for yield in bioenergy Populus: identifying GxE interactions from growth at three contrasting sites. *Tree Gen Genom* 2008;**4**:97–112.

78. Ranjan P, Yin TM, Zhang XY, Kalluri UC, Yang XH, Jawdy S, et al. Bioinformatics-based identification of candidate genes from QTLs associated with cell wall traits in. *Popul Bioenerg Res* 2010;**3**:172–82.

79. Penning B, Hunter CT, Tayengwa R, Eveland E, Dugard CK, Olek A, et al. Genetic resources for maize cell wall biology. *Plant Physiol* 2009;**151**:1703–28.

80. Kelley SS, Rials TG, Snell R, Groom LH, Sluiter A. Use of near infrared spectroscopy to measure the chemical and mechanical properties of solid wood. *Wood Sci Technol* 2004;**38**:257–76.

81. Monrroy M, Mendonca RT, Ruiz J, Baeza J, Freer J. Estimating glucan, xylan, and methylglucuronic acids in kraft pulps of *Eucalyptus globulus* using FT-NIR spectroscopy and multivariate analysis. *J Wood Chem Technol* 2009;**29**:150–63.
82. Liu L, Ye XP, Womac AR, Sokhansanj S. Variability of biomass chemical composition and rapid analysis using FT-NIR techniques. *Carbohydr Polym* 2010;**81**:820–9.
83. Pattathil S, Avci U, Baldwin D, Swennes AG, McGill JA, Popper Z, et al. A comprehensive toolkit of plant cell wall glycan-directed monoclonal antibodies. *Plant Physiol* 2010;**153**:514–25.
84. Mansfield SD, Kim H, Lu FC, Ralph J. Whole plant cell wall characterization using solution-state 2D NMR. *Nat Protoc* 2012;**7**:1579–89.
85. Donohoe BS, Decker SR, Tucker MP, Himmel ME, Vinzant TB. Visualizing lignin coalescence and migration through maize cell walls following thermochemical pretreatment. *Biotech Bioeng* 2008;**101**:913–25.
86. Selig MJ, Tucker MP, Law C, Doeppke C, Himmel ME, Decker SR. High throughput determination of glucan and xylan fractions in lignocelluloses. *Biotechnol Lett* 2011;**33**:961–7.
87. Penning BW, Sykes RW, Babcock NC, Dugard CK, Held MA, Klimek JF, et al. Genetic determinants for enzymatic digestion of lignocellulosic biomass are independent of those for lignin abundance in a maize recombinant inbred population. *Plant Physiol* 2014;**165**(4):1475–87.
88. Bradbury PJ, Zhang Z, Kroon DE, Casstevens TM, Ramdoss Y, Buckler ES. TASSEL: software for association mapping of complex traits in diverse samples. *Bioinformatics* 2007;**23**:2633–5.
89. Tian F, Bradbury PJ, Brown PJ, Hung H, Sun Q, Flint-Garcia S, et al. Genome-wide association study of leaf architecture in the maize nested association mapping population. *Nat Genet* 2011;**43**:159–62.
90. Li X, Zhu CS, Yeh CT, Wu W, Takacs EM, Petsch KA, et al. Genic and nongenic contributions to natural variation of quantitative traits in maize. *Genome Res* 2012;**22**:2436–44.
91. Morgan JLW, Strumillo J, Zimmer J. Crystallographic snapshot of cellulose synthesis and membrane translocation. *Nature* 2013;**493**:181–6.
92. Giddings Jr TH, Brower DL, Staehelin LA. Visualization of particle complexes in the plasma membrane of *Micrasterias denticulata* associated with the formation of cellulose fibrils in primary and secondary cell walls. *J Cell Biol* 1980;**84**:327–39.
93. Mueller SC, Brown Jr RM. Evidence for an intramembrane component associated with a cellulose microfibril-synthesizing complex in higher plants. *J Cell Biol* 1980;**4**:315–26.
94. Delmer DP. Cellulose biosynthesis: exciting times for a difficult field of study. *Annu Rev Plant Physiol Plant Mol Biol* 1999;**50**:245–76.
95. Kurek I, Kawagoe Y, Jacob-Wilk D, Doblin M, Delmer D. Dimerization of cotton fiber cellulose synthase catalytic subunits occurs via oxidation of the zinc-binding domains. *Proc Nat Acad Sci USA* 2002;**99**:11109–14.
96. Kennedy CJ, Cameron GJ, Sturcová A, Apperley DC, Altaner C, Wess TJ, et al. Microfibril diameter in celery collenchyma cellulose: X-ray scattering and NMR evidence. *Cellulose* 2007;**14**:235–46.
97. Fernandes AN, Thomas LH, Altaner CM, Callow P, Forsyth VT, Apperley DC, et al. Nanostructure of cellulose microfibrils in spruce wood. *Proc Natl Acad Sci USA* 2011;**108**:E1195–203.
98. Thomas LH, Forsyth VT, Sturcová A, Kennedy CJ, May RP, Altaner CM, et al. Structure of cellulose microfibrils in primary cell walls from collenchyma. *Plant Physiol* 2013;**161**:465–76.
99. Ding S-Y, Himmel ME. The maize primary cell wall microfibril: a new model derived from direct visualization. *J Agric Food Chem* 2006;**54**:597–606.
100. Vergara CE, Carpita NC. β-D-Glycan synthases and the CesA gene family: lessons to be learned from the mixed-linkage (1→3),(1→4)-β-D-glucan synthase. *Plant Mol Biol* 2001;**47**:145–60.
101. Sethaphong L, Haigler CH, Kubicki JD, Zimmer J, Bonetta D, Debolt S, et al. Tertiary model of a plant cellulose synthase. *Proc Natl Acad Sci USA* 2013;**110**:7512–7.
102. Olek AT, Rayon C, Makowski L, Kim HR, Ciesielski P, Badger J, et al. The structure of the catalytic domain of a plant cellulose synthase and its assembly into dimers. *Plant Cell* 2014;**26**:2996–3009.
103. Amor Y, Haigler CH, Johnson S, Wainscott M, Delmer DP. A membrane-associated form of sucrose synthase and its potential role in synthesis of cellulose and callose in plants. *Proc Natl Acad Sci USA* 1995;**92**:9353–7.

104. Barratt DHP, Derbyshire P, Findlay K, Pike M, Wellner N, Lunn J, et al. Normal growth of Arabidopsis requires cytosolic invertase but not sucrose synthase. *Proc Natl Acad Sci USA* 2009;**106**:13124–9.

105. Coleman HD, Yan J, Mansfield SD. Sucrose synthase affects carbon partitioning to increase cellulose production and altered cell wall ultrastructure. *Proc Natl Acad Sci USA* 2009;**106**:13118–23.

106. Crowell EF, Gonneau M, Stierhof Y-D, Höfte H, Vernhettes S. Regulated trafficking of cellulose synthases. *Curr Opin Plant Biol* 2010;**13**:700–5.

107. Schindelman G, Morikami A, Jung J, Baskin TI, Carpita NC, Derbyshire P, et al. COBRA encodes a putative GPI-anchored protein, which is polarly localized and necessary for oriented cell expansion in Arabidopsis. *Genes Dev* 2001;**15**:1115–27.

108. Roudier F, Fernandez AG, Fujita M, Himmelspach R, Borner GHH, Schindelman G, et al. COBRA, an Arabidopsis extracellular glycosyl-phosphatidyl inositol-anchored protein, specifically controls highly anisotropic expansion through its involvement in cellulose microfibril orientation. *Plant Cell* 2005;**17**:1749–63.

109. Sato K, Suzuki R, Nishikubo N, Takenouchi S, Ito S, Nakano Y, et al. Isolation of a novel cell wall architecture mutant of rice with defective Arabidopsis COBL4 ortholog BC1 required for regulated deposition of secondary cell wall components. *Planta* 2010;**232**:257–70.

110. Sindhu A, Langewisch T, Olek A, Multani DS, McCann MC, Vermerris W, et al. Maize *Brittle stalk2* encodes a COBRA-like protein expressed in early organ development but required for tissue flexibility at maturity. *Plant Physiol* 2007;**145**:1444–59.

111. Peña MJ, Zhong R, Zhou GK, Richardson EA, O'Neill MA, Darvill AG, et al. Arabidopsis *irregular xylem8* and *irregular xylem9*: implications for the complexity of glucuronoxylan biosynthesis. *Plant Cell* 2007;**19**:549–63.

112. York WS, O'Neill MA. Biochemical control of xylan biosynthesis: which end is up? *Curr Opin Plant Biol* 2008;**11**:258–65.

113. Brown DM, Goubet F, Wong VW, Goodacre R, Stephens E, Dupree P, et al. Comparison of five xylan synthesis mutants reveals new insight into the mechanisms of xylan synthesis. *Plant J* 2007;**52**:1154–68.

114. Brown DM, Zhang Z, Stephens E, Dupree P, Turner SR. Characterization of IRX10 and IRX10-like reveals an essential role in glucuronoxylan biosynthesis in Arabidopsis. *Plant J* 2009;**57**:732–46.

115. Brown D, Wightman R, Zhang Z, Gomez LD, Atanassov I, Bukowski J-P, et al. Arabidopsis genes *IRREGULAR XYLEM (IRX15)* and *IRX15L* encode DUF579-containing proteins that are essential for normal xylan deposition in the secondary cell wall. *Plant J* 2011;**66**:401–13.

116. Jensen JK, Kim H, Cocuron JC, Orler R, Ralph J, Wilkerson CG. The DUF579 domain containing proteins IRX15 and IRX15-L affect xylan synthesis in Arabidopsis. *Plant J* 2011;**66**:387–400.

117. Scheller HV, Ulvskov P. Hemicelluloses. *Annu Rev Plant Biol* 2010;**61**:263–89.

118. Mortimer JC, Miles GP, Brown DM, Zhang Z, Segura MP, Weimar T, et al. Absence of branches from xylan in Arabidopsis *gux* mutants reveals potential for simplification of lignocellulosic biomass. *Proc Natl Acad Sci USA* 2010;**107**:17409–14.

119. Bromley JR, Busse-Wicher M, Tryfona T, Mortimer JC, Zhang Z, Brown DM, et al. GUX1 and GUX2 glucuronyltransferases decorate distinct domains of glucuronoxylan with different substitution patterns. *Plant J* 2013;**74**:423–34.

120. Nieduszynski IA, Marchessault RH. Structure of β-D-(1→4)-xylan hydrate. *Biopolymers* 1972;**11**:1335–44.

121. Gabbay SM, Sundararajan PR, Marchessault RH. X-ray and stereochemical studies on xylan diacetate. *Biopolymers* 1972;**11**:79–94.

122. Yui T, Imada K, Shibuya N, Ogawa K. Conformation of an arabinoxylan isolated from the rice endosperm cell-wall by X-ray-diffraction and a conformational-analysis. *Biosci Biotechnol Biochem* 1995;**59**:965–8.

123. Zeng W, Jiang N, Nadella R, Killen TL, Nadella V, Faik A. A glucurono(arabino)xylan synthase complex from wheat contains members of the GT43, GT47, and GT75 families and functions cooperatively. *Plant Physiol* 2010;**154**:78–97.

124. Lee C, Teng Q, Zhong R, Ye Z-H. The four Arabidopsis reduced wall acetylation genes are expressed in secondary wall-containing cells and required for the acetylation of xylan. *Plant Cell Physiol* 2011;**52**:1289–301.

125. Manabe Y, Nafisi M, Verhertbruggen Y, Orfila C, Gille S, Rautengarten C, et al. Loss-of-function mutation of reduced wall acetylation in Arabidopsis leads to reduced cell wall acetylation and increased resistance to *Botrytis cinerea*. *Plant Physiol* 2011;**155**:1068–78.

126. Chuck GS, Tobias C, Sun L, Kraemer F, Li CL, Dibble D, et al. Overexpression of the maize *Corngrass1* microRNA prevents flowering, improves digestibility, and increases starch content of switchgrass. *Proc Natl Acad Sci USA* 2011;**108**:17550–5.

127. Demura T, Ye Z-H. Regulation of plant biomass production. *Curr Opin Plant Biol* 2010;**13**:299–304.

128. Zhong R, Lee C, Ye Z-H. Evolutionary conservation of the transcriptional network regulating secondary cell wall biosynthesis. *Trends Plant Sci* 2010;**15**:625–32.

129. Zhao Q, Dixon RA. Transcriptional networks for lignin biosynthesis: more complex than we thought? *Trends Plant Sci* 2011;**16**:227–33.

130. Kubo M, Udagawa M, Nichikubo N, Horiguchi G, Yamaguchi M, Ito J, et al. Transcription switches for protoxylem and metaxylem vessel formation. *Genes Dev* 2005;**19**:1855–60.

131. Zhong R, Demura T, Ye J-H. SND1, a NAC domain transcription factor, is a key regulator of secondary wall synthesis in fibers of Arabidopsis. *Plant Cell* 2006;**18**:3158–70.

132. Mitsuda N, Iwase A, Yamamoto H, Yoshida M, Seki M, Shinozaki K, et al. NAC transcription factors, NST1 and NST3, are key regulators of the formation of secondary walls in woody tissues of Arabidopsis. *Plant Cell* 2007;**19**:270–80.

133. Nakano Y, Nishikubo N, Goué N, Yamaguchi M, Katayama Y, Demura T. MYB transcription factors orchestrate the developmental program of xylem vessels in Arabidopsis roots. *Plant Biotechnol* 2010;**27**:267–72.

134. McCarthy RL, Zhong R, Ye Z-H. MYB83 is a direct target of SND1 and acts redundantly with MYB46 in the regulation of secondary wall biosynthesis in Arabidopsis. *Plant Cell* 2009;**50**:1950–64.

135. Ohtani M, Nishikubo N, Xu B, Yamaguchi M, Mitsuda N, Goué N, et al. A NAC domain protein family contributing to the regulation of wood formation in poplar. *Plant J* 2011;**67**:499–512.

136. Yamaguchi M, Mitsuda N, Ohtani M, Ohme-Takagi M, Kato K, Demura T. Vascular-related nac-domain directly regulates the expression of a broad range of genes for xylem vessel formation. *Plant J* 2011;**66**:579–90.

137. Bosch M, Mayer C-D, Cookson A, Donnison IS. Identification of genes involved in cell wall biogenesis in grasses by differential gene expression profiling of elongating and non-elongating maize internodes. *J Exp Bot* 2011;**62**:3545–61.

138. Zhong R, Lee C, McCarthy RL, Reeves CK, Jones EG, Ye Z-H. Transcriptional activation of secondary wall biosynthesis by rice and maize NAC and MYB transcription factors. *Plant Cell Physiol* 2011;**52**:1856–71.

139. Zhou J, Lee C, Zhong R, Ye Z-H. MYB58 and MYB63 are transcriptional activators of the lignin biosynthetic pathway during secondary cell wall formation in *Arabidopsis*. *Plant Cell* 2009;**21**:248–66.

140. Yan L, Xu CH, Kang YL, Gu TW, Wang DX, Zhao SY, et al. The heterologous expression in *Arabidopsis thaliana* of sorghum transcription factor *SbbHLH1* downregulates lignin synthesis. *J Exp Bot* 2013;**64**:3021–32.

141. Zhao Q, Wang H, Yin Y, Xu Y, Chen F, Dixon RA. Syringyl lignin biosynthesis is directly regulated by a secondary cell wall master switch. *Proc Natl Acad Sci USA* 2010;**107**:14496–501.

142. Zhao Q, Gallego-Giraldo L, Wang HZ, Zeng YN, Ding SY, Chen F, et al. An NAC transcription factor orchestrates multiple features of cell wall development in *Medicago truncatula*. *Plant J* 2010;**63**:100–14.

143. Wang H, Avci U, Nakshima J, Hahn MG, Chen F, Dixon RA. Mutation of WRKY transcription factors initiates pith secondary wall formation and increases stem biomass in dicotyledonous plants. *Proc Natl Acad Sci USA* 2010;**107**:22338–43.

144. Ambavaram MMR, Krishnan A, Trijatmiko KR, Pereira A. Coordinated activation of cellulose and repression of lignin biosynthetic pathways in rice. *Plant Physiol* 2011;**155**:916–31.

CHAPTER 5

Processive Cellulases

David B. Wilson

Department of Molecular Biology and Genetics, Cornell University, Ithaca, NY, USA

Structural and functional studies of cellulases have identified three functionally different classes of cellulases, all of which are capable of acting synergistically with cellulases from other classes on crystalline cellulose: endoglucanases, processive endoglucanases, and exocellulases.[1] Known cellulase sequences are listed in the CAZy web site.[2] The extreme diversity of cellulases, probably results from the enormous diversity of plant cell walls.[3] Consistent with this, some cellulase sequences show evidence for positive selection.[4] Endoglucanases are the most abundant class, and they occur in 11 of the 12 cellulase families. They possess open active site clefts that can bind to any accessible site along a cellulose chain. These enzymes make one or a few cleavages around their binding site and then dissociate. Because cellulose oligosaccharides longer than cellohexose are insoluble, most endocellulases produce about 30% insoluble reducing ends and 70% soluble reducing sugars, which are mainly cellobiose, because larger soluble oligosaccharides are rapidly cleaved to cellobiose by these enzymes. An endocellulase that is preferentially bound near the end of a cellulose chain could produce mainly soluble reducing sugars, but few if any such enzymes are known.

The first processive endocellulase to be well characterized was *Thermobifida fusca* Cel9A, which produces about 87% soluble reducing ends.[5] This enzyme contains a weakly binding family 3 carbohydrate binding module (CBM), which is rigidly attached to the family 9 catalytic domain (CD). A cellulose chain modeled into its active site also lies along the binding surface of the CBM.[6] Thus this CBM is an integral part of the catalytic site of this enzyme. Removal of the family 3 CBM converted this enzyme into a nonprocessive endocellulase that produced only 44% soluble reducing ends, which makes it the least processive endoglucanase characterized so far.[7] Another unusual property of this enzyme is that in its processive mode, it initially produces cellotetraose, while most processive cellulases produce mainly cellobiose. As digestion proceeds, the dominant product is cellobiose, because cellotetraose is readily cleaved to cellobiose by this enzyme. It appears that Cel9A, which has an open active site cleft, initially binds to any accessible site along a cellulose chain like other endocellulase, but after it cleaves the chain, the fragment bound to the weakly binding CBM can rebind into the active site, allowing the enzyme to processively cleave cellotetraose from the nonreducing end of the cellulose chain. This process continues until the enzyme finally dissociates. Similar enzymes

Direct Microbial Conversion of Biomass to Advanced Biofuels. http://dx.doi.org/10.1016/B978-0-444-59592-8.00005-1

have been found in many cellulolytic bacteria, including anerobic bacteria-producing cellulo-somes. In fact, it appears that in cellulosomes this type of cellulase has replaced the nonreducing end specific family 6 exocellulase.[8] There is a good reason for this, because it appear that anaerobic bacteria need to produce oligosaccharides with an average length of four glucose residues to obtain enough energy for growth on cellulose, and an exocellulase mainly produces cellobiose, while the processive endoglucanase produces mainly cellotetraose.[9] Another class of processive endoglucanases, which are in family 5, were found in *Saccharophagus degradans*, in the brown rot basidiomycete, *Gloeophyllum trabeum*, and in the mushroom, *Volvariella volvacea*.[10–12] While their mechanism is not well studied, they do not require a CBM for processivity, suggesting that their processivity is determined by the affinities of the glucose-binding subsites in their CD. This type of processivity is well studied in certain chitinases.[13]

Exocellulases are the largest class of processive cellulases and were the first ones to be identified. There are two types of exocellulases, one type attacks the nonreducing end of a cellulose chain, cleaving off cellobiose, and these enzymes are all in family 6.[14] *Trichoderma reesei* Cel6A was the first cellulase CD to have its structure determined, and its active site is present in a tunnel, as are those of every known exocellulase.[15] The other type of exocellulase attacks the reducing end of cellulose chains, again processively cleaving off cellobiose residues. It is surprising that most fungal reducing end exocellulases are in family 7, while all bacterial reducing end specific exocellulases are in family 48. Anaerobic fungi, which are present in the rumen, contain family 48 exocellulases that are similar to those in rumen bacteria, suggesting that they were acquired by horizontal gene transfer.[16] It is not clear why there are two structurally different families of reducing end specific exocellulases and only one family of nonreducing end specific exocellu-lases. However, one possible explanation is that family 7 exocellulases might not be able to fold up in an anaerobic environment, so that anaerobic bacteria evolved family 48 exocellulases. However, this would not explain why all aerobic bacteria produce family 48 exocellulases, although family 7 exocellulases are significantly more active than family 48 exocellulases.

It has been shown that a family 7 exocellulase can synergize with both purified and crude *T. fusca* cellulases, and all the mixtures showed higher activity than the corresponding mixtures containing *T. fusca* Cel48A.[7] Exocellulases do not appear to cause internal cleavage of cellulose, as shown by their very high production of soluble reducing ends (93–95%) and their inability to reduce the viscosity of Carboxymethylcellulose (CMC), although there are claims in the literature that exocellulases will occasionally make such cleavages.[17,13] The low level of insoluble reducing ends produced by an exocellulase (5–7%) may result from the enzyme exposing preexisting reducing ends that were hidden underneath cellulose chains that it hydrolyzed.

A key limitation in studying cellulase processivity is the lack of a simple direct assay for determining how processive an enzyme is. Processivity is defined as the average number of cleavages made by a cellulase on cellulose chains before it dissociates. The determination of the number of soluble reducing ends versus the number of insoluble ends produced by a cellulase gives a rough measure of processive, but does not give an accurate value for the processivity of the enzyme. This assay is most simply run on filter paper discs in which the disc can be removed after it is incubated with an enzyme, rinsed, and the amount of insoluble ends in it can be determined by a reducing sugar assay. The supernatant can be assayed for reducing sugars to determine the amount of soluble reducing ends.[7] A more quantitative assay for exocellulases is to incubate the enzyme with a cellulose substrate and determine the ratio of cellobiose to cellotriose produced by HPLC. This assay assumes that the initial positioning of a cellulose chain either results in the first cleavage product being cellobiose or cellotriose and that half of the cellulose chains will bind in each way. This occurs because every other glucose residue in cellulose has a different stereochemistry, and each glucose subsite in a cellulase is specific for one stereochemistry. After the first cleavage, every subsequent cleavage will produce cellobiose, so that the ratio of cellobiose to cellotriose gives the processivity of the enzyme. This assumes that the enzyme does not cleave cellotriose. If it does cleave cellotriose the amount of glucose will determine how much cellotriose was cleaved, so that the processivity is the amount of cellobiose—glucose/cellotriose + glucose. This assay relies on a number of assumptions that have not been tested and probably underestimates processivity. Processivity also has been assayed by modifying the reducing ends in the cellulose substrate and then dividing the amount of soluble reducing sugar produced by the enzyme by the amount of insoluble reducing ends it produces.[19] The most direct measure of processivity is to follow the movement of individual cellulases on a cellulose fiber using scanning atomic force microscopy.[20] This assay has been carried out for *T. reesei* Cel7A and Cel6B, and the results were very different, as Cel7A was very processive, while Cel6A did not show processivity at the resolution of this measurement.[20] The distance moved by the measured Cel7A molecules gave a processivity of about 20–40, but there could be some molecules that only move a short distance that would not have been seen due to the limited resolution of the technique. This would reduce the bulk processivity measured by the other assays, which measure the average processivity of a large population of molecules. Another approach to study the processivity of individual cellulase molecules is to label them with a fluorescent group and observe them by confocal imaging.[21] A study of three cellulases, Cel5A, an endocellulase, Cel6B, an exocellulase, and the processive endocellulase, Cel9A, did not see evidence for processive movement of any of these enzymes, possibly because of its low resolution (about 20 nm). Each of the enzymes showed different behavior on the cellulose, but there was no evidence of surface diffusion for any of them. Cel5A was more mobile than the other two enzymes, possibly because its CD dissociates faster because it is not processive.[21] Surface diffusion had been reported for a *Cellulomonas fimi* endocellulase by photobleaching of fluorescent cellulase, but it appears that the restoration of fluorescence was caused by reattachment of unbound cellulases, not surface diffusion.[22]

An important property of cellulases acting on crystalline cellulose is synergism, because some cellulase mixtures have significantly higher activity than the sum of the activities of each individual enzyme. The specific activity of an effective mixture of multiple cellulases can be more than 10 times that of any individual cellulase. Synergism is rarely seen in mixtures of two cellulases in the same class, i.e., two different endocellulases. It is seen in mixtures of an endocellulase and an exocellulase, mixtures of a processive endocellulase and any other class of cellulase, and mixtures of a reducing end attacking exocellulase and a nonreducing end attacking exocellulase.[23] At this time, mixtures of family 5 and family 9 processive endocellulases have not been tested for synergism. While there is still a great deal to be learned about cellulase synergism, it appears that an important cause is the limited amounts of accessible substrate for individual cellulases. Thus, if two cellulases attack different sites on the cellulose particle and create new substrates for each other, they will give synergism. For example, an endocellulase will create more ends for either type of exocellulase, and an exocellulase may create new accessible sites for an endocellulase by modifying the cellulose surface as it moves along a cellulose chain. It is interesting that cellulose incubated with an endocellulase, which is then removed, is a better substrate for an exocellulase then untreated cellulose, because it contains more chain ends.[24] However the reverse is not true, although studies of synergism in a mixture of an endocellulase and an exocellulase show that both enzymes were stimulated.[7] A processive endocellulase is both creating new ends for exocellulases and disrupting the neighboring chains during its processive stage, thus allowing it to synergize with both endocellulases and exocellulases. *T. fusca* Cel9A has the highest activity on crystalline cellulose of any cellulase, possibly because it can synergize with itself.

Exocellulases are significantly less active than good endocellulases on most substrates and somewhat less active on crystalline cellulose, but they are the most abundant cellulases in most cellulolytic microorganisms grown on cellulose. For *T. reesei*, an active celluloytic fungus, the reducing end specific exocellulase, Cel7A, makes up nearly 70% of the total cellulase protein,[25] while for the bacterium *T. fusca*, the comparable enzyme Cel48A makes up about 35%, and its other exocellulase is another 35% of the total cellulase protein.[26] The difference between the two organisms may reflect the fact that Cel7A is significantly more active than Cel48A. Inactivation of either *T. fusca* exocellulase in *T. fusca* crude cellulase by treatment with a specific antiserum causes a major loss of activity on bacterial cellulose, showing that both exocellulases are responsible for a significant fraction of the activity of *T. fusca* crude cellulase.

It appears that the rate-limiting step for crystalline cellulose hydrolysis is the placement of a segment of a cellulose molecule into the active site of the cellulase. The change in cellulose hydrolysis rate with temperature is too high for this step to simply be diffusion of the cellulase to an accessible site on the substrate, so that the enzyme must participate in placing the cellulose molecule into the active site. To design cellulases with improved activity on crystalline cellulose, it is necessary to identify the residues in the cellulase that help place a cellulose molecule into its active site. It seems likely that in an exocellulase, the residues on the surface of the cellulase close to the entrance to the active site tunnel will carry out this step. As shown in Figure 1, there are a limited number of such residues at the entrance of the tunnels for the three family 48 exocellulases: *T. fusca* Cel48A, *Clostridium thermocellum* CelS, and *Clostridium cellulolyticum* CelF, in which three dimensional structures are known. All of the tunnel entrance residues in all three enzymes potentially can bind to cellulose, and there is about a three-fold enrichment of aromatic residues among these residues for all three enzymes, as compared to the total enzyme surface. However, only one of the four aromatic residues in *T. fusca* Cel48A are conserved in all family members. We have mutated each of the aromatic residues in *T. fusca* Cel48A to Ala, and in each case the mutant enzyme had reduced activity on crystalline cellulose and normal activity on the soluble substrates cellohexose and cellopentose, showing that each one of these residues is specifically required for crystalline cellulose activity.[27] Mutating either of two nonaromatic residues to Ala did not change activity on crystalline cellulose. These results suggest that it is the aromatic residues near the tunnel entrance that function to place a cellulose end into the active site tunnel, but we do not know exactly how this would occur. It is interesting that the four aromatic residues at the tunnel entrance are in a line.

There is an excellent review about the role of processivity in the hydrolysis of crystalline substrates that is focused on chitin hydrolysis, but also extensively discusses cellulose hydrolysis.[28]

There are several important questions about exocellulase processivity that still have not been answered. One is how the cellulose chain, which remains in the active site tunnel after the

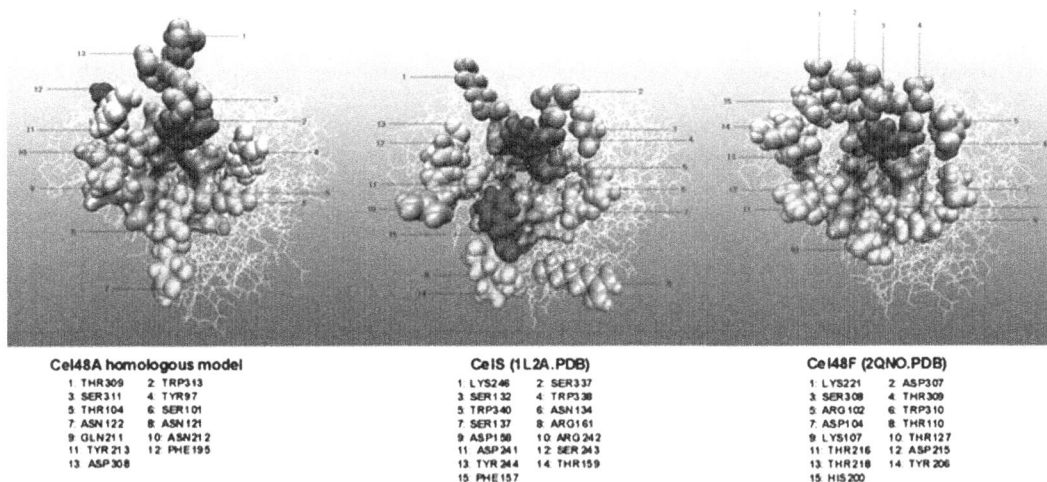

Cel48A homologous model		CelS (1L2A.PDB)		Cel48F (2QNO.PDB)	
1. THR309	2. TRP313	1. LYS246	2. SER337	1. LYS221	2. ASP307
3. SER311	4. TYR97	3. SER132	4. TRP338	3. SER309	4. THR309
5. THR104	6. SER101	5. TRP340	6. ASN134	5. ARG102	6. TRP310
7. ASN122	8. ASN121	7. SER137	8. ARG161	7. ASP104	8. THR110
9. GLN211	10. ASN212	9. ASP158	10. ARG242	9. LYS107	10. THR127
11. TYR213	12. PHE195	11. ASP241	12. SER243	11. THR216	12. ASP215
13. ASP308		13. TYR244	14. THR159	13. THR218	14. TYR206
		15. PHE157		15. HIS200	

Figure 1
Surface residues near the tunnel entrance of three family 48 cellulases.

cleaved cellobiose has dissociated, is able to be moved to fill the empty subsites, given the many interactions that exist between the cellulose chain and the enzyme? Another is what causes the enzyme to dissociate from a cellulose chain, given the many interactions between them? Studies of the affinity of cellulases and inactive mutant cellulases to cellulose, both with and without their CBM, show that Wildtype (WT) CDs have a low affinity, while the inactive mutant CDs have a high affinity to cellulose, and this is true of both endocellulases and exocellulases.[28] The low affinity of exocellulase CDs to cellulose seems surprising, given that even after cleavage there are many interactions between the CD and the cellulose chain, and the CD is processive so that it should not dissociate after each cleavage event. At present, the reasons for these findings are not known. Finally, the way in which exocellulases transiently disrupt a cellulose substrate so that they increase the activity of endocellulases in synergistic mixtures is not known.

Acknowledgments

This work was sponsored by the BioEnergy Science Center (BESC). BESC is a U.S. Department of Energy Bioenergy Research Center supported by the Office of Biological and Environmental Research in the DOE Office of Science.

References

1. Wilson DB. Processive and nonprocessive cellulases for biofuel production—lessons from bacterial genomes and structural analysis. *Appl Microbiol Biotechnol* 2012;**93**:497–502.
2. Henrissat B, Bairoch A. Updating the sequence-based classification of glycosyl hydrolases. *Biochem J* 1996;**316**:695–6.
3. Lee KJ, Marcus SE, Knox JP. Cell wall biology: perspectives from cell wall imaging. *Mol Plant* 2011;**4**:212–9.
4. Mayer WE, Schuster LN, Bartelmes G, Dieterich C, Sommer RJ. Horizontal gene transfer of microbial cellulases into nematode genomes is associated with functional assimilation and gene turnover. *BMC Evol Biol* 2011;**11**:13.
5. Irwin D, Shin D-H, Zhang S, Barr BK, Sakon J, Karplus PA, et al. Roles of the catalytic domain and two cellulose binding domains of *Thermomonospora fusca* E4 in cellulose hydrolysis. *J Bacteriol* 1998;**180**: 1709–14.
6. Sakon J, Irwin D, Wilson DB, Karplus PA. Structure and mechanism of endo/exocellulase E4 from *Thermomonospora fusca*. *Nat Struct Biol* 1997;**4**:810–8.
7. Irwin DC, Spezio M, Walker LP, Wilson DB. Activity studies of eight purified cellulases: specificity, synergism, and binding domain effects. *Biotechnol Bioeng* 1993;**42**:1002–13.
8. Caspi J, Barak Y, Haimovitz R, Gilary H, Irwin DC, Lamed R, et al. *Thermobifida fusca* exoglucanase Cel6B is incompatible with the cellulosomal mode in contrast to endoglucanase Cel6A. *Syst Synth Biol* 2010;**4**: 193–201.
9. Lynd LR, Weimer PJ, van Zyl WH, Pretorius IS. Microbial cellulose utilization: fundamentals and biotechnology. *Microbiol Mol Biol Rev* 2002;**66**:506–77.
10. Watson BJ, Zhang H, Longmire AG, Moon YH, Hutcheson SW. Processive endoglucanases mediate degradation of cellulose by *Saccharophagus degradans*. *J Bacteriol* 2009;**191**:5697–705.
11. Cohen R, Suzuki MR, Hammel KE. Processive endoglucanase active in crystalline cellulose hydrolysis by the brown rot basidiomycete *Gloeophyllum trabeum*. *Appl Environ Microbiol* 2005;**71**:2412–7.

12. Zheng F, Ding S. Processivity and enzymatic mode of a glycoside hydrolase family 5 endoglucanase from *Volvariella volvacea*. *Appl Environ Microbiol* 2013;**79**:989–96.

13. Vaaje-Kolstad G, Horn SJ, Sørlie M, Eijsink VG. The chitinolytic machinery of *Serratia marcescens*—a model system for enzymatic degradation of recalcitrant polysaccharides. *FEBS J* 2013;**280**(13):3028–49.

14. Barr BK, Hsieh Y-L, Ganem B, Wilson DB. Identification of two functionally different classes of exocellulases. *Biochemistry* 1996;**35**:586–92.

15. Rouvinen J, Bergfors T, Teeri T, Knowles JK, Jones TA. Three-dimensional structure of cellobiohydrolase II from *Trichoderma reesei*. *Science* 1990;**249**:380–6.

16. Youssef NH, Couger MB, Struchtemeyer CG, Liggenstoffer AS, Prade RA, Najar FZ, et al. The genome of the anaerobic fungus *Orpinomyces* sp. strain C1A reveals the unique evolutionary history of a remarkable plant biomass degrader. *Appl Envir Microbiol* 2013;**79**:4620–34.

17. Boisset C, Fraschini C, Schülein M, Henrissat B, Chanzy H. Imaging the enzymatic digestion of bacterial cellulose ribbons reveals the endo character of the cellobiohydrolase Cel6A from *Humicola insolens* and its mode of synergy with cellobiohydrolase Cel7A. *Appl Environ Microbiol* 2000;**66**:1444–52.

18. Ståhlberg J, Johansson G, Pettersson G. *Trichoderma reesei* has no true exo-cellulase: all intact and truncated cellulases produce new-reducing end groups on cellulose. *Biochem Biophys Acta* 1993;**1157**:107–13.

19. Horn SJ, Sørlie M, Vårum KM, Väljamäe P, Eijsink VG. Measuring processivity. *Methods Enzymol* 2012;**510**:69–95.

20. Imai T, Boisset C, Samejima M, Igarashi K, Sugiyama J. Unidirectional processive action of cellobiohydrolase Cel7A on Valonia cellulose microcrystals. *FEBS Lett* 1998;**432**:113–6.

21. Moran-Mirabal JM, Bolewski JC, Walker LP. *Thermobifida fusca* cellulases exhibit limited surface diffusion on bacterial micro-crystalline cellulose. *Biotechnol Bioeng* 2013;**110**:47–56.

22. Jervis EJ, Haynes CA, Kilburn DG. Surface diffusion of cellulases and their isolated binding domains on cellulose. *J Biol Chem* 1997;**272**:24016–23.

23. Vuong TV, Wilson DB. Processivity, synergism, and substrate specificity of *Thermobifida fusca* Cel6B. *Appl Environ Microbiol* 2009;**75**:6655–61.

24. Väljamäe P, Sild V, Nutt A, Pettersson G, Johansson G. Acid hydrolysis of bacterial cellulose reveals different modes of synergistic action between cellobiohydrolase I and endoglucanase I. *Eur J Biochem* 1999;**266**:327–34.

25. Teeri TT, Koivula A, Linder M, Wohlfahrt G, Divne C, Jones TA. *Trichoderma reesei* cellobiohydrolases: why so efficient on crystalline cellulose? *Biochem Soc Trans* 1998;**26**:173–8.

26. Spiridonov NA, Wilson DB. Regulation of biosynthesis of individual cellulases in *Thermomonospora fusca*. *J Bacteriol* 1998;**180**:3529–32.

27. Kostylev M, Alahuhta M, Chen M, Brunecky R, Himmel ME, Lunin VV, et al. Cel48A from *Thermobifida fusca*: structure and site directed mutagenesis. *Biotechnol Bioeng* 2014;**111**(4):664–73.

28. Payne CM, Baban J, Horn SJ, Backe PH, Arvai AS, Dalhus B, et al. Hallmarks of processivity in glycoside hydrolases from crystallographic and computational studies of the *Serratia marcecscens* chitinases. *J Biol Chem* 2012;**287**:36322–30.

CHAPTER 6

Bacterial AA10 Lytic Polysaccharide Monooxygenases Enhance the Hydrolytic Degradation of Recalcitrant Substrates

Nathan Kruer-Zerhusen, David B. Wilson

Department of Molecular Biology and Genetics, Cornell University, Ithaca, NY, USA

Substrate Recalcitrance and Cellulase Mixtures

Polysaccharide macromolecules are produced by numerous organisms for structural and energy storage roles. Structural cellulose is the most abundant of these structural polysacchrides, comprising up to 50% of plant dry weight.[1] Biomass represents a sustainable supply of feedstock material for bioconversion into liquid fuel and other materials such as plastics. Cellulose is advantageous for fermentative liquid fuel production because saccharification results in pure glucose, which can be used directly for many bioconversion applications. The chemical composition of cellulose is simple, being made up of linear glucose chains, but degrading it into monomeric components is hindered by its structural complexity within biomass. Lignocellulosic plant biomass is composed of cellulose microfibrils forming layers crosslinked by hemicellulose and embedded in lignin. This complex structure limits the efficiency of saccharification to usable components and increases the expense of plant bimass feedstock conversion to liquid fuels.[2]

Cellulose is made up of repeating cellobiose units of glucose monomers linked by β-1,4 glycosidic bonds.[1] Long cellulose polysaccharides can have a very high degree of polymerization (DP), with polymers made up of thousands of glucose monomers. These polysaccharides stabilize each other within the cellulose structure, with the equatorial oxygen atoms of glucose rings forming inter- and intramolecular hydrogen bonds. Noncovalent bonds between adjacent chains form a cellulose sheet that interacts with adjacent sheets through van der Waals forces and planar stacking interactions.[3] Ordered cellulose polysaccharide chains are stabilized by this network of interactions and form the crystalline core of microfibrils whereas surface chains are more solvent exposed and may have a more amorphous structural character.

Direct Microbial Conversion of Biomass to Advanced Biofuels. http://dx.doi.org/10.1016/B978-0-444-59592-8.00006-3

Within plant biomass, many cellulose microfibrils overlap to form a matrix of fibers. The molecular interactions forming stable networks on multiple scales allow cellulose to provide structural rigidity to plant cell walls. In addition, these interactions cause cellulose to be very recalcitrant and resist degradation by chemical and enzymatic means.[3] The tight structure of cellulose microfibers resists depolymerization by preventing enzymes from accessing digesible material. This provides cellulose with very high stability over time, with the glycosidic bonds of intact cellulose having a hypothesized half-life of 5 million years at neutral pH in the absence of catalysts.[4]

Chitin is the second most abundant crystalline biopolysaccharide, found as structural material in many insects and crustaceans. It is similarly structurally resistant to degradation, but it is less densely packed because of the presence of an acetamido group at the 2-hydroxy position.[5]

Biomass produced by plants for structural and storage roles characteristically has a high degree of recalcitrance to chemical and enzymatic depolymerization. Biomass recalcitrance is defined as the chemical and physical properties preventing the degradation of lignocellulose into components via microbial or enzymatic means. This recalcitrance is a major limiting factor in insoluble polysaccharide enzymatic saccharification.[2] In addition to the structure of crystalline cellulose, hemicellulose and lignin contribute to biomass recalcitrance by restricting the access to cellulose. By covering cellulose microfibers, hemicellulose and lignin present a physical barrier to chemical and enzymatic substrate access.

Hemicelluloses are more structurally complex than cellulose. They are made up of polymeric branching macromolecules of saccharide monomers, such as xylans, mannans, and glucans. Hemicellulose serves to interact directly with cellulose microfibers and form connections between adjacent fibers, with pectin cross-linking them to other matrix components.[2] Hemicellulose has less crystallinity and is more readily digestible than cellulose, although structural branching and acetylation can increase the difficulty.[6] Microbes secrete multiple enzymes as part of their system of degradative enzymes to digest these cross-linking co-polymers.[7]

Lignin is a more complex polymer composed of differing aromatic subunits. Lignin surrounds the cellulose microfibers and covalently links the associated hemicellulose to form a hydrophobic encasing material. Lignin provides structural strength to biomass through these inter- and intramolecular covalent linkages.[8] The insoluble polyphenolic structure of lignin makes it very difficult to degrade, requiring specialized oxidative enzymes or harsh chemicals. The enzymatic mechanisms of microbial lignin depolymerization are still not fully characterized.

Biomass pretreatment is used to remove the hemicelluloses and lignin, which blocks enzyme access to cellulose microfibrils. The first step for biomass saccharification is to reduce the substrate recalcitrance by performing a physical or chemical pretreatment. This pretreatment improves enzymatic hydrolysis by removing noncellulosic material to expose the crystalline cellulose microfibers to enzymatic digestion. The crystalline nature of cellulose microfibers themselves causes a high degree of recalcitrance. Saccharification of this cellulose to free sugars is still limited by the crystalline structure.

Model polysaccharide substrates for investigating recalcitrance and enzymatic degradation include Avicel, phosphoric acid swollen cellulose (PASC), filter paper, bacterial cellulose (BC), and pretreated lignocellulosic biomass. Depending on the degree and type of mechanical and chemical changes during preprocessing, these substrates can have different amorphous regions, crystallinity, DP, and productive binding surfaces. In particular, PASC is dramatically affected structurally during processing, with acid treatment reducing cellulose chain order. This enables investigation into enzyme function on a substrate with significantly reduced crystallinity compared with other cellulose forms.[9]

In addition to plants, crystalline cellulose is produced by the bacterium *Acetobacter xylinum*. BC is a useful model substrate for cellulose digestion because it is composed of pure cellulose microfibers without inhibiting hemicellulose or lignin. BC is structurally analogous to the crystalline core of plant microfibrils, being made up of rectangular fibers packed into ribbons, which then overlap. This ribbon structure gives BC high crystallinity while also having high surface area. Compared with other cellulose model substrates, such as filter paper and Avicel, BC has a significantly lower DP.[1]

A theoretical cellulose microfiber can be perfectly crystalline in packing, but actual cellulose substrates are structurally heterogeneous. Features in the cellulose such as pits, surface micropores, and capillaries give cellulose a higher total surface area compared with an ideal microfiber of the same size.[1] Heterogeneous substrate regions affect cellulase kinetics on chemically homogenous crystalline cellulose substrates such as BC and Avicel. The heterogeneity of cellulose crystallinity is thought to be the major contributing factor to the nonlinear kinetics of cellulases.[10]

Some microbes are able to degrade biomass to liberate usable sugar compounds. By secreting cellulases, hemicellulases, and lignin oxidases, these microbes efficiently degrade the heterogenous co-polymer composition of plant cell walls.[7] *Thermobifida fusca* is a ubiquitous cell wall-degrading gram-positive actinomycete found in aerobic soils.[11] Multiple secreted enzymes enable *T. fusca* growth on crystalline cellulose as the sole source of carbon, and this system has been extensively studied.[12] Many biomass-degrading enzymes are similar between the cellulases system of *T. fusca* and those secreted by the industrially favored filamentous fungus *Hypocrea jecorina* (formerly *Trichoderma reesei*). *T. fusca* serves as an effective model system for the more complex set of cellulases produced by filamentous fungi.

Chitin is similarly used as a source of energy for microbial growth, with secreted chitinases produced by many marine and soil bacteria. Because chitin is less crystalline, chitinase enzyme systems tend to be less complex. The secreted chitinases system of the model chitinolytic bacterium *Serratia marcescens* requires only four enzymes to convert β-chitin into usable saccharides.[13]

Cellulases and chitinases are a diverse group of glycoside hydrolases that degrade their polysaccharide substrates using multiple mechanisms.[14] Cellulases are expressed by a small fraction of microorganisms as well as several animals such as sea squirts, nematodes, and insects.[15,16] The Carbohydrate Active Enzymes database (www.CAZy.org) groups cellulases and other biomass active enzymes into families based on their structural folds. Cellulases currently form 14 distinct families, which can each include examples of different modes of activity.[17]

The general classes of cellulases are exocellulases, endocellulases, and processive endocellulases. These cellulases use acid-base chemistry to catalyze the addition of water to the β-1,4 glycosidic bond of polysaccharides, resulting in hydrolytic cleavage.[18] The active site of hydrolytic cellulases can only accommodate a single polysaccharide chain, making the separation and binding of a chain from crystalline cellulose a key step in cellulose hydrolysis. The active site can be within a tunnel, as for exocellulases, or inside of a structural cleft for endocellulases.[19] Within the active site, conserved catalytic residues can interact with the positioned polysaccharide chain. Active site glutamate or aspartic acid is used to catalyze the cleavage reaction.[20] Hydrogen bonding networks in the cellulase active site adjust the pKa of the catalytic residue side chains for optimal activity.[21]

Many cellulases have binding domains separated from their catalytic domain by flexible linkers. These carbohydrate-binding modules (CBMs) are globular domains with aromatic surface residues that enable tight binding to crystalline polysaccharides.[22] CBMs primarily function to anchor the catalytic domain and provide increased access to the insoluble substrate.[23] There are different conserved structural folds forming multiple CAZy families of CBMs. CBMs are present as domains of many cell wall-degrading enzymes, with varying binding specificity.

Cellulose-directed CBMs typically have a set of aromatic residues on the binding surface positioned so as to form hydrophobic stacking interactions with the glucose rings on the substrate surface. CBM binding is tight but reversible, contributing to the net binding of the cellulase along with the innate binding ability of the catalytic domain.[24] Because CBMs are smaller and more compact than catalytic domains, they can bind to surface-exposed regions of the insoluble substrate and they can enter pores within substrates that the catalytic domain cannot.[25] Overlapping fibers, or ribbons in the case of BC, form interstitial spaces to create a more complex structural environment for binding.

Some evidence points to noncatalytic disruption of cellulose by the binding of CBMs. Tightly associated CBMs such as the family 3c CBM of *T. fusca* Cel9A are thought to contribute to the extraction of individual cellulose chains for positioning in the active site, but CBMs themselves are not thought to significantly affect microfibril structure or crystallinity.[26]

An unconfirmed but possible role of CBMs is the disruption of noncovalently bound cellulose regions, such as separation of microfibers resulting in increased net surface area. Some results support this notion; for example, pretreatment with a CBM led to morphological changes of ramie cotton fiber surfaces when observed by microscopy and enabled other CBMs to penetrate deeper into the fiber structure.[27]

There are claims of CBM enhancement of crystalline cellulose digestion through substrate disruption such as substrate swelling of loosely bound regions or surface defibrillation, but this effect is still controversial. Different groups have observed different effects, possibly because of the different enzyme systems studied or substrates assayed. CBMs are confirmed to act to target catalytic domains to different regions of cellulose and to increase the affinity and binding to the insoluble substrate.[24]

Expansins, a recently characterized family of plant proteins, cause a loosening of plant cell walls to enable plant cell growth. Expansins have two polysaccharide binding domains, each with an open binding surface. Addition of plant expansins to cellulase mixtures enhances crystalline cellulose degradation, with a greater effect at low cellulase loading.[5] Expansin-like proteins are expressed by cellulolytic fungi, such as the swollenin family of *H. jecorina*. Swollenins enhance cellulase activity on recalcitrant cellulose through an unknown nonhydrolytic mechanism. The stimulation despite lack of hydrolysis suggests a change to the cellulose substrate to reduce recalcitrance and enable increased cellulase activity.[28]

Substrate recalcitrance significantly affects the rate of cellulase activity on polysaccharide substrates. *T. fusca* cellulases are well modeled by classical Michaelis–Menten kinetics as long as the substrate is soluble and homogenously mixed. Short oligosaccharides and colorimetric analogs such as 2,4-dinitrophenyl-β-d-cellobioside can be used to measure cellulase kinetic parameters such as K_m and V_{max}.[29] The application of classical kinetics suggests that cellulases behave as optimized enzymes and can rapidly perform the bond hydrolysis step of substrate digestion when the substrate is soluble.

The activity of cellulases on insoluble polysaccharide substrates has been consistently observed to be nonlinear with respect to time. The cause of the declining rate constant in cellulase reactions has not been conclusively established. A common position is the accessible substrate changes over time, rather than cellulases becoming inactivated, substrate inhibition, or product inhibition.[10] Because cellulases are described by classical kinetics on soluble oligosaccharide substrates, another step likely represents the rate-limiting step for the degradation of crystalline cellulose.

Insoluble substrates are not as simple as soluble analogs and present a heterogeneous material with inconsistent recalcitrance, binding sites, and physical barriers. The application of classical Michaelis–Menten kinetics to cellulase activity on insoluble substrates has been attempted multiple times, but it is complicated by the rate constant declining over time.[30]

The assumptions made by Michaelis–Menten kinetics do not allow for changing rate constants; therefore, other approaches must be used to characterize cellulase kinetics. Changing rate constants can be accommodated by "fractal-like" kinetics, which can be used to describe systems with diffusion over inconsistent three-dimensional substrates such as cellulose.[10] Kinetic models that incorporate fractal-like kinetics may better describe cellulase–substrate interactions and their change over the course of digestion.[31,32]

Because cellulase active sites can only accommodate a single chain, they are limited in their accessible substrate when digesting recalcitrant material. Because cellulose has a high degree of recalcitrance and heterogeneity, multiple cellulases can interact with the substrate to perform different roles. Cellulase synergism is the effect of multiple cellulases in a mixture having activity that is higher than the sum of individual cellulases alone.[33] The most effective microbial and industrial cellulase systems depend on mixtures of highly synergistic cellulases. Synergism between cellulases is only observed on more crystalline and recalcitrant material and does not require the cellulases to physically interact. An exception to the requirement of crystalline substrates for synergism is β-glucosidase and xylosidases, which relieve the product inhibition caused by the accumulation of released disaccharides.[34]

The enhanced cellulose saccharification observed by synergistic mixtures of cellulases has been explained mechanistically by several models. Each situation depends on cellulases creating substrate for another to attack. The magnitude of synergism decreases as protein loading increases, with the highest effect at protein loading far below substrate saturation.[35] The endo-exo model described by Wood and McCrae describes an endocellulase making cleavages on surface-exposed polysaccharide chains. The new ends generated by these endocellulase cleavages can then be bound by exocellulases to make processive cleavages.[36] In the exo–exo model, exocellulases with different substrate preferences form a synergistic combination. Two exocellulases directed toward opposite ends of cellulose chains operate simultaneously, attacking different regions of the same substrate.[37]

More recently, an additional model of cellulase synergism on crystalline cellulose has been proposed that involves cellulose surface changes. In this model, a processive endocellulase rapidly cleaves an insoluble cellulose surface into soluble products, and the resulting surface grooves are cleared away by a processive exocellulase.[34] This mechanism for synergism is highly efficient on crystalline cellulose because of the high ability of the processive cellulase to degrade crystalline surfaces. Each of these hydrolase models describe increased activity due to different substrate preference or enhanced substrate availability by the action of one or both of the cellulases in the mixture. The enhancement of recalcitrant substrate saccharification by cellulase synergistic mixtures is an important basis for understanding the mechanistic basis for cellulose and biomass digestion.

Because cellulases can only accommodate a single cellulose chain in the active site, chain separation from the insoluble substrate has been proposed as the rate-limiting step of cellulose degradation.[38] Crystallinity of the substrate increases as cellulose becomes more digested; therefore, changes to the substrate that enable chain separation would be advantageous for synergistic mixtures. Coughlan defined "amorphogenesis" to describe the change to an insoluble substrate that leads to reduced crystallinity and increased substrate accessibility.[39] The chemical composition of polysaccharide chains remains unchanged as a result of amorphogenesis, but hydrogen bonding among chains is reduced and they become more physically separated. Processes that contribute to amorphogenesis include delamination, dispersion, and swelling of crystalline cellulose regions. This change to the substrate enables increased cellulase access over a greater fraction of the insoluble substrate and a corresponding increase in the extent of digestion.

The underlying concept of amorphogenesis resulting from the disruption of crystalline cellulose regions was first proposed by Reese in the C_1/C_x system. The unidentified "swelling factor," or C_1, of microbial cellulase systems would induce amorphogenesis of crystalline cellulose regions to enable greater substrate access for the other cellulases.[40] Different proteins have been investigated for this role of initial crystalline substrate disruption, but the identity of the Reese C_1 swelling factor has not been discovered.

Lytic Polysaccharide Monooxygenases

Once exposed via biomass pretreatment, crystalline cellulose microfibers still present significant recalcitrance to digestion. This recalcitrance is addressed to some degree by synergistic combinations of hydrolytic cellulases on the basis of the observation of increased percentage conversion of cellulase mixtures. The cost of biomass saccharification is largely based on the expense of industrial cellulase mixtures and their capacity to digest a high percentage of the biomass substrate. The cost of industrial cellulase mixtures has dropped threefold in recent years, coinciding with investigation into a new class of oxidative enzymes aiding in the breakdown of plant cell walls (Novazyme, personal communication).

For many years, the enzymes responsible for most cellulose digestion were thought to be limited to hydrolases, such as cellulases, hemicellulases, and esterases.[41] Auxiliary proteins expressed and secreted by microbes were known to use oxidative mechanisms to act on insoluble lignin. Secreted proteins of multiple enzyme families have recently been characterized as stimulating hydrolysis without having an obvious hydrolase active site structure or generating typical hydrolysis products. Originally thought to be binding modules, and then shown to have weak endocellulase-like activity, family 33 proteins were characterized to stimulate chitin hydrolysis by aerobic bacteria.[13] Cellulolytic fungal proteins were similarly classified as weak endocellulases in the glycoside hydrolase 61 family (GH61), but they later were shown to lack the typical cleft active site when crystal structures were obtained.[42]

Both of these enzyme families have been given novel classifications in the CAZy database on the basis of their ability to synergize with hydrolytic cellulases. The auxiliary activity (AA) family includes all of the oxidative enzymes acting on lignin and other portions of hemicellulose in addition to these lytic polysaccharide monooxygenases. As with traditional CAZy families, the AA families are based on a well-characterized founding member and others with sequence similarities indicating similar three-dimensional structures. The goal of this categorization is to provide a global list of carbohydrate oxidases from bacteria, fungi, and others. The former fungal GH61 family of lytic polysaccharide monooxygenases has been reclassified as AA9, whereas the bacterial family 33 CBMs have become CAZy family AA10.[17] These auxiliary enzymes have been found to play a significant role in the breakdown of lignocellulose and biomass saccharification. The members of the AA family all perform some redox role associated with polymeric substrate degradation. Bacterial AA10 enzymes so far characterized appear to contribute to cellulase mixtures by reduction of substrate recalcitrance or increased substrate accessibility for hydrolytic cellulases, with relatively minimal soluble product release on their own.[43] These auxiliary enzymes are abundant in fungal and bacterial cellulase systems, with genomes commonly containing multiple genes encoding AA enzymes.[17]

The discovery of the oxidative activity of bacterial AA enzymes was preceded by investigation into a group of related chitin binding proteins (CHBs) in the late 1990s. Much work was done investigating the extracellular enzymes of gram-positive soil bacteria such as *Streptomyces* spp. with similar phenotypes, ecological roles, and crystalline polysaccharide substrate utilization, such as the gram-positive model organism *T. fusca*. The CHBs originally under investigation from gram-positive soil bacteria included proteins with varying size and substrate specificities. These small binding proteins are secreted in large amounts by gram-positive bacteria during growth on crystalline polysaccharides.

CHB1 of *Streptomyces olivaceoviridis*, an 18.7 kDa protein induced by growth on chitin, was the first to be characterized. CHB1 binds only α-chitin, mediated in part by a conserved tryptophan on the basis of spectroscopy and mutagenesis.[44] CHB2 of *Streptomyces reticuli* is 18.6 kDa and also binds only α-chitin, including fungal structures containing chitin.[45] CHB3 from *Streptomyces coelicolor* is a smaller CHB at 14.9 kDa that is more flexible in its binding, being able to bind both α and β forms of chitin as well as soluble chitosan. CHB3 is related to the other CHBs, but with less sequence identity compared with the others.[46] CHbB of *Bacillus amyloliquefaciens*, 19.8 kDa, is somewhat similar to CHB3 in its binding capacity because it binds α- and β-chitin (with a preference for the β form), but it can bind with some affinity to cellulose but not to soluble chitosan.[47]

CBP21 of the gram-negative soil bacterium *S. marcessens* is similar to the others despite being phylogenetically distinct from *Streptomyces* spp. CBP21 is a similar size at 18.8 kDa, and it shows high sequence identity with CHB1. The characterized crystalline chitin binding properties are also similar, although CBP21 binds preferentially to β-chitin over other forms.[48] CBP21 was the first CHB of this family of secreted binding proteins for which the structure was solved, revealing a fibronectin-like fold with a bud-like extension.[49] The structure revealed that many of the conserved aromatic residues thought to play a similar role as those found on the binding surface of CBMs were actually buried deep in the core of the protein. As will be discussed later in this section, polar surface residues important for binding were investigated through site-directed mutagenesis.

These CBPs are similar in upregulation, secretion, and binding with the much smaller antifungal protein Afp1 of *Streptomyces tendae*. Afp1 can bind to α-chitin and chitosan but not cellulose, akin to CHB3 of *S. coelicolor*.[50] However, the antifungal proteins are distinct from CBPs, with Afp1 having much smaller size at 9.8 kDa, different structural fold (having more sequence homology with hydrolase CBMs), and inherent antifungal ability, which the other CHBs lack.[51] Afp1 is smaller than the CHBs, but it is not as small as the even smaller antifungal chitin binding lectins secreted by gram-positive bacteria.[50]

The mechanism of antifungal protein activity is not confirmed, but expression of CHBs is induced by living or autoclaved fungal chitin. CHBs such as CHB1 bind α-chitin and cover the fungal chitin to form a "glue-like' covering," which may enable physical contact interactions between bacterial and fungal hyphae. The *S. olivaceoviridis* CHB1 was originally proposed to act antifungally through binding to the chitin at growing hyphal tips. The proposed role was nonhydrolytic disruption through binding to underlying microfibrils to release additional surface chains for exochitinase attack, similar to the nonhydrolytic disruption effect proposed for CBMs.[44] Bacterial exochitinases alone have been observed to effectively destroy chitinous fungal hyphae; therefore, it is unknown whether any additional disruption would be beneficial.[52]

It is unclear whether antifungal proteins are actually involved in antifungal host defense, or bacterial antifungal CHBs may be more a byproduct of nutritional processes. Inhibitory effects appear to be due to their ability to bind chitin at the hyphal apex, where chitin is most accessible and blocks cell wall assembly. This leads to the observed aberrant hyphal growth and morphological effects such as branching and hyphal swelling. Inhibition is highly dependent on environmental conditions, and it is specific to species, suggesting different binding targets for different CHBs. It is interesting to note that the antifungal activity of CHBs is antagonized by divalent cations such as Ca^{2+}.[53]

All of these CBPs show similarities: small in size, compact structural folds, and high secretion in response to growth on insoluble substrates. Despite similar sequences, their varied binding properties show a fundamental diversity among the group. Each protein is specific but limited in its substrate binding, typically predominantly targeted toward one form of crystalline chitin and most unable to bind to cellulose or soluble chitosan. SoCHB1, SrCHB2, ScCHB3, BaCHBB, and SmCBP21 are all now categorized as AA10 proteins, whereas StAfp1 is not in that family (http://www.cazy.org, 2013).

Plants produce numerous pathogenesis-related (PR) proteins, including the PR-4 family of small chitin-binding antifungal proteins.[53] The mechanism of fungal inhibition is unknown, but it has been hypothesized to involve cell wall weakening at the fungal hyphal akin to the hypothesized bacterial CHB mechanism.[54] Plant PR-4 proteins are similar in size but are not structurally related to bacterial antifungal CBPs and likely rely on different activity to achieve this fungal inhibition.[55]

AA10 proteins are found within insect virus genomes, classified as the GP37 family in baculoviruses and fusolins in entomopoxvirus (EPV). Although not essential, these proteins act as virulence factors to increase the infectivity of an insect virus on the host. Both groups are multidomain proteins with a highly conserved N-terminal chitin binding domain recently classified as members of the AA10 family (http://www.cazy.org, 2013). In addition to viruses, multidomain virulence factors containing N-terminal AA10 domains are found in bacteria, such as the *Vibrio cholera* colonization factor GbpA.[56]

The viral virulence factor fusolin forms crystalline protein structures referred to as spindles, which do not contain virions but enhance infectivity of EPV and nucleopolyhedroviruses up to 10^6-fold. The N-terminal AA10 domain was found to be essential for the dramatic increase in nucleopolyhedrovirus infectivity, with the smaller C-terminal region being critical for spindle formation but entirely dispensable for activity. Glycosylation at several locations provides enhanced binding and stability when exposed to insect proteases within the gut. The AA10 domain contains a completely conserved N-terminal HGY motif, a common motif among AA10 sequences. A conserved histidine corresponding to His114 of CBP21 was investigated by mutation to alanine. This mutation led to retention of chitin binding but loss of all virus-enhancing ability.[57]

The peritrophic membrane (PM) within the insect digestive tract is a network of chitin and proteins that forms a barrier to viral infection of midgut epithelial cells.[58] The PM can be disintegrated by the addition of EPV fusolin in the form of spindles. The enhancement of viral infection is hypothesized to be due to structural changes to the PM chitin network caused by the AA10 domain of fusolin, leading to structural disruption and increased chitinase activity.[57]

Baculovirus GP37 and fusolin protein sequences cluster separately and are distinct from bacterial AA10 proteins via phylogenetic analysis, suggesting they are not directly related. They may conceivably share a common ancestor because baculoviruses are known to have acquired the chitinase *v-chiA* gene from bacteria via horizontal transfer, but this has not yet been established.[59]

Several AA enzymes from bacterial and fungal systems have been recently characterized, with multiple crystal structures solved. CBP21 of *S. marcescens* was the first aerobic bacterial AA enzyme to be investigated for its ability to synergize with chitinases.[13] CBP21 was found to have oxidative activity when a reducing agent was included, generating oxidized oligosaccharide products.[60] The substrate preference of CBP21 oxidative activity is limited to crystalline forms of β-chitin, showing no activity or stimulation on chitooligomers or chitosan, the soluble form of chitin. This result led to the reclassification from CBP with weak endochitinase activity to lytic polysaccharide monooxygenase.[60]

Most species of aerobic bacteria that digest insoluble substrates have genes encoding either one or two AA10 proteins. These AA10 proteins likely catalyze oxidative cleavage of insoluble polysaccharides, but the majority of these identified genes have not yet been characterized. In contrast with aerobic bacteria encoding only one or a pair of polysaccharide monooxygenase genes, AA9 genes tend to be abundant within the genomes of cellulolytic fungi.[17] These often contain many copies with slight variations or different oxidized product production.[61] The presence of multiple similar protein sequences within genomes has been discussed in the literature, but it represents a trait of these enzyme families that has not been fully explored.

Many accessory proteins are upregulated when cellulolytic organisms are grown on cellulose substrates, but in general these proteins are poorly explored.[62] The expression of AA10 enzymes by *T. fusca* is controlled by the carbon source and appears to be similar to the regulation of cellulases and hemicellulases. This substrate-dependent control enables optimal secretion of enzymes against a given substrate for maximum activity. *T. fusca* AA10 enzymes are upregulated when grown on cellulose-containing substrates, especially the pure cellulose substrate bacterial microcrystalline cellulose (BMCC), as measured based on mRNA concentrations via microarray. The genes encoding AA10A and AA10B are upregulated between 4.8- and 13-fold during growth on BC, an upregulation in mRNA levels comparable to other *T. fusca* cellulases.[63]

Both *T. fusca* AA10 sequences contain signal peptides targeting them for secretion.[64] Investigation of the *T. fusca* secretome on different substrates compared with glucose controls showed that AA10A and AA10B are abundantly secreted when grown on cellulose. AA10A enrichment was twice the next most enriched cellulase when growth on cellulose was compared with glucose.[63] This enrichment was confirmed using iTraq tandem mass spectrometry (MS/MS) quantification on different substrates, with AA10 enzymes being enriched, especially when *T. fusca* was cultured on pure cellulose.[7] Most of the proteins secreted by *T. fusca*, at least 180, are not yet characterized and their role in cellulose degradation is not known.[63]

AA10 proteins can consist of an AA10 catalytic domain alone as well as occasionally a catalytic domain with an additional family 2 CBM domain. A fibronectin type-III (FNIII) module with uncharacterized function is often found between the catalytic domain and CBM of AA10 enzymes.[17] The additional CBM domain may direct the AA10 protein to different regions of the crystalline substrate because of the different binding mode relying more on aromatic residues compared with the mostly polar residues involved in AA10 catalytic domain binding interactions. This is one potential explanation for the simultaneous production of multiple AA10 forms during crystalline cellulose digestion.

The chitinolytic CBP21 of *S. marcescens* was the first bacterial AA10 enzyme for which the structure was solved by X-ray crystallography.[49] SmCBP21 and *T. fusca* AA10A have significant sequence similarity, suggesting structural and functional similarity.[64] The structure of CBP21 resembles a globular binding module with a foot-like extension. The flat surface formed by the core fold and this extension forms a binding surface found to be responsible for substrate interaction.[49]

The CBP21 structure establishes several structural features that appear to be consistent among bacterial AA10 enzymes. The structures solved to date all have a core fibronectin type-III/immunoglobulin-like β-sandwich fold. Multiple conserved aromatic residues are found in the core of this β-sandwich, which provide structural stability through buried hydrophobic interactions. Multiple loops and helices extend from the β-sandwich core to form the foot-like extension. The binding surface is enriched in polar residues, with few aromatic residues in important positions. This composition of mostly polar residues is distinct from the aromatic residues that contribute to the binding ability of most CBMs.[22] The single aromatic residue of CBP21 is in an important position and contributes significantly to surface binding.[49]

The most significant feature of AA10 enzymes is the absence of a binding pocket or cleft that could accommodate a cellulose chain. Two histidine residues on the binding surface, one being the N-terminal residue, are absolutely conserved among AA10 enzymes. These are responsible for chelating a metal ion, likely to be copper in all AA10 enzymes on the basis of convincing results of EfCBM33A and BaCBM33A binding kinetics.[65,66] The copper atom is coordinated by multiple residues resulting in a "histidine brace"-type planar binding pocket, which creates a type II copper site. Imidazole rings such as the histidine side chain are common coordinating side chain residues for type II copper sites.[67] Through nuclear magnetic resonance spectroscopy, CBP21 was found to be very structurally rigid except for several residues surrounding the metal ion binding pocket.[68]

T. fusca secretes two AA10 enzymes: AA10A (formerly E7) and AA10B (formerly E8). AA10A is the smaller of the two at 21.3 kDa and consists of only the catalytic domain. AA10B has a catalytic domain of similar size, but with an additional CBM2 domain separated by an FNIII-like domain.[64] Although they have low sequence identity (13%), they are predicted to have similar structure of their catalytic domains and behave similarly in most assays. Both exhibit the oxidative activity in the presence of reducing agents, which is emblematic of the AA family (Kruer-Zerhusen, unpublished).

The critical structural features of *T. fusca* AA10 proteins can be inferred by homology modeling of the AA10A structure on the basis of high sequence similarity with the existing crystal structures of bacterial AA10 enzymes. There is a stretch of AA10A sequence that does not align with the CBP21 sequence, mapping to the foot-like extension of the homology model. These additional residues likely contribute to an extension of the binding surface, similar to the extended flat surface observed in the structure of *Bacillus amyloliquefaciens* BaCBM33 (formerly characterized as ChbB).[66]

The flat surface of AA10 enzymes contains multiple residues shown to be significant for CBP21 binding to crystalline chitin.[49] It is interesting to note that the aromatic residue located on this extension found to be most important for CBP21 binding, Y54, is not found in the sequence of TfAA10A. BaCBM33 has a tyrosine at residue 57, but this appears to be buried in the structure. BaCBM33 has a tryptophan near this position, W50, which appears more available to contribute to surface binding. Although TfAA10A lacks the tyrosine around residue 57, it shares a similar potentially surface-exposed tryptophan with BaCBM33.

Several other polar residues found to be important for CBP21 binding are not conserved on the modeled surface of TfAA10A. These include two significant glutamate residues, E55 and E60. AA10A instead has E112 and N79, which may be located in similar positions on the binding surface. Binding surface residues likely dictate the differing binding preferences observed for AA10 proteins, and their investigation should elucidate this specificity.

It is interesting that both AA10 proteins produced by *T. fusca* show similar binding affinity to BC crystalline cellulose, but TfAA10B shows higher binding to filter paper and α-chitin. The multidomain TfAA10B has only slightly tighter binding to cellulose, where it would be expected to perform much better than the single-domain form on the basis of the presence of the additional tight binding CBM2.[64] The catalytic domain of TfAA10A and TfAA10B may already be optimized for cellulose binding; thus, the additional CBM only changes the accessible binding regions. The varied binding specificity of similar AA proteins targeting to different substrate regions may be an explanation for the multiple gene copies observed in many genomes.

T. fusca AA10A binds to cellulose and chitin and can stimulate chitinases as well as cellulases.[64] However, CBP21 is limited to binding to chitin and stimulating chitinases and does not bind cellulose.[49] AA10 proteins capable of binding and cleaving crystalline chitin are abundant in marine bacteria, where chitin from marine animals is a source of energy and nitrogen. BLAST results show different groups of AA10 enzymes depending on the source of the target protein sequence. When CBP21 is used, the homologous sequences are derived from chitin-utilizing bacteria whereas for TfAA10A the homologues are all from potential cellulolytic bacteria (http://blast.ncbi.nlm.nih.gov, 2013). Because there are few copies of AA10 genes within bacterial genomes, this suggests that the AA10 proteins have differing substrate specificities corresponding to the ecological role of the host organism.

CBP21 was the first AA10 enzyme shown to have oxidative cleavage ability. The stimulation of chitinases by CBP21 was dramatic when external reducing agents were present, an effect that required molecular oxygen.[60] This stimulation was only found on crystalline substrates, and there was no reaction with hexameric *N*-acetylglucosamine under the same conditions. To date, no AA10 enzymes have been shown to have activity on soluble oligosaccharides, which is in agreement with the proposed multichain binding orientation of polysaccharide monooxygenase enzymes.[61] The products observed from CBP21 oxidative cleavage using matrix-assisted laser desorption/ionization-time-of-flight mass spectrometry were aldonic acids, the open lactone ring formed through oxidation of the C1 position.[4] Products included native and oxidized sugars with a range of DP. Ethylenediaminetetraacetic acid is able to completely inhibit AA10 oxidative activity, which is recoverable by the addition of copper in sufficient concentration.[65]

Cellulose oxidative cleavage was first shown using the AA10 CelS2 from *S. coelicolor* A3, which is a multidomain AA10 similar in domain structure to TfAA10B. When incubated on cellulose, aldonic cello-oligosaccharides were produced, agreeing with the C1 oxidation observed previously on chitin.[43] Digestion of Avicel released native and oxidized oligosaccharide products, compared with significantly fewer soluble products resulting from filter paper digestion. The difference in DP between the two substrates may explain this result because of the significantly higher DP of filter paper.[69] The oxidative cleavages made on crystalline surfaces may result in products that remain insoluble, complicating their detection.

TfAA10A and TfAA10B produce oxidized products when incubated with crystalline cellulose (BC) in the presence of reducing agents. TfAA10B appears to have higher activity than TfAA10A, releasing significantly more soluble oxidized oligosaccharides when assayed alone on BMCC. Oxidized oligosaccharide products produced by TfAA10A have been confirmed to be lactone aldonic acids by liquid chromatography-MS/MS (Kruer-Zerhusen, unpublished). These oxidized products have a range of DP when measured by high-performance anion-exchange chromatography and compared with oxidized oligosaccharide standards prepared using the iodine oxidation method used by Forsberg et al.

The AA10 enzymes of *T. fusca* similarly require reducing agents for activity, although the structure of the reducing agent is not restrictive. Reducing agents that enable oxidative cleavage include ascorbate, glutathione, and nicotinamide adenosine dinucleotide (NADH). It is interesting to note that the small molecule dehydroascorbate also appears to stimulate activity by AA10A, although through an unknown mechanism. The oxidized form of glutathione, glutathione disulfide, does not stimulate oxidation, and dehydroascorbate should similarly lack the reducing power provided by the other small-molecule reducing agents (Kruer-Zerhusen, unpublished).

Both AA10 enzymes of *T. fusca* show significant synergism with almost all of the hydrolytic cellulases of the *T. fusca* system. TfAA10A has been shown to give a substantial synergistic effect in binary assays with the exocellulases Cel48A and Cel6B and the endocellulases Cel6A and Cel5A. Curiously, TfAA10A does not synergize with the processive endocellulase Cel9A either in a binary mixture or when included in the established Cel9A-Cel48A synergistic combination. There may be some overlapping mechanistic activity on crystalline cellulose that would explain this result (Kruer-Zerhusen, unpublished).

The stimulation of hydrolytic activity by *T. fusca* AA10s occurs predominantly during the beginning of hydrolysis. The extent of digestion is improved significantly for exocellu- lases, which find recalcitrant crystalline material particularly difficult. The stimulation provided by AA10 monooxygenases appears to stimulate cellulases by a significant increase in the b parameter in the $f(t) = At^b$, which is a nonlinear fit applied to time course data.[32] It is unclear whether the decrease in stimulation over time is due to substrate change, enzyme inactivation, reducing agent depletion, or some other factor. The addition of reducing agent halfway through the experiment does not provide additional stimulation, suggesting that the decline in stimulation is due to some other factor (Kruer-Zerhusen, unpublished).

A proposed mechanism for the oxidative cleavage performed by lytic polysaccharide mono- oxygnases has been described. The generation of a superoxide radical at the copper active site results from the transition of the bound Cu(I) to Cu(II). This radical oxygen is positioned by the monooxygenase surface to abstract a hydrogen from a particular side of the substrate β-1,4 glycosidic bond depending on the enzyme's catalytic preference. An external electron is required to reduce the superoxide for the simultaneous release of a molecule of water and generation of an oxo radical. The oxo radical is temporarily stabilized by the copper ion and specifically reacts with the substrate radical to form a covalent bond and result in hydroxyl- ation at the substrate carbon. The oxidation of carbon adjacent to the β-1,4 glycosidic bond results in spontaneous elimination of the adjacent chain, resulting in an aldonolactone (or 4-ketoaldose at the C4 position) at the end of the polysaccharide chain. The external reducing agent then reduces the Cu(II) of the lytic polysaccharide monooxygnases to reset the enzyme for an additional catalytic cycle.[70]

Fungal AA9 lytic polysaccharide monooxygneases, formerly classified as the GH61 family, have been investigated as components of fungal cellulase system for many years. Investiga- tion into the AA9 (formerly family GH61) proteins from fungi have shown that they also act as copper monooxygenases. The AA9 proteins of *Tribulus terrestris* were characterized as being able to stimulate the activity of *H. jecorina* cellulase mixtures, although only when acting on pretreated biomass and not on pure crystalline substrates.[62] The reducing power allowing for AA9 stimulation on pretreated corn stover was found to be gallate, a soluble small molecule cofactor released from the biomass substrate.[4] In the presence of reducing agents such as gallate and ascorbate, AA9 are able to perform oxidative cleavages on cellu- lose similar to the AA10 family.[4] AA9 monooxygenases act on the same cellulose substrates as the bacterial AA10 ScCelS2 and similarly require an external source of reducing power. Either a small molecule reducing agent such as ascorbate, lignin components of pretreated biomass, or a protein partner such as the flavoprotein cellulose dehydrogenase (CDH) can serve as reducing agents.[71,72] CDH is produced by most cellulolytic fungi, and its role in cellulose cleavage is still being explored.

The first structure of a fungal GH61 (now AA9) solved was Cel61B, the most abundantly secreted GH61 of *H. jecorina*.[42] Although it had low sequence identity with CBP21, the structure shows obvious similarities in the flat binding surface and conserved metal ion binding pocket configuration. It was found that including transition metals was critical for obtaining crystals, suggesting a role in maintaining surface chain stability in AA9 that is not as critical in AA10 enzymes. Similar to previous results for family 33 CBMs, low endoglucanase activity was observed in the absence of an external reducing agent, which led to their initial characterization.[42]

An interesting structural difference is the methylation of the absolutely conserved N-terminal histidine residue in AA9 proteins, although this modification has not been found in any of the bacterial AA10 proteins studied so far.[61] It will be interesting to see if replacing the methylated histidine ring with an unmodified histidine side chain changes the activity of a family 61 protein.

There is evidence that some family 61 proteins have greater diversity in their oxidized products with a proposed separation into distinct subfamilies.[73] Some AA9 monooxygenases create oxidized oligosaccharide products that are oxidized at the C4 position whereas others produce products oxidized at the C1 position. The functional significance of different oxidation positions is still being investigated. Oxidation at other positions has been proposed, but it has not been fully explored.

In contrast to cellulolytic or chitinolytic bacteria containing one or two AA10 genes, most cellulolytic fungi contain multiple genes encoding AA9 monooxygenases.[17] Searching for homologous sequences, it was observed that AA9 are abundant in fungal genomes, but they are limited only to eukaryotes. This limitation of the family to eukaryotes is an unusual property for cellulase families.[42]

Conclusion

It has been proposed that lytic polysaccharide monooxygenases of bacterial and fungal systems perform substrate disruption activity on crystalline substrates. This substrate amorphogenesis synergistically stimulates the activity of hydrolytic cellulases. The introduction of a carboxylic acid at the reducing end (or the nonreducing end for some AA9 oxidases) may disrupt the normal conformation of the sugar ring and introduce new charge interactions that conflict with neighboring cellulose chains, thus causing disruption and increasing substrate accessibility.[60] Although a chemical reaction mechanism has been proposed for the cleavage of insoluble polysaccharide substrates, the mechanism explaining lytic polysachharide monooxygenase synergistic stimulation of hydrolases has not been confirmed.[70]

There remain many questions regarding the activity and role of lytic polysaccharide monooxygenases. The source of reducing power required for activity in vitro is still unconfirmed. Are small molecules from substrate disruption or protein partners such as CDH responsible? What structural features dictate the diversity of substrate binding observed in bacterial AA10 enzymes? What is the role of lytic polysaccharide monooxygenases in other systems not involved in crystalline substrate degradation? The investigation of these novel enzymes should provide insights into how recalcitrant substrates are overcome, and specifically the mechanism by which hydrolytic cellulases work synergistically with auxiliary enzymes to digest plant biomass.

Acknowledgments

This work was sponsored by the BioEnergy Science Center (BESC). BESC is a U.S. Department of Energy (DOE) Bioenergy Research Center supported by the Office of Biological and Environmental Research in the DOE Office of Science.

References

1. Lynd LR, Weimer PJ, Willem HVZ, Pretorius IS. Microbial cellulose utilization: fundamentals and biotechnology. *Microbiol Mol Biol Rev* 2002;**66**:506–77.
2. Himmel ME, Ding SY, Johnson DK, Adney WS, Nimlos MR, Brady JW, et al. Biomass recalcitrance: engineering plants and enzymes for biofuels production. *Science* 2007;**315**:804–7.
3. Chundawat SPS, Bellesia G, Uppugundla N, da Costa Sousa L, Gao D, Cheh AM, et al. Restructuring the crystalline cellulose hydrogen bond network enhances its depolymerization rate. *J Am Chem Soc* 2011;**133**:11163–74.
4. Quinlan RJ, Sweeney MD, Lo Leggio L, Otten H, Poulsen JCN, Johansen KS, et al. Insights into the oxidative degradation of cellulose by a copper metalloenzyme that exploits biomass components. *Proc Natl Acad Sci USA* 2011;**108**:15079–84.
5. Arantes V, Saddler JN. Access to cellulose limits the efficiency of enzymatic hydrolysis: the role of amorphogenesis. *Biotechnol Biofuels* 2010;**3**:2–11.
6. Horn SJ, Vaaje-Kolstad G, Westereng B, Eijsink VG. Novel enzymes for the degradation of cellulose. *Biotechnol Biofuels* 2012;**5**:45.
7. Adav SS, Ng CS, Arulmani M, Sze SK. Quantitative iTRAQ secretome analysis of cellulolytic *Thermobifida fusca*. *J Proteome Res* 2010;**9**:3016–24.
8. Zhao X, Zhang L, Liu D. Biomass recalcitrance. Part I: the chemical compositions and physical structures affecting the enzymatic hydrolysis of lignocellulose. *Biofuels, Bioprod Biorefining* 2012;**6**:465–82.
9. Zhang Y, Cui J, Lynd L, Kuang L. A transition from cellulose swelling to cellulose dissolution by o-phosphoric acid: evidence from enzymatic hydrolysis and supramolecular structure. *Biomacromolecules* 2006;**7**:644–8.
10. Zhang S, Wolfgang DE, Wilson DB. Substrate heterogeneity causes the nonlinear kinetics of insoluble cellulose hydrolysis. *Biotechnol Bioeng* 1999;**66**:35–41.
11. Lykidis A, Mavromatis K, Ivanova N, Anderson I, Land M, DiBartolo G, et al. Genome sequence and analysis of the soil cellulolytic actinomycete *Thermobifida fusca* YX. *J Bacteriol* 2007;**189**:2477–86.
12. Wilson DB. Studies of *Thermobifida fusca* plant cell wall degrading enzymes. *Chem Rec* 2004;**4**:72–82.
13. Vaaje-Kolstad G, Horn SJ, van Aalten DMF, Synstad B, Eijsink VGH. The non-catalytic chitin-binding protein CBP21 from *Serratia marcescens* is essential for chitin degradation. *J Biol Chem* 2005;**280**:28492–7.
14. Vuong TV, Wilson DB. Glycoside hydrolases: catalytic base/nucleophile diversity. *Biotechnol Bioeng* 2010;**107**:195–205.

15. Mba Medie F, Davies GJ, Drancourt M, Henrissat B. Genome analyses highlight the different biological roles of cellulases. *Nat Rev Microbiol* 2012;**10**:227–34.

16. Smant G, Stokkermans JP, Yan Y, de Boer JM, Baum TJ, Wang X, et al. Endogenous cellulases in animals: isolation of beta-1,4-endoglucanase genes from two species of plant-parasitic cyst nematodes. *Proc Natl Acad Sci USA* 1998;**95**:4906–11.

17. Levasseur A, Drula E, Lombard V, Coutinho PM, Henrissat B. Expansion of the enzymatic repertoire of the CAZy database to integrate auxiliary redox enzymes. *Biotechnol Biofuels* 2013;**6**:41.

18. Walker LP, Wilson DB. Enzymatic hydrolysis of cellulose: an overview. *Bioresour Technol* 1991;**36**:3–14.

19. Breyer WA, Matthews BW. A structural basis for processivity. *Protein Sci: A Publ Protein Soc* 2001;**10**:1699–711.

20. Zechel DL, Withers SG. Glycosidase mechanisms: anatomy of a finely tuned catalyst. *Acc Chem Res* 2000;**33**:11–8.

21. Zhou W, Irwin DC, Escovar-Kousen J, Wilson DB. Kinetic studies of *Thermobifida fusca* Cel9A active site mutant enzymes. *Biochemistry* 2004;**43**:9655–63.

22. Hashimoto H. Recent structural studies of carbohydrate-binding modules. *Cell Mol Life Sci* 2006;**63**:2954–67.

23. Wilson DB. Three microbial strategies for plant cell wall degradation. *Ann N Y Acad Sci* 2008;**1125**:289–97.

24. Jung H, Wilson DB, Walker LP. Binding and reversibility of *Thermobifida fusca* Cel5A, Cel6B, and Cel48A and their respective catalytic domains to bacterial microcrystalline cellulose. *Biotechnol Bioeng* 2003;**84**:151–9.

25. Jung H, Wilson DB, Walker LP. Binding mechanisms for *Thermobifida fusca* Cel5A, Cel6B, and Cel48A cellulose-binding modules on bacterial microcrystalline cellulose. *Biotechnol Bioeng* 2002;**80**:380–92.

26. Sakon J, Irwin D, Wilson DB, Karplus PA. Structure and mechanism of endo/exocellulase E4 from *Thermomonospora fusca*. *Nat Struct Biol* 1997;**4**:810–8.

27. Din N, Gilkes N, Tekant B. Non-Hydrolytic disruption of cellulose fibres by the binding domain of a bacterial cellulase. *Nat Biotechnol* 1991;**9**:1096–9.

28. Baker JO, King MR, Adney WS, Decker SR, Vinzant TB, Lantz SE, et al. Investigation of the cell-wall loosening protein expansin as a possible additive in the enzymatic saccharification of lignocellulosic biomass. *Appl Biochem Biotechnol* 2000;**84–86**:217–23.

29. Vuong TV, Wilson DB. Processivity, synergism, and substrate specificity of *Thermobifida fusca* Cel6B. *Appl Environ Microbiol* 2009;**75**:6655–61.

30. Bansal P, Hall M, Realff MJ, Lee JH, Bommarius AS. Modeling cellulase kinetics on lignocellulosic substrates. *Biotechnol Adv* 2009;**27**:833–48.

31. Väljamäe P, Kipper K, Pettersson G, Johansson G. Synergistic cellulose hydrolysis can be described in terms of fractal-like kinetics. *Biotechnol Bioeng* 2003;**84**:254–7.

32. Kostylev M, Wilson D. Two-parameter kinetic model based on a time-dependent activity coefficient accurately describes enzymatic cellulose digestion. *Biochemistry* 2013;**52**:5656–64.

33. Wilson D, Kostylev M. Cellulase processivity. In: Himmel ME, editor. *Biomass conversion: methods and protocols*, vol. 908. Totowa, NJ: Humana Press; 2012. p. 93–9.

34. Kostylev M, Wilson D. Synergistic interactions in cellulose hydrolysis. *Biofuels* 2012;**3**:61–70.

35. Watson DL, Wilson DB, Walker LP. Synergism in binary mixtures of *Thermobifida fusca* cellulases Cel6B, Cel9A, and Cel5A on BMCC and Avicel. *Appl Biochem Biotechnol* 2002;**101**:97–111.

36. Wood TM, McCrae SI. Synergism between enzymes involved in the solubilization of native cellulose. In: Brown R, editor. *Hydrolysis of cellulose: mechanisms of enzymatic and acid catalysis*. Washington, DC: American Chemical Society; 1979.

37. Tomme P, Heriban V, Claeyssens M. Adsorption of two cellobiohydrolases from *Trichoderma reesei* to Avicel: evidence for "exo-exo" synergism and possible "loose complex" formation. *Biotechnol Lett* 1990;**12**:525–30.

38. Wilson DB. Cellulases and biofuels. *Curr Opin Biotechnol* 2009;**20**:295–9.

39. Coughlan MP. The properties of fungal and bacterial cellulases with comment on their production and application. *Biotechnol Genet Eng Rev* 1985;**3**:39–109.

40. Reese ET, Siu RGH, Levinson HS. The biological degradation of soluble cellulose derivatives and its relationship to the mechanism of cellulose hydrolysis. *J Bacteriol* 1950;**59**:485–97.

41. Wyman CE, Decker SR, Himmel ME, Brady JW, Skopec CE. Hydrolysis of cellulose and hemicellulose. In: Dumitriu S, editor. *Polysaccharides: structural diversity and functional versatility.* 2nd ed. CRC Press; 2004.

42. Karkehabadi S, Hansson H, Kim S, Piens K, Mitchinson C, Sandgren M. The first structure of a glycoside hydrolase family 61 member, Cel61B from *Hypocrea jecorina*, at 1.6 A resolution. *J Mol Biol* 2008;**383**:144–54.

43. Forsberg Z, Vaaje-Kolstad G, Westereng B, Bunæs AC, Stenstrøm Y, MacKenzie A, et al. Cleavage of cellulose by a CBM33 protein. *Protein Sci* 2011;**20**:1479–83.

44. Schnellmann J, Zeltins A, Blaak H, Schrempf H. The novel lectin-like protein CHB1 is encoded by a chitin-inducible *Streptomyces olivaceoviridis* gene and binds specifically to crystalline alpha-chitin of fungi and other organisms. *Mol Microbiol* 1994;**13**:807–19.

45. Kolbe S, Fischer S, Becirevic A, Hinz P. The *Streptomyces reticuli* α-chitin-binding protein CHB2 and its gene. *Microbiology* 1998;**144**:1291–7.

46. Saito A, Miyashita K, Biuković G, Schrempf H. Characteristics of a *Streptomyces coelicolor* A3(2) extracellular protein targeting chitin and chitosan. *Appl Environ Microbiol* 2001;**67**:1268–73.

47. Chu HH, Hoang V, Hofemeister J, Schrempf H. A *Bacillus amyloliquefaciens* ChbB protein binds beta- and alpha-chitin and has homologues in related strains. *Microbiology* 2001;**147**:1793–803.

48. Suzuki K, Suzuki M, Taiyoji M, Nikaidou N, Watanabe T. Chitin binding protein (CBP21) in the culture supernatant of *Serratia marcescens* 2170. *Biosci Biotechnol Biochem* 1998;**62**:128–35.

49. Vaaje-Kolstad G, Houston DR, Riemen AHK, Eijsink VGH, van Aalten DMF. Crystal structure and binding properties of the *Serratia marcescens* chitin-binding protein CBP21. *J Biol Chem* 2005;**280**:11313–9.

50. Bormann C, Baier D, Hörr I, Raps C, Berger J, Jung G, et al. Characterization of a novel, antifungal, chitin-binding protein from *Streptomyces tendae* Tü901 that interferes with growth polarity. *J Bacteriol* 1999;**181**:7421–9.

51. Campos-Olivas R, Hörr I, Bormann C, Jung G, Gronenborn AM. Solution structure, backbone dynamics and chitin binding of the anti-fungal protein from *Streptomyces tendae* TU901. *J Mol Biol* 2001;**308**:765–82.

52. Schrempf H. Recognition and degradation of chitin by streptomycetes. *Anton Leeuw Int J Gen Mol Microbiol* 2001;**79**:285–9.

53. Theis T, Stahl U. Antifungal proteins: targets, mechanisms and prospective applications. *Cell Mol Life Sci CMLS* 2004;**61**:437–55.

54. Selitrennikoff C. Antifungal proteins. *Appl Environ Microbiol* 2001;**67**:2883–94.

55. Bertini L, Proietti S, Aleandri MP, Mondello F, Sandini S, Caporale C. Modular structure of HEL protein from Arabidopsis reveals new potential functions for PR-4 proteins. *Biol Chem* 2012;**393**:1–14.

56. Wong E, Vaaje-Kolstad G, Ghosh A, Hurtado-Guerrero R, Konarev PV, Ibrahim AFM, et al. The *Vibrio cholerae* colonization factor GbpA possesses a modular structure that governs binding to different host surfaces. *PLoS Pathog* 2012;**8**:e1002373.

57. Takemoto Y, Mitsuhashi W, Murakami R, Konishi H, Miyamoto K. The N-terminal region of an entomopoxvirus fusolin is essential for the enhancement of peroral infection, whereas the C-terminal region is eliminated in digestive juice. *J Virol* 2008;**82**:12406–15.

58. Li Z, Li C, Yang K, Wang L, Yin C, Gong Y, et al. Characterization of a chitin-binding protein GP37 of *Spodoptera litura* multicapsid nucleopolyhedrovirus. *Virus Res* 2003;**96**:113–22.

59. Salvador R, Ferrelli ML, Berretta MF, Mitsuhashi W, Biedma ME, Romanowski V, et al. Analysis of EpapGV gp37 gene reveals a close relationship between granulovirus and entomopoxvirus. *Virus Genes* 2012;**45**:610–3.

60. Vaaje-Kolstad G, Westereng B, Horn SJ, Liu Z, Zhai H, Sørlie M, et al. An oxidative enzyme boosting the enzymatic conversion of recalcitrant polysaccharides. *Science* 2010;**330**:219–22.

61. Li X, Beeson WT, Phillips CM, Marletta MA, Cate JHD. Structural basis for substrate targeting and catalysis by fungal polysaccharide monooxygenases. *Structure* 2012;**20**:1051–61.

62. Harris PV, Welner D, McFarland KC, Re E, Navarro Poulsen JC, Brown K, et al. Stimulation of lignocellulosic biomass hydrolysis by proteins of glycoside hydrolase family 61: structure and function of a large, enigmatic family. *Biochemistry* 2010;**49**:3305–16.

63. Chen S, Wilson DB. Proteomic and transcriptomic analysis of extracellular proteins and mRNA levels in *Thermobifida fusca* grown on cellobiose and glucose. *J Bacteriol* 2007;**189**:6260–5.

64. Moser F, Irwin D, Chen S, Wilson DB. Regulation and characterization of *Thermobifida fusca* carbohydrate-binding module proteins E7 and E8. *Biotechnol Bioeng* 2008;**100**:1066–77.

65. Vaaje-Kolstad G, Bøhle LA, Gåseidnes S, Dalhus B, Bjørås M, Mathiesen G, et al. Characterization of the chitinolytic machinery of Enterococcus faecalis V583 and high-resolution structure of its oxidative CBM33 enzyme. *J Mol Biol* 2012;**416**:239–54.

66. Hemsworth GR, Taylor EJ, Kim RQ, Gregory RC, Lewis SJ, Turkenburg JP, et al. The copper active site of CBM33 polysaccharide oxygenases. *J Am Chem Soc* 2013;**135**:6069–77.

67. Lu Y, Roe JA, Bender CJ, Peisach J, Banci L, Bertini I, et al. New type 2 copper-cysteinate proteins. Copper site histidine-to-cysteine mutants of yeast copper-zinc superoxide dismutase. *Inorg Chem* 1996;**35**:1692–700.

68. Aachmann FL, Sørlie M, Skjåk-Bræk G, Eijsink VGH, Vaaje-Kolstad G. NMR structure of a lytic polysaccharide monooxygenase provides insight into copper binding, protein dynamics, and substrate interactions. *Proc Natl Acad Sci USA* 2012;**109**.

69. Kleman-leyer KM, Gilkes NR, Miller Jr RC, Kirk TK. Changes in the molecular-size distribution of insoluble cellulose by the action of recombinant *Cellulomonas fimi* cellulases. *Biochem J* 1994;**302**:463–9.

70. Phillips CM, Beeson WT, Cate JH, Marletta MA. Cellobiose dehydrogenase and a copper-dependent polysaccharide monooxygenase potentiate cellulose degradation by *Neurospora crassa*. *ACS Chem Biol* 2011;**6**:1399–406.

71. Cannella D, Hsieh CWC, Felby C, Jørgensen H. Production and effect of aldonic acids during enzymatic hydrolysis of lignocellulose at high dry matter content. *Biotechnol Biofuels* 2012;**5**:26.

72. Langston JA, Shaghasi T, Abbate E, Xu F, Vlasenko E, Sweeney MD. Oxidoreductive cellulose depolymerization by the enzymes cellobiose dehydrogenase and glycoside hydrolase 61. *Appl Environ Microbiol* 2011;**77**:7007–15.

73. Beeson WT, Phillips CM, Cate JHD, Marletta MA. Oxidative cleavage of cellulose by fungal copper-dependent polysaccharide monooxygenases. *J Am Chem Soc* 2012;**134**:890–2.

CHAPTER 7

New Insights into Microbial Strategies for Biomass Conversion

Sarah E. Hobdey, Bryon S. Donohoe, Roman Brunecky, Michael E. Himmel, Yannick J. Bomble

Biosciences Center, National Renewable Energy Laboratory (NREL), Golden, CO, USA

Introduction

Biomass is the most abundant and renewable source of carbon on our planet. With the right technologies and increased conversion yields, biomass-derived fuels have the potential to supply a significant fraction of current liquid fuels in the United States. However, a major obstacle in the conversion of biomass to fuel is the high capital cost of chemical pretreatment required to overcome the innate recalcitrance of plant cell walls.[1,2] Current efforts to reduce these costs by substantially reducing the severity of, or eliminating entirely, chemical pretreatment leave a more intact and thus complex cell wall material containing more of the hemicellulosic polymers intact. In this process scenario, there is a greater need for a biological approach for breaking down these complex hemicellulosic materials into usable sugars. Plant cell walls contain primarily four carbon-rich polymers: cellulose, hemicellulose, pectin, and lignin.[3] Cellulose microfibrils provide the fibers that, when combined with the other cell wall components, create a fiber reinforced matrix material that serves as a protective barrier from mechanical, chemical, enzymatic, and microbial degradation. Cellulose is composed of β-1,4-linked glucose polymers; in the cell wall, these glucose polymers are organized in directional cellodextrin chains that assemble into larger, longer microfibrils that can be highly compact and crystalline, greatly restricting accessibility to the interior chains of cellulose mirofibrils.[4] Cellulose is synthesized by a large macromolecular protein complex called the cellulose synthase complex (CSC). CSCs form in the plant cell membrane and are responsible for polymerization of glucose into β-1,4-linked glucan and translocation of glucan across the plasma membrane, where aggregation of glucan chains occurs to form cellulose microfibrils.[5] In contrast to cellulose, hemicelluloses are noncrystalline polysaccharides composed mostly of a five- or six-carbon sugar backbone physically intermingled with cellulose microfibrils and held by noncovalent interactions. Hemicellulose chains are more accessible to chemical and enzymatic degradation than cellulose and therefore are not as recalcitrant. Hemicellulose is synthesized in the Golgi apparatus of the plant cell and is secreted by vesicular transport

Direct Microbial Conversion of Biomass to Advanced Biofuels. http://dx.doi.org/10.1016/B978-0-444-59592-8.00007-5

into the plant cell wall.[6] Hemicelluloses are a heterogeneous class of branched polysaccharides, such as glucuronoxylan in hardwoods, in which β-D-xylan polymers are decorated with acetate and 4-*O*-methyl glucuronate side chains. Soft wood hemicelluloses are primarily galactoglucomannans and arabinoglucuronoxylans.[7]

In nature, bacteria and fungi have evolved multiple strategies to disrupt and deconstruct plant cell walls for their own use. These methods use one of the following broad strategies for biomass deconstruction: monofunctional enzymes, multifunctional enzymes, and highly aggregated (cellulosomal) enzymes[8–11], all of which include enzymes such as cellulases, hemicellulases, and other polysaccharide-degrading enzymes. Each of these enzyme systems relies on the synergistic interaction between the enzymes or functional domains of the enzymes to enhance the rate of biomass conversion into monomeric sugar. In 1950, Elwyn T. Reese was the first researcher to report synergistic effects by mixing cellulase fractions.[12] In this keystone report, Reese et al. described this behavior, which is now known as endocellulase and exocellulase synergism. Today, intermolecular synergy has been reported for almost every known cellulolytic fungal and bacterial system studied in detail. In fungal, and some bacterial, systems, cellulases are secreted as independent, monofunctional molecules with functionalities that complement each other to create an intermolecular synergistic enzyme cocktail.[13] In contrast to the monofunctional cellulase system, some cellulolytic anaerobic bacteria and fungi use cellulases that self-assemble onto a protein scaffold attached to the cell surface. In this case, intermolecular synergy relies on the proximity of enzymes with complementary functionalities.[14,15] Finally, with the recent discovery of multifunctional enzymes, scientists have also described intramolecular synergy, in which multiple catalytic domains of different functions are combined into a single molecule.[11,16] This chapter discusses the three distinct paradigms of cellulase synergy, illustrates examples of their performance, and postulates on the potential for novel cocktail designs.

Distinct Enzyme Synergy Paradigms in Cellulolytic Microorganisms
Free Enzyme Systems

Cellulolytic bacteria and fungi that secrete free enzymes depend on the hydrolysis of lignocellulose into usable sugars by enzymes with specific substrate specificities. In general, there is believed to be cooperative or synergistic action between at least three classes of free enzymes.[13]

- Endo-β-(1,4)-glucanases (EC 3.2.1.4) hydrolyze soluble and insoluble β-(1,4)-glucan substrates both internally and from reducing and nonreducing ends in a nonprocessive or processive manner.
- Exo-β-(1,4)-D-glucanases (EC 3.2.1.74 and EC 3.2.1.91) liberate D-glucose or D-cellobiose from soluble and insoluble β-(1,4)-D-glucan processively from the reducing or nonreducing end.
- β-D-glucosidases and cellobiases (EC 3.2.1.21); act on cellobiose and less commonly, soluble cellodextrins to free D-glucose.[17]

In conjunction with the three primary cellulase functionalities, there are also glycoside hydrolases that help breakdown hemicellulose, such as xylanases. Hemicellulases can be classified into one of three categories:

- Endo-acting hemicellulases that cleave polysaccharide chains internally.
- Exo-acting hemicellulases that cleave polysaccharides processively from the reducing or nonreducing ends.
- Accessory enzymes that cleave unique backbone linkages and side chains on decorated polysaccharides.

Cellulases and hemicellulases are glycoside hydrolases (GHs), which hydrolyze the glycosidic bond between two or more carbohydrates or between a carbohydrate and non-carbohydrate moiety. GHs are classified into families based on protein sequence and predicted peptide folds and not the specific activity; currently, there are 133 families of GHs (cazy.org). In addition to GHs, there are several other carbohydrate-active enzyme families that use other activities to cleave cellulose and hemicelluloses. These enzymes are classified into three groups: polysaccharide lyases (PLs), carbohydrate esterases (CEs), and auxiliary activities (AAs) families. Another important group of proteins commonly associated with carbohydrate-active enzymes are the carbohydrate binding modules (CBMs). There are currently 68 CBM families listed in CAZY, also classified by sequence and fold, which recognize crystalline cellulose, amorphous cellulose, and hemicellulose chemistries and morphologies, as well as other carbohydrates not discussed in this chapter. CBMs are important for localizing the correct carbohydrate-active enzyme to the correct substrate for hydrolysis.[18,19]

Intermolecular synergy

In the most fundamental classification, cellulases in the monofunctional enzyme system are defined as the cellobiohydrolases (CBHs; exo-β-(1,4)-D-glucanases), endoglucanases (EGs; endo-β-(1,4)-glucanases), and β-D-glucosidases. CBHs are processive enzymes that hydrolyze cellulose from chain ends, whereas EGs typically hydrolyze cellulose chains nonprocessively at random positions internally. Cellobiases hydrolyze cellobiose, the final product of most CBHs, to glucose, thereby preventing end product inhibition of the CBHs.

The monofunctional enzyme system of *Hypocrea jecorena* (*Trichoderma reesei*), for example, secretes large quantities of processive CBHs (Table 1). These processive cellulases have been classified into distinct families of enzymes that can hydrolyze cellulose from either the reducing or nonreducing ends.[20] For example, the CBHs from GH family 7 CBH, Cel7A, processively hydrolyze cellulose from the reducing end, whereas GH family 6, Cel6A, processively hydrolyses cellulose from the nonreducing end. Processive enzymes are supplemented by the EGs, such as Cel5A and Cel7B from *T. reesei*, which cleave internal cellulose chains to create additional reducing and nonreducing ends for CBHs to begin processive hydrolysis. EGs have a further benefit in that they also create shorter cellulose chains to

Table 1: Representative GHs, functions and architecture.

	Organism	Protein	Enzyme activity	Domains
Aerobic	T. reesei (fungus)	Cel7A	Cellobiohydrolase	CBM1-linker-GH7
		Cel6A	Cellobiohydrolase	CBM1-linker-GH6
		Cel7B	Endoglucanase	CMB1-linker-GH7
		Cel5A	Endoglucanase	CMB1-linker-GH5
		Bgl1	β-glucosidase	GH3
	T. fusca[26,49] (bacterium)	Cel48A	Cellobiohydrolase	CBM2-linker-GH48[50]
		Cel6B	Cellobiohydrolase	CBM2-linker-GH6
		Cel9B	Endoglucanase	CMB3c-linker-GH9
		Cel6A	Endoglucanase	CMB2-linker-GH6
		BglC	β-glucosidase	GH1
	S. degradans[51] (bacterium)	Cel6A	Cellobiohydrolase	CBM2-linker-CBM2-linker-GH6
		Cel9A	Endoglucanase	GH9
		Cel5A	Endoglucanase	GH5-CBM6-CBM6-CBM6-GH5
		Bgl1A[a]	β-glucosidase	GH1
Anaerobic	P. equi[52-55] (fungus)	Cel5A	Endoglucanase	Doc-GH5-GH5-GH5-Doc-Doc
		Cel6A	Cellobiohydrolase	Doc-Doc-GH6
		Cel45A		Doc-Doc-Doc-GH45
		Cel45A		DocII-DocIII-GH45
		Cel3A	β-glucosidase	GH3-aux-Doc-Doc-Doc
	C. thermocellum[8,56,57] (bacterium)	CipA	Scaffoldin	CohI-CohI-CBM3-CohI-CohI-CohI-CohI-CohI-CohI-CohI-DocII
		CelA	Endoglucanase	GH8-DocI
		CelS	Cellobiohydrolase	GH48-DocI
		BglA[a]	β-glucosidase	GH1
		CelJ	Endo- and exo-glucanase	CBM30-GH9-GH44-DocI-CBM44
	R. flavefaciens[58] (bacterium)	CelB[b]	Cellobiohydrolase	GH5-unknown-linker-Doc
		EndB	Endoglucanase	GH44-CBM-Doc
		ScaA	Scaffoldin	Coh-Coh-Coh-XDoc
		ScaB	Scaffoldin	Coh-Coh-Coh-Coh-Coh-Coh-Coh-XDoc
		BglB[b]	β-glucosidase	
	C. bescii[11,59] (bacterium)	CelA	Endo- and exo- glucanase	GH9-CBM3c-CBM3b-CBM3b-GH48
		CelC	Endoglucanase/ mannanase	GH9-CBM3-CBM3-CBM3-GH5
			Exoglucanase/ xylanase	GH10-CBM3-CBM3-GH48
		CelB	Xylanase/ endoglucanase	GH10-CBM3-CBM3-CBM3-GH5
		ManA	Mannanase	GH5-CBM3-CBM3-GH44

[a]No cellular secretion signal.
[b]From related species found in CAZY database.

facilitate the release of processive enzymes (discussed below). Combining the activities of EG and CBHs releases large quantities of soluble cellodextrins, in particular cellobiose. β-D-glucosidases hydrolyze these smaller glucose dimers to glucose monomers, which are then used by the fungal (or bacterial) cell for metabolism.

Similar to the monofunctional enzyme system of fungi, bacteria, such as *Thermobifida fusca*, use a parallel synergistic system.[21] *T. fusca* is an aerobic, soil bacterium that thrives in elevated temperatures over a large range of pH. Like *T. reesei*, *T. fusca* secretes endocellulases and exocellulases along with many hemicellulases for plant cell wall deconstruction[22–25] (Table 1). Furthermore, cellobiose is transported into the bacterial cell for hydrolysis by an intracellular β-D-glucosidase.[26] Like the fungal cellulase systems, these bacterial endocellulases and exocellulases have CBMs that assist in substrate recognition and enzyme localization.

In practice, these enzymes exhibit intermolecular synergy in that the secretome of these three types of enzymes have a total calculated effect much higher than that of each individual enzyme. Thus, for complete cellulose digestion, enzyme cocktails and commercial formulations developed for biomass conversion must contain each of these activities. Two examples of modern commercially available cellulase enzyme formulations are CTec2 (Novozymes) Accelerase (DuPont).

Three-dimensional cellulase structure

Glycoside hydrolase crystal structures have been critical for elucidating the function of these enzymes. In particular, most endocellulases and exocellulases in the monofunctional enzyme system consist of at least two domains, a CBM and a catalytic domain (CD). These two domains are often connected by a nonstructured linker peptide that aids in protein flexibility and/or cellulose binding.[27,28]

CBMs are noncatalytic, functionally independent domains connected to the catalytic domain of many GHs. There are currently 68 families of CBMs classified by fold and carbohydrate recognition (www.cazy.org). The family 7 CBH (Cel7A) from *T. reesei*, for example, is attached to a family 1CBM. This CBM is thought to bind to cellulose fibers to effectively increase the local concentration of the catalytic modules on crystalline cellulosic substrates, target specific carbohydrate structures, and possibly disrupt lignocellulose.[29,30]

The cellulase catalytic domain contains the enzyme active site and therefore must be able to bind, orient, and hydrolyze the β-(1,4)-glycosidic bond. Interestingly, it is the tunnel-like structure around the active site that seems to be important in determining an enzyme's processivity.[31] Cel7A, for example, is a highly processive enzyme with an enclosed "tunnel" that accommodates a single cellulose chain that is threaded through the active site. Less processive enzymes or EGs do not have such extensive tunnels around their active sites, which allows these enzymes to bind and release cellulose chains at random. Consequently, it is thought that EGs are not only synergistic with CBHs, in that they create initiation sites for CBHs, but they also create release sites for highly processive enzymes with highly enclosed active sites.

Self-Assembling, Highly Aggregated Enzyme Systems

An alternative plant cell wall-degrading enzyme paradigm that evolved in some bacteria and fungi involves large, multienzyme complexes that are tethered together and commonly bound to the cell surface by long linker peptides and large glycoproteins, called scaffoldins. In contrast to monofunctional enzyme systems, a multitude of enzymes are non-covalently linked to a scaffoldin to create a gigantic, multifunctional mega-Dalton-sized complex, called the cellulosome. Similar to the *T. reesei* cellulase secretome, cellulosomes incorporate processive and nonprocessive cellulases, hemicellulases, and other carbohydrate-degrading enzymes for synergistic action on biomass.[32]

The cellulosome complex is assembled by the high-affinity, type- and species-specific, noncovalent interaction between the cohesin modules of the scaffoldin and the enzyme-associated dockerin modules.[33,34] Primary scaffoldins typically contain multiple cohesin modules, which can bind directly to dockerins-associated enzymes or to secondary scaffoldins. CipA, the primary scaffoldin from *Clostridum thermocellum*, for example, contains type I and type II cohesin modules that can bind to dockerin-associated enzymes or to secondary scaffoldins, respectively, thereby enabling many different types of enzymes to be assembled into the final cellulosome complex.[34] In contrast to the monofunctional enzyme system, CBMs are less often associated with each individual enzyme, but are typically a component of the scaffoldin protein (Table 1). In this case, the scaffoldin-associated CBM(s) are responsible for maintaining the cellulosome proximity to the lignocellulosic substrate for efficient saccharification.

Intermolecular versus intramolecular synergy

Compared to the monofunctional enzyme systems, which exhibit strictly intermolecular synergy, cellulosomes can be considered to have both intermolecular and intramolecular synergy. Since the scaffoldin brings the catalytic modules of independent function into close physical association with each other, two enzymes can be thought of as separate and having intermolecular synergy. However, considering that it is the scaffoldin that is associated with all the enzymes making up a single cellulosome, the cellulosome can also be thought of as displaying intramolecular synergy as well.

Both the monofunctional enzyme and cellulosomal systems use the same families of cellulases and hemicellulases, with the exception of GH 6, 7 and 45, which are present only in noncellulosomal fungal systems. The synergistic interaction between endoglucanases and exoglucanases is; therefore, similar to the mono-functional enzymes system. However, because the enzymes decorating cellulosomes are all part of a large, mega-Dalton assembly, the flexible movement of the scaffoldin is the likely mechanism by which enzymes of complementary catalytic activity are brought in close proximity to cell wall substrates.

Designer cellulosomes

Producing artificial, multienzyme complexes or designer cellulosomes is one method researchers have used recently to optimize biomass deconstruction. Designer cellulosomes comprise a chimeric scaffoldin with a CBM and various cohesin modules taken from different cellulosome-producing microbial species. The approach permits control of the composition and arrangement of individual cellulases on the engineered cellulosome.[35] Structure-function analysis of cellulosomes and their associated enzymes has uncovered different types of cohesin and dockerin associations. These associations are specific to different microbial species, in which some type I cohesins will bind type I dockerins of some species, but not others; i.e., type II cohesins of *C. thermocellum* will bind to type II dockerins of *A. celluloyticus*; however, type I cohesins of *C. celluloyticum* cannot bind the type I dockerins of *C. thermocellum*.[36] Designer cellulosomes can be engineered from components of native and non-native cellulosomes, as well as chimeric monofunctional enzymes containing relevant dockerin modules. The inclusion of selected enzymes is controlled in order of addition and type of enzyme activity to form highly specific cellulolytic complexes.[37] The ordered incorporation of selected enzymes into designer cellulosomes has been shown to improve synergism between cellulases by specific substrate targeting by the CBM, by the proximity of the cellulases in the complex, and by flexibility of the scaffoldin.[15,38,39]

Multifunctional Enzyme Systems

The third emerging paradigm that can be viewed as intermediate between that of monofunctional enzymes and highly aggregated enzymes (cellulosomes) is that of the multifunctional enzymes. Multifunctional enzymes are single gene products composed of two or more catalytic activities.[11,16,40–42] These enzymes are usually of high molecular weight and have one or several CBMs. In nature, multifunctional enzymes exist in both monofunctional enzyme systems and cellulosomal systems (Table 1). The classification of multifunctional enzymes is based on substrate specificities and, therefore, can be grouped into one of four different classes: cellulase—cellulase, cellulase–hemicellulase, hemicellulase–hemicellulase, and hemicellulase–carbohydrate esterase systems.[8]

Intramolecular synergy

The presence of two different enzymes in the same polypeptide chain suggests that the proximity of the designated catalytic modules necessitates concerted action on a given portion of the lignocellulosic substrate, and, indeed, it has been shown that each class of multifunctional enzymes display enhanced activity when compared to their individual modules.[16,43]

One well-characterized example of this cellulase paradigm is the multifunctional cellulase–cellulase enzyme from the thermophilic, cellulolytic, bacterium *Caldicellulosiruptor bescii*, CelA. CelA contains an *N*-terminal endoglucanase module, GH9, and a *C*-terminal exoglucanase module, GH48, which are separated by three family 3 CBMs.[11,16] When these modules are separated, but in the same cocktail, they exhibit synergistic digestion of cellulose; however, when tested as a single molecule, this synergism is substantially enhanced.[43] Having both endoglucanse and exoglucanase activity, CelA is capable of hydrolyzing microcrystalline cellulose alone and much more efficiently than the model endoglucanases and exoglucanses (*A. cellulolyticus* E1 and *T. reesei* Cel7A) from free enzyme systems.[11] This further suggests that the proximity, or local concentration, of synergistic activities is important for increased digestion of cellulose.

Along with cellulase–cellulase multifunctional enzymes, hemicellulase–hemicellulase multifunctional enzymes commonly have two different activities in the same protein sequence. For example, GH26 and GH10 display primarily mannanase and xylanase activities, respectively.[42] CBMs of these enzymes usually target a single polysaccharide chain to localize the associated enzymes to relevant portions of the substrate, either before or during the processes of degradation, as newly exposed sites on the substrates become accessible. Interestingly, some of these multifunctional hemicellulases also have CBM3 modules, which are known to exclusively bind cellulose.

The evolution of multifunctional enzymes is likely to have occurred from the fusion of various catalytic modules with the benefit of performing their respective functions in the hydrolysis of plant cell walls at close proximity.[8] However, it is not unreasonable to believe that researchers can use this "natural" strategy to construct artificial multifunctional enzyme chimeras based on functionality and substrate configuration in an attempt to optimize saccharification levels. Indeed, Xu et al. recently showed that designed, chimeric multifunctional cellulases could be created from *C. thermocellum* components.[35] This multifunctional cellulase linked together two important *C. thermocellum* cellulases, CbhA and CelA, and exhibited higher activity than the combination of its individual components.

New Cellulose Digestion Strategies Promoting Interspecies Synergism

For years, the primary digestion strategy observed among most cellulolytic enzymes was observable at the gross microfibril level. The first such observation (1985) was made by Chanzy and coworkers[44] using electron microscopy to observe the unidirectional action of Cel6A from *T. reesei* acting on Valonia microcrystalline cellulose. It is now thought that the processive action of exo-acting cellulases produces sharpened tip morphologies at primarily the reducing ends of cellulose microfibrils.[45] These authors suggested that the synergy between these cellulases caused the formation of sharpened cellulose microcrystals at the reducing ends. This endocellulase/exocellulase synergistic cellulose tip sharpening mechanism was further confirmed by several research groups and most recently observed for the Cel7A rich commercial preparation Cellic CTec2 (Novozymes).[10] Moreover, several recent studies have uncovered novel digestion strategies used by less common cellulolytic systems, such the cellulosomal bacterium *C. thermocellum* and the hyperthemophilic bacterium, *C. bescii*. These systems bring new insights to the natural paradigms of cellulose digestion.

Cellulose Deconstruction by Cellulosomes: An Efficient and Complementary Deconstruction Mechanism

The specific, non-covalent attachment and organization of the cellulosome enables enzyme co-localization, which results in importantly different enzymatic performances compared to the monofunctional enzymes. A recent study comparing digestion of various substrates by the commercial fungal preparation (CTec2) and purified, isolated cellulosomes from *C. thermocellum*, found that purified cellulosomes degrade Avicel and filter paper cellulose faster than the monofunctional enzymes, but was less effective than monofunctional enzymes on pretreated biomass.[10] This result was attributed to the size and mobility of monofunctional enzymes compared to the cellulosomes. More interestingly, Transmission Electron Microscopy (TEM) micrographs of Avicel particles partially digested by cellulosomes revealed irregular, splayed-out cellulose ends. This observation is in stark contrast to monofunctional enzymes, which seem to have a tapering effect that results in a sharpening of the particle end.[4] By splaying out the ends of crystalline cellulose, the cellulosomes effectively increase the surface area that is accessible to enzymatic digestion by roughly two-fold.[10] This observation likely reflects the basic enzyme differences for each paradigm: where monofunctional enzymes are able to find, bind, hydrolyze, and release substrate based on individual enzyme function; cellulosomes having many catalytic units per complex tend to find, bind, and digest the entirety of the substrate before dissociation.[10]

In conjunction with different cellulose fibril digestion morphologies, the authors also noted different localization patterns of cellulosomes and free cellulases during digestion of switch grass. Cellulosomes were found near fractures in the plant cell wall, whereas free enzymes were able to penetrate into the secondary cell wall,[10] indicating that, in real plant cell wall material, cellulosomes are too large to penetrate the porosity of pretreated biomass cell walls, emphasizing the benefit of splaying cellulose fibers to increase accessibly for the much larger, complexed enzymes. While the proposed mechanism displays the benefits of splaying, these results also provide insights to the limitations that cellulosomes experience digesting biomass compared to their action on pure cellulose.

Considering the substantial difference between the ablative mode of action of monofunctional enzymes and the cellulose splaying strategy used by cellulosomes, it is seems obvious that great synergy could exist between these two cellulolytic systems and that this is possibly why

C. thermocellum naturally uses both the aggregated cellulosomal systems and an extensive monofunctional enzyme system.[46] Indeed, it has been shown that the combination of purified cellulosomes and back addition of a commercial free cellulase system had higher activity on Avicel, reaching 100% conversion in 24 h, compared to 70% with cellulosomes or free enzymes.[10] Two possible synergistic methods can be surmised: (1) the cellulosome embarks on a microfibril bundle, splaying it out to effectively increase the surface area accessible for monofunctional enzymes to degrade; and (2) in the case of natural substrates, monofunctional enzymes penetrate through the primary and secondary plant cell wall, increasing the porosity of the wall material allowing cellulosomes to more quickly splay and penetrate the material.

The Hyperthermophilic Cellulase from Caldicellulosiruptor bescii CelA Degrades Cellulose by Several Complementary Mechanisms

As mentioned previously, cellulases from the emerging CBP microorganism *C. bescii* are of great interest due to their extremely high efficiency, provided by the combination of complementary catalytic domains within the same gene product.[16,43] The cellulase, CelA, from *C. bescii* was shown to convert 100% of pure crystalline cellulose within seven days in the presence of a thermostable β-D-glucosidase at a temperature of 85 °C.[11] CelA also dramatically outperformed a mixture of the exocellulases/endocellulases Cel7A/Cel5A from the aerobic fungus, *T. reesei,* and bacterium, *A. cellulolyticus.*[11] Beyond the hyperactivity of CelA, the comparison of this bacterial cellulase to its fungal counterparts reveals distinct digestion strategies never reported for any previously characterized cellulolytic systems (Figure 1).

Avicel particles partially digested by CelA exhibited narrowed, irregular, but not finely tapered morphology on one end and an irregular, scalloped angled morphology on the opposite end. However, in addition to the ablation activity often attributed to exoglucanases, CelA was also able to burrow into Avicel particles and create cavities. Evidence of these cavities was clear from TEM, in which CelA digested particles showed irregularly spaced cavities along the length of the particles (Figure 2). The size of these particles varied from 15 to 30 nm in length and up to 150 nm in depth. These cavities were wide enough to accommodate one or several CelA molecules given the size distribution of the cellulase inferred by molecular dynamics. This finding suggests that the smaller cavities may be the result of individual CelA molecules, whereas the larger cavities may be the result of mature cavities merging or multiple enzymes working in the same, enlarged cavity. The presence of several binding modules most certainly restricts the processive behavior of this cellulase and increases its residence time on a particular location, leading to the formation of cavities. It should also be mentioned that in addition to this cavity-forming mechanism, there are likely other, less well-characterized mechanisms at play, as suggested by the two dramatically different end morphologies previously mentioned. This new digestion strategy could lead to increased synergy with more traditional processive cellulases, because it generates more accessible surface area and makes more cellulose ends available.

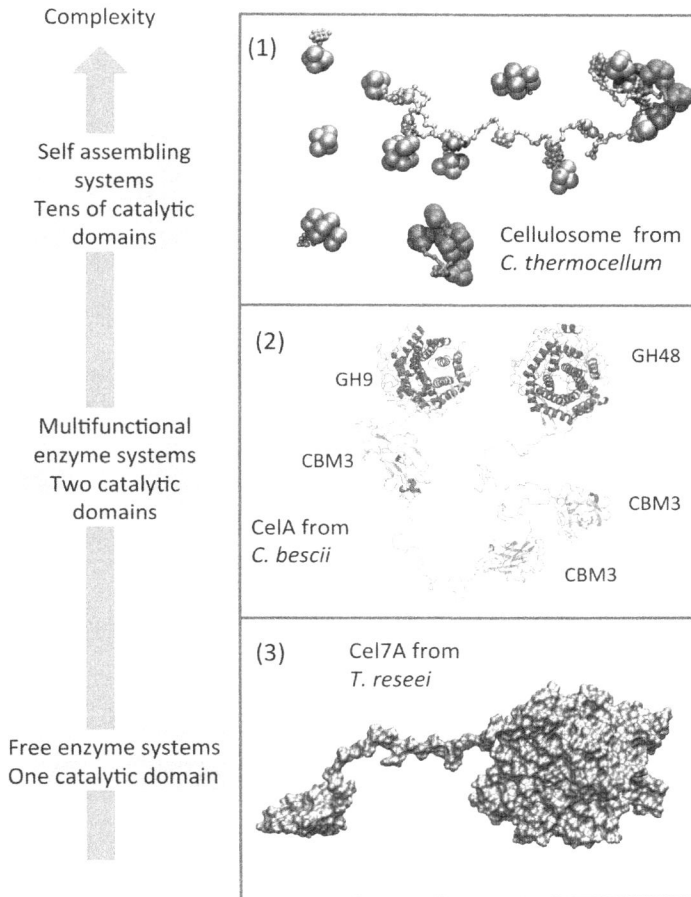

Figure 1
Complexity of the three cellulase systems used in nature by microorganisms to degrade lignocellulose. *Ref. 11.*

However, due to the differences in optimal temperature with most highly active exoglucanases, this synergistic effect has yet to be observed (Figures 3 and 4).

Future Perspective

As new cellulolytic microorganisms are discovered through bio-prospecting and molecular characterization, there are opportunities to discover new biomass degradation paradigms. The picture of biomass degradation is increasingly more complex, in which multifunctional cellulases from many thermophilic organisms are now being considered as potential game changers and templates for more efficient chimeric cellulases. The recent discovery of the high cellulolytic activity of *C. bescii* and its most abundant cellulase, CelA, has remotivated

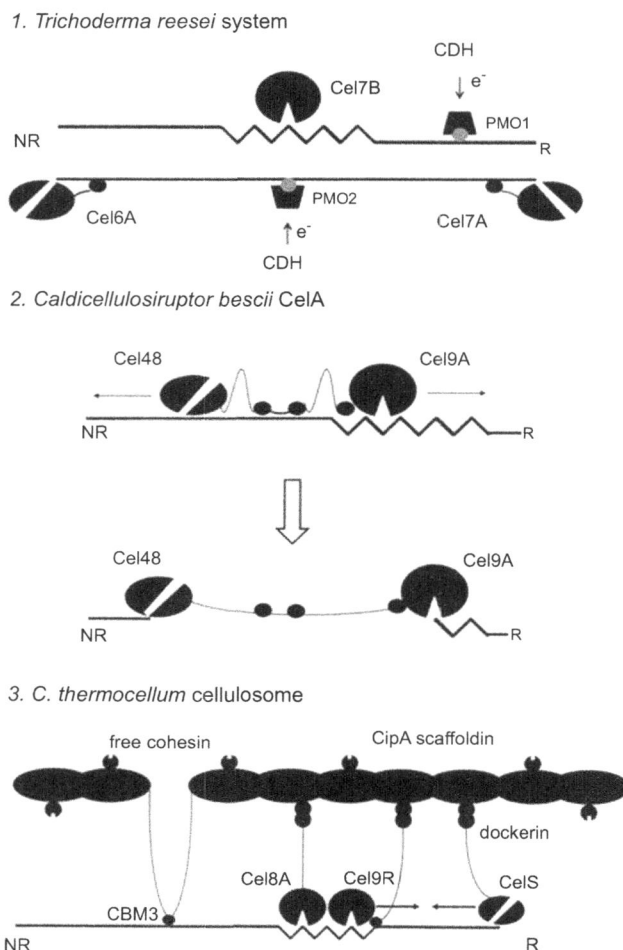

Figure 2
Depiction of the mechanism of action of the three natural cellulose degrading systems known today. The (1) mono-functional system deployed by most filamentous fungi (*T. reesei*) and bacteria, (2) the multi-functional system used by some bacteria (*C. bescii*), and the (3) highly aggregated system used by very few bacteria (*C. thermocellum*) and fungi.

researchers to search for similar microbes in thermophilic environments. Unlike the relatively new multifunctional systems, self-assembling systems have been extensively studied and well characterized biochemically, but there still remains much to be understood about the specific strategies they use to degrade biomass. The recent discovery of the high synergistic activity of cellulosomes from *C. thermocellum* and fungal cellulase-based formulation CTec2 shows that much remains to be understood. Additionally, in both cases, advanced imaging of digested substrates has proven to be a great tool for providing new insights into the mechanisms by which these cellulolytic systems degrade biomass.

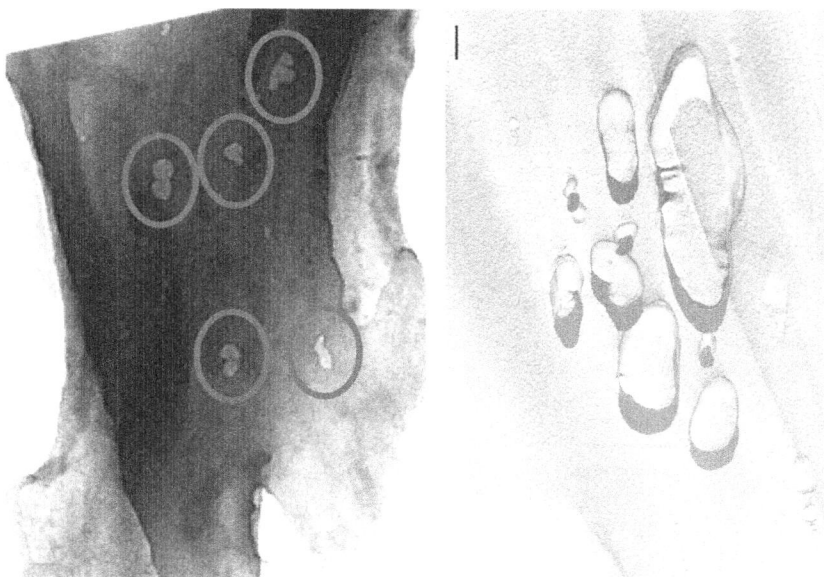

Figure 3

TEM micrographs of small Avicel PH101 particles from digestions at ~65% conversion. CelA digested particles display cavities of various sizes on the surface (green (light gray in print versions)). Plate on the right is one tilt image from a tomographic study of a cavity field. The length insert bar indicates 100 nm.

Understanding these newly discovered biomass digestion strategies is an important step toward understanding biomass deconstruction in nature. However, they also highlight how little is known to date about the variety of biomass digestion mechanisms that exists in the biosphere, for example from a consortia of organisms. There exist great opportunities for the discovery of new enzymes and microorganisms that can synergistically degrade biomass or be used to augment current enzyme formulations. Indeed, because most synergistic action of microorganisms or enzymes on biomass is known from prospecting, researchers often look at microbes that have evolved in the same environment and benefit from cooperative degradation of biomass. This is the case for microbes found in hot springs, the rumen, or soil. Whereas this strategy has enabled significant advances in the biofuels industry, it appears that a better approach to enhance existing formulations or CBP microbes would be to take advantage of the different digestion strategies that exist in different ecosystems. Indeed, as mentioned above, the recent study by Resch et al. shows that two enzymatic systems from two different ecosystems can synergistically degrade biomass because of the complementary nature of their biomass digestion mechanisms.

Another area for potential mechanistic synergy is illustrated by CelA from *C. bescii* and its cavity-forming mechanism recently discovered by Brunecky et al.[11] The formation of cavities and splaying of end surfaces observed on CelA digested samples indicates the available

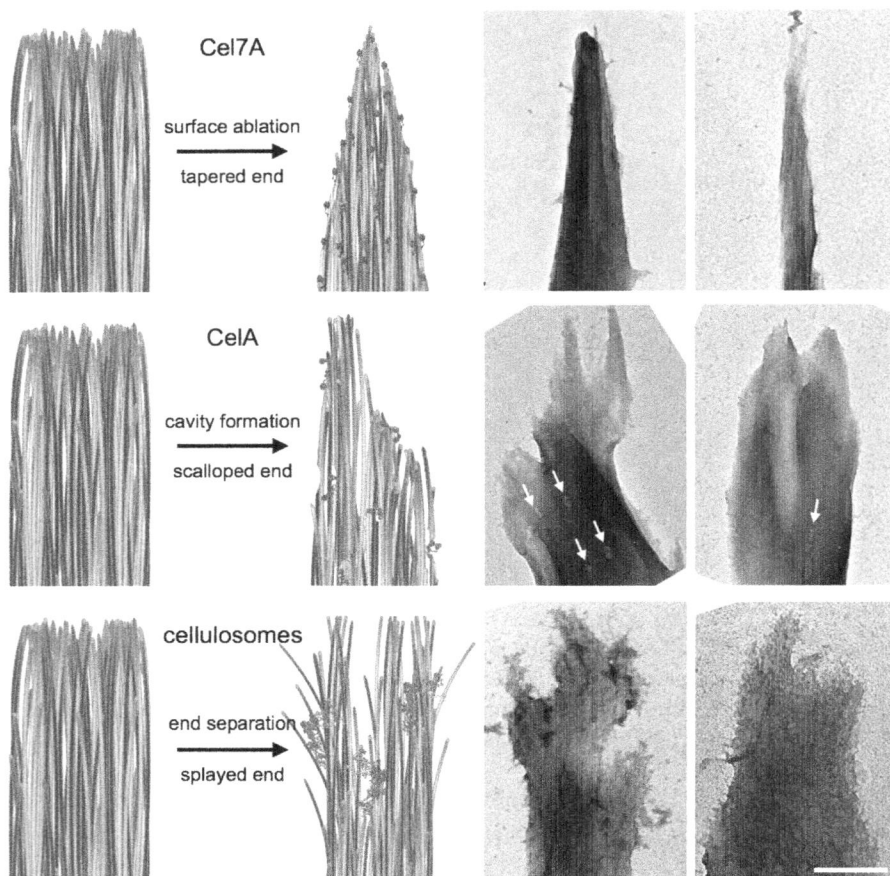

Figure 4
Illustrations and transmission electron micrographs of digested Avicel particles. Imaging of these substrates has revealed new insights into different strategies used by cellulase systems to deconstruct biomass. In the monofunctional enzyme system, in which individual cellulases are limited to ablating the surface, the particle obtains a smooth, tapered end. The tapering of one end of the particle can be understood by the abundance of the reducing end oriented exoglucanase, Cel7A, in CTec2 and that the microfibrils that make up these small Avicel particles must be oriented in parallel. In the CelA digested particles, the same end displays a more irregular, scalloped morphology. In addition, a unique feature of CelA digested particles is the appearance of excavated cavities (white arrows). Cellulosome digested Avicel particles display separation of individual cellulose microfibrils creating more splayed particle end morphology. Scale bar = 500 μm. *Modified from Refs 10,11.*

surface area of the biomass is being increased by the action of the CelA enzyme. Given that surface area and available binding area are limiting factors in free enzyme cellulase systems and that significant enhancements in hydrolysis rates can be achieved with ball milling,[47,48] it follows that the conventional free enzyme systems, which function in a surface ablative manner, should be able to take advantage of the increased accessibility to the substrate provided by the CelA cavity forming mechanism to increase their rate of hydrolysis.

Acknowledgments

The fungal topics were supported by the US Department of Energy (DOE) Energy Efficiency and Renewable Energy (EERE) Bioenergy Technologies Office (BETO), and the bacterial topics were supported by the BioEnergy Science Center (BESC). BESC is a US DOE Bioenergy Research Center supported by the Office of Biological and Environmental Research in the US DOE Office of Science.

References

1. Himmel ME, Ding SY, Johnson DK, Adney WS, Nimlos MR, Brady JW, et al. Biomass recalcitrance: engineering plants and enzymes for biofuels production. *Science* 2007;**315**:804–7.
2. Chundawat SP, Beckham GT, Himmel ME, Dale BE. Deconstruction of lignocellulosic biomass to fuels and chemicals. *Annu Rev Chem Biomol Eng* 2011;**2**:121–45.
3. Brunecky R, Selig MJ, Wei H, Resch M, Himmel ME. In: Goldman SL, editor. *Compendium of bioenergy plants corn*. CRC press; 2014. p. 33–77.
4. Somerville C. Cellulose synthesis in higher plants. *Annu Rev Cell Dev Biol* 2006;**22**:53–78.
5. Delmer DP, Amor Y. Cellulose biosynthesis. *Plant Cell* 1995;**7**:987–1000.
6. Sandhu AP, Randhawa GS, Dhugga KS. Plant cell wall matrix polysaccharide biosynthesis. *Mol Plant* 2009;**2**:840–50.
7. Naran R, Black S, Decker S, Azadi P. Extraction and characterization of native heteroxylans from delignified corn stover and aspen. *Cellulose* 2009;**16**:661–75.
8. Himmel ME, Xu Q, Luo Y, Ding S-Y, Lamed R, Bayer EA. Microbial enzyme systems for biomass conversion: emerging paradigms. *Biofuels* 2010;**1**(2):323–41.
9. Morais S, Barak Y, Lamed R, Wilson DB, Xu Q, Himmel ME, et al. Paradigmatic status of an endo- and exoglucanase and its effect on crystalline cellulose degradation. *Biotechnol Biofuels* 2012;**5**:78.
10. Resch MG, Donohoe BS, Baker JO, Decker SR, Bayer EA, Beckham GT, et al. Fungal cellulases and complexed cellulosomal enzymes exhibit synergistic mechanisms in cellulose deconstruction. *Energ Environ Sci* 2013;**6**:1858–67.
11. Brunecky R, Alahuhta M, Xu Q, Donohoe BS, Crowley MF, Kataeva IA, et al. Revealing nature's cellulase diversity: the digestion mechanism of *Caldicellulosiruptor bescii* CelA. *Science* 2013;**342**:1513–6.
12. Reese ET, Siu RG, Levinson HS. The biological degradation of soluble cellulose derivatives and its relationship to the mechanism of cellulose hydrolysis. *J Bacteriol* 1950;**59**:485–97.
13. Stickel Jonathan J, Philips RB, Elander Rischard T, McMillan James. In: Bisaria Virendra S, Kondo A, editors. *Bioprocessing of renewable resources to commodity bioproducts*. Wiley; 2014. p. 77–97.
14. Caspi J, Barak Y, Haimovitz R, Irwin D, Lamed R, Wilson DB, et al. Effect of linker length and dockerin position on conversion of a *Thermobifida fusca* endoglucanase to the cellulosomal mode. *Appl Environ Microbiol* 2009;**75**:7335–42.
15. Fierobe HP, Bayer EA, Tardif C, Czjzek M, Mechaly A, Belaich A, et al. Degradation of cellulose substrates by cellulosome chimeras. Substrate targeting versus proximity of enzyme components. *J Biol Chem* 2002;**277**:49621–30.
16. Zverlov V, Mahr S, Riedel K, Bronnenmeier K. Properties and gene structure of a bifunctional cellulolytic enzyme (CelA) from the extreme thermophile "*Anaerocellum thermophilum*" with separate glycosyl hydrolase family 9 and 48 catalytic domains. *Microbiology* 1998;**144**(Pt 2):457–65.
17. Beguin P, Aubert JP. The biological degradation of cellulose. *FEMS Microbiol Rev* 1994;**13**:25–58.
18. Boraston AB, Bolam DN, Gilbert HJ, Davies GJ. Carbohydrate-binding modules: fine-tuning polysaccharide recognition. *Biochem J* 2004;**382**:769–81.

19. Guillen D, Sanchez S, Rodriguez-Sanoja R. Carbohydrate-binding domains: multiplicity of biological roles. *Appl Microbiol Biotechnol* 2010;**85**:1241–9.
20. Divne C, Stahlberg J, Reinikainen T, Ruohonen L, Pettersson G, Knowles JK, et al. The three-dimensional crystal structure of the catalytic core of cellobiohydrolase I from *Trichoderma reesei*. *Science* 1994; **265**:524–8.
21. Lykidis A, Mavromatis K, Ivanova N, Anderson I, Land M, DiBartolo G, et al. Genome sequence and analysis of the soil cellulolytic actinomycete *Thermobifida fusca* YX. *J Bacteriol* 2007;**189**:2477–86.
22. Irwin DC, Spezio M, Walker LP, Wilson DB. Activity studies of eight purified cellulases: specificity, synergism, and binding domain effects. *Biotechnol Bioeng* 1993;**42**:1002–13.
23. Ghangas GS, Wilson DB. Cloning of the *Thermomonospora fusca* endoglucanase E2 gene in *Streptomyces lividans*: affinity purification and functional domains of the cloned gene product. *Appl Environ Microbiol* 1988;**54**:2521–6.
24. Zhang S, Lao G, Wilson DB. Characterization of a *Thermomonospora fusca* exocellulase. *Biochemistry* 1995;**34**:3386–95.
25. Wilson DB. Studies of *Thermobifida fusca* plant cell wall degrading enzymes. *Chem Rec* 2004;**4**:72–82.
26. Spiridonov NA, Wilson DB. Cloning and biochemical characterization of BglC, a beta-glucosidase from the cellulolytic actinomycete *Thermobifida fusca*. *Curr Microbiol* 2001;**42**:295–301.
27. Payne CM, Resch MG, Chen L, Crowley MF, Himmel ME, Taylor 2nd LE, et al. Glycosylated linkers in multimodular lignocellulose-degrading enzymes dynamically bind to cellulose. *Proc Natl Acad Sci USA* 2013;**110**:14646–51.
28. Sammond DW, Payne CM, Brunecky R, Himmel ME, Crowley MF, Beckham GT. Cellulase linkers are optimized based on domain type and function: insights from sequence analysis, biophysical measurements, and molecular simulation. *PloS One* 2012;**7**:e48615.
29. Nidetzky B, Steiner W, Hayn M, Claeyssens M. Cellulose hydrolysis by the cellulases from *Trichoderma reesei*: a new model for synergistic interaction. *Biochem J* 1994;**298**(Pt 3):705–10.
30. Carrard G, Koivula A, Soderlund H, Beguin P. Cellulose-binding domains promote hydrolysis of different sites on crystalline cellulose. *Proc Natl Acad Sci USA* 2000;**97**:10342–7.
31. Divne C, Stahlberg J, Teeri TT, Jones TA. High-resolution crystal structures reveal how a cellulose chain is bound in the 50 A long tunnel of cellobiohydrolase I from *Trichoderma reesei*. *J Mol Biol* 1998;**275**:309–25.
32. Bayer EA, Belaich J-P, Shoham Y, Lamed R. The cellulosomes: multienzyme machines for degradation of plant cell wall polysaccharides. *Annu Rev Microbiol* 2004;**58**:521–54.
33. Pagès S, Bélaïch A, Bélaïch J-P, Morag E, Lamed R, Shoham Y, et al. Species-specificity of the cohesin-dockerin interaction between *Clostridium thermocellum* and *Clostridium cellulolyticum*: prediction of specificity determinants of the dockerin domain. *Proteins: Struct Funct Bioinforma* 1997;**29**:517–27.
34. Mechaly A, Fierobe H-P, Belaich A, Belaich J-P, Lamed R, Shoham Y, et al. Cohesin-dockerin interaction in cellulosome assembly: a single hydroxyl group of a dockerin domain distinguishes between nonrecognition and high affinity recognition. *J Biol Chem* 2001;**276**:9883–8.
35. Xu Q, Ding S-Y, Brunecky R, Bomble Y, Himmel M, Baker J. Improving activity of minicellulosomes by integration of intra- and intermolecular synergies. *Biotechnol Biofuels* 2013;**6**:126.
36. Fontes CM, Gilbert HJ. Cellulosomes: highly efficient nanomachines designed to deconstruct plant cell wall complex carbohydrates. *Annu Rev Biochem* 2010;**79**:655–81.
37. Ding SY, Xu Q, Crowley M, Zeng Y, Nimlos M, Lamed R, et al. A biophysical perspective on the cellulosome: new opportunities for biomass conversion. *Curr Opin Biotechnol* 2008;**19**:218–27.
38. Morais S, Barak Y, Hadar Y, Wilson DB, Shoham Y, Lamed R, et al. Assembly of xylanases into designer cellulosomes promotes efficient hydrolysis of the xylan component of a natural recalcitrant cellulosic substrate. *mBio* 2011;**2**.
39. Vazana Y, Barak Y, Unger T, Peleg Y, Shamshoum M, Ben-Yehezkel T, et al. A synthetic biology approach for evaluating the functional contribution of designer cellulosome components to deconstruction of cellulosic substrates. *Biotechnol Biofuels* 2013;**6**:182.
40. Najmudin S, Guerreiro CI, Ferreira LM, Romao MJ, Fontes CM, Prates JA. Overexpression, purification and crystallization of the two C-terminal domains of the bifunctional cellulase ctCel9D-Cel44A from *Clostridium thermocellum*. *Acta Crystallogr Sect F, Struct Biol Cryst Commun* 2005;**61**:1043–5.

41. Sakka K, Yoshikawa K, Kojima Y, Karita S, Ohmiya K, Shimada K. Nucleotide sequence of the *Clostridium stercorarium* xylA gene encoding a bifunctional protein with beta-D-xylosidase and alpha-L-arabinofuranosidase activities, and properties of the translated product. *Biosci Biotechnol Biochem* 1993;**57**:268–72.

42. Dam P, Kataeva I, Yang S-J, Zhou F, Yin Y, Chou W, et al. Insights into plant biomass conversion from the genome of the anaerobic thermophilic bacterium *Caldicellulosiruptor bescii* DSM 6725. *Nucleic Acids Res* 2011;**39**:3240–54.

43. Yi Z, Su X, Revindran V, Mackie RI, Cann I. Molecular and biochemical analyses of CbCel9A/Cel48A, a highly secreted multi-modular cellulase by *Caldicellulosiruptor bescii* during growth on crystalline cellulose. *PloS One* 2013;**8**:e84172.

44. Chanzy H, Henrissat B, Vuong R, Schulein M. Undirectional degradation of valonia cellulose microcrystals subjected to cellulase action. *FEBS Lett* 1985;**184**:285–8.

45. Boisset C, Fraschini C, Schulein M, Henrissat B, Chanzy H. Imaging the enzymatic digestion of bacterial cellulose ribbons reveals the endo character of the cellobiohydrolase Cel6A from *Humicola insolens* and its mode of synergy with cellobiohydrolase Cel7A. *Appl Environ Microb* 2000;**66**:1444–52.

46. Berger E, Zhang D, Zverlov VV, Schwarz WH. Two noncellulosomal cellulases of *Clostridium thermocellum*, Cel9I and Cel48Y, hydrolyse crystalline cellulose synergistically. *FEMS Microbiol Lett* 2007;**268**:194–201.

47. Mais U, Esteghlalian AR, Saddler JN, Mansfield SD. Enhancing the enzymatic hydrolysis of cellulosic materials using simultaneous ball milling. *Appl Biochem Biotechnol* 2002;**98**:815–32.

48. Walker LP, Wilson DB. Enzymatic hydrolysis of cellulose: an overview. *Bioresour Technol* 1991;**36**:3–14.

49. Wilson DB. Studies of *Thermobifida fusca* plant cell wall degrading enzymes. *Chem Rec* 2004;**4**:72–82.

50. Irwin DC, Zhang S, Wilson DB. Cloning, expression and characterization of a family 48 exocellulase, Cel48A, from *Thermobifida fusca*. *Eur J Biochem/FEBS* 2000;**267**:4988–97.

51. Taylor 2nd LE, Henrissat B, Coutinho PM, Ekborg NA, Hutcheson SW, Weiner RM. Complete cellulase system in the marine bacterium *Saccharophagus degradans* strain 2-40T. *J Bacteriol* 2006;**188**:3849–61.

52. Eberhardt RY, Gilbert HJ, Hazlewood GP. Primary sequence and enzymic properties of two modular endoglucanases, Cel5A and Cel45A, from the anaerobic fungus *Piromyces equi*. *Microbiology* 2000;**146** (Pt 8):1999–2008.

53. Harhangi HR, Freelove AC, Ubhayasekera W, van Dinther M, Steenbakkers PJ, Akhmanova A, et al. Cel6A, a major exoglucanase from the cellulosome of the anaerobic fungi *Piromyces* sp. E2 and *Piromyces equi*. *Biochim Biophys Acta* 2003;**1628**:30–9.

54. Nagy T, Tunnicliffe RB, Higgins LD, Walters C, Gilbert HJ, Williamson MP. Characterization of a double dockerin from the cellulosome of the anaerobic fungus *Piromyces equi*. *J Mol Biol* 2007;**373**:612–22.

55. Steenbakkers PJ, Harhangi HR, Bosscher MW, van der Hooft MM, Keltjens JT, van der Drift C, et al. Beta-glucosidase in cellulosome of the anaerobic fungus *Piromyces* sp. strain E2 is a family 3 glycoside hydrolase. *Biochem J* 2003;**370**:963–70.

56. Bayer EA, Shimon LJW, Shoham Y, Lamed R. Cellulosomes—structure and ultrastructure. *J Struct Biol* 1998;**124**:221–34.

57. Grabnitz F, Seiss M, Rucknagel KP, Staudenbauer WL. Structure of the beta-glucosidase gene Bgla of *Clostridium-Thermocellum*—sequence-analysis reveals a superfamily of cellulases and beta-glycosidases including human lactase phlorizin hydrolase. *Eur J Biochem* 1991;**200**:301–9.

58. Miller MEB, Antonopoulos DA, Rincon MT, Band M, Bari A, Akraiko T, et al. Diversity and strain specificity of plant cell wall degrading enzymes revealed by the draft genome of *Ruminococcus flavefaciens* FD-1. *PloS One* 2009;**4**.

59. Dam P, Kataeva I, Yang SJ, Zhou F, Yin Y, Chou W, et al. Insights into plant biomass conversion from the genome of the anaerobic thermophilic bacterium *Caldicellulosiruptor bescii* DSM 6725. *Nucleic Acids Res* 2011;**39**:3240–54.

CHAPTER 8

New Paradigms for Engineering Plant Cell Wall Degrading Enzymes

Sarah Moraïs[1], Michael E. Himmel[2], Edward A. Bayer[1]
[1]*Department of Biological Chemistry, The Weizmann Institute of Science, Rehovot, Israel;*
[2]*Biosciences Center, National Renewable Energy Laboratory (NREL), Golden, CO, USA*

Introduction

The dominant paradigms for plant cell wall degradation include free enzyme systems (Figure 1(a)), cellulosomes (Figure 1(b)), multifunctional enzyme systems (Figure 1(c)), and cell-anchored enzyme systems (Figure 2).[1]

Over half of a century ago, Mandels and Reese initiated extensive research on glycoside hydrolases involved in plant cell wall degradation while studying cellulases and their regulation in *Trichoderma viride*.[2,3] Since then, glycoside hydrolases have been classified thus far in more than 130 different families[4] on the basis of their sequence homologies. The glycoside hydrolases are modular enzymes that contain a catalytic module that cleaves the glycoside bond and (frequently) a carbohydrate-binding module (CBM) that targets the enzyme to the polysaccharide substrate (Figure 1(a)).

Cellulosomes were discovered in 1983 by Bayer and Lamed[5,6] in the anaerobic thermophilic bacterium *Clostridium thermocellum*. These enzymatic complexes comprise dockerin-containing enzymes, a "primary scaffoldin" and "anchoring scaffodins." The primary scaffoldin is the backbone of the complex. It is composed of multiple cohesin modules that serve to integrate the dockerin-containing enzymes and a CBM for targeting of the complex to the cellulose substrate. The anchoring scaffoldins bind to the primary scaffoldin via a special type of cohesin/ dockerin interaction in which an X-dockerin modular dyad on the primary scaffoldin attaches to one or more cohesins on corresponding anchoring scaffoldins. Thus, the cellulosomes are attached to the bacterial cell surface via an SLH module (S-layer homology) located on the C-terminus of the various anchoring scaffoldins. In subsequent work, more extensive variability in the organization of cellulosome complexes revealed other bacterial strains. For example, in *Acetivibrio cellulolyticus*, the number of enzymes incorporated into the

Direct Microbial Conversion of Biomass to Advanced Biofuels. http://dx.doi.org/10.1016/B978-0-444-59592-8.00008-7

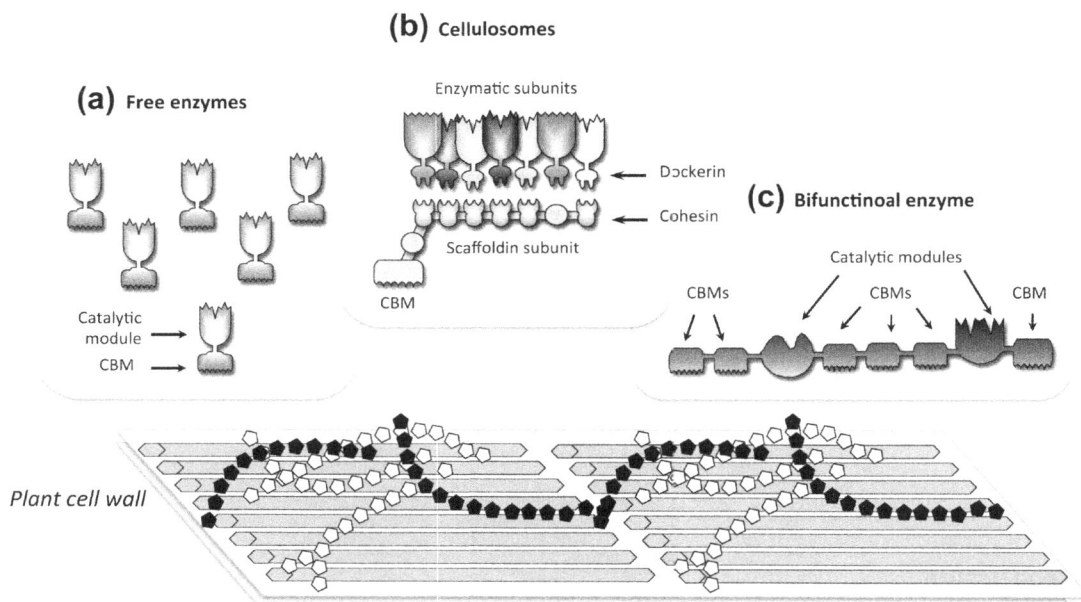

Figure 1
Major enzymatic paradigms for plant cell wall degradation: (a) free enzyme systems,
(b) cellulosomes, and (c) multifunctional enzymes.

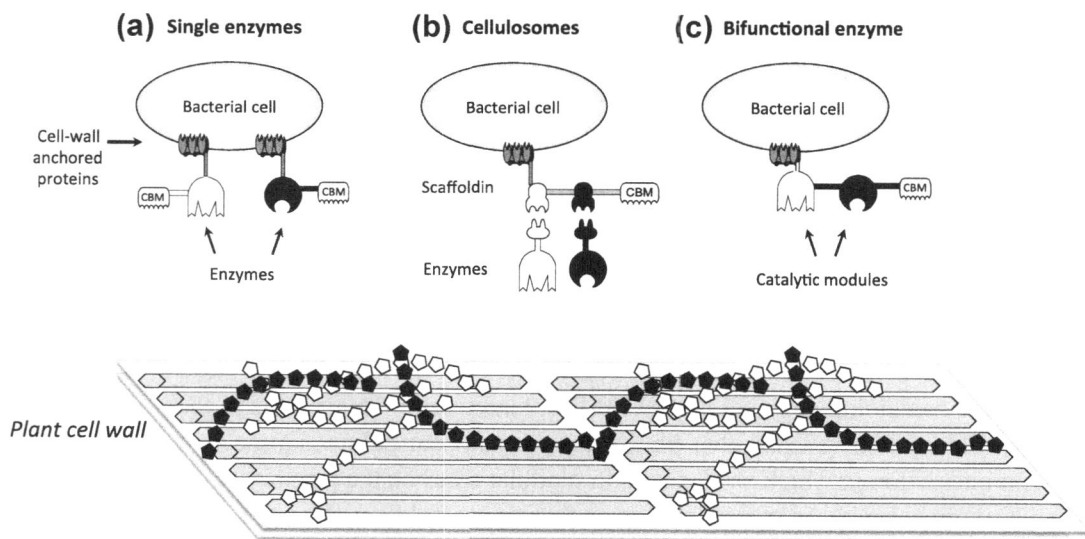

Figure 2
Cell-surface display of enzymes and complexed enzymes: (a) single enzymes, (b) cellulosomes,
and (c) bifunctional enzymes.

cellulosome complex was found to be amplified by the involvement of an "adaptor scaffoldin," which mediates between the primary and anchoring scaffoldins.[7,8] It is clear that the organization of enzymes into a cellulosome concentrates them and perhaps positions them in a suitable orientation with respect to each other and to the cellulosic substrate. Thus, this grouping results in an enzyme proximity effect and a common targeting of the enzymes to the substrate that is believed to render the cellulosome more effective than free enzymes in degrading recalcitrant cellulose substrates.[5,6] Moreover, the fact that the complex is attached to the substrate and the cell results in a minimal diffusion loss of enzymes and hydrolytic products, with the latter also entering the cells directly without inhibiting the enzymes.

A third enzymatic paradigm is that of the multifunctional enzymes. The first evidence for the existence of bifunctional enzymes was published in 1990 by Bergquist and colleagues,[9] in which a bifunctional exo/endoglucanase from the extreme thermophile *Caldocellum saccharolyticum* (*Caldicellulosiruptor saccharolyticus*) was characterized. Although bifunctional enzymes are commonly found in hyperthermophilic bacteria, they have also been discovered in mesophiles.[10] Subsequent descriptions of multifunctional enzymes were published mainly because of genome sequencing.[11] These high-molecular-weight enzymes are composed of several catalytic modules involved in the degradation of plant cell walls (cellulase–cellulase, hemicellulase–hemicellulase, hemicellulase–cellulose, and hemicellulase–carbohydrate esterase systems) together with one or several CBMs.[1,11] Some also contain dockerin modules for incorporation in cellulosome complexes. The presence of several catalytic modules in the same polypeptide chain would seem to indicate that their "enforced proximity" would account for enhanced concerted action on cellulosic substrates.

Single enzymes, cellulosomes, and multifunctional enzymes are either found as "free" systems or anchored to the microbial cell wall via an SLH module or alternative cell-anchoring process, such as the sortase-mediated attachment associated with some rumen bacteria.[12] The proximity of the enzymes to the bacterial cell is believed to minimize diffusion loss of hydrolytic products and allow direct utilization of produced soluble sugars. The cell-anchored enzyme systems, whether in the free, the cellulosomal, or the multifunctional mode together embody a fourth paradigm.

These enzymatic paradigms have been the subject of extensive research and engineering to augment the action of natural systems in the intricate degradation of plant cell walls.[13]

Engineering of Single Enzymes

Since the discovery of glycosides hydrolases and the prospect of potential applications of cellulases in deconstructing cellulosic biomass toward biofuel production, significant improvements have been achieved in cellulase engineering in efficiency and cost reduction.

Thermostable cellulases are particularly attractive because they offer many advantages in the bioconversion process, which include higher specific activity and stability, microbial growth inhibition, increase in mass transfer rate due to lower fluid viscosity, and greater flexibility in the bioprocess.[14] Two decades ago, disulfide bridges were inserted into the enzyme to confer stability, thermoprotection, and prevent denaturation.[15] Since then, directed evolution has successfully led to significant increases in thermostability but also in enzymatic activity or pH stability,[14] although the exact features that confer thermostability to proteins are still unknown. Enzymes with improved thermal stability were also obtained by SCHEMA, a structure-guided approach that produces chimeric proteins by interchanging contiguous blocks of amino acids that include recombination processes.[16,17] Very recently, this method was used to synthesize a diverse set of Cel48 exoglucanase chimaeras on the basis of three native Cel48 enzymes from mesophilic and thermophilic *Clostridia*. As many as 60 active chimaeras were characterized, significantly expanding knowledge on sequence–function relationships within the important GH48 family.[18] The activities of these Cel48 exoglucanase derivatives were established alone and not in the presence of other enzymes, notably endoglucanases. It still remains to be determined whether these exocellulases will work synergistically in the context of multienzyme complexes and whether the resultant cocktails will exhibit enhanced cellulolytic activities.

A recent overview[19] of various methods for engineering cellulases (and glycoside hydrolases) cited rational design (based on the structure/function of the protein[20]), directed evolution (using error-prone polymerase chain reaction[20–22]), and knowledge-based library designs based on multiple sequence alignment (consensus approach[23]). Additional methods including degenerate oligonucleotide gene shuffling were described, which can lead, for example, to enhanced performance of xylanases.[24,25] Some methods can also be combined as the association of rational design, and random mutagenesis served to significantly stabilize a large cellobiose phosphorylase from *C. thermocellum*.[26]

Cellulases, hemicellulases, and other glycoside hydrolases are not only engineered for biomass degradation but also for additional applications[27] (e.g., improved thermal stability for cellulase applications as detergents,[28] improving xylanase activity under alkaline conditions for pulp and paper industry,[29,30] or using β-glucanases to depolymerize cereal glucans in the brewing and animal feedstuff industries[31]).

Cellulosome Engineering
Mini-Cellulosomes

In mini-cellulosomes, a recombinant truncated form of the wild-type scaffoldin is used, which thus contains a smaller and manageable number of cohesins (Figure 3(a)). Therefore, the cohesins exhibit uniform specificity characteristics and recognize the dockerin of the same species, resulting in cellulosome complexes of heterogeneous content. Mini-cellulosomes

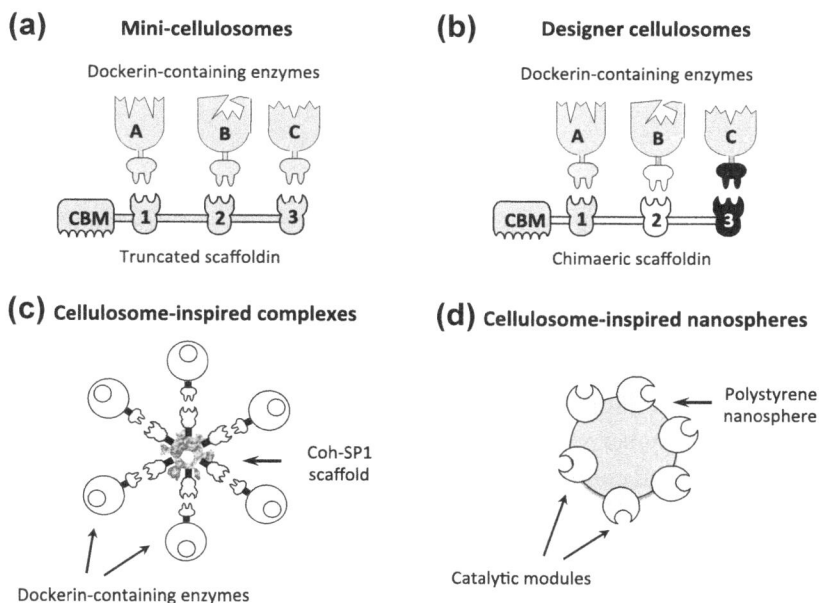

Figure 3
Engineering of cellulosomes: (a) mini-cellulosomes; (b) designer cellulosomes; and (c and d) cellulosome-inspired complexes of Heyman,[114] Morais,[61] and Blanchette.[65]

were first constructed in 2002.[32] In this study, mini-cellulosomes comprised copies of a recombinant cellulosomal endoglucanase and a truncated scaffoldin from *Clostridium cellulovorans*, and synergistic activity on cellulosic substrates was demonstrated. Later on, it was suggested that in *C. cellulovorans*, the cohesin–dockerin interaction might be more selective than originally believed.[33]

In another article, *Bacillus subtilis* was transformed with *C. cellulovorans* enzymes and a truncated scaffoldin, and the resultant strain produced mini-cellulosomes complexes with cellulolytic activity.[34] Increasing enzyme copies in mini-cellulosome complexes enhanced the synergistic effect between the catalytic modules and served to demonstrate the importance of the clustering effect (physical enzyme proximity) of the enzymes within the mini-cellulosome complex for efficient degradation of several plant cell wall substrates.[35] Another research effort provided evidence that *Saccharomyces cerevisiae* cells can be engineered to produce mini-cellulosomes (composed of a chimeric endoglucanase originated from *C. thermocellum* and a *C. cellulovorans* mini-scaffoldin), and the resultant complex degraded cellulosic substrates for ethanol production.[36] Likewise, *Corynebacterium glutamicum* has been transformed with the same mini-cellulosome components for amino-acid production.[37] In these two works, it is unclear why the authors would include a thermophilic enzyme in a mesophilic organis, and the enzymatic activity of the mini-cellulosomes would certainly have been increased by selecting a mesophilic endoglucanase. Another research effort indicated that mixing diverse types of enzymes (i.e., mannanase and endoglucanase) in mini-cellulosomes served to increase the synergistic action of the enzymes.[38]

Recently, the complete in vitro reconstitution of the *C. thermocellum* cellulosome was achieved.[39,40] In an earlier article, the authors confirmed the key role of CipA scaffoldin in cellulosome assembly and efficiency toward crystalline cellulose degradation by a series of mutations in *CipA* gene.[39,40] A *C. thermocellum* mutant was obtained with a completely defective CipA scaffoldin. Nevertheless, the mutant strain produced the traditional set of cellulosomal enzymes in the free form. The latter cellulase system was 15-fold less active in degrading crystalline cellulose than the wild-type bacterium. In the subsequent study,[39,40] the authors simply mixed the culture supernatant of the mutant *C. thermocellum* containing the free dockerin-bearing enzymes with a purified form of the full-length recombinant CipA scaffoldin. The reassociated cellulosome exhibited a 12-fold enhancement as compared with free enzymes on microcrystalline cellulose degradation and was at least 80% as efficient as the native cellulosomes. This study highlighted the essential function of the CipA scaffoldin in assembling an enzymatic complex allowing enhanced enzyme synergy.

Designer Cellulosomes

Designer cellulosomes have been proposed as a tool for understanding the structure–function relationship of cellulosome components and for subsequent biotechnological application in waste management and biofuel production.[41–43] In designer cellulosomes, each chimeric enzyme is appended with a dockerin of divergent specificity that binds specifically to a matching cohesin of a chimeric scaffoldin. Thus, in contrast with mini-cellulosomes, designer cellulosomes allow precise incorporation of the different enzymes into the chimeric scaffoldin, and the composition of designer cellulosomes is homogeneous with respect to the enzyme content and the exact location of the enzymes within the complex (Figure 3(b)). The first demonstrations in the construction and use of artificial cellulosomes were reported in 2001.[44] In this work, divalent designer cellulosomes were assembled with components of *Clostridium cellulolyticum*, and the complex exhibited enhanced degradation of microcrystalline cellulose. Two cohesins originating from different bacterial species and exhibiting divergent specificities were fused into a single polypeptide chain together with a CBM for targeting of the enzymes to the cellulosic substrate, thus forming the chimeric scaffoldin. Chimeric enzymes that contained matching dockerins were constructed in parallel, enabling their precise incorporation into the designer cellulosome complex.

Two factors that serve to enhance deconstruction of recalcitrant cellulosic substrates were defined: the enzyme targeting to the substrate surface via the CBM of the scaffoldin and the physical proximity effect of the enzyme components.[45] In addition, the resulting enhancement in substrate deconstruction was shown to increase with the recalcitrance of the cellulosic substrate.[46] Also, for more complex lignocellulosic substrates (wheat straw), the contribution of a large spectrum of enzymes (from different glycoside hydrolase families) specialized for the different subcomponents of the substrate was demonstrated.

The designer cellulosome approach also enabled fabrication of novel and inventive cellulosome geometries, and their activities on crystalline cellulosic substrates were compared with those of more conventional designer cellulosomes.[47] This study established the negative influence of multiple CBMs in designer-cellulosome complexes in cellulose degradation, thus corroborating the results of a previous study,[45] and further indicated that increased architectural restrictions and elevated levels of rigidity appeared to decrease the activity of the resultant designer cellulosomes. In one case, a family 6 fungus-derived cellulase was included into designer cellulosome modes together with standard cellulosomal enzymes.[48] In this study, the two factors—targeting effect and proximity effect—were observed to occur separately and not in combination. The authors suggested that the origin of the enzymes from the different microbial systems may have been responsible for the apparent antagonism between the proximity and CBM targeting effects and that the benefit of combined effects may occur in designer cellulosomes composed only of bacterial enzymes. In fact, family 6 enzymes have not been observed to be a component of native cellulosomes. It is interesting to note that two family 6 enzymes—an endoglucanase and an exoglucanase—derived from the aerobic bacterium *Thermobifidia fusca*, were incorporated into designer cellulosomes.[49] The endoglucanase performed well in the cellulosome mode, but the family 6 exoglucanase exhibited an "antiproximity" effect and was inappropriate for use as a component in designer cellulosomes.

The designer cellulosome approach was also used to examine the interplay of prominent cellulosomal and noncellulosomal cellulases from *C. thermocellum* on crystalline cellulose.[50] In this case, the targeting effect was found to be the major factor responsible for the synergism among the enzyme combinations whereas the proximity effect appeared to play a negligible role. Thus, designer cellulosome complexes may exhibit both of these effects, either singly or in combination, depending on the characteristics (specific enzymes, composition and organization of scaffoldin, linker regions, etc.) of the individual system and its relationship to the status of the substrate. The phenomena that cause the synergistic effect seem to depend on the characteristics of the specific enzyme combination used to fabricate the designer cellulosome and the properties of the component parts vary with each study.

In 2006, the complete conversion of the free enzyme system of the aerobic thermophile bacterium *T. fusca* into the cellulosomal mode was initiated. This highly cellulolytic bacterium possesses a set of only six cellulases and four xylanases. This finite and manageable panel of enzymes allows the very attractive possibility of converting the entire enzymatic system into the cellulosomal mode, which eliminates the difficulties in selecting enzymes from a highly diverse set for inclusion into designer cellulosomes. At first, the cellulases were engineered into chimeric cellulosomal enzymes by replacing their native CBM with a dockerin of divergent specificity. Several designer cellulosome complexes exhibited enhanced cellulose-degrading activity as compared with the free wild-type enzyme degradation.[49,51–53] The significance of linker length and dockerin position in enzyme design was examined,[51] and it was established that linker length had apparently no influence on the activity of the chimaeras. However, the position of the dockerin in the chimeric enzymes appeared to be an important parameter.

The combined action of cellulases and xylanases together in the same designer cellulosome complex served to enhance the combined synergistic activities of the enzymes toward a natural complex wheat straw substrate.[54] While preparing different classes of designer cellulosomes— those that contain only cellulases, those that contain only hemicellulases, and those of mixed composition—the advantages of using the mixture of enzymes in a single cellulosome for degradation of the wheat straw substrate were demonstrated, suggesting a strong proximity effect among cellulases and xylanases.[55] Thus, the entire xylanolytic system of *T. fusca* was assembled into a defined designer cellulosome complex and its combined saccharolytic activity was compared versus that of the free xylanase system. The data demonstrated enhanced synergistic activities for the xylanolytic designer cellulosomes on the natural recalcitrant wheat straw substrate degradation.[56] In parallel, another article reported the constructions of several divalent designer "xylanosomes" that performed higher xylanolytic activities on arabinoxylan and destarched corn bran when compared with that of the free enzymes.[57]

Very recently, the designer cellulosome technique was pushed to its apparent limit (six different chimeric dockerin-bearing enzymes from *T. fusca* bound at specific locations onto a hybrid scaffoldin).[58] The artificial designer cellulosome complexes obtained were comparable in size with natural cellulosomes. Evidence for proper assembly and stability was provided, and their enzymatic activity on raw substrates (pretreated on not) was compared with those of the free enzyme system and of natural cellulosomes. The action of these designer cellulosomes on untreated wheat straw exhibited a 1.6-fold enhancement toward the combination of wild-type enzymes and was 33–42% as efficient as the natural cellulosomes of *C. thermocellum*. The reduction of substrate complexity by pretreatment of the wheat straw substrate allowed complete conversion of the substrate into soluble saccharides by native cellulosomes. However, the pretreatment removed the advantage of the designer cellulosomes because the free enzymes displayed higher levels of activity, indicating that enzyme proximity between these selected enzymes was less significant on pretreated substrates.

An overview of the methodologies essential for designing and examining cellulosome complexes was published recently.[59]

Cellulosome-Inspired Complexes

In 2007, the assembly of a *T. fusca* enzyme, endoglucanase Cel5A, on a designed ring-shaped scaffold termed Coh-SP1 was studied.[40] For this purpose, the gene of a *C. thermocellum* cohesin was fused in frame to the SP1 protein,[60] and, upon expression of the chimeric gene, 12 Coh-SP1 self-assembled to form a circular scaffold (Figure 3(c)). The Coh-SP1 scaffold successfully allowed the incorporation of an estimated average of 10.5 endoglucanase molecules harboring the matching dockerin. These nanobioreactors were shown to be significantly more efficient for cellulose degradation than equivalent amounts of the free enzyme. Later on, a chimeric exoglucanase, Cel6B, derived also from *T. fusca*, was subjected to interaction with the Coh-SP1 scaffold, and full incorporation of the exoglucanase (12 copies) was observed.[61] The complexation of the exoglucanase on the scaffoldin resulted in a dramatic decrease in enzymatic activity. However, when the complexed exoglucanase was combined with a relatively low concentration of wild-type Cel5A endoglucanase, a marked enhancement of cellulolytic activity over that of the combined free, uncomplexed enzymes was observed. To account for this surprising result, a synergistic mechanism was proposed in which the endoglucanase cleaves internal sites of the cellulose chains, and the new chain ends of the substrate are now readily accessible to the scaffold-borne exoglucanase, thereby resulting in highly effective, synergistic degradation of cellulosic substrates.

In parallel, another scaffold from the hyperthermo-acidophilic archaeon *Sulfolobus shibatae*, comprising 18 cohesin subunits was engineered.[62] This scaffold allowed the incorporation of a mixture of cellulosomal enzymes from *C. thermocellum* and the enzymatic complex was termed "rosettazyme." These complexes exhibited increased cellulose-degrading activity compared with the activity of the free enzymes in solution.

Recently, Kim and colleagues proposed a new design for artificial cellulosome complexes.[63,64] In this effort, biotinylated forms of the catalytic module and CBM are independently clustered to streptavidin and then associated with inorganic nanoparticles. These complexes allowed increasing valence of the CBM, which appeared to be beneficial for degradation of insoluble cellulosic substrates. Very recently, *T. viride* cellulases were conjugated to polystyrene nanospheres (Figure 3(d)). Whereas the complexed enzymes exhibited similar levels of cellulolytic activity on carboxymethyl cellulose, they achieved enhanced levels of degradation on microcrystalline cellulose and natural cellulosic substrates.[65]

Cellulosome-inspired complexes were also grafted onto the cell surface of *Lactococcus lactis*. Truncated CipA scaffoldins of *C. thermocellum* were functionally displayed on the bacterial cell surface and interacted with two reporter enzymes harboring the matching dockerin from *C. thermocellum*.[66] Later on, a chimeric scaffoldin comprising type I and II cohesins from *C. thermocellum* was displayed on the *L. lactis* surface, and dockerin-containing chimeric reporter enzymes, produced in *Escherichia coli*, were assembled ex vivo on the scaffoldin. The sequential binding of the two enzymes suggested that parameters such as protein size and position within the scaffold affect assembly of the designer cellulosome complex.[67]

Multifunctional Enzyme Design

Bifunctional enzymes have been proposed for various biotechnological uses,[68,69] and there are numerous reports in the literature that document attempts to engineer multifunctional enzyme chimaeras.[70]

The first end-to-end fusions of glycoside hydrolases were reported in 1987.[71] The resultant chimeric enzyme was composed of an exoglucanase and an endoglucanase from *Cellulomonas fimi*. Both enzymatic activities were retained, but the binding abilities to microcrystalline cellulose were abolished upon fusion. In 1994, a fusion between a *B. subtilis* xylanase and a *C. fimi* cellulase with an internal CBM was reported that retained parental degrading activities on cellulose and xylan substrates.[72]

End-to-end fusions were also demonstrated to depend on the design of the protein chimaeras. Thus, a xylanase/cellulase fusion using a *C. thermocellum* xylanase and a cellulase from *Pectobacterium chrysanthemi* (*Erwinia chrysanthemi*) resulted in a chimaera able to degrade xylan and cellulose. However, the reverse fusion cellulase/xylanase lost both enzymatic activities.[73] In addition, the effect on increased linker lengths between the two fused modules was shown to be beneficial to the enzymatic activity of the resultant bifunctional enzyme. As in some other research efforts, it is unknown why the authors chose to mix enzymes from thermophilic and mesophilic bacteria, and studies of this nature would be more meaningful if the scientists involved would stay true to the inherent characteristics of the enzymes they choose to study. Likewise, a bifunctional cellulases/xylanase enzyme was constructed from two genes of *Thermotoga maritima*, and the resultant chimaera exhibited enzymatic activity on cellulase and xylan. However, when reversing the order of the catalytic modules in the chimaera, the enzyme lost its activities, probably because of protein misfolding.[74] Another fusion protein was prepared using *T. maritima* genes, a bifunctional cellulase/β-glucosidase, and the same trend concerning the reverse chimaera activity was observed. In addition, lower specific activities were obtained for the chimeric enzymes than for the wild-type enzymes.[75] Fusions between an *Aspergillus niger* xylanase and a *T. maritima* glucanase served to demonstrate that fusion of a large catalytic module at the C-terminus contributes more to enzyme catalytic activity, whereas fusion of a large catalytic module at the N-terminus disturbs substrate binding affinity of the enzyme.[76]

In parallel, another research was published on an end-to-end fusion of a glucanase and a xylanase from a different *Bacillus* species. In this research, the obtained fusion enzyme exhibited increased glucanase activity and decreased xylanase activity as compared with the parental enzymes.[77] Later on, the same research group highlighted the role of proper linker peptides between the two genes of a bifunctional enzyme chimaera. In this context, a linker peptide was inserted in between the two genes (glucanase and xylanase) to reduce protein-folding interference and allow for optimal function of the two enzymes, either independently or in concert, on the substrate, and the chimaera with the extended peptide spacer resulted in enhancement of glucanase and xylanase activities as compared with the wild-type enzymes.[78]

Multifunctional enzyme conjugates (up to three fused catalytic modules) have been reported recently to have increased enzymatic activities and synergy compared with a simple combination of their parental enzymes.[79,80] These authors observed enhanced degradation of natural substrates upon fusing two or three complementary hemicellulases (xylanase, arabinofurano-sidase, and xylosidase) into the same polypeptide chain.

In view of the above works and difficulties in engineering a fully functional and optimized chimaera, later studies engineered several fusions of enzymes in an attempt to obtain at least one fused enzyme with enhanced degrading activities of its catalytic modules. In 2011, four fusion models between a cellulase and a xylanase from a *Paenibacillus* strain isolated from an insect gut were designed, and their predicted three-dimensional structures were analyzed using circular dichroism spectroscopy. The optimal fusion in terms of structural features (i.e., approximating the wild-type enzymes) was cloned and characterized, and the chimaera exhibited increased enzymatic activities for both catalytic modules.[81] The same year, another article reported the construction of six different chimaeras between cellulases and β-glucosidase from *C. thermocellum*. The most active chimaera on amorphous cellulose also demonstrated enhanced thermostability.[82]

In another study, a *B. subtilis* laccase and xylanase were combined in two different bifunctional enzyme constructs. Laccase activity was superior to the enzymatic activity of the parental enzyme whereas xylanase activity was similar to that of the parental enzyme.[83] The chimeric enzymes exhibited strong activity on kraft pulp cellulose and hold potential for biobleaching applications.

A recent publication reports the fusion of a cellulase to a β-glucosidase from a *Paenibacillus* strain in an attempt to relieve feedback inhibition of the cellulase by the enhanced action of the second enzyme, which degrades the inhibitory cellobiose to noninhibitory glucose units.[84] One of the six bifunctional enzymes designed exhibited enhanced degradation on pretreated rice straw and cellulose and achieved significant synergism with a commercial enzyme preparation on pretreated biomass.

In 2012, four enzymes fusions between an exoglucanase and an endoglucanase from *T. fusca* were designed (Figure 4) and characterized.[85] Thus, an inhibitory effect on cellulose degradation was observed when two copies of the family 2 CBM were present on the bifunctional enzyme. In addition, the position of the various modules on the polypeptide chain appeared to be of critical importance to the activity of the enzyme. Nevertheless, the most active bifunctional chimaera achieved exhibited reduced levels of cellulose degradation compared with the combination of the wild-type enzymes or their inclusion into the cellulosomal mode.

Cell Wall-Anchored Paradigms

The three enzymatic paradigms cited above are naturally found either as free enzymatic systems or as cell wall-anchored enzymes. Numerous attempts in transforming bacterial cells, fungi, or yeast with cellulases have been successful (Figure 2(a)). In initial attempts, cellulases have been displayed in *E. coli* and *B. subtilis* by creating cellulase fusion proteins that

Wild-type Enzymes

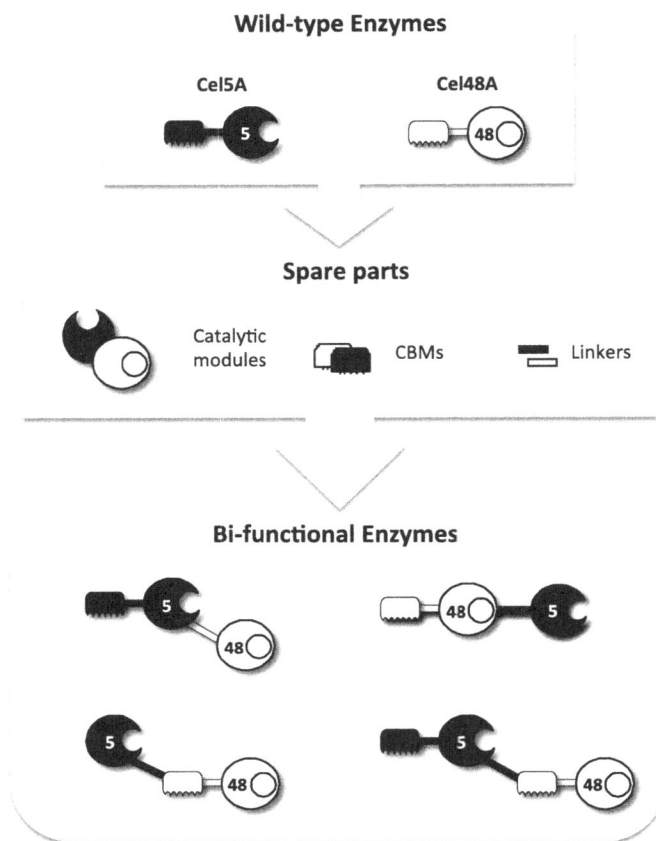

Figure 4
Design of several architectures of bifunctional enzymes based on the gene fusion of an
endoglucanase and an exoglucanase.[85]

are associated with the membrane.[86–88] Another research demonstrated that yeast cells
expressing endo- and exoglucanase from *Trichoderma reseei* and a β-glucosidase from
Aspergillus aculeatus at their cell surface exhibited higher ethanol yields than those that
secreted these enzymes.[89–92] Yeasts cells displaying a *Paenibacillus polymyxa* xylanase have
also been reported.[93] The focus of most researches is to develop a consolidated bioprocessing
(CBP) organism, and bacteria such as *Zymobacter palmae* have been transformed with a
Cellulomonas endoglucanase.[94] Toward this goal, a library of 35 fungal β-glucosidases was
screened for their ability to be both displayed on the *S. cerevisiae* cell surface and functional,
and the most active enzyme was selected to develop a cellobiose-fermenting strain.[95]

Means for attaching cellulosomes to cell surfaces have been considered (Figure 2(b)).
Research groups have created *S. cerevisiae* strains that display designer mini-cellulosomes
using a covalent link to the β-1,6-glucan within the cell wall using a glycosyl

phosphatidylinositol signal motif to display the mini-scaffoldin on the yeast cell surface.[96–98] Assembly of mini-cellulosomes was achieved either by incubating the yeast with purified dockerin-containing cellulases[96,97] or directly in vivo by co-expressing the enzymes.[98] These authors established that it was possible to produce yeast strains that could produce ethanol from cellulose, a step toward the construction of a CBP microorganism that could produce high levels of ethanol directly from biomass. Another research effort demonstrated the efficiency of mini-cellulosomes composed of three complementary hemicellulases displayed on the *S. cerevisiae* surface for the single-step conversion of xylan to ethanol.[99] Very recently, another publication demonstrated that it is possible to use sortase enzymes to attach a mini-cellulosome to the surface of *B. subtilis*.[100] An additional strategy has been recently used to display a mini-scaffoldin on the cell surface of *B. subtilis* using a cell wall-binding module from a cell wall hydrolase of the bacterium.[101] In the two latter studies, the addition of purified dockerin-bearing enzymes to the bacterial cells led to the proper assembly of mini-cellulosomes and enhanced cellulose degradation was demonstrated.

Lately, strategies for augmenting the copies of cell surface-displayed cellulases in *S. cerevisiae* were developed because efficient display of numerous copies of enzymes or complexed enzymes in the form of designer or mini-cellulosome is challenging.[102,103]

To date, no attempt in attaching a bifunctional enzyme to a microorganism cell surface has been reported (Figure 2(c)). Research in this direction could also lead to efficient microorganisms able to deconstruct cellulosic biomass into soluble sugars.

In addition, cellulolytic activity of these engineered strains is still significantly lower than the activity of cellulosome-displaying bacteria; thus, improved methods to efficiently degrade cellulose are required. Producing designer cellulosomes in *C. thermocellum* should be considered because it is currently the most efficient system for degrading cellulosic biomass, and a recent publication demonstrated that its efficiency could be increased in vivo by integration of a β-glucosidase to the complex.[104]

Reflections and Perspectives

Improvement of enzymatic paradigms for plant cell wall degradation has progressed significantly during the last several decades. However, improving the activity profiles of a single enzyme or groups of individual enzymes will not necessarily lead to significant improvements in the overall degradation of plant-derived cellulosic biomass. Indeed, how a modified enzyme works in concert with other enzymes in an enzymatic cocktail to efficiently degrade recalcitrant cellulosic substrates is the real challenge.

In nature, cellulosome-producing bacteria have intricate enzymatic and regulatory systems in which the different paradigms are all represented and even mixed. For example, the premier cellulosome-producing bacteria, *C. thermocellum* and *C. cellulovorans*, are known to also produce free enzymes.[105–109] In addition, some multifunctional enzymes of *C. thermocellum, A. cellulolyticus* and *Ruminococcus flavefaciens*, bear dockerin modules enabling their integration in the respective cellulosome systems.[1,110] Moreover, in addition to cell wall-anchored cellulosomes, *C. thermocellum* displays single enzymes on its cell surface via SLH modules.[111] This implies that the concerted action of the different enzymatic paradigms benefits the overall degradation of plant cell wall polysaccharides by certain cellulolytic bacteria. Research in this direction could provide novel mechanistic insight into the efficient synergistic activity of enzymatic paradigms on the deconstruction of plant cell wall substrates into soluble sugars.

It is clear from the accumulated research efforts of numerous groups over the years that several major phenomena are crucial for efficient degradation of crystalline cellulosic substrates: (1) the selective targeting of the enzymes to the surface of the substrate, (2) packaging of the enzymes into a multienzyme format in which their physical proximity facilitates enhanced synergistic action, and (3) substrate channeling.[112,113] In addition to these major features, additional characteristics of the system can contribute to enhanced cellulolysis. These may include approaches for enriching product formation, removal of toxic substances and inhibitors of enzyme action, cell uptake of the sugar products, etc. For enzymes either integrated into cellulosomes or consolidated into multifunctional chimaeras, the precise contribution of the intermodular linkers and the configuration and tertiary status of the different components vis-à-vis each other are still obscure factors. In the future, novel information regarding the connection among sequence, structure, and activity is still indispensable for a clear understanding of the improved action of cellulosomes and other multienzyme systems on crystalline cellulosic substrates. Such information will be vital for our future capacity to formulate better enzyme cocktails and to improve individual enzymes and/or enzyme systems, such as designer cellulosomes. Knowledge into the fine structure of these enzyme systems and their action on these particularly recalcitrant, but especially desirable, resources is the key to our future success in overcoming the barriers presented by cellulosic biomass en route to cost-effective production of biofuels.

One major consideration is whether processing of cellulosic biomass will be best served by future development of better enzyme cocktails or through the action of improved cellulase- or cellulosome-producing microbes. In using the bacteria and fungi for this purpose, the major advantage is a natural and viable cell-based system that innately replenishes and adjusts itself to the fluctuating nature of its substrate through cell propagation. Nevertheless, their cellulase-and/or cellulosome systems have been honed by evolution to deal with lignocellulosic materials in their natural state, and Nature's wants and needs are not necessarily compatible with human desire. In the deconstruction of complex carbohydrates to monosaccharides as an interim feedstock for biofuel production, we require an even more rapid and heightened degradation of cellulosic feedstocks than Nature can provide. Therefore, we will be compelled to engineer improved microbes by genetic and metabolic engineering approaches, which is one of the major scientific and engineering efforts today. The microbial cell-based approach is certainly attractive in many respects, but the control of bacterial or fungal cultures during massive deconstruction of lignocellulosic materials is a foreboding venture. If co-cultures and mixed cultures are used, then the possible complications are accordingly magnified.

The conventional approach for enhanced processing of cellulosic biomass has been attempted over the past half century or so and involves the production of high-activity enzyme cocktails that contain cellulases, hemicellulases, and other plant cell wall polysaccharide-degrading enzymes that would produce soluble sugars that could serve as substrates for fermentation of ethanol or other types of biofuel. The challenge is to improve the activities and longevity of the numerous enzymes necessary and to improve their production capacity within the framework of ensuring their combined synergistic action. There is no question that this has been an especially daunting challenge throughout the past decades. Employment of native bacteria and fungi for this purpose has fallen far short of the goal. Although high-production mutants have shown great improvement in this direction, they are still (currently) inadequate as a final commercial solution. Conventional engineering efforts have failed to provide a cost-effective process, and there is a sense that the passage from science to engineering was premature. It is still an open question whether the purported enhanced cellulolytic/hemicellulolytic properties of native and designer cellulosomes will serve to resolve this challenge in the future. The fact that cellulosome-producing anaerobes also produce additional free enzymes, which further enhance polysaccharide degradation, suggests that mixed systems may be considered for additional process enhancement.

A second approach for industrial conversion of cellulosic biomass to biofuels is to first use cellulolytic bacteria and fungi for the production of soluble sugars, followed by a subsequent process involving the fermentation of an ethanologenic microbe, such as yeast, or a solventogenic bacterium that could produce, for example, butanol. This approach has the explicit advantage that the strain or strains used, together with their engineered cellulolytic and biofuel-producing features, would be self-propagating. The more contemporary approach is to combine the two processes into a single microbe, commonly termed the CBP approach. Unfortunately, Mother Nature has not deemed it fit to evolve such an organism for large-scale production of ethanol or solvents. Although some microbes can degrade cellulosic materials in a moderately efficient way and produce ethanol as an end product, the amount of ethanol produced falls far short of that required for cost-effective production. The alternative is even more discouraging because microbes that are naturally capable of producing large amounts of ethanol do not utilize lignocellulosics. Therefore, in the CBP approach, we would have to either engineer a cellulolytic microbe to increase ethanol production or provide the appropriate enzyme systems, likely involving dozens of genes encoding for enzymes and ancillary functions, to an ethanologenic microbe. Finally, a CBP microorganism can be engineered from a host that lacks both cellulolytic and ethanologenic/solventogenic features. Although this may appear to present a formidable double barrier, this might depend on the auxiliary features (physical, chemical, and enzymatic) of the microbe chosen as host. If the microorganism is resistant to high levels of the desired biofuel product, particularly amenable to genetic changes, and the tools and other components are readily available, then this approach may indeed prove valid in the future. In any case, the lack of initial cellulose-degrading or ethanol/solvent-producing properties should not be considered grounds for preclusion.

Another alternative exists by which the natural or improved functions of a CBP microbe can be supported by addition of extraneous enzymes for further process enhancement. In the final analysis, consideration of overall production and/or processing schemes will all boil down to whether the final price of enzyme production and/or cell-based processing is cost-effective and competitive for replacement of fossil fuels. In past decades, interest in biomass-to-biofuel conversion from the public, industrial, and governmental sectors has been directly proportional to the price of gasoline. In the relatively short period of time since the recent convergence of environmental and political factors into the equation, much progress has been accomplished in reducing the costs of biofuel production from cellulosic biomass. However, there is still much to do in this context, and the question remains whether the global forces at play will continue to exhibit patience and support until the remaining scientific challenges will be resolved.

Acknowledgments

This work was sponsored by the BioEnergy Science Center (BESC). BESC is a U.S. Department of Energy (DOE) Bioenergy Research Center supported by the Office of Biological and Environmental Research in the DOE Office of Science.

References

1. Himmel M, Xu Q, Luo Y, Ding S, Lamed R, Bayer E. Microbial enzyme systems for biomass conversion: emerging paradigms. *Biofuels* 2010;**1**:323–41.
2. Mandels M, Reese E. Induction of cellulase in *Trichoderma viride* as influenced by carbon source and metals. *J Bacteriol* 1957;**73**:269–78.
3. Mandels M, Reese E. Induction of cellulase in fungi by cellobiose. *J Bacteriol* 1960;**79**:816–26.
4. Cantarel BL, Coutinho PM, Rancurel C, Bernard T, Lombard V, Henrissat B. The Carbohydrate-Active EnZymes database (CAZy): an expert resource for Glycogenomics. *Nucleic Acids Res* 2009;**37**:D233–8.
5. Lamed R, Setter E, Bayer EA. Characterization of a cellulose-binding, cellulase-containing complex in *Clostridium thermocellum*. *J Bacteriol* 1983;**156**:828–36.
6. Lamed R, Setter E, Kenig R, Bayer EA. The cellulosome – a discrete cell surface organelle of *Clostridium thermocellum* which exhibits separate antigenic, cellulose-binding and various cellulolytic activities. *Biotechnol Bioeng Symp* 1983;**13**:163–81.
7. Ding S-Y, Bayer EA, Steiner D, Shoham Y, Lamed R. A novel cellulosomal scaffoldin from *Acetivibrio cellulolyticus* that contains a family-9 glycosyl hydrolase. *J Bacteriol* 1999;**181**:6720–9.
8. Ding S-Y, Lamed R, Bayer EA, Himmel ME. In: Setlow JK, editor. *Genetic engineering: principles and methods*, vol. 25. New York: Kluwer Academic Publishers; 2003. p. 209–26.
9. Saul DJ, Williams LC, Grayling RA, Chamley LW, Love DR, Bergquist PL. celB, a gene coding for a bifunctional cellulase from the extreme thermophile "*Caldocellum saccharolyticum*". *Appl Environ Microbiol* 1990;**56**:3117–24.

10. Flint HJ, Martin J, McPherson CA, Daniel AS, Zhang JX. A bifunctional enzyme, with separate xylanase and beta(1,3-1,4)-glucanase domains, encoded by the xynD gene of *Ruminococcus flavefaciens. J Bacteriol* 1993;**175**:2943–51.

11. Xu Q, Luo Y, Bu L, Ding S-Y, Lamed R, Bayer EA, et al. In: Moo-Young M, Butler M, Webb C, Moreira A, Bai F, editors. *Comprehensive biotechnology*, vol. 3. Amsterdam: Elsevier B.V.; 2011. 2nd ed. p. 15–25.

12. Rincon MT, Cepeljnik T, Martin JC, Lamed R, Barak Y, Bayer EA, et al. Unconventional mode of attachment of the *Ruminococcus flavefaciens* cellulosome to the cell surface. *J Bacteriol* 2005;**187**:7569–78.

13. Elkins JG, Raman B, Keller M. Engineered microbial systems for enhanced conversion of lignocellulosic biomass. *Curr Opin Biotechnol* 2010;**21**:657–62.

14. Yeoman CJ, Han Y, Dodd D, Schroeder CM, Mackie RI, Cann IK. Thermostable enzymes as biocatalysts in the biofuel industry. *Adv Appl Microbiol* 2010;**70**:1–55.

15. Wakarchuk WW, Sung WL, Campbell RL, Cunningham A, Watson DC, Yaguchi M. Thermostabilization of the *Bacillus circulans* xylanase by the introduction of disulfide bonds. *Protein Eng* 1994;**7**:1379–86.

16. Meyer MM, Hochrein L, Arnold FH. Structure-guided SCHEMA recombination of distantly related beta-lactamases. *Protein Eng Des Sel* 2006;**19**:563–70.

17. Heinzelman P, Snow CD, Wu I, Nguyen C, Villalobos A, Govindarajan S, et al. A family of thermostable fungal cellulases created by structure-guided recombination. *Proc Natl Acad Sci USA* 2009;**106**:5610–5.

18. Smith MA, Rentmeister A, Snow CD, Wu T, Farrow MF, Mingardon F, et al. A diverse set of family 48 bacterial glycoside hydrolase cellulases created by structure-guided recombination. *FEBS J* 2012; **279**:4453–65.

19. Anbar M, Bayer EA. Approaches for improving thermostability characteristics in cellulases. *Methods Enzymol* 2012;**510**:261–71.

20. Percival Zhang YH, Himmel ME, Mielenz JR. Outlook for cellulase improvement: screening and selection strategies. *Biotechnol Adv* 2006;**24**:452–81.

21. Anbar M, Lamed R, Bayer E. Thermostability enhancement of *Clostridium thermocellum* cellulosomal endoglucanase Cel8A by a single glycine substitution. *ChemCatChem* 2010;**2**:997–1003.

22. Liu W, Zhang XZ, Zhang Z, Zhang YH. Engineering of *Clostridium phytofermentans* endoglucanase Cel5A for improved thermostability. *Appl Environ Microbiol* 2010;**76**:4914–7.

23. Anbar M, Gul O, Lamed R, Sezerman UO, Bayer EA. Improved thermostability of *Clostridium thermocellum* endoglucanase Cel8A by using consensus-guided mutagenesis. *Appl Environ Microbiol* 2012;**78**:3458–64.

24. Bergquist PL, Reeves RA, Gibbs MD. Degenerate oligonucleotide gene shuffling (DOGS) and random drift mutagenesis (RNDM): two complementary techniques for enzyme evolution. *Biomol Eng* 2005;**22**:63–72.

25. Bergquist PL, Gibbs MD. Degenerate oligonucleotide gene shuffling. *Methods Mol Biol* 2007;**352**:191–204.

26. Ye X, Zhang C, Zhang YH. Engineering a large protein by combined rational and random approaches: stabilizing the *Clostridium thermocellum* cellobiose phosphorylase. *Mol Biosyst* 2012;**8**:1815–23.

27. Cherry JR, Fidantsef AL. Directed evolution of industrial enzymes: an update. *Curr Opin Biotechnol* 2003; **14**:438–43.

28. Sandgren M, Gualfetti PJ, Shaw A, Gross LS, Saldajeno M, Day AG, et al. Comparison of family 12 glycoside hydrolases and recruited substitutions important for thermal stability. *Protein Sci* 2003;**12**:848–60.

29. Gibbs MD, Nevalainen KM, Bergquist PL. Degenerate oligonucleotide gene shuffling (DOGS): a method for enhancing the frequency of recombination with family shuffling. *Gene* 2001;**271**:13–20.

30. Prade RA. Xylanases: from biology to biotechnology. *Biotechnol Genet Eng Rev* 1996;**13**:101–31.

31. Planas A. Bacterial 1,3-1,4-beta-glucanases: structure, function and protein engineering. *Biochim Biophys Acta* 2000;**1543**:361–82.

32. Murashima K, Chen CL, Kosugi A, Tamaru Y, Doi RH, Wong SL. Heterologous production of *Clostridium cellulovorans engB*, using protease-deficient *Bacillus subtilis*, and preparation of active recombinant cellulosomes. *J Bacteriol* 2002;**184**:76–81.

33. Koukiekolo R, Cho HY, Kosugi A, Inui M, Yukawa H, Doi RH. Degradation of corn fiber by *Clostridium cellulovorans* cellulases and hemicellulases and contribution of scaffolding protein CbpA. *Appl Environ Microbiol* 2005;**71**:3504–11.

34. Cho HY, Yukawa H, Inui M, Doi RH, Wong SL. Production of minicellulosomes from *Clostridium cellulovorans* in *Bacillus subtilis* WB800. *Appl Environ Microbiol* 2004;**70**:5704–7.
35. Cha J, Matsuoka S, Chan H, Yukawa H, Inui M, Doi RH. Effect of multiple copies of cohesins on cellulase and hemicellulase activities of *Clostridium cellulovorans* mini-cellulosomes. *J Microbiol Biotechnol* 2007;**17**: 1782–8.
36. Hyeon JE, Yu KO, Suh DJ, Suh YW, Lee SE, Lee J, et al. Production of minicellulosomes from *Clostridium cellulovorans* for the fermentation of cellulosic ethanol using engineered recombinant *Saccharomyces cerevisiae*. *FEMS Microbiol Lett* 2010;**310**:39–47.
37. Hyeon JE, Jeon WJ, Whang SY, Han SO. Production of minicellulosomes for the enhanced hydrolysis of cellulosic substrates by recombinant *Corynebacterium glutamicum*. *Enzyme Microb Technol* 2011;**48**:371–7.
38. Jeon SD, Yu KO, Kim SW, Han SO. A celluloytic complex from *Clostridium cellulovorans* consisting of mannanase B and endoglucanase E has synergistic effects on galactomannan degradation. *Appl Microbiol Biotechnol* 2011;**90**:565–72.
39. Krauss J, Zverlov VV, Schwarz WH. In vitro reconstitution of the complete *Clostridium thermocellum* cellulosome and synergistic activity on crystalline cellulose. *Appl Environ Microbiol* 2012;**78**:4301–7.
40. Zverlov VV, Klupp M, Krauss J, Schwarz WH. Mutations in the scaffoldin gene, cipA, of *Clostridium thermocellum* with impaired cellulosome formation and cellulose hydrolysis: insertions of a new transposable element, IS1447, and implications for cellulase synergism on crystalline cellulose. *J Bacteriol* 2008;**190**:4321–7.
41. Bayer EA, Lamed R. The cellulose paradox: pollutant par excellence and/or a reclaimable natural resource? *Biodegradation* 1992;**3**:171–88.
42. Bayer EA, Lamed R, Himmel ME. The potential of cellulases and cellulosomes for cellulosic waste management. *Curr Opin Biotechnol* 2007;**18**:237–45.
43. Bayer EA, Morag E, Lamed R. The cellulosome – a treasure-trove for biotechnology. *Trends Biotechnol* 1994;**12**:378–86.
44. Fierobe H-P, Mechaly A, Tardif C, Belaich A, Lamed R, Shoham Y, et al. Design and production of active cellulosome chimeras: selective incorporation of dockerin-containing enzymes into defined functional complexes. *J Biol Chem* 2001;**276**:21257–61.
45. Fierobe H-P, Bayer EA, Tardif C, Czjzek M, Mechaly A, Belaich A, et al. Degradation of cellulose substrates by cellulosome chimeras: substrate targeting versus proximity of enzyme components. *J Biol Chem* 2002;**277**:49621–30.
46. Fierobe H-P, Mingardon F, Mechaly A, Belaich A, Rincon MT, Lamed R, et al. Action of designer cellulosomes on homogeneous versus complex substrates: controlled incorporation of three distinct enzymes into a defined tri-functional scaffoldin. *J Biol Chem* 2005;**280**:16325–34.
47. Mingardon F, Chanal A, Tardif C, Bayer EA, Fierobe H-P. Exploration of new geometries in cellulosome-like chimeras. *Appl Environ Microbiol* 2007;**73**:7138–49.
48. Mingardon F, Chanal A, López-Contreras AM, Dray C, Bayer EA, Fierobe H-P. Incorporation of fungal cellulases in bacterial minicellulosomes yields viable, synergistically acting cellulolytic complexes. *Appl Environ Microbiol* 2007;**73**:3822–32.
49. Caspi J, Barak Y, Haimovitz R, Gilary H, Irwin D, Lamed R, et al. *Thermobifida fusca* exoglucanase Cel6B is incompatible with the cellulosomal mode in contrast to endoglucanase Cel6A. *Syst Synth Biol* 2010;**4**:193–201.
50. Vazana Y, Moraïs S, Barak Y, Lamed R, Bayer EA. Interplay between *Clostridium thermocellum* family-48 and family-9 cellulases in the cellulosomal versus non-cellulosomal states. *Appl Environ Microbiol* 2010;**76**:3236–43.
51. Caspi J, Barak Y, Haimovitz R, Irwin D, Lamed R, Wilson DB, et al. Effect of linker length and dockerin position on conversion of a *Thermobifida fusca* endoglucanase to the cellulosomal mode. *Appl Environ Microbiol* 2009;**75**:7335–42.
52. Caspi J, Irwin D, Lamed R, Fierobe H-P, Wilson DB, Bayer EA. Conversion of noncellulosomal *Thermobifida fusca* free exoglucanases into cellulosomal components: comparative impact on cellulose-degrading activity. *J Biotechnol* 2008;**135**:351–7.
53. Caspi J, Irwin D, Lamed R, Shoham Y, Fierobe H-P, Wilson DB, et al. *Thermobifida fusca* family-6 cellulases as potential designer cellulosome components. *Biocatal Biotransformation* 2006;**24**:3–12.

54. Moraïs S, Barak Y, Caspi J, Hadar Y, Lamed R, Shoham Y, et al. Contribution of a xylan-binding module to the degradation of a complex cellulosic substrate by designer cellulosomes. *Appl Environ Microbiol* 2010;**76**:3787–96.

55. Moraïs S, Barak Y, Caspi J, Hadar Y, Lamed R, Shoham Y, et al. Cellulase-xylanase synergy in designer cellulosomes for enhanced degradation of a complex cellulosic substrate. *mBio* 2010;**1**:e00285–00210.

56. Morais S, Barak Y, Hadar Y, Wilson DB, Shoham Y, Lamed R, et al. Assembly of xylanases into designer cellulosomes promotes efficient hydrolysis of the xylan component of a natural recalcitrant cellulosic substrate. *mBio* 2011;**2**:e00233–00211.

57. McClendon SD, Mao Z, Shin HD, Wagschal K, Chen RR. Designer xylanosomes: protein nanostructures for enhanced xylan hydrolysis. *Appl Biochem Biotechnol* 2012;**167**:395–411.

58. Morais S, Morag E, Barak Y, Goldman D, Hadar Y, Lamed R, et al. Deconstruction of lignocellulose into soluble sugars by native and designer cellulosomes. *mBio* 2012;**3**:e00508–00512.

59. Vazana Y, Morais S, Barak Y, Lamed R, Bayer EA. Designer cellulosomes for enhanced hydrolysis of cellulosic substrates. *Methods Enzymol* 2012;**510**:429–52.

60. Wang WX, Pelah D, Alergand T, Shoseyov O, Altman A. Characterization of SP1, a stress-responsive, boiling-soluble, homo-oligomeric protein from aspen. *Plant Physiol* 2002;**130**:865–75.

61. Moraïs S, Heyman A, Barak Y, Caspi J, Wilson DB, Lamed R, et al. Enhanced cellulose degradation by nano-complexed enzymes: synergism between a scaffold-linked exoglucanase and a free endoglucanase. *J Biotechnol* 2010;**147**:205–11.

62. Mitsuzawa S, Kagawa H, Li Y, Chan SL, Paavola CD, Trent JD. The rosettazyme: a synthetic cellulosome. *J Biotechnol* 2009;**143**:139–44.

63. Kim D, Nakazawa H, Umetsu M, Matsuyama T, Ishida N, Ikeuchi A, et al. A nanocluster design for the construction of artificial cellulosomes. *Catal Sci Technol* 2012;**2**:499–503.

64. Kim DM, Umetsu M, Takai K, Matsuyama T, Ishida N, Takahashi H, et al. Enhancement of cellulolytic enzyme activity by clustering cellulose binding domains on nanoscaffolds. *Small* 2011;**7**:656–64.

65. Blanchette C, Lacayo CI, Fischer NO, Hwang M, Thelen MP. Enhanced cellulose degradation using cellulase-nanosphere complexes. *PLoS One* 2012;**7**:e42116.

66. Wieczorek AS, Martin VJ. Engineering the cell surface display of cohesins for assembly of cellulosome-inspired enzyme complexes on *Lactococcus lactis*. *Microb Cell Fact* 2010;**9**:69.

67. Wieczorek AS, Martin VJ. Effects of synthetic cohesin-containing scaffold protein architecture on binding dockerin-enzyme fusions on the surface of *Lactococcus lactis*. *Microb Cell Fact* 2012;**11**:160.

68. Khandeparker R, Numan MT. Bifunctional xylanases and their potential use in biotechnology. *J Indus Microbiol Biotechnol* 2008;**35**:635–44.

69. Nixon AE, Ostermeier M, Benkovic SJ. Hybrid enzymes: manipulating enzyme design. *Trends Biotechnol* 1998;**16**:258–64.

70. Rizk M, Antranikian G, Elleuche S. End-to-end gene fusions and their impact on the production of multifunctional biomass degrading enzymes. *Biochem Biophys Res Commun* 2012;**428**:1–5.

71. Warren RA, Gerhard B, Gilkes NR, Owolabi JB, Kilburn DG, Miller Jr RC. A bifunctional exoglucanase-endoglucanase fusion protein. *Gene* 1987;**61**:421–7.

72. Tomme P, Gilkes NR, Miller Jr RC, Warren AJ, Kilburn DG. An internal cellulose-binding domain mediates adsorption of an engineered bifunctional xylanase/cellulase. *Protein Eng* 1994;**7**:117–23.

73. An JM, Kim YK, Lim WJ, Hong SY, An CL, Shin EC, et al. Evaluation of a novel bifunctional xylanase–cellulase constructed by gene fusion. *Enz Microb Technol* 2005;**36**:989–95.

74. Hong SY, Lee JS, Cho KM, Math RK, Kim YH, Hong SJ, et al. Assembling a novel bifunctional cellulase-xylanase from *Thermotoga maritima* by end-to-end fusion. *Biotechnol Lett* 2006;**28**:1857–62.

75. Hong SY, Lee JS, Cho KM, Math RK, Kim YH, Hong SJ, et al. Construction of the bifunctional enzyme cellulase-beta-glucosidase from the hyperthermophilic bacterium *Thermotoga maritima*. *Biotechnol Lett* 2007;**29**:931–6.

76. Liu L, Wang L, Zhang Z, Guo X, Li X, Chen H. Domain-swapping of mesophilic xylanase with hyper-thermophilic glucanase. *BMC Biotechnol* 2012;**12**:28.

77. Lu P, Feng MG, Li WF, Hu CX. Construction and characterization of a bifunctional fusion enzyme of *Bacillus*-sourced beta-glucanase and xylanase expressed in *Escherichia coli*. *FEMS Microbiol Lett* 2006;**261**:224–30.
78. Lu P, Feng MG. Bifunctional enhancement of a beta-glucanase-xylanase fusion enzyme by optimization of peptide linkers. *Appl Microbiol Biotechnol* 2008;**79**:579–87.
79. Fan ZM, Wagschal K, Chen W, Montross MD, Lee CC, Yuan L. Multimeric hemicellulases facilitate biomass conversion. *Appl Environ Microbiol* 2009;**75**:1754–7.
80. Fan ZM, Wagschal K, Lee CC, Kong Q, Shen KA, Maiti IB, et al. The construction and characterization of two xylan-degrading chimeric enzymes. *Biotechnol Bioeng* 2009;**102**:684–92.
81. Adlakha N, Rajagopal R, Kumar S, Reddy VS, Yazdani SS. Synthesis and characterization of chimeric proteins based on cellulase and xylanase from an insect gut bacterium. *Appl Environ Microbiol* 2011; **77**:4859–66.
82. Lee HL, Chang CK, Teng KH, Liang PH. Construction and characterization of different fusion proteins between cellulases and beta-glucosidase to improve glucose production and thermostability. *Bioresour Technol* 2011;**102**:3973–6.
83. Ribeiro LF, Furtado GP, Lourenzoni MR, Costa-Filho AJ, Santos CR, Nogueira SC, et al. Engineering bifunctional laccase-xylanase chimeras for improved catalytic performance. *J Biol Chem* 2011;**286**:43026–38.
84. Adlakha N, Sawant S, Anil A, Lali A, Yazdani SS. Specific fusion of beta-1,4-endoglucanase and beta-1,4-glucosidase enhances cellulolytic activity and helps in channeling of intermediates. *Appl Environ Microbiol* 2012;**78**:7447–54.
85. Morais S, Barak Y, Lamed R, Wilson DB, Xu Q, Himmel ME, et al. Paradigmatic status of an endo- and exoglucanase and its effect on crystalline cellulose degradation. *Biotechnol Biofuels* 2012;**5**:78.
86. Francisco JA, Stathopoulos C, Warren RAJ, Kilburn DG, Georgiou G. Specific adhesion and hydrolysis of cellulose by intact *Escherichia coli* cells expressing surface-anchored cellulase or cellulose-binding domains. *Biotechnology* 1993;**11**:491–5.
87. Kim JH, Park IS, Kim BG. Development and characterization of membrane surface display system using molecular chaperon, prsA, of *Bacillus subtilis*. *Biochem Biophys Res Commun* 2005;**334**:1248–53.
88. Kim YS, Jung HC, Pan JG. Bacterial cell surface display of an enzyme library for selective screening of improved cellulase variants. *Appl Environ Microbiol* 2000;**66**:788–93.
89. Fujita Y, Ito J, Ueda M, Fukuda H, Kondo A. Synergistic saccharification, and direct fermentation to ethanol, of amorphous cellulose by use of an engineered yeast strain codisplaying three types of cellulolytic enzyme. *Appl Environ Microbiol* 2004;**70**:1207–12.
90. Yanase S, Yamada R, Kaneko S, Noda H, Hasunuma T, Tanaka T, et al. Ethanol production from cellulosic materials using cellulase-expressing yeast. *Biotechnol J* 2010;**5**:449–55.
91. Matano Y, Hasunuma T, Kondo A. Display of cellulases on the cell surface of *Saccharomyces cerevisiae* for high yield ethanol production from high-solid lignocellulosic biomass. *Bioresour Technol* 2012;**108**:128–33.
92. Matano Y, Hasunuma T, Kondo A. Cell recycle batch fermentation of high-solid lignocellulose using a recombinant cellulase-displaying yeast strain for high yield ethanol production in consolidated bioprocessing. *Bioresour Technol* 2012;**135**:403–9.
93. Yeasmin S, Kim CH, Park HJ, Sheikh MI, Lee JY, Kim JW, et al. Cell surface display of cellulase activity-free xylanase enzyme on *Saccharomyces cerevisiae* EBY100. *Appl Biochem Biotechnol* 2011;**164**:294–304.
94. Kojima M, Akahoshi T, Okamoto K, Yanase H. Expression and surface display of *Cellulomonas* endoglucanase in the ethanologenic bacterium *Zymobacter palmae*. *Appl Microbiol Biotechnol* 2012;**96**:1093–104.
95. Wilde C, Gold ND, Bawa N, Tambor JH, Mougharbel L, Storms R, et al. Expression of a library of fungal beta-glucosidases in *Saccharomyces cerevisiae* for the development of a biomass fermenting strain. *Appl Microbiol Biotechnol* 2012;**95**:647–59.
96. Lilly M, Fierobe HP, van Zyl WH, Volschenk H. Heterologous expression of a *Clostridium* minicellulosome in *Saccharomyces cerevisiae*. *FEMS Yeast Res* 2009;**9**:1236–49.
97. Tsai SL, Oh J, Singh S, Chen R, Chen W. Functional assembly of minicellulosomes on the *Saccharomyces cerevisiae* cell surface for cellulose hydrolysis and ethanol production. *Appl Environ Microbiol* 2009;**75**:6087–93.

98. Wen F, Sun J, Zhao H. Yeast surface display of trifunctional minicellulosomes for simultaneous saccharification and fermentation of cellulose to ethanol. *Appl Environ Microbiol* 2010;**76**:1251–60.

99. Sun J, Wen F, Si T, Xu JH, Zhao H. Direct conversion of xylan to ethanol by recombinant *Saccharomyces cerevisiae* strains displaying an engineered minihemicellulosome. *Appl Environ Microbiol* 2012;**78**:3837–45.

100. Anderson TD, Robson SA, Jiang XW, Malmirchegini GR, Fierobe HP, Lazazzera BA, et al. Assembly of minicellulosomes on the surface of *Bacillus subtilis*. *Appl Environ Microbiol* 2011;**77**:4849–58.

101. You C, Zhang XZ, Sathitsuksanoh N, Lynd LR, Zhang YH. Enhanced microbial utilization of recalcitrant cellulose by an ex vivo cellulosome-microbe complex. *Appl Environ Microbiol* 2012;**78**:1437–44.

102. Yang J, Dang H, Lu JR. Improving genetic immobilization of a cellulase on yeast cell surface for bioethanol production using cellulose. *J Basic Microbiol* 2012;**53**(4):381–9.

103. Han Z, Zhang B, Wang YE, Zuo YY, Su WW. Self-assembled amyloid-like oligomeric-cohesin Scaffoldin for augmented protein display on the *Saccharomyces cerevisiae* cell surface. *Appl Environ Microbiol* 2012;**78**:3249–55.

104. Gefen G, Anbar M, Morag E, Lamed R, Bayer EA. Enhanced cellulose degradation by targeted integration of a cohesin-fused beta-glucosidase into the *Clostridium thermocellum* cellulosome. *Proc Natl Acad Sci USA* 2012;**109**:10298–303.

105. Doi RH, Park JS, Liu CC, Malburg LM, Tamaru Y, Ichiishi A, et al. Cellulosome and noncellulosomal cellulases of *Clostridium cellulovorans*. *Extremophiles* 1998;**2**:53–60.

106. Kosugi A, Murashima K, Doi RH. Characterization of two noncellulosomal subunits, ArfA and BgaA, from *Clostridium cellulovorans* that cooperate with the cellulosome in plant cell wall degradation. *J Bacteriol* 2002;**184**:6859–65.

107. Han SO, Cho HY, Yukawa H, Inui M, Doi RH. Regulation of expression of cellulosomes and noncellulosomal (hemi)cellulolytic enzymes in *Clostridium cellulovorans* during growth on different carbon sources. *J Bacteriol* 2004;**186**:4218–27.

108. Berger E, Zhang D, Zverlov VV, Schwarz WH. Two noncellulosomal cellulases of *Clostridium thermocellum*, Cel9I and Cel48Y, hydrolyse crystalline cellulose synergistically. *FEMS Microbiol Lett* 2007;**268**:194–201.

109. Tamaru Y, Miyake H, Kuroda K, Nakanishi A, Matsushima C, Doi RH, et al. Comparison of the mesophilic cellulosome-producing *Clostridium cellulovorans* genome with other cellulosome-related clostridial genomes. *Microb Biotechnol* 2011;**4**:64–73.

110. Bharali S, Purama RK, Majumder A, Fontes CM, Goyal A. Functional characterization and mutation analysis of family 11, Carbohydrate-Binding Module (CtCBM11) of cellulosomal bifunctional cellulase from *Clostridium thermocellum*. *Indian J Microbiol* 2007;**47**:109–18.

111. Fuchs KP, Zverlov VV, Velikodvorskaya GA, Lottspeich F, Schwarz WH. Lic16A of *Clostridium thermocellum*, a non-cellulosomal, highly complex endo-beta-1,3-glucanase bound to the outer cell surface. *Microbiology* 2003;**149**:1021–31.

112. Zhang Y. Substrate channeling and enzyme complexes for biotechnological applications. *Biotechnol Adv* 2011;**29**:715–25.

113. You C, Myung S, Zhang Y. Facilitated substrate channeling in a self-assembled trifunctional enzyme complex. *Angew Chem Int Ed Engl* 2012;**51**:8787–90.

114. Heyman A, Barak Y, Caspi J, Wilson DB, Bayer EA, Shoseyov O. Multiple display of catalytic domain modules on a protein scaffold, nano fabrication of enzyme particles. *J Biotechnol* 2007;**131**:433–9.

PART 3

Fuels from Fungi and Yeast

CHAPTER 9

Expression of Fungal Hydrolases in Saccharomyces cerevisiae

W.H. van Zyl[1], R. den Haan[2], S.H. Rose[1], D.C. la Grange[3]
[1]Department of Microbiology, University of Stellenbosch, Stellenbosch, South Africa; [2]Department of Biotechnology, University of the Western Cape, Bellville, South Africa; [3]Department of Biochemistry, Microbiology and Biotechnology, University of Limpopo, Sovenga, South Africa

Introduction

Lignocellulosic biomass is an abundant, renewable feedstock that can be used for the sustainable production of biofuels and commodity chemicals if an economically viable technology can be developed to overcome its recalcitrance. Conversion of biomass to ethanol or other commodities via a biological route is initiated with a pretreatment process to render the polymeric fractions more accessible to enzymatic hydrolysis.[1] In the case of ethanol production, four biologically mediated events then convert pretreated lignocellulose to ethanol: (1) production of depolymerizing enzymes; (2) hydrolysis of the polysaccharide components of biomass; (3) fermentation of the resulting hexose; and (4) pentose sugars present in the broth. Improvements on this process flow generally involve merging two or more of these steps. Polysaccharide hydrolysis and sugar fermentation steps are combined in simultaneous saccharification and fermentation (SSF) of hexoses or simultaneous saccharification and co-fermentation (SSCF) of both hexoses and pentoses, assuming a suitable fermentative organism is available.[2,3] Fermentation of the sugars produced by the action of the hydrolytic enzymes in SSF or SSCF avoids the feedback inhibition effect of these sugars on the enzymes. However, when a mesophilic process organism is used in SSF or SSCF processes, it requires lowering the operating temperature to a level that is suboptimal for enzymatic activity. The final goal would be one-step consolidated bioprocessing (CBP), requiring direct microbial conversion of pretreated lignocellulose to bioethanol in a single reactor without the requirement for addition of exogenous enzymes. This would signify a breakthrough for low-cost biomass processing owing to the economic benefits of process integration and avoiding the high costs of enzymes that make the biological conversion route unattractive.[4–7] The costs of current biomass conversion technologies therefore would be significantly reduced by organisms that simultaneously hydrolyze the cellulose and hemicellulose in biomass and produce a commodity product such as ethanol at a high rate and titer.[3]

Direct Microbial Conversion of Biomass to Advanced Biofuels. http://dx.doi.org/10.1016/B978-0-444-59592-8.00009-9

The yeast *Saccharomyces cerevisiae* has long been used for the production of ethanol at industrial scale.[8,9] Characteristics that make it suitable for industrial ethanol production include a high rate of ethanol production from glucose (3.3 g/L/h), high ethanol tolerance, and a generally regarded as safe (GRAS) status. It is also relatively easy to manipulate the *S. cerevisiae* genome. However, this yeast species has a number of shortcomings in terms of being used for direct microbial conversion of biomass, such as its inability to hydrolyze cellulose and hemicellulose or use the pentose sugars available in lignocellulosic biomass. A number of research groups have been working on widening the substrate range of *S. cerevisiae* through genetic engineering to include monomeric forms of the sugars contained in plant biomass, including xylose,[5,8] arabinose,[10] and cellobiose.[11] Over the past three decades, several researchers have attempted the expression of genes encoding lignocellulolytic hydrolases in *S. cerevisiae*. Some have sought to produce these enzymes in an organism that would not yield interfering activities so as to gain insight into its mechanism,[12] whereas others have sought to enable the yeast to hydrolyze nonnative polymeric substrates.[13] This chapter will highlight some of the major successes and struggles in attempts to express fungal hydrolases in the yeast *S. cerevisiae*.

Cellulose and Hemicellulose Structure and Hydrolysis

Lignocellulosic plant biomass consists of 40%–55% cellulose, 25%–50% hemicellulose, and 10%–40% lignin, depending on the source.[14] The main polysaccharide is crystalline cellulose consisting of β-1,4 linked glucose residues, which also represents the major fraction of fermentable sugars. Enzymatic hydrolysis of crystalline cellulose requires synergistic interaction of three major types of enzymatic activity, collectively referred to as cellulases: (1) endo-glucanases (EGs) (EC 3.2.1.4) that act on amorphous regions of cellulose releasing cellodextrins and providing free chain ends; (2) exoglucanases, including cellodextrinases and cellobiohydrolases (CBHs) (EC 3.2.1.91) that act on the crystalline part of cellulose in a processive manner from free chain ends and release mainly cellobiose; and (3) β-glucosidases (BGLs) (EC 3.2.1.21) that convert cellobiose and small cello-oligosaccharides to glucose[15] (Figure 1).

Although cellulose is the major polymer in plant material, hemicelluloses can make up as much as 35% of the dry weight in certain species.[16,17] In 1891, Schulze called the easily extractable, low-molecular-weight polysaccharides that occur together with cellulose and lignin in plant tissues hemicellulose.[18] Unlike cellulose, hemicellulose is far more complex, and it does not have a homogeneous chemical composition. Hemicelluloses are heteroglucans, and the name given to a particular hemicellulose molecule depends on the monomers that predominate its backbone. Thus, the backbone of xylan, mannan, galactan, and arabinan consist of mainly D-xylose, D-mannose, D-galactose, or L-arabinose residues, respectively. After glucose, xylose is the most abundant sugar present in plant material.

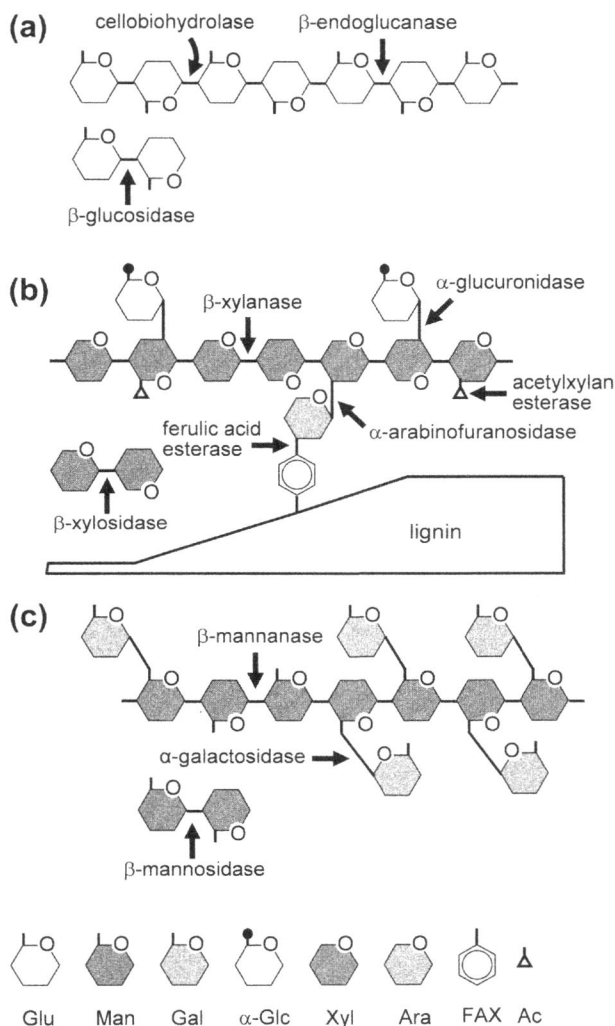

Figure 1
Diagram illustrating the complexity of cellulose and hemicellulose structure and the enzymes required for their hydrolysis. Cellulose (a) and hemicellulose structures for arabinoxylan (b) and galactomannans (c) depicting the different side-chains present on each. Hexoses are distinguished from pentoses by the presence of a protruding line from the cyclic hexagon (pyranose ring), depicting the CH_2OH group. Hydrolase enzymes and the bonds targeted for cleavage in the four polysaccharide structures are indicated by arrows[38].

The main chain of xylan is analogous to that of cellulose, but is composed of xylose monomers linked with β-1,4 bonds.[19] The xylan backbone can be substituted with other saccharide units without substantially changing the basic backbone conformation.[20] The solubility of xylan is directly proportional to the number of substituents. The frequency and composition of substituents in xylan differ, depending on the plant origin. The xylopyranose units of the xylan main chain can be substituted at the C-2 and/or C-3 positions. Acetyl, arabinosyl, and glucuronosyl residues are the most common substituents found. In species that contain arabinose side chains, they may be esterified with *p*-coumaric acid or ferulic acid. It is usually these ferulic acid substituents that engage in covalent cross-linking of xylan molecules with lignin or with other xylan molecules.

The xylopyranose units of xylan from hardwood are substituted at the C-2 and/or C-3 position with acetic acid and with 4-*O*-methyl-α-D-glucuronic acid at the C-2 position.[18,21] The acetyl content in the cell walls of higher plants can constitute up to 2% of the plant's dry mass. Softwood xylans are substituted with 4-*O*-methyl-α-D-glucuronic acid and L-arabinose. The D-xylosyl residues in the xylan main chain can be substituted at the C-2 position with 4-*O*-methyl-α-D-glucuronic acid and with arabinose at the C-3 position. The xylans from grasses have a large content of arabinofuranosyl side-chains linked to C-2 and/or C-3 of the β-D-xylanopyranose residue in the main chain. In addition, such xylans contain 2–5% by weight of *O*-acetyl groups linked to C-2 or C-3 of the xylopyranose units. Cell walls of grasses also contain 1–2% phenolic acid (*p*-coumaric acid and ferulic acid) substituents attached to position five of the arabinose side-chains.[18,22]

Because the structure of xylan is variable, a more complex assembly of enzymes is required than for cellulose hydrolysis.[21,23] Degradation of the β-1,4-xylan backbone requires the action of endo-β-1,4-xylanases (EC 3.2.1.8) and β-xylosidases (EC 3.2.1.37); the former are generally considered to be those enzymes that hydrolyze the xylan backbone, whereas the latter are those that hydrolyze xylo-oligomers produced through the action of β-xylanases. To achieve complete degradation of complex substituted xylans, a series of accessory debranching enzymes, namely α-D-glucuronidases (EC 3.2.1.139), α-L-arabinofuranosidases (EC 3.2.1.55), acetyl xylan esterases (EC 3.1.1.72) and feruloyl esterases, are also needed (Figure 1).

For cost-effective conversion of lignocellulose to biocommodities, the hydrolysis of all polysaccharides, including mannan, is of interest. Glucomannan typically contributes ~5% of the dry weight in plants; however, galactoglucomannans can make up as much as 25% of the dry weight in softwood species.[24] The mannose content in spruce and pine is higher than the xylose content, implying that mannanases are as important as xylanases in terms of bioethanol production from these species. Mannan hydrolysis and mannose fermentation has been largely overlooked. Low molecular mass (>30 kDa) mannans act as structural polysaccharides that cross-link cellulose microfibrils, whereas high molecular mass mannans are used as storage carbohydrates in seeds.[25] The backbone of linear mannans consist of β-1,4 linked mannose units, which can be interrupted by D-glucose units (glucomannan). Branched mannans contain 1,6-linked D-galactose, resulting in galactomannan and galactoglucomannan. The mannan and glucose residues can be acetylated at the C-2 or C-3 positions. A low degree of α-1,6-linked D-galactose substitution results in densely packed granular and crystalline structures or microfibrils, similar to cellulose microfibrils.[26]

Microbial hydrolysis of mannan is initiated by the action of the endo-β-1,4-mannanases (EC 3.2.1.78), which cleave the mannoglycosidic linkages in the mannan backbone of glucomannan, galactomannan, and galactoglucomannan.[27] Degradation is affected by the type and extent of substitution of the backbone. In general, the main end-products of mannan hydrolysis by β-mannanases are mannobiose and mannotriose. In rare cases, β-mannanases have been reported to produce limited amounts of mannose in addition to mannobiose and mannotriose.[28] The β-mannosidases (EC 3.2.1.25) cleave the terminal linkage at the nonreducing end of the oligosaccharides (with a degree of polymerization of up to six), releasing mannose. Only a few β-mannosidases are able to release mannose from the mannan polymer.[24] Glucopyranose units at the nonreducing end of the oligosaccharides, derived from the hydrolysis of glucomannan and galactoglucomannan are released by β-glucosidases. Side-chain substitutions require the action of an α-galactosidase (EC 3.2.1.22) and an acetyl mannan esterase for the removal of the galactopyranosyl and acetyl groups, respectively.

Expression of Fungal Cellulases in Saccharomyces cerevisiae

Successful expression of fungal cellulases in yeasts dates back to the late 1980s when reports of the expression of EGs and CBHs in *S. cerevisiae* first appeared.[12,29] It was soon surmised that due to protein processing problems such as aberrant folding and hyperglycosylation and the relatively low titer production of the heterologous cellulases, the usefulness of producing these proteins in yeast both as a tool to allow their characterization or to broaden the yeast's substrate range would be limited. However, in the interim several fungal cellulases have been expressed at relatively high levels (Table 1), allowing the yeast to assimilate cellulosic substrates with the concomitant production of ethanol.[30]

Wild type *S. cerevisiae* is unable to grow on the disaccharide cellobiose, the soluble main product of the action of cellobiohydrolases. However, the successful heterologous expression and secretion of a *Saccharomycopsis fibuligera* β-glucosidase (BGL) enabled a recombinant strain to grow on and ferment cellobiose at roughly the same rate as on glucose in anaerobic conditions.[11] The heterologous production and secretion of BLGs from several other sources in yeast also allowed the strains to grow on and ferment cellobiose.[31] The secreted β-glucosidase hydrolyzes cellobiose outside of the cell and transport as well as catabolism of the resultant glucose products can then occur as per usual for *S. cerevisiae*. Immobilization of the secreted β-glucosidase onto the cell wall may hold the advantage that released glucose molecules are in the immediate vicinity of the cell wall and available for immediate uptake.[32,33]

As an alternative, intracellular cellobiose utilization was also engineered. The *Neurospora crassa* high affinity cellodextrin transport system was reconstituted into *S. cerevisiae*.[34] This enabled a recombinant strain also producing an intracellular β-glucosidase to grow on cellodextrins up to cellotetraose. A xylose fermenting strain was subsequently engineered to also produce the high affinity cellodextrin transporter and intracellular β-glucosidase and used to

Table 1: Various fungal cellulose hydrolyzing enzymes heterologously expressed in *S. cerevisiae*

Organism	Gene	GH family	Promoter	Extracellular activity	References
β-Glucosidase					
Aspergillus aculeatus	*bgl1*	3	*GAPDH*	21.3 U/g DCW (pNPG)	54
Saccharomycopsis fibuligera	*bgl1*	3	*PGK1*	800 U/g DCW (pNPG)	52
Endomyces fibuliger	*bgl1*	3	Native	2023 U/g DCW (C2)	106
Candida wickerhamii	*bglB*	3	*ADH1*	0.298 U/L (pNPG)	107
Phanerochaete chrysosporium	*bglB*	1	*ENO1*	420 U/L (pNPG)	31
Endoglucanase					
Trichoderma reesei	*eg1*	7	*PGK1*	72 U/g DCW (HEC)	108
			ENO1	450 U/g DCW (CMC)	52
	eg2	5	*GAPDH*	3.64 U/g DCW (AC)	54
Aspergillus niger	*eng1*	5	*GAPDH*	574 U/L (CMC)	109
Aspergillus aculeatus	*eg3*	12	*GAP*	60 U/L (CMC)	110
Cryptococcus flavus	*cmc1*	5	*GAP*	12 500 U/L (CMC)	111
Cellobiohydrolase 1					
T. reesei	*cbh1*	7	*ENO1*	0.6 U/L (MUL) 3% (Avicel)	40
A. niger	*cbhB*	7	*ENO1*	0.035 U/L (AC)	41
P. chrysosporium	*cbh1-4*	7	*PGK1*	0.035 U/L (AC)	112
H. grisea	*cbh1*	7	*ENO1*	3.3 U/L (MUL), 9% (Avicel)	40
T. emersoni	*cbh1*	7	*ENO1*	145 U/L (MUL), 7% (Avicel)	40
T. emersonii & T. reesei	*cbh1-CCBM*	7	*ENO1*	84 U/L (MUL), 11% (Avicel)	40
Cellobiohydrolase 2					
T. reesei	*cbh2*	6	*ENO1*	0.14 U/L (AC)	41
			PGK1	6% (Avicel)	40
A. bisporus	*cel3*	6	*TPI*	0.06 U/g DCW (AC)	113
C. heterostrophus	*cbh2*	6	*PGK1*	6% (Avicel)	40
C. lucknowense	*cbh2b*	6	*PGK1*	9% (Avicel)	40

U = micromole substrate released/min, pNPG = *p*-nitrophenol glucopyranoside, C2 = cellobiose, HEC = hydroxyethyl cellulose, CMC = carboxymethyl cellulose, AC = amorphous cellulose, MUL = methylumbelliferyl lactoside, % Avicel refers to the amount of Avicel hydrolysed in the assay in 48 h. *GAPDH* = glyceraldehyde 3-phosphate dehydrogenase, *PGK1* = phosphoglycerate kinase, *ADH1* = alcohol dehydrogenase, *ENO1* = enolase, *GAP* = glyceraldehyde 3-phosphate dehydrogenase, *TPI* = triose phosphate isomerase.

co-ferment mixtures of xylose and cellobiose.[35] It was demonstrated that intracellular hydro-lysis of cellobiose minimized glucose repression of xylose fermentation, allowing co-con-sumption of cellobiose and xylose and improving ethanol yields. This was partly due to avoiding the competition between xylose and glucose for transport into the cell. Sadie et al.[36] showed that expression of the gene encoding lactose permease of *Kluyveromyces lactis* (*lac12*) also facilitated transport of cellobiose into a recombinant *S. cerevisiae* strain and that combined expression with a *Clostridium stercorarium* cellobiose phosphorylase (*cepA*) allowed phosphorolysis of cellobiose enabling growth on cellobiose as sole carbohydrate source.

Expression of fungal endoglucanase (EG) encoding genes representing various CAZy Glyco-side Hydrolase (GH) families (http://www.cazy.org) in *S. cerevisiae* was initially more successful than CBH production. Relatively high percentages of total cell protein were reported for some EGs.[29,37,38] High activity of recombinant EGs on synthetic and amorphous cellulose substrates such as phosphoric acid swollen cellulose (PASC) and carboxymethyl cellulose (CMC) were also reported. This was not a surprising result considering the often two to three order of magnitude higher specific activity of EG enzymes on the substrates they are measured on relative to CBHs and their substrates. While recombinant EGs are often hyperglycosylated, several have been produced with enzyme activity and stability properties comparable to those of the native proteins. Sufficiency analysis showed that EG and BGL expression in *S. cerevisiae* will not be a limiting step in cellulase system reconstruction.[38]

CBHs have been successfully produced and secreted by *S. cerevisiae* and other yeasts and were tested for their activity on a variety of substrates ranging from small synthetic molecules such as *p*-nitrophenyl-β-D-cellobioside (*p*NPC) and *p*-methylumbelliferyl-β-D-lactoside (MULac) to amorphous and crystalline forms of cellulose.[39] A general feature among most reports of heterologous CBH production in *S. cerevisiae* is that a relatively low titer of secreted CBH was found, although the range of reported values is large — 0.002 to >1% of total cell protein (tcp).[38,40] Low protein titer coupled with the low specific activity of CBHs on polymeric substrates has led to the identification of CBH expression as a restrictive step for developing *S. cerevisiae* for direct microbial conversion of biomass to commodity prod-ucts.[3] However, the amount of CBH1 required to enable growth on crystalline cellulose was calculated to be within the capacity of heterologous protein production in *S. cerevisiae* in terms of total cellular protein, *i.e.*, 1–10% of the tcp.[41–43] Recently, the expression of rela-tively high levels of CBHs in *S. cerevisiae* was reported.[40,44] CBH production levels of up to 4% tcp were reported, which met the calculated levels sufficient for growth on cellulose at the rate required for an industrial process.[30]

CBHs produced in *S. cerevisiae* consistently displayed high and variable levels of glycosyl-ation.[40,41,45] CBH1 enzymes originating from *Thermoascus aurantiacus*, *Talaromyces emer-sonii*, *Neosartorya fischeri*, and *Trichoderma reesei* and CBH2s originating from *Chrysosporium lucknowense*, *Acremonium cellulolyticus*, and *T. reesei* all showed significant levels of mainly *N*-attached hyperglycosylation. Generally, it was found that the fundamental attributes of the enzymes, such as their activity on crystalline cellulose substrates and, for example, the thermostability of the *T. aurantiacus* enzyme, remained unchanged. Some authors reported decreased specific activity for certain heterologous CBHs on polymeric substrates, presumably as a result of hyperglycosylation, although this was not always the case.[12,41,46,47] The *N*-glycans added by *S. cerevisiae* to recombinant *T. emersonii* CBH1 seemed to improve the stability of the enzyme and the activity on crystalline cellulose (at 70 °C) to some extent; however, its ability to bind Avicel decreased.

More than 20 CBH encoding genes of fungal origin were expressed in the studies of Ilmén et al.[40] and Heinzelman et al.[48] Remarkable variation was found in the secreted protein levels and activities measured between the different recombinant strains. This is noteworthy, given the fact that the overall sequence homology of the GH7 proteins and the GH6 proteins produced were both more than 60% and that the genes were expressed under the control of identical promoters on identical episomal plasmids. For example, activities of the CBH1-expressing strains on the soluble substrate MULac, ranged over three orders of magnitude. The highest secreted activity was obtained from a strain producing the *T. emersonii* CBH1, an enzyme that lacks a carbohydrate binding module (CBM).[40] Addition of a *C*- or *N*-terminal CBM, originating from various fungal CBH1 sources, to the *T. emersonii* CBH1 increased the specific activity of this enzyme on crystalline cellulose. However, the production level was reduced relative to that of the enzyme with no CBM, suggesting that the presence of a CBM increased the complexity of CBH production in yeast.

In some cases, such as for recombinant *T. emersonii* and *Acremonium thermophilum* CBH1 enzymes, the estimations of the protein concentrations based on total protein and the estimation of the protein concentration of active CBH1 based on its specific activity on MULac were fairly consistent.[40] In comparison, the enzymatic activities of the *T. reesei* and *C. thermophilum* CBH1s were disproportional to the amount of secreted protein measured, suggesting that only a fraction of these secreted enzyme pools were enzymatically active. This was similar to the results obtained when *T. reesei* CBH1 was expressed in *Pichia pastoris*, in which circular dichroism assays indicated that the lack of active enzymes was due to the improper formation of disulfide bonds.[49,50] Expression of the most successfully produced CBH1s and CBH2s in the same host cell lead to Avicel conversion efficiencies that exceeded that of the corresponding strains expressing only one enzyme—likely due to a synergistic effect.[40] However, lower MULac activities in these strains indicated that CBH1 production was lower than when it was produced alone; this effect varied depending on the co-expressed partner.

To ascertain why highly homologous CBHs were secreted at vastly different levels and why co-expression altered the production levels of CBHs in comparison to single expression, differences in *cbh* mRNA levels, gene copy number, and secretion stress-induced responses were investigated in strains with both high and low cellulase production.[40] It was demonstrated that strains producing highly secreted CBHs had much higher messenger RNA (mRNA) levels, higher vector copy numbers, and higher levels of MULac activity than strains producing poorly expressed CBHs. The same criteria were also greater for the strain producing the *T. emersonii* CBH1 than those for strains co-producing this enzyme, or its CBM-attached derivatives with any CBH2. In co-expressing strains, mRNA levels of both *cbh1* and *cbh2* were reduced when compared with the corresponding strains expressing only one *cbh* gene, and this was consistent with lower vector copy numbers. Furthermore, spliced *HAC1* mRNA coding for the unfolded protein response (UPR)-inducing transcription factor, could be detected in all CBH-producing strains, indicating that the UPR was induced. Strains producing efficiently secreted enzymes had relatively low levels of the spliced *HAC1* transcript, suggesting that the expression of these proteins were less stressful for the cell's secretion machinery. Conversely, the *T. reesei* CBH1 caused a stronger UPR induction and far less active enzyme was secreted. Overall, these results showed that there are certain gene candidates that were more compatible with expression in yeast than others. However, the features that lead to incompatibility, marked by low levels of mRNA, low episomal plasmid copy number, and low levels of secreted protein and a strong induction of the UPR, are difficult to define.

There have been numerous reports showing production of combinations of cellulases in *S. cerevisiae* specifically with the aim of enabling the organism to grow on or convert polymeric substrates. Cho et al.[51] showed that a strain co-producing a β-glucosidase and an exocellulase/endocellulase could grow to some extent on cellodextrins. An *S. cerevisiae* strain co-secreting the *T. reesei* EG1 (*cel7B*) and the *S. fibuligera* β-glucosidase (*cel3A*) and releasing the recombinant cellulases into the surrounding media was able to grow on and produce 1.0 g/L ethanol from 10 g/L PASC.[52] Jeon et al.[53] constructed a similar strain that produced significantly more EG activity and notably improved conversion of PASC to ethanol was achieved.

Fujita et al.[54,55] reported co-expression of up to three cellulases in *S. cerevisiae*. The secreted cellulases were tethered to the cell surface by producing them as fusion proteins with a section of the C-terminal end of the yeast cell surface protein α-agglutinin. High cell density suspensions of a strain displaying the *T. reesei* EG2, CBH2, and the *Aspergillus aculeatus* β-glucosidase could convert 10 g/L PASC to approximately 3 g/L ethanol. Yamada et al.[56] developed a resourceful method to integrate a number of various surface-tethered cellulase encoding genes through multicopy δ-integration, to optimize expression levels. Cellulase expression cassettes encoding the three main cellulase activities were integrated into *S. cerevisiae* chromosomes in a single step, and strains expressing optimum ratios of these cellulases were selected by growth on media containing PASC as sole carbon source. Although the overall integrated gene copy numbers of an efficient "cocktail" δ-integrated strain was roughly half that of a conventional δ-integrant strain, the PASC degradation activity (64.9 mU/g-wet cells) was higher than that of a conventional strain (57.6 mU/g-wet cells). This suggested that optimization of the cellulase expression ratio improved PASC degradation activity more than simply achieving higher cellulase gene copy numbers.

Matano et al.[57] enhanced cellulase activities on a recombinant *S. cerevisiae* yeast cell surface displaying *T. reesei* EG2 and CBH2 and *A. aculeatus* BGL1 by integrating additional *eg2* and *cbh2* gene cassettes into the recombinant strain. As a result, 43.1 g/L ethanol was produced from 200 g-dry weight/L pretreated rice straw after performing 2-h liquefaction and a subsequent 72-h fermentation in the presence of commercial cellulase loaded to 10 FPU/g-biomass. Ethanol yield by the recombinant strain reached 89% of the theoretical maximum, which was 1.4-fold higher than the strain without additional gene copies. In SSF on 100 g/L Avicel with 1.0 FPU/mL cellulase, a wild-type yeast strain yielded 79.7% of the theoretical maximum ethanol yield in 96 h.[57] A recombinant yeast strain displaying a CBH, EG, and BGL yielded 87.3% of the theoretical maximum ethanol yield with significantly less residual substrate. Higher ethanol production may be attributed to higher efficiency of the cellulases displayed on the cell surface. It is postulated that the close proximity of the enzymes on the cell surface enables synergistic hydrolysis of the cellulosic substrate.[32] Furthermore, reutilization of the yeast cells in subsequent processing runs enables reuse of the enzymes displayed on their cell surface without reproduction of yeast cells, reducing the cost of yeast propagation and enzyme addition.

The most efficient cellulose hydrolyzing organisms are those found in the rumen of certain herbivores and produce an enzyme complex called a cellulosome that contains all the enzymes needed to completely break down cellulose and hemicellulose.[58] The enzymes are all attached to a noncatalytic, cell surface bound protein called a scaffoldin. The scaffoldin contains cohesins that interact specifically with dockerins attached to all cellulosomal enzymes. The advantage of this arrangement is that the enzymes are all in close proximity, and the sugars produced through the combined action of these enzymes can be taken up by the cell.[59] Several groups have attempted to reconstruct a minicellulosome on the *S. cerevisiae* cell surface.[60–63] Recently, an *S. cerevisiae* strain was engineered to display a trifunctional minicellulosome. It consisted of a miniscaffoldin containing a cellulose-binding domain and three cohesin modules, anchored to the cell surface and three types of cellulases, *T. reesei* EG2 and CBH2 and *A. aculeatus* BGL1, each bearing a C-terminal dockerin.[63] This strain was able to hydrolyze and ferment PASC to ethanol with a titer of 1.8 g/L.

The cellulase co-producing strains described above were active on amorphous cellulose substrates, but generally fared poorly on crystalline cellulose substrates in the absence of added commercial enzyme preparations, likely due to the low titers of secreted CBHs. Because exoglucanase activity is required for the hydrolysis of crystalline cellulose, high-level expression of cellobiohydrolases in these strains could enable conversion of crystalline cellulose to ethanol. As mentioned previously, the CBH expression levels achieved by Ilmén et al.[40] meets the calculated levels required for growth on cellulose at rates required for an industrial process.[30] Using these exoglucanases, a strain was constructed that was able to convert most of the glucan residue available in paper sludge to ethanol.[44] The strain was also able to displace 60% of the commercial enzymes preparation required to convert the sugars available in pretreated hardwood to ethanol in an SSF configuration. A similar strain secreting the *T. reesei* EG2, CBH2 and the *A. aculeatus* BGL1 produced ethanol in one step from pretreated corn stover without the addition of exogenously produced enzymes. This strain fermented 63% of the available cellulose in 96 h to 2.6% v/v ethanol.[64] These results demonstrate that cellulolytic *S. cerevisiae* strains can be constructed and used as a platform for developing an economical advanced biofuel process.

Expression of Xylan Hydrolases in Saccharomyces cerevisiae

S. cerevisiae cannot hydrolyze xylan, nor can it use the resulting monomeric pentose sugars, therefore extensive genetic engineering is necessary to enable it to convert xylan to ethanol. The first two reports of the successful expression of xylanases in yeast were published separately in 1992. The *Cryptococcus albidus*[65,66] and the *Aspergillus kawachii*[67] GH11 enzymes expressed were cloned from complementary DNA. These and subsequent studies were primarily undertaken because of the industrial need for xylanases completely free of cellulases used for biotransformations.[19,68]

These initial studies were followed by reports on the expression of xylanases from *A. kawachii*,[69] *Aspergillus niger*,[70] *Aureobasidium pullulans*,[71] and *T. reesei*[72] (Table 2). *A. pullulans* is a yeast-like organism well known for its ability to break down xylan and grow on xylose. Expression of the *A. pullulans xynA* gene under the control of the *GAL1* promoter in *S. cerevisiae* produced 28.6 U/mL of xylanase activity 6 h after galactose induction. Fair levels of xylanase activity were obtained with the *A. kawachii* (18 U/mL) and *T. reesei* (9 U/mL) genes under the control of the constitutive *PGK1* promoter and terminator. In both cases, *FUR1* autoselection was used to maintain the episomal plasmid during growth on rich medium.[72,73] The highest xylanase activity in *S. cerevisiae* was obtained using the same strategy, but gene expression was under the control of the derepressible *ADH2* promoter.[72] Using this promoter, the authors reported 71 U/ml after 70 h of growth in rich medium. In 2002, Fujita et al.[54] reported anchoring the *T. reesei xyn2* on the surface of *S. cerevisiae* using a C-terminal section of α-agglutinin as anchor. Cell surface display concentrated enzymes on the cell surface, allowing the cell to be used as whole-cell biocatalysts for the production of xylo-oligosaccharides from xylan.

Expression of a xylanase gene in *S. cerevisiae* will enable it to hydrolyze xylan to short xylo-oligosaccharides, mainly xylobiose and xylotriose.[74,75] Further hydrolysis to D-xylose requires the action of a β-xylosidase. The first report of xylosidase activity in *S. cerevisiae* was in 1997.[76] This was through the expression of the *Bacillus pumilus xynB*. Xylosidase activity was very low; however, the same group subsequently reported the expression of the *A. niger xlnD* gene in *S. cerevisiae*.[75] This enzyme was expressed significantly better, and 0.3 U/mL of xylosidase activity on *p*-nitrophenyl-β-D-xylopyranoside (pNPX) was reported. Co-expression of the *T. reesei xyn2* with the *A. niger xlnD* yielded a yeast capable of breaking

Table 2: Various hemicellulose hydrolyzing enzymes heterologously expressed in *S. cerevisiae*

Organism	Gene	GH family	Promoter	Extracellular activity (IU/ml)	Reference
Xylanase					
Cryptococcus flavus	*Cfxyn1*	11	*PGK1*	2.5	65
Aspergillus niger	*xyn4*	11	*ADH2*	5.4	70
	xyn5	11	*ADH2*	4.3	
Thermomyces lanuginosus	*xynA*	11	*PGK1*	13.2	114
Aspergillus kawachii	*xynC*	11	*PGK1*	18	69
Pyrenophora tritici-repentis	*xyl1*	43	*ENO1*	3.0	77
Aureobasidium pullulans	*xynA*	11	*GAL1*	28.6	71
Aureobasidium pullulans	*Xyn10*	10		5	77
Trichoderma reesei	*xyn2*	11	*ADH2*	71	72
			PGK1	9	
			ENO1	12	77
Xylosidase					
Aspergillus niger	*xlnD*	3	*ADH2*	0.3 (pNPX)	75
Trichoderma reesei	*bxl1*	3	*PGK1*	1 (pNPX)	115
Cochliobolus carbonum	*xyl1*	43	*ENO1*	0.12 (pNPX)	77
Pyrenophora tritici-repentis	*xyl1*	43	*ENO1*	0.2 (pNPX)	77
Aspergillus oryzae	*xylA*	43	*GAPDH*	Displayed on the cell surface	88
α-L-Arabinofuranosidase					
Aspergillus niger	*abfB*	54	*PGK1*	1.4 (pNPA)	100
Trichoderma reesei	*abf1*	54	*PGK1*	10 (pNPA)	115
Aspergillus awamori	*arfB*		*PGK1*		116
α-Glucuronidase					
Aureobasidium pullulans	*aguA*	67	*ADH2*		97
Pichia stipitis		115	*PGK1*		117
Mannanase					
Agaricus bisporus	*cel*	5	*TPI*	2.5×10^{-3}	118
Aspergillus aculeatus	*man1*	5	*PGK1*	22.9	28
			ADH2	31.4	28
			GAL1		119

Table 2: Various hemicellulose hydrolyzing enzymes heterologously expressed in
S. cerevisiae—cont'd

Organism	Gene	GH family	Promoter	Extracellular activity (IU/ml)	Reference
Orpinomyces sp	*manA*	5	*GAL1*	1.15	120
Trichoderma reesei	*man1*	5	*PGK1*	0.013	121
α-Galactosidase					
Trichoderma reesei	*agl1*	27	*PGK1*	25.9	101
Trichoderma reesei	*agl2*	36	*PGK1*	1.04	101
Trichoderma reesei	*agl3*	27	*PGK1*	0.06	101
β-Galactosidase					
Aspergillus niger	*lacA*		*ADH1*	7.5	122
Kluyveromyces lactis	*lac4*			1.7	102

U = micromole substrate released/min, pNPX = *p*-nitrophenol xylopyranoside, pNPA = *p*-nitrophenol arabinofuranoside, *PGK1* = phosphoglycerate kinase, *ADH2* = alcohol dehydrogenase 2, *ADH1* = alcohol dehydrogenase 1, *ENO1* = enolase, *GAL1* = galactokinase, *GAPDH* = glyceraldehyde 3-phosphate dehydrogenase, *TPI* = triose phosphate isomerase.

down the xylan backbone to its monomeric constituents D-xylose, achieving 57% conversion.[75] Brevnova et al.[77] expressed the *Pyrenophora tritici-repentis* GH 43 xylosidase in *S. cerevisiae*. This enzyme did not only yielded 6.9-fold higher xylosidase activity than the recombinant *A. niger xlnD* xylosidase when the genes were expressed on identical expression vectors, but also exhibited significant xylanase activity on birchwood xylan.

S. cerevisiae has been genetically engineered to grow on xylose with two different metabolic pathways by independent groups. The xylose reductase (XR, *Pichia stipitis XYL1*) and xylitol dehydrogenase (XDH, *P. stipitis XYL2*) pathway was the first introduced.[78] This pathway uses an XR to convert xylose to xylitol, with mainly NADPH as co-factor. The resultant xylitol is then converted to xylulose, with NAD^+ as co-factor. Xylulose then enters the pentose phosphate pathway and is ultimately converted to ethanol under anaerobic conditions. A number of attempts to express an active xylose isomerase (XI) in *S. cerevisiae* were met with limited success for several years.[79–82] Eventually, a xylose isomerase with sufficient activity to support growth on xylose was obtained from the anaerobic fungus *Piromyces* sp E2.[83] To obtain rapid fermentation of xylose under anaerobic conditions, a xylulokinase (XK, *Pichia stipitis XYL3*) was also overexpressed, as were a number of enzymes in the pentose phosphate pathway. To minimize the accumulation of xylitol, *GRE3* encoding an aldose reductase was deleted. Recently the *Clostridium phytofermentans* XI was successfully expressed in *S. cerevisiae*.[84] It was subsequently shown that the activity and kinetic parameters of the *C. phytofermentans* XI is comparable to that of the *Piromyces* XI, but it is significantly less sensitive to xylitol inhibition.[85] This gene was also instrumental in engineering a strain that could directly convert xylose to isobutanol. In a paper comparing the two pathways, the XR-XDH pathway was superior in terms of xylose consumption, specific ethanol productivity and final ethanol concentration; however, the pathway caused a co-factor imbalance, leading to an accumulation of xylitol.[86,87] The XI, on the other hand, does not require any co-factors, and xylose is converted to xylulose in a single step. The highest ethanol yield therefore was observed in the yeast strain using the XI pathway.

Strains of *S. cerevisiae* were now available than could convert xylan to xylose, and yet other strains were engineered to ferment xylose to ethanol. Katahira et al.[88] combined the two capabilities into a single yeast strain. The authors displayed the *T. reesei xyn2* xylanase and the *A. oryzae xylA* xylosidase on the surface of *S. cerevisiae* using the α-agglutinin anchor.[74] This strain was able to produce 0.31 g/L xylose from birchwood xylan. The XR-XDH pathway was introduced to enable it to convert the liberated xylose to 7.1 g/L ethanol. Ethanol was produced at a rate of 0.13 g/L/h, and the yield was 0.30 g/g sugar consumed.

As mentioned earlier, unlike cellulose, xylan is not a homopolymer and contains side-chains that hinder enzymatic attack on the xylan backbone. Most of the reported xylanases expressed in *S. cerevisiae* up to 2012 are GH11 enzymes. These are fairly small proteins, typically around 21–25 kDa, which is perhaps part of the reason why they express well in *S. cerevisiae*. The β-xylanases of GH10 are capable of cleaving glycosidic linkages in the xylan main chain closer to the substituents, such as 4-*O*-methyl-D-glucuronic acid, acetic acid, and L-arabinose.[89,90] The *A. pullulans xyn10* GH10 xylanase was expressed in *S. cerevisiae*, yielding 5 U/mL of activity on birchwood xylan.[77]

More complete hydrolysis of the xylan backbone can be obtained by removing the acetyl, arabinose, and glucuronic acid side-chains. Few published studies report on the expression of debranching enzymes in *S. cerevisiae* (Table 2). Recombinant expression of the *T. reesei* arabinofuranosidase lead to the production of an enzyme that was not only capable of hydrolyzing the artificial chromophoric substrate *p*-nitrophenyl-α-L-arabinofuranoside (pNPA), but also released arabinose from arabinoxylans.[91] Expressing an arabinofuranosidase in a xylose fermenting yeast also capable of fermenting L-arabinose[92–96] should lead to improved conversion of arabinoxylan to ethanol. The α-glucuronidase expressed by De Wet et al.[97] was active on a series of substrates from aldobiouronic acid to aldopentaouronic acid. *S. cerevisiae* thus is capable of producing these debranching enzymes in an active form that would allow removal of these side-chains from the xylan backbone. Brevnova et al.[77] reported the expression of acetyl xylan esterases. The authors showed that the presence of the *N. fischeri* or *T. reesei* acetyl xylan esterases produced by *S. cerevisiae* increased the yield of xylose from pretreated hardwood during treatment with xylanase and xylosidase.

The construction of a mini-hemicellulosome/xylanosome in *S. cerevisiae* was also reported.[98,99] Sun et al.[99] used fungal enzymes that were successfully expressed in

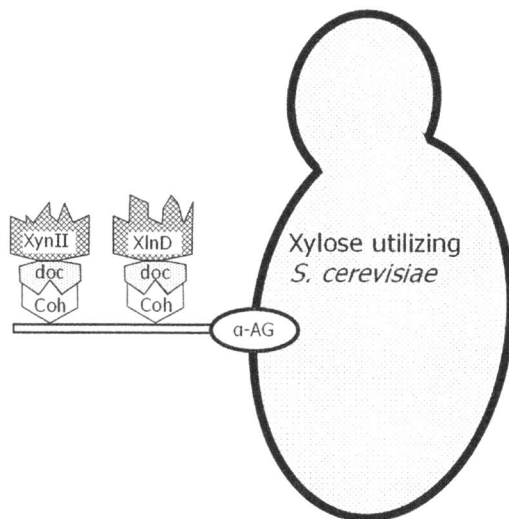

Figure 2
Mini-hemicellulosome constructed by Sun et al.[99] to create an *S. cerevisiae* strain capable of fermenting birchwood xylan directly to ethanol. XynII = *T. reesei* xylanase 2 (XynII), XlnD = *A. niger* xylosidase (XlnD), coh = *C. thermocellum* cohesin, doc = *C. thermocellum* dockerin, αAG = the *S. cerevisiae* α-agglutinin cell wall anchor.

S. cerevisiae previously. For the construction of a mini-hemicellulosome they attached a dockerin module, from the thermophilic anaerobic bacterium *Clostridium thermocellum*, to a *T. reesei* xylanase[72] and the *A. niger* xylosidase[75] and arabinofuranosidase.[100] Only part of the *C. thermocellum* CipA scaffoldin containing three cohesins, were used for the mini-hemicellulosome. The α-agglutinin anchor was used to display the mini-CipA on the cell surface of *S. cerevisiae*. The mini-CipA and dockerin containing hemicellulases were all expressed in an *S. cerevisiae* strain containing the XR-XDH pathway for xylose use (Figure 2). Unfortunately, the arabinofuranosidase in the trifunctional mini-hemicellulosome did not show activity on arabinoxylan. However, yeast displaying a bifunctional mini-hemicellulosome CipA3-XynII-XlnD simultaneously hydrolyzed and fermented birchwood xylan to ethanol with a yield of 0.31 g/g of sugar consumed.

Expression of Mannan Hydrolases in Saccharomyces cerevisiae

S. cerevisiae cannot degrade mannan, but is able to utilise the mannose monomer through glycolysis.[24] Heterologous production of mannanases, mannosidases and α-galactosidases was reviewed by Chauhan et al.[25] and Van Zyl et al.[26] Only a few of the genes were expressed in *S. cerevisiae* as host (Table 2). Setati et al.[28] was the first to report high level production of an endomannanase in *S. cerevisiae*. This endomannanase was also able to produce low levels of mannose from ivory nut mannan. Of the three *T. reesei* α-galactosidases encoded by *agl1*, *agl2* and *agl3*, only *agl1* could be expressed at high levels.[101] Low levels of expression of the *Kluyveromyces lactis* β-galactosidase were obtained with expression in *S. cerevisiae*.[102] Yet, this was significant since the native Lac4 of *K. lactis* is produced in both dimeric and tetrametic forms with the monomeric peptide being 124 kDa in size. This makes the Lac4 probably the largest functional heterologous protein secreted by *S. cerevisiae*.

β-mannosidases have been expressed successfully in *Pichia pastoris*, *A. niger*, and *Aspergillus oryzae*,[26] yet no successful expression has been reported in *S. cerevisiae*. Overexpression of the *Cellvibrio mixtus Man5A* (synthetic codon optimized), *A. niger* (*mndA*), *Aspergillus tubingensis* (*AtmndA*) in *S. cerevisiae* proved to be unsuccessful (WH van Zyl, unpublished results). Since β-mannosidases are involved in the removal of mannose units on glycosylated proteins,[103] it was proposed that overexpression of β-mannosidases may lead to interference of the glycosylation process, which leads to down regulation of its expression.

Discussion

The expression and secretion of various cellulases, xylanases and mannanases have been successfully demonstrated and represent milestones towards the development of *S. cerevisiae* for CBP conversion of lignocellulosic biomass to ethanol. However, the secretion of certain hydrolases, such as CBHs, has proved to be challenging. Screening of large collections of fungal genes encoding CBHs belonging to glycoside hydrolases families six and seven yielded recombinant enzyme candidates more compatible with expression in *S. cerevisiae*.[40] These enzymes were not only expressed at higher titers, but their production proved less stressful to the cell. Using CBHs as an example, a dedicated bioinformatics study trying to link specific features of the best secreted enzymes that leads to efficient folding secretion by yeast could be prudent to better understand and engineer efficient enzyme production by *S. cerevisiae*. This may require solving the protein crystal structures of a selection of well and poorly expressed enzymes from the same GH family. Using a similar approach of screening a large collection of other glycoside hydrolases in *S. cerevisiae* could not only yield high titers of xylanases, etc., but broaden the bioinformatics data set trying to identify universal protein features for efficient production.

Stable cellulolytic yeasts have been produced by various groups by integrating multicopy genes of one or more GH enzymes into the yeast genome. However, the optimal co-expression of different enzymes for other activities, including enzymes that are active on hemicelluloses, is still required to enable CBP yeasts to completely convert pretreated lignocellulosic feedstocks. Optimizing gene dosage for specific feedstocks will also be required for optimal substrate conversion in a CBP process. This might also require optimizing yeast strains for specific feedstocks and pretreatment processes or, vice versa, optimizing the pretreatment process to best suit the CBP yeast. This might be one of the limitations of CBP versus SSF or SSCF in which adjusting enzyme mixtures for different feedstock and pretreatment processes will be easier than re-engineering special CBP yeast strains for each feedstock/pretreatment process. It should be noted that addition of auxiliary enzymes to the CBP process will in all likelihood be a prerequisite in the foreseeable future due to the plethora of enzymes required for efficient biomass conversion. A further obstacle in the development of CBP yeasts may be the inability to produce high enough titers of the different enzymes.

Apart from optimizing the co-expression of glycoside hydrolases, further engineering and optimization of appropriate CBP yeast host strains will also be required. A limited number of industrial strains for the bioconversion of lignocellulosics are currently available. Further improvements in the level of secreted heterologous proteins in industrial strains will be required to make CBP more efficient and economically viable. The secretion machinery of yeasts is a multistep process, and a vast array of protein targets can be manipulated in isolation or in concert to enhance heterologous protein secretion. The potential of "-omics" data and other postgenomic technologies to identify possible gene candidates or pathways to be used in strain engineering strategies is clearly illustrated by the transcriptomics-based work of Gasser et al.[104] The advantages of identifying secretion-enhancing genes and overexpressing them individually or in combination were eloquently demonstrated by Kroukamp et al.[105] In this study, the authors demonstrated a synergetic 4.5-fold increase in *Saccharomycopsis fibuligera* BGL1 production with the overexpression of native *S. cerevisiae PSE1* and *SOD1*. They also demonstrated how some genes might enhance secretion in combination with other effectors, but not in isolation. One drawback is the observation that most of the positive effects observed thus far have been highly protein-specific, signifying that the effects should be assessed on a case-by-case basis for different proteins of interest. However, it is predicted that the identification of compatible gene candidates for all required activities, and the combination of these genes in a strain engineered for their optimal secretion, will enable the construction of ideal CBP yeast strains. Apart from more efficient production of glycoside hydrolases, the engineering of robust yeast hosts that cope better with fermentation inhibitors are crucial if CBP on both the soluble and insoluble fractions of pre-treated lignocellulosics are considered. Similar approaches as described for enhanced protein production can be designed to identify target genes to enhance these multi-gene traits. In conclusion, good progress toward producing cellulases at high titers in *S. cerevisiae* has been demonstrated. Moreover, combining the search and expression of new glycoside hydrolases, while simultaneously improving different favorable features of CBP yeasts, as highlighted in this review, will be fundamental in bringing direct microbial conversion with *S. cerevisiae* to commercial reality.

References

1. Van Zyl WH, Den Haan R, La Grange DC. Developing cellulolytic organisms for consolidated bioprocessing of lignocellulosics. In: Gupta VK, Tuohy MG, editors. *Biofuel technologies: recent developments.* New York: Springer Verlag; 2013. p. 189–220.
2. Lynd LR, Weimer PJ, Van Zyl WH, Pretorius IS. Microbial cellulose utilization: fundamentals and biotechnology. *Microbiol Mol Biol Rev* 2002;**66**:506–77.
3. Lynd LR, Van Zyl WH, McBride JE, Laser M. Consolidated bioprocessing of cellulosic biomass: an update. *Curr Opin Biotechnol* 2005;**16**:577–83.

4. Anex RP, Aden A, Kazi FK, Fortman J, Swanson RM, Wright MM, et al. Techno-economic comparison of biomass-to-transportation fuels via pyrolysis, gasification, and biochemical pathways. *Fuel* 2010;**89** (Suppl. 1):S29–35.

5. Hahn-Hägerdal B, Karhumaa K, Fonseca C, Spencer-Martins I, Gorwa-Grauslund MF. Towards industrial pentose-fermenting yeast strains. *Appl Microbiol Biotechnol* 2007;**74**:937–53.

6. Hamelinck CN, Van Hooijdonk G, Faaij APC. Ethanol from lignocellulosic biomass: techno-economic performance in short-, middle- and long-term. *Biomass Bioenerg* 2005;**28**:384–410.

7. Van Zyl WH, Chimphango AFA, Den Haan R, Gorgens JF, Chirwa PWC. Next-generation cellulosic ethanol technologies and their contribution to a sustainable Africa. *Interface Focus* 2011;**1**:196–211.

8. Kuyper M, Hartog MMP, Toirkens MJ, Almering MJH, Winkler AA, Van Dijken JP, et al. Metabolic engineering of a xylose-isomerase-expressing *Saccharomyces cerevisiae* strain for rapid anaerobic xylose fermentation. *FEMS Yeast Res* 2005;**5**:399–409.

9. Van Dijken JP, Bauer J, Brambilla L, Duboc P, Francois JM, Gancedo C, et al. An interlaboratory comparison of physiological and genetic properties of four *Saccharomyces cerevisiae* strains. *Enzyme Microb Technol* 2000;**26**:706–14.

10. Karhumaa K, Wiedemann B, Hahn-Hagerdal B, Boles E, Gorwa-Grauslund MF. Co-utilization of L-arabinose and D-xylose by laboratory and industrial *Saccharomyces cerevisiae* strains. *Microb Cell Fact* 2006;**5**:18–28.

11. van Rooyen R, Hahn-Hagerdal B, La Grange DC, Van Zyl WH. Construction of cellobiose-growing and fermenting *Saccharomyces cerevisiae* strains. *J Biotechnol* 2005;**120**:284–95.

12. Penttilä ME, Andre L, Lehtovaara P, Bailey M, Teeri TT, Knowles JK. Efficient secretion of two fungal cellobiohydrolases by *Saccharomyces cerevisiae*. *Gene* 1988;**63**:103–12.

13. Stambuk BU, Alves Jr SL, Hollatz C, Zastrow CR. Improvement of maltotriose fermentation by *Saccharomyces cerevisiae*. *Lett Appl Microbiol* 2006;**43**:370–6.

14. Sun Y, Cheng J. Hydrolysis of lignocellulosic materials for ethanol production: a review. *Bioresour Technol* 2002;**83**:1–11.

15. Zhang YH, Lynd LR. Toward an aggregated understanding of enzymatic hydrolysis of cellulose: noncomplexed cellulase systems. *Biotechnol Bioeng* 2004;**88**:797–824.

16. Suurnäkki A, Tenkanen M, Buchert J, Viikari L. Hemicellulases in the bleaching of chemical pulps. In: Scheper T, editor. *Advances in biochemical engineering*. Berlin: Springer Verlag; 1997. p. 261–87.

17. Thomson JA. Molecular biology of xylan degradation. *FEMS Microbiol Rev* 1993;**104**:65–82.

18. Eriksson K-EL, Blanchette RA, Ander P. *Microbial and enzymatic degradation of wood and wood components*. New York: Springer-Verlag; 1990.

19. Biely P. Microbial xylanolytic systems. *Trends Biotechnol* 1985;**3**:286–90.

20. Hazlewood GP, Gilbert HJ. Molecular biology of hemicellulases. In: Coughlan MP, Hazlewood GP, editors. *Hemicellulose and hemicellulases*. London: Portland Press; 1993. p. 103–26.

21. Puls J, Schuseil J. Chemistry of hemicelluloses: relationship between hemicellulose structure and enzymes required for hydrolysis. In: Coughlan MP, Hazlewood GP, editors. *Hemicellulose and hemicellulases*. London: Portland Press; 1993. p. 1–27.

22. Mueller-Harvey I, Hartley RD, Harris PJ, Curzon EH. Linkage of *p*-coumaroyl and feruloyl groups to cell-wall polysaccharides of barley straw. *Carbohydr Res* 1986;**148**:71–85.

23. Zhang J, Siika-aho M, Tenkanen M, Viikari L. The role of acetyl xylan esterase in the solubilization of xylan and enzymatic hydrolysis of wheat straw and giant reed. *Biotechnol Biofuels* 2011;**4**(1):60.

24. Girio FM, Fonseca C, Carvalheiro F, Duarte LC, Marques S, Bogel-Lukasik R. Hemicelluloses for fuel ethanol: a review. *Bioresour Technol* 2010;**101**:4775–800.

25. Chauhan PS, Puri N, Sharma P, Gupta N. Mannanases: microbial sources, production, properties and potential biotechnological applications. *Appl Microbiol Biotechnol* 2012;**93**:1817–30.

26. Van Zyl WH, Rose SH, Trollope K, Gorgens JF. Fungal β-mannanases: mannan hydrolysis, heterologous production and biotechnological applications. *Process Biochem* 2010;**45**:1203–13.

27. Moreira LR, Filho EX. An overview of mannan structure and mannan-degrading enzyme systems. *Appl Microbiol Biotechnol* 2008;**79**:165–78.

28. Setati ME, Ademark P, Van Zyl WH, Hahn-Hagerdal B, Stalbrand H. Expression of the *Aspergillus aculeatus* endo-beta-1,4-mannanase encoding gene (*man1*) in *Saccharomyces cerevisiae* and characterization of the recombinant enzyme. *Protein Expr Purif* 2001;**21**:105–14.

29. Penttilä ME, Andre L, Saloheimo M, Lehtovaara P, Knowles JK. Expression of two *Trichoderma reesei* endoglucanases in the yeast *Saccharomyces cerevisiae. Yeast* 1987;**3**:175–85.

30. Olson DG, McBride JE, Shaw AJ, Lynd LR. Recent progress in consolidated bioprocessing. *Curr Opin Biotechnol* 2012;**23**:1–11.

31. Njokweni A, Rose S, Van Zyl WH. Fungal β-glucosidase expression in *Saccharomyces cerevisiae. J Ind Microbiol Biotechnol* 2012;**39**:1445–52.

32. Hasunuma T, Sanda T, Yamada R, Yoshimura K, Ishii J, Kondo A. Metabolic pathway engineering based on metabolomics confers acetic and formic acid tolerance to a recombinant xylose-fermenting strain of *Saccharomyces cerevisiae. Microb Cell Fact* 2011;**10**:2.

33. Hasunuma T, Kondo A. Development of yeast cell factories for consolidated bioprocessing of lignocellulose to bioethanol through cell surface engineering. *Biotechnol Adv* 2012;**30**:1207–18.

34. Galazka JM, Tian C, Beeson WT, Martinez B, Glass NL, Cate JH. Cellodextrin transport in yeast for improved biofuel production. *Science* 2010;**330**:84–6.

35. Ha S-J, Galazka JM, Kim SR, Choi J-H, Yang X, Seo J-H, et al. Engineered *Saccharomyces cerevisiae* capable of simultaneous cellobiose and xylose fermentation. *Proc Natl Acad Sci USA* 2011;**108**:504–9.

36. Sadie CJ, Rose SH, Den Haan R, Van Zyl WH. Co-expression of a cellobiose phosphorylase and lactose permease enables intracellular cellobiose utilisation by *Saccharomyces cerevisiae. Appl Microbiol Biotechnol* 2011;**90**:1373–80.

37. La Grange DC, Den Haan R, Van Zyl WH. Engineering cellulolytic ability into bioprocessing organisms. *Appl Microbiol Biotechnol* 2010;**87**:1195–208.

38. Van Zyl WH, Lynd LR, Den Haan R, McBride JE. Consolidated bioprocessing for bioethanol production using *Saccharomyces cerevisiae. Adv Biochem Eng Biotechnol* 2007;**108**:205–35.

39. Den Haan R, Kroukamp H, Van Zyl JH, Van Zyl WH. Cellobiohydrolase secretion by yeast: current state and prospects for improvement. *Process Biochem* 2013;**148**:1–12.

40. Ilmén M, Den Haan R, Brevnova E, McBride J, Wiswall E, Froehlich A, et al. High level secretion of cellobiohydrolases by *Saccharomyces cerevisiae. Biotechnol Biofuels* 2011;**4**(1):30.

41. Den Haan R, McBride JE, La Grange DC, Lynd LR, Van Zyl WH. Functional expression of cellobiohydrolases in *Saccharomyces cerevisiae* towards one-step conversion of cellulose to ethanol. *Enzyme Microb Technol* 2007;**40**:1291–9.

42. Park EH, Shin YM, Lim YY, Kwon TH, Kim DH, Yang MS. Expression of glucose oxidase by using recombinant yeast. *J Biotechnol* 2000;**81**:35–44.

43. Romanos MA, Scorer CA, Clare JJ. Foreign gene expression in yeast: a review. *Yeast* 1992;**8**:423–88.

44. McBride JE, Brevnova E, Ghandi C, Mellon M, Froehlich A, Delaault K, et al. Yeast expressing cellulases for simultaneous saccharification and fermentation using cellulose. PCT/US2009/065571. 5-27-2010.

45. Hong J, Tamaki H, Yamamoto K, Kumagai H. Cloning of a gene encoding thermostable cellobiohydrolase from *Thermoascus aurantiacus* and its expression in yeast. *Appl Microbiol Biotechnol* 2003;**63**:42–50.

46. Reinikainen T, Ruohonen L, Nevanen T, Laaksonen L, Kraulis P, Jones TA, et al. Investigation of the function of mutated cellulose-binding domains of *Trichoderma reesei* cellobiohydrolase I. *Proteins* 1992;**14**:475–82.

47. Voutilainen SP, Murray PG, Tuohy MG, Koivula A. Expression of *Talaromyces emersonii* cellobiohydrolase Cel7A in *Saccharomyces cerevisiae* and rational mutagenesis to improve its thermostability and activity. *Protein Eng Des Sel* 2010;**23**:69–79.

48. Heinzelman P, Komor R, Kanaan A, Romero P, Yu X, Mohler S, et al. Efficient screening of fungal cellobiohydrolase class I enzymes for thermostabilizing sequence blocks by SCHEMA structure-guided recombination. *Protein Eng Des Sel* 2010;**23**:871–80.

49. Boer H, Teeri TT, Koivula A. Characterization of *Trichoderma reesei* cellobiohydrolase Cel7A secreted from *Pichia pastoris* using two different promoters. *Biotechnol Bioeng* 2000;**69**:486–94.

50. Godbole S, Decker SR, Nieves RA, Adney WS, Vinzant TB, Baker JO, et al. Cloning and expression of *Trichoderma reesei* cellobiohydrolase I in Pichia pastoris. *Biotechnol Prog* 1999;**15**:828–33.

51. Cho KM, Yoo YJ, Kang HS. δ-Integration of endo/exo-glucanase and β-glucosidase genes into the yeast chromosomes for direct conversion of cellulose to ethanol. *Enzyme Microb Technol* 1999;**25**:23–30.
52. Den Haan R, Rose SH, Lynd LR, Van Zyl WH. Hydrolysis and fermentation of amorphous cellulose by recombinant *Saccharomyces cerevisiae*. *Metab Eng* 2007;**9**:87–94.
53. Jeon E, Hyeon J-E, Suh DJ, Suh Y-W, Kim SW, Song KH, et al. Production of cellulosic ethanol in *Saccharomyces cerevisiae* heterologous expressing *Clostridium thermocellum* endoglucanase and *Saccharomycopsis fibuligera* beta-glucosidase genes. *Mol Cells* 2009;**28**:369–73.
54. Fujita Y, Takahashi S, Ueda M, Tanaka A, Okada H, Morikawa Y, et al. Direct and efficient production of ethanol from cellulosic material with a yeast strain displaying cellulolytic enzymes. *Appl Environ Microbiol* 2002;**68**:5136–41.
55. Fujita Y, Ito J, Ueda M, Fukuda H, Kondo A. Synergistic saccharification, and direct fermentation to ethanol, of amorphous cellulose by use of an engineered yeast strain codisplaying three types of cellulolytic enzyme. *Appl Environ Microbiol* 2004;**70**:1207–12.
56. Yamada R, Taniguchi N, Tanaka T, Ogino C, Fukuda H, Kondo A. Cocktail δ-integration: a novel method to construct cellulolytic enzyme expression ratio-optimized yeast strains. *Microb Cell Fact* 2010;**9**:32.
57. Matano Y, Hasunuma T, Kondo A. Display of cellulases on the cell surface of *Saccharomyces cerevisiae* for high yield ethanol production from high-solid lignocellulosic biomass. *Bioresour Technol* 2012;**108**:128–33.
58. Devaux M. The cellulosome of *Clostridium cellulolyticum enzyme*. *Microb Technol* 2004;**37**:373–85.
59. Fierobe H-P, Bayer EA, Tardif C, Czjzek M, Mechaly A, Belaich A, et al. Degradation of cellulose substrates by cellulosome chimeras. *J Biol Chem* 2008;**277**:49621–30.
60. Ito J, Kosugi A, Tanaka T, Kuroda K, Shibasaki S, Ogino C, et al. Regulation of the display ratio of enzymes on the *Saccharomyces cerevisiae* cell surface by the immunoglobulin G and cellulosomal enzyme binding domains. *Appl Environ Microbiol* 2009;**75**:4149–54.
61. Lilly M, Fierobe HP, Van Zyl WH, Volschenk H. Heterologous expression of a *Clostridium* minicellulosome in *Saccharomyces cerevisiae*. *FEMS Yeast Res* 2009;**9**:1236–49.
62. Tsai SL, Oh J, Singh S, Chen R, Chen W. Functional assembly of minicellulosomes on the *Saccharomyces cerevisiae* cell surface for cellulose hydrolysis and ethanol production. *Appl Environ Microbiol* 2009;**75**:6087–93.
63. Wen F, Sun J, Zhao H. Yeast surface display of trifunctional minicellulosomes for simultaneous saccharification and fermentation of cellulose to ethanol. *Appl Environ Microbiol* 2010;**76**:1251–60.
64. Khramtsov N, McDade L, Amerik A, Yu E, Divatia K, Tikhonov A, et al. Industrial yeast strain engineered to ferment ethanol from lignocellulosic biomass. *Bioresour Technol* 2011;**102**:8310–3.
65. Parachin NS, Siqueira S, de Faria FP, Torres FAG, De Moraes LMP. Xylanases from *Cryptococcus flavus* isolate I-11: enzymatic profile, isolation and heterologous expression of CfXYN1 in *Saccharomyces cerevisiae*. *J Mol Catal B Enzym* 2009;**59**:52–7.
66. Moreau A, Durand S, Morosoli R. Secretion of a *Cryptococcus albidus* xylanase in *Saccharomyces cerevisiae*. *Gene* 1992;**116**:109–13.
67. Ito K, Ikemasu T, Ishikawa T. Cloning and sequencing of the xynA gene encoding xylanase A of *Aspergillus kawachii*. *Biosci Biotech Biochem* 1992;**56**:906–12.
68. Beg QK, Kapoor M, Mahajan L, Hoondal GS. Microbial xylanases and their industrial applications: a review. *Appl Microbiol Biotechnol* 2001;**56**:326–38.
69. Crous JM, Pretorius IS, Van Zyl WH. Cloning and expression of an *Aspergillus kawachii* endo-1,4-β-xylanase gene in *Saccharomyces cerevisiae*. *Curr Genet* 1995;**28**:467–73.
70. Luttig M, Pretorius IS, Zyl WH. Cloning of two β-xylanase-encoding genes from *Aspergillus niger* and their expression in *Saccharomyces cerevisiae*. *Biotechnol Lett* 1997;**19**:411–5.
71. Li XL, Ljungdahl LG. Expression of *Aureobasidium pullulans xynA* in, and Secretion of the Xylanase from, *Saccharomyces cerevisiae*. *Appl Envir Microbiol* 1996;**62**:209–13.
72. La Grange DC, Pretorius IS, Van Zyl WH. Expression of a *Trichoderma reesei* β-xylanase gene (*XYN2*) in *Saccharomyces cerevisiae*. *Appl Envir. Microbiol* 1996;**62**:1036–44.
73. Kern L, De Montigny J, Jund R, Lacroute F. The *FUR1* gene of *Saccharomyces cerevisiae*: cloning, structure and expression of wild-type and mutant alleles. *Gene* 1990;**88**:149–57.

74. Fujita Y, Katahira S, Ueda M, Tanaka A, Okada H, Morikawa Y, et al. Construction of whole-cell biocatalyst for xylan degradation through cell-surface xylanase display in *Saccharomyces cerevisiae*. *J Mol Catal B-Enzym* 2002;**17**:189–95.

75. La Grange DC, Pretorius IS, Claeyssens M, Van Zyl WH. Degradation of xylan to D-xylose by recombinant *Saccharomyces cerevisiae* coexpressing the *Aspergillus niger* β-xylosidase (*xlnD*) and the *Trichoderma reesei* xylanse II (*xyn2*) genes. *Appl Environ Microbiol* 2001;**67**:5512–9.

76. La Grange DC, Pretorius IS, Van Zyl WH. Cloning of the *Bacillus pumilus* β-xylosidase gene (*xynB*) and its expression in *Saccharomyces cerevisiae*. *Appl Microbiol Biotechnol* 1997;**47**:262–6.

77. Brevnova E, McBride J, Wiswall E, Wegner KS, Caiazza N, Hau H, et al. Yeast expressing saccharolytic enzymes for consolidated bioprocessing using starch and cellulose. PCT/US2011/039192[WO2011/153516], 6-3-2011.

78. Walfridsson M, Anderlund M, Bao X, Hahn-Hägerdal B. Expression of different levels of enzymes from the *Pichia stipitis* XYL1 and XYL2 genes in *Saccharomyces cerevisiae* and its effects on product formation during xylose utilization. *Appl Microbiol Biotechnol* 1997;**48**:218–24.

79. Walfridsson M, Bao X, Anderlund M, Lilius G, Bulow L, Hahn-Hägerdal B. Ethanolic fermentation of xylose with *Saccharomyces cerevisiae* harboring the *Thermus thermophilus xylA* gene, which expresses an active xylose (glucose) isomerase. *Appl Environ Microbiol* 2013;**62**:4648–51.

80. Gardonyi M, Hahn-Hägerdal B. The *Streptomyces rubiginosus* xylose isomerase is misfolded when expressed in *Saccharomyces cerevisiae*. *Enzyme Microb Technol* 2003;**32**:252–9.

81. Amore R, Wilhelm M, Hollenberg CP. The fermentation of xylose—an analysis of the expression of Bacillus and Actinoplanes xylose isomerase genes in yeast. *Appl Microbiol Biotechnol* 1989;**30**:351–7.

82. Sarthy AV, McConaughy BL, Lobo Z, Sundstrom JA, Furlong CE, Hall BD. Expression of the *Escherichia coli* xylose isomerase gene in *Saccharomyces cerevisiae*. *Appl Envir Microbiol* 1987;**53**:1996–2000.

83. Kuyper M, Harhangi HR, Stave AK, Winkler AA, Jetten MS, De Laat WTAM, et al. High-level functional expression of a fungal xylose isomerase: the key to efficient ethanolic fermentation of xylose by *Saccharomyces cerevisiae*? *FEMS Yeast Res* 2003;**4**:69–78.

84. Brat D, Boles E, Wiedemann B. Functional expression of a bacterial xylose isomerase in *Saccharomyces cerevisiae*. *Appl Envir Microbiol* 2009;**75**:2304–11.

85. Brat D, Boles E. Isobutanol production from D-xylose by recombinant *Saccharomyces cerevisiae*. *FEMS Yeast Res* 2012;**13**:241–4.

86. Olofsson K, Runquist D, Hahn-Hägerdal B, Liden G. A mutated xylose reductase increases bioethanol production more than a glucose/xylose facilitator in simultaneous fermentation and co-fermentation of wheat straw. *Amb Express* 2011;**1**:1–4.

87. Karhumaa K, Sanchez RG, Hahn-Hägerdal B, Gorwa-Grauslund M-F. Comparison of the xylose reductase-xylitol dehydrogenase and the xylose isomerase pathways for xylose fermentation by recombinant. *Saccharomyces cerevisiae*. *Microb Cell Fact* 2007;**6**:1–10.

88. Katahira S, Fujita Y, Mizuike A, Fukuda H, Kondo A. Construction of a xylan-fermenting yeast strain through codisplay of xylanolytic enzymes on the surface of xylose-utilizing *Saccharomyces cerevisiae* cells. *Appl Envir Microbiol* 2004;**70**:5407–14.

89. Biely P, Tenkanen M. Enzymology of hemicellulose degradation. In: Harman, Kubicek CP, editors. *Trichoderma and gliocladium*. Taylor and Francis Publishers; 1999. p. 25–47.

90. Biely P, Vrsanská M, Tenkanen M, Kluepfel D. Endo-β-1,4-xylanase families: differences in catalytic properties. *J Biotechnol* 1997;**57**:151–66.

91. Margolles-Clark E, Tenkanen M, Nakari-Setalä T, Penttilä M. Cloning of genes encoding α-L-arabinofuranosidase and β-xylosidase from *Trichoderma reesei* by expression in *Saccharomyces cerevisiae*. *Appl Envir Microbiol* 1996;**62**:3840–6.

92. Richard P, Verho R, Putkonen M, Londesborough J, Penttilä M. Production of ethanol from L-arabinose by *Saccharomyces cerevisiae* containing a fungal L-arabinose pathway. *FEMS Yeast Res* 2007;**3**:185–9.

93. Becker J, Boles E. A modified *Saccharomyces cerevisiae* strain that consumes L-arabinose and produces ethanol. *Appl Environ Microbiol* 2003;**69**:4144–50.

94. Wisselink HW, Toirkens MJ, Wu O, Pronk JT, van Maris AJA. Novel evolutionary engineering approach for accelerated utilization of glucose, xylose and arabinose mixtures by engineered *Saccahromyces cerevisiae* strains. *Appl Environ Microbiol* 2009;**75**:907–14.

95. Wisselink HW, Toirkens MJ, Berriel MDRF, Winkler AA, van Dijken JP, et al. Engineering of *Saccharomyces cerevisiae* for efficient anaerobic alcoholic fermentation of L-arabinose. *Appl Envir Microbiol* 2007;**73**:4881–91.

96. Garcia Sanchez R, Karhumaa K, Fonseca C, Sànchez Nogué V, Almeida JRM, Larsson CU, et al. Improved xylose and arabinose utilization by an industrial recombinant *Saccharomyces cerevisiae* strain using evolutionary engineering. *Biotechnol Biofuels* 2010;**3**:1–13.

97. de Wet BJM, Van Zyl WH, Prior BA. Characterization of the Aureobasidium pullulans alpha-glucuronidase expressed in *Saccharomyces cerevisiae*. *Enzyme Microb Technol* 2006;**38**:649–56.

98. McCledon SD, Mao Z, Shin HD, Wagschal K, Chen RR. Designer xylanosomes: protein nanostructures for enhanced xylan hydrolysis. *Appl Biochem Biotechnol* 2012;**167**:395–411.

99. Sun J, Wen F, Si T, Xu JH, Zhao H. Direct conversion of xylan to ethanol by recombinant *Saccharomyces cerevisiae* strains displaying an engineered minihemicellulosome. *Appl Envir Microbiol* 2012;**78**:3837–45.

100. Crous JM, Pretorius IS, Van Zyl WH. Cloning and expression of the α-L-arabinofuranosidase gene (*ABFB*) of *Aspergillus niger* in *Saccharomyces cerevisiae*. *Appl Microbiol Biotechnol* 1996;**46**:256–60.

101. Margolles-Clark E, Ilmén M, Penttilä M. Expression patterns of ten hemicellulase genes of the filamentous fungus *Trichoderma reesei* on various carbon sources. *J Biotechnol* 1997;**57**:167–79.

102. Becerra M, Diaz Prado S, Cerdaín E, Gonzalez Siso MI. Heterologous *Kluyveromyces lactis* β-galactosidase secretion by *Saccharomyces cerevisiae* super-secreting mutants. *Biotechnol Lett* 2001;**23**:33–40.

103. Wildt S, Gerngross TU. The humanization of *N*-glycosylation pathways in yeast. *Nat Rev Microbiol* 2005;**3**:119–28.

104. Gasser B, Sauer M, Maurer M, Stadlmayr G, Mattanovich D. Transcriptomics-based identification of novel factors enhancing heterologous protein secretion in yeasts. *Appl Envir Microbiol* 2007;**73**:6499–507.

105. Kroukamp H, Den Haan R, Van Wyk N, Van Zyl WH. Overexpression of native *PSE1* and *SOD1* in *Saccaromyces cerevisiae* improved heterologous cellulase production. *Appl Energy* 2013;**102**:150–6.

106. Van Rensburg P, Van Zyl WH, Pretorius IS. Engineering yeast for efficient cellulose degradation. *Yeast* 1998;**14**:67–76.

107. Skory CD, Freer SN, Bothast RJ. Expression and secretion of the *Candida wickerhamii* extracellular beta-glucosidase gene, *bglB*, in *Saccharomyces cerevisiae*. *Curr Genet* 1996;**30**:417–22.

108. Zurbriggen BD, Penttila ME, Viikari L, Bailey MJ. Pilot scale production of a *Trichoderma reesei* endo-beta-glucanase by brewer's yeast. *J Biotechnol* 1991;**17**:133–46.

109. Hong J, Tamaki H, Akiba S, Yamamoto K, Kumagai H. Cloning of a gene encoding a highly stable endo-beta-1,4-glucanase from *Aspergillus niger* and its expression in yeast. *J Biosci Bioeng* 2001;**92**:434–41.

110. Ooi T, Minamiguchi K, Kawaguchi T, Okada H, Murao S, Arai M. Expression of the cellulase (FI-CMCase) gene of *Aspergillus aculeatus* in *Saccharomyces cerevisiae*. *Biosci Biotechnol Biochem* 1994;**58**:954–6.

111. Mochizuki D, Miyahara K, Hirata D, Matsuzaki H, Hatano T, Fukui S, et al. Overexpression and secretion of cellulolytic enzymes by δ-sequence-mediated multicopy integration of heterologous DNA sequences into the chromosomes of *Saccharomyces cerevisiae*. *J Ferment Bioeng* 1994;**77**:468–73.

112. Petersen SH, Van Zyl WH, Pretorius IS. Development of a polysaccharide degrading strain of *Saccharomyces cerevisiae*. *Biotechnol Tech* 1998;**12**:615–9.

113. Chow CM, Yague E, Raguz S, Wood DA, Thurston CF. The *cel3* gene of *Agaricus bisporus* codes for a modular cellulase and is transcriptionally regulated by the carbon source. *Appl Envir Microbiol* 1994;**60**:2779–85.

114. Mchunu NP. Expression of a modified xylanase in yeast [MSc thesis]. Durban University of Technology; 2009.

115. Margolles-Clark E, Tenkanen M, Nakari-Setala T, Penttila M. Cloning of genes encoding α-L-arabinofuranosidase and β-xylosidase from *Trichoderma reesei* by expression in *Saccharomyces cerevisiae*. *Appl Envir Microbiol* 1996;**62**:3840–6.

116. Zietsman AJJ, de Klerk D, Van Rensburg P. Coexpression of α-L-arabinofuranosidase and β-glucosidase in *Saccharomyces cerevisiae*. *FEMS Yeast Res* 2010;**11**:88–103.

117. Gomes K. Enzymatic modification of the functional properties of xylan from lignocellulose feedstocks [MSc thesis]. University of Stellenbosch; 2012.

118. Tang CM, Waterman LD, Smith MH, Thurston CF. The *cel4* gene of *Agaricus bisporus* encodes a β-mannanase. *Appl Envir Microbiol* 2001;**67**:2298–303.

119. Christgau S, Kauppinen S, Vind J, Kofod LV, Dalboge H. Expression cloning, purification and characterization of a β-1,4-mannanase from Aspergillus aculeatus Biochem. *Mol Biol Int* 1994;**33**:917–25.

120. Ximenes EA, Chen H, Kataeva IA, Cotta MA, Felix CR, Ljungdahl LG, et al. A mannanase, ManA, of the polycentric anaerobic fungus *Orpinomyces* sp. strain PC-2 has carbohydrate binding and docking modules. *Can J Microbiol* 2005;**51**:559–68.

121. Stalbrand H, Saloheimo A, Vehmaanpera J, Henrissat B, Penttila M. Cloning and expression in *Saccharomyces cerevisiae* of a *Trichoderma reesei* β-mannanase gene containing a cellulose binding domain. *Appl Envir Microbiol* 1995;**61**:1090–7.

122. Domingues L, Oliveira C, Castro I, Lima N, Teixeira JA. Production of β-galactosidase from recombinant *Saccharomyces cerevisiae* grown on lactose. *J Chem Technol Biotechnol* 2004;**79**:908–15.

CHAPTER 10

Identification of Genetic Targets to Improve Lignocellulosic Hydrocarbon Production in Trichoderma reesei *Using Public Genomic and Transcriptomic Datasets*

Shihui Yang[1], Wei Wang[2], Hui Wei[2], Michael E. Himmel[2], Min Zhang[1]

[1]*National Bioenergy Center, National Renewable Energy Laboratory (NREL), Golden, CO, USA;*
[2]*Biosciences Center, National Renewable Energy Laboratory (NREL), Golden, CO, USA*

Background

Bacteria, yeasts and fungi can naturally synthesize fatty acids, isoprenoids, or polyalkanoates for energy storage. These compounds have high energy densities and are compatible with current fuel infrastructure, permitting their exploitation for hydrocarbon fuel production.[1–4] Many microorganisms are being developed as potential biofuel production strains, yeast strains are currently the leading industrial biocatalyst microorganisms,[5] and engineered bacteria such as *Escherichia coli*, *Zymomonas mobilis*, *Corynebacterium glutamicum*, and *Bacillus subtilis* are also being developed and deployed to address commercially important inoculum requirements.[6–10] However, all have limitations for economic advanced biofuel production in terms of robustness, substrate use, productivity, and yield.

Consolidated bioprocessing (CBP) is a promising strategy for economic lignocellulosic biofuel production, which integrates all steps of enzyme production, saccharification, and fermentation biologically. A CBP microbial biocatalyst should therefore produce and secret cellulytic enzymes to solubilize lignocellulosic biomass substrates into simple sugars and, at the same time, produce desired chemicals efficiently with high yield and titer. Despite recent exploration on developing microbial consortia as CBP biocatalysts,[11–13] it will still be challenging to overcome the complexity of CBP consortia to meet the needs for commercial production of biofuels at an acceptable cost. The classical CBP strain development strategies focusing on a single microorganism are still the better approach for industrial biotechnology applications, which include two strategies termed "native strategy" to increase productivity of a native cellulolytic microorganism and "recombinant strategy" to enable lignocellulosic biomass use capability of a microbial biocatalyst with excellent productivity.[14,15] Despite a lot of efforts being devoted to developing a promising CBP strain in the direction of

Direct Microbial Conversion of Biomass to Advanced Biofuels. http://dx.doi.org/10.1016/B978-0-444-59592-8.00010-5

recombinant strategy to enable lignocellulosic assimilation capability of classical microbial biocatalysts such as yeast *Saccharomyces cerevisiae*, oleaginous yeast *Yarrowia lipolytica*, and bacterial species of *E. coli* and *Z. mobilis*, no obvious champion could be developed yet. *Clostridium* species are the models for native strategy, receiving intensive attention, but a lot of barriers still need to be overcome to increase productivity for commercial production.[16–19] Fungal CBP is a novel concept proposed by National Renewable Energy Laboratory scientists recently, and the feasibility and strategies to develop a model fungal cellulase producer, *T. reesei*, as the fungal CBP platform was discussed, which opens a new direction toward CBP development for industrial application.[20]

The filamentous fungus *T. reesei* is one of the main producers of cellulases and hemicellulases for commercial lignocellulosic bioethanol production, with efficient systems for cellulase induction and nutrient transportation.[21] It possesses at least three classical enzymes of exoglucanases (syn. cellobiohydrolases), endoglucanases, and β-glucosidases, with new players involved in cellulose degradation identified recently, such as GH61 polysaccharide monooxygenases (PMOs), expansin-like proteins swollenin (SWOI), and expansin/family 45 endoglucanase-like proteins (EEL1, EEL2, and EEL3).[22–24] The main cellulase production in *T. reesei* is regulated sophisticatedly, and various transcriptional factors (TFs) controlling cellulase gene expression have been discovered, such as XYR1, CRE1, ACE1, ACE2, AreA, BglR, and HAP2/3/5 complex.[23,25–32] In addition, the environmental conditions for cellulase induction have also been investigated, with several inducers identified, including cellulose, disaccharides of cellobiose, lactose and sophorose, and low-molecular weight compounds such as L-arabitol and L-sorbose.[33,34]

Recent technical breakthroughs in next-generation sequencing (NGS), systems biology, and synthetic biology[35–38] have significantly increased the amount of data available on the metabolism and regulation of biofuel-producing organisms, leading to a potential paradigm shift in industrial biocatalyst development.[3,4,38–50] As one of the major producers of cellulases and hemicellulases at the commercial scale, *T. reesei* has also been widely studied using genomic and systems biology approaches. For example, the genome sequence of *T. reesei* has been reported and annotated.[24] Comparative genomic studies between wild type and mutant strains with improved cellulase production (e.g., QM9414, RUT C30) were carried out, and the genetic loci associated with excellent cellulase production were identified and characterized.[51,52] In addition, quite a few studies using transcriptomic and proteomic approaches have been reported.[23,53–58] Although nearly all of them were focused on cellulase induction, and there is no systems biology study reported yet for *T. reesei* as a CBP candidate for biofuel production, these systems biology datasets still can help us understand the global transcriptional profiles of *T. reesei* in different environmental conditions. In this chapter, we will show the promise of using public genomic and transcriptomic datasets to help us identify genetic targets and to guide metabolic engineering practice of improving hydrocarbon production in *T. reesei*.

Materials and Methods

Trichoderma reesei *Protein Function Annotation and Pathway Reconstruction*

Although the 34-Mb genome sequence of *T. reesei* has been reported and annotated,[24] the annotation has not been systematically conducted since its first release to reflect the recent exponential explosion of the genomic information. To identify the proteins related to hydrocarbon (e.g., terpenoid and fatty acid) biosynthesis, metabolism, and regulation, the protein sequences of *T. reesei* has been extracted and reannotated functionally. In brief, 9143 protein sequences containing all manually curated and automatically annotated models chosen from the filtered model sets representing the best gene model of each locus (TreeseiV2_Frozen-GeneCatalog20081022.proteins.fasta) were downloaded from the JGI website (http://genome.jgi-psf.org/Trire2/Trire2.download.ftp.html) and imported into CLC Genomics Workbench (V7.0) as the reference protein sequences for Blast search. In addition, the protein sequences were also imported into Blast2GO for the functional annotation and CAZYmes Analysis Toolkit (CAT) for analysis and annotation of CAZYmes (Carbohydrate Active enZYmes),[59,60] which was then compared to a recent reannotated CAZy genes of *T. reesei*.[22] The KEGG pathways were extracted from annotation result, as was the information of KOG, enzyme code, and the reaction substrate(s) and product(s). The potential homologous gene(s) in *T. reesei* were identified by reiterated BlastP searches. The information of protein product and conserved domains were examined, and the pathway was reconstructed with the enzyme and pathway information from literature search (Figure 1).

Trichoderma reesei *Microarray and RNA Sequencing Dataset Collection and Transcriptomic Analyses*

Currently, 21 RNA sequencing (RNA-Seq) runs are available at National Center for Biotechnology Information Sequence Read Archive (NCBI SRA) database from three studies.[56,57] The first study (SRP018852) used ABI SOLiD four system, including eight samples of wild-type QM6a strain from three different conditions: (1) 48-h growth in glucose-based minimal media (three runs, 48Gluc); (2) 24-h growth after transferring mycelia from glucose-based media into media containing wheat straw as a sole carbon source (three runs, 24Straw); and (3) 5-h growth after addition of glucose to straw cultures (two runs, 5Gluc).[57] The second study (SRP024316) used Illumina Hiseq2000 to investigate the cellulase induction mechanism of wild-type strain TU-6 with four samples: WT_Avicel, WT_No Carbon, WT_glucose, and stp1 deletion mutant-Avicel. The third study (SRP034709) also used Illumina Hiseq2000 to investigate the cellulase formation mechanism by growing QM9414 strain in Mandels-Andreotti medium supplemented with either 1% cellulose, or 2% glucose, or 1 mM sophorose with three biological replicates.[56]

The SRA files of two RNA-Seq datasets (SRP018852 and SRP024316) were downloaded from NCBI and then converted into fastq files using the NCBI sratoolkit 2.3.3–4, which were

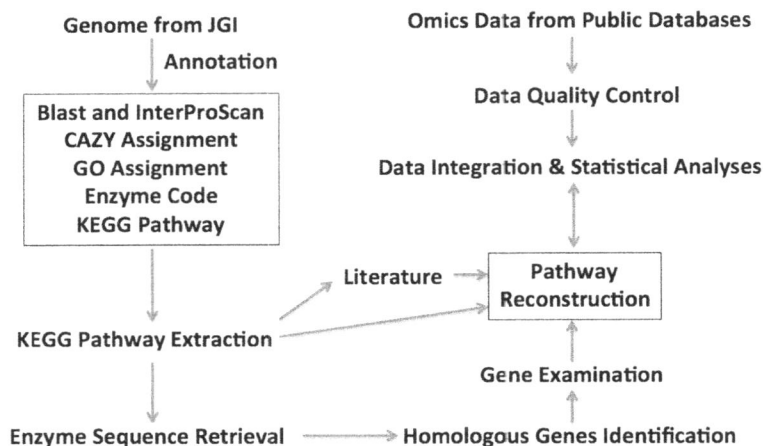

Figure 1
Flowchart of pathway reconstruction and omics data integration for this study.

then checked by FastQC software for read quality assurance before importing into CLC Genomics Workbench. These fastq files were then individually mapped to 9143 reference transcript sequences (TreeseiV2_FrozenGeneCatalog20081022). RNA-Seq analyses were then carried out to generate the reads per kilobase transcript per million reads (RPKM) values for all transcripts. In addition, the normalized counts in the file "GSE53629_QM9414_GLU_SPH_CEL_normalized.counts.txt" of the third study (SRP034709) for every transcript were downloaded, and the RPKM values were then calculated based on the formula of RPKM = $10^9 \times C/(N \times L)$, in which C is the number of mappable reads that fell onto the genes, N is the total number of mappable reads in the experiment, and L is the sum of the exons in base pairs. The RPKM values of all three RNA-Seq studies were then $\log_2$ transformed and merged as one file, and the average $\log_2$-based RPKM values for each condition were calculated from all available biological replicates, except one biological replicate of 2% glucose supplementation condition (GSM1297504) in the third study (SRP034709) was discarded due to its large variation with other two biological replicates (data not shown).

The distributions of $\log_2$-based RPKM values in each condition were then calculated, and the $\log_2$-based RPKM values for transcripts identified in the glycolysis/fermentation and terpenoid biosynthesis pathways were extracted, and the ratios for two RNA-Seq studies with replicates (SRP018852 and SRP034709) were calculated to compare the effect of other carbon sources (straw biomass in SRP018852, whereas sophorose and cellulose in SRP034709) with glucose. The gene expression strength and differential expression comparing glucose with other carbon sources were then color-coded and reflected on the metabolic pathways.

Within the NCBI GEO database, 20 studies were deposited for 166 samples. We looked into these studies and selected eight microarray studies (GSE53874, GSE46155, GSE39276,

GSE39111, GSE36448, GSE27471, GSE27581, GSE22687) for 43 different conditions of strains, carbon sources, culture conditions, or growth phases. Since these arrays used different platforms, the annotation of each platform was also downloaded, and the gene expression values for each condition were merged into one table based on common gene identification or sequence name. Finally, these $\log_2$-based array data were further merged with the RNA-Seq dataset, which were then imported into JMP Genomics 6.0 (SAS Inc., NC) for statistical analyses.

Results and Discussions

Identify Target Genes for Metabolic Engineering by Genomic Metabolic Pathway Analysis

We have completed the protein annotation for *T. reesei* using Blast2GO and identified the potential carbohydrate-active enzymes (CAZymes). In brief, 1658 proteins have been assigned to 115 Kyoto Encyclopedia of Genes and Genomes (KEGG) pathways, and 218 proteins are homologous to 83 CAZyme families, with GH28 the most abundant one (29 enzymes). In addition, the KEGG pathways of glycolysis, fermentation, and terpenoid biosynthesis have been extracted from the KEGG results with the literature reports; we reconstructed these two specific pathways by identifying the homologous genes in *T. reesei* through Blast and by manually checking the information of the conserved domains and chemical reactions that these enzymes are involved in (Figure 2). In summary, *T. reesei* has complete glycolysis, fermentation, and terpenoid biosynthesis pathways, although phosphate acetyltransferase (PAT), pyruvate formate-lyase (PFL), and farnesene biosynthesis genes were not identified.

Trichoderma reesei does not possess the methylerythritol phosphate (MEP) pathway, but uses mevalonate pathway for terpenoid biosynthesis (Figure 2). Although several proteins homologous to β-farnesene synthase have been identified, including Trire2|56,966, Trire2|53,079, Trire2|75,468, and Trire2|37,950, considering their relatively low similarity (1.52E-33, 4.26E-33, 9.0E-28, and 3.38E-27 respectively) and the fact that the parental strain QM6a does not produce farnesene, heterogonous farnesene synthase gene metabolic engineering is required for farnesene production in the *T. reesei*. In addition, the detection of significant amount of nerolidol within parental *T. reesei* strain indicates the existence of active pathway for nerolidol production (data not shown). Protein Trire2|112,028 is homologous to (3S,6E)-nerolidol synthase with low E-value of 2.3E-14, and Trire2|53,079 is homologous to (3R,6E)-nerolidol synthase (4.2.3.49) with an E-value of 2.0E-28, which may contribute to the nerolidol production in *T. reesei*. (3R,6E)-nerolidol synthase belongs to acyclic sesquiterpene synthase involved in sesquiterpenoid and triterpenoid biosynthesis and is homologous to β-farnesene synthase (4.2.3.47) for β-farnesene biosynthesis and farnesyl diphosphatase (3.1.7.6) for (2E,6E)-farnesol biosynthesis (Figure 2). Although the similarity of

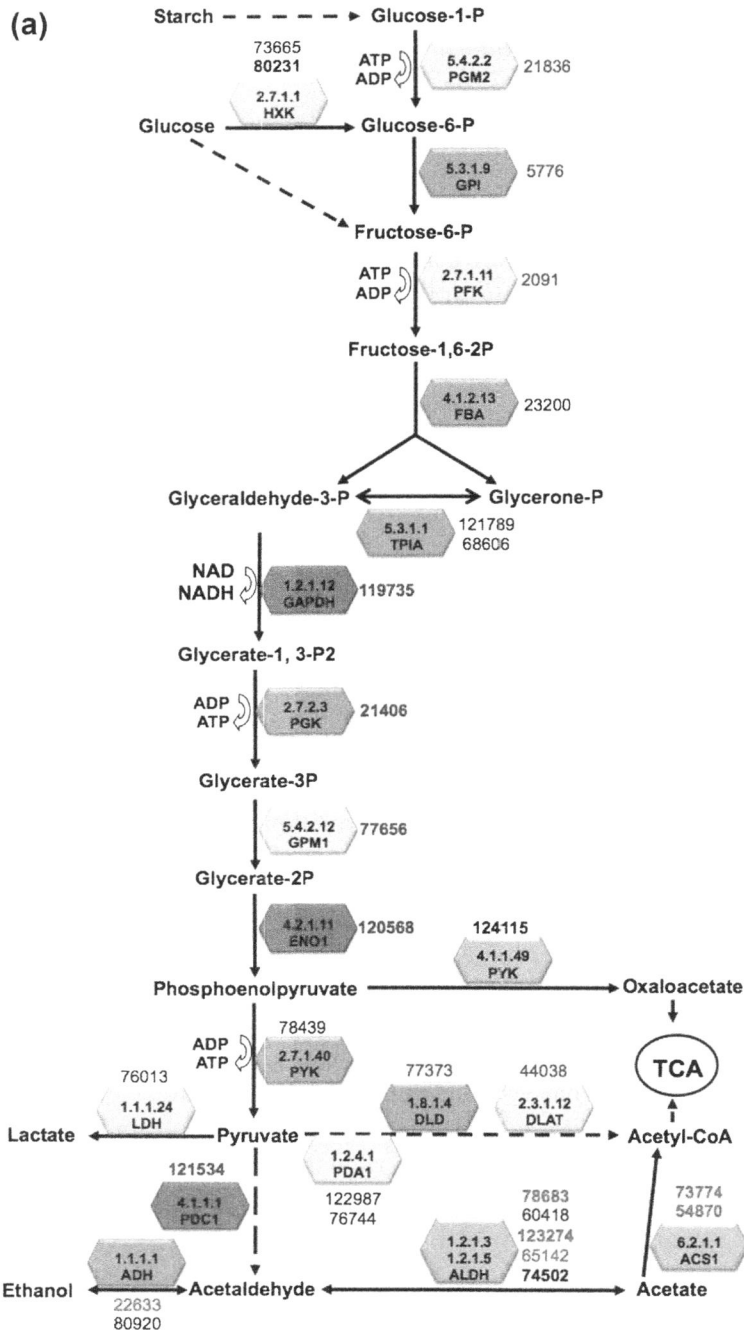

Figure 2

The reconstructed glycolysis and fermentation pathway (**a**) and terpenoid biosynthesis pathway (**b**). Enzymes are represented by gray diamond boxes. The ones filled with red, salmon, or canta-loupe colors are those with transcriptional expression strength among the top 2.5%, 10%, or 25% of all transcripts respectively. The one filled with blue color is among the lowest 2.5%. The number(s) adjacent to enzyme represent the homologous proteins ID number identified in *T. reesei* with yellow highlight indicating reasonable similarity to homologous proteins. Multiple numbers were included for enzyme with multiple homologues. Red font indicates biomass induced compared to glucose, whereas blue indicates downregulated. Bold font indicates at least 2-fold changes.

(b)

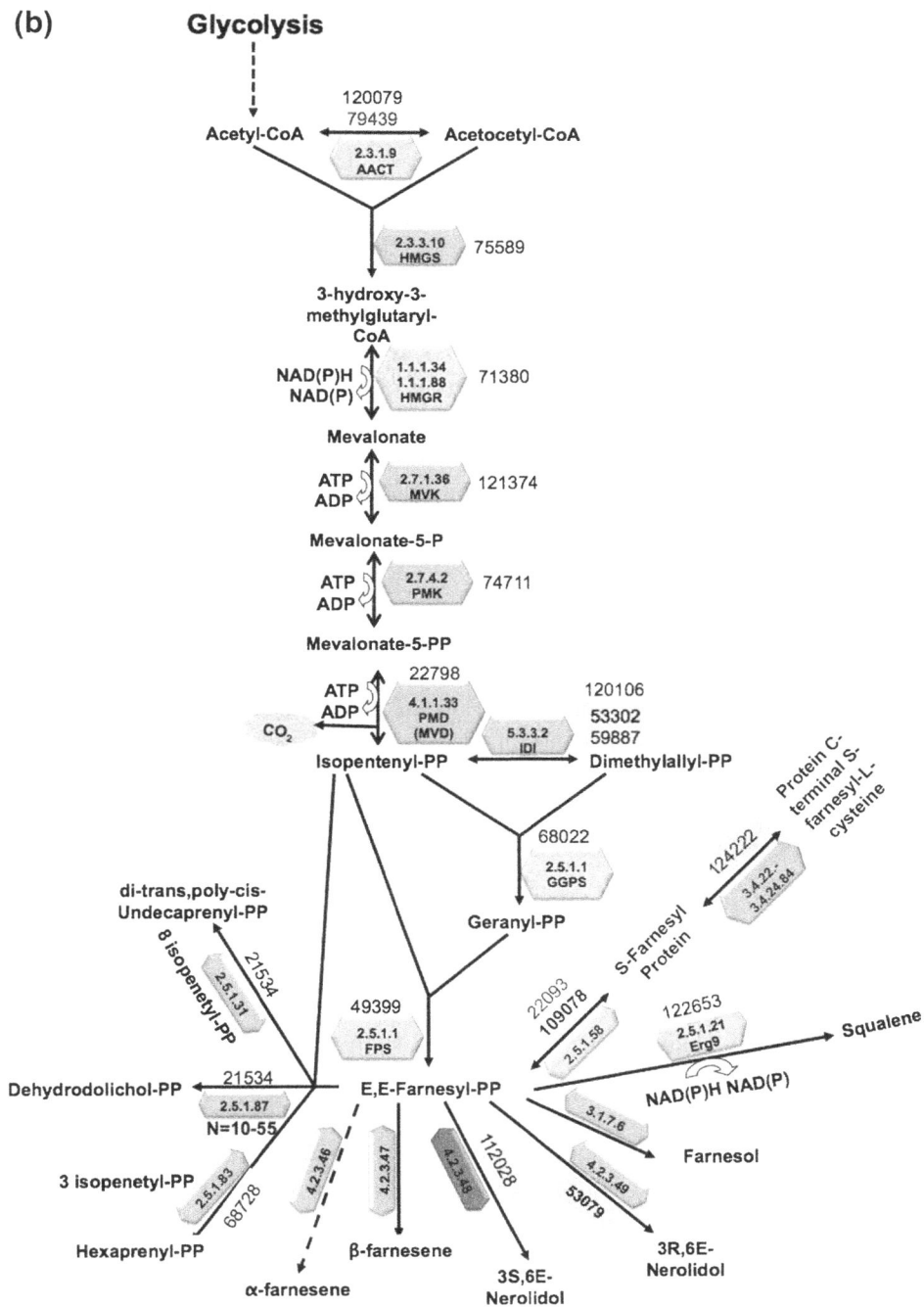

Figure 2—cont'd

Trire2|53,079 to β-farnesene synthase is slightly higher than to (3R,6E)-nerolidol synthase, Trire2|53,079 most likely is a nerolidol synthase instead of β-farnesene synthase, because nerolidol is detected in *T. reesei*, but not farnesene. Therefore, the knockout of enzyme involved in nerolidol production (e.g., Trire2|53,079 and Trire2|112,028) may help direct the carbon flow to the farnesene production because these enzymes are competing for the same substrate of farnesyl diphosphate (Figure 2).

In addition, multiple hits from Blast result were identified for several enzymes (Figure 2). Further characterization of these multiple hits for one reaction indicated that most of them actually are potential paralogues with common conserved functional domains (e.g., HXK2, TPIA, ADH, ACS1, ALDH, and AACT). However, some of them are subunits of a functional enzyme complex with different functional domains. For example, Trire2|122,987 and Trire2|76,744 are the alpha and beta subunits, respectively, of a potential pyruvate dehydrogenase E1 component (PDA1) in the fermentation pathway, and Trire2|22,093 and Trire2|109,078 are the alpha and beta subunits, respectively, of a potential farnesyltransferase in the terpenoid biosynthesis pathway (Figure 2).

Identify Target Genes for Metabolic Engineering by Transcriptomic Analysis

Genomic annotation and pathway analysis can provide guidance on metabolic engineering. For example, it is a straightforward approach to include all genes encoding the enzyme subunits for metabolic engineering. However, it will be challenging to pick the genetic candidate for a specific enzyme with several paralogous genes. In this section, we will discuss the integration of omics datasets, such as transcriptomic data to facilitate the selection of genetic targets among multiple paralogues and to identify key pathway genes for metabolic engineering.

Two different kinds of information were obtained through transcriptomic data integration and statistical analyses, the gene expression strength in terms of the $\log_2$-based RPKM values and the differential gene expression pattern. Results based on RNA-Seq were primarily used in this work because one RNA-Seq study (SRP018852) used QM6α wild type and two RNA-Seq studies of SRP018852 and SRP034709 contain RNA-Seq runs of multiple biological replicates for statistical analyses. Because only raw sequencing run data are available for one RNA-Seq study (SRP024316), we downloaded the SRA files and generated our in-house RPKM values for this RNA-Seq study using our RNA-Seq data analysis pipeline.[61] To validate our RNA-Seq analysis results for this study, we compared the RPKM values generated similarly for another RNA-Seq study (SRP018852), using its raw sequencing SRA datasets with the published RPKM values from this study (GEO# GSE44648, GSM1088358_5Gluc2.htseq.rpkm.txt). Our in-house RNA-Seq data analysis results have high correlation among biological replicates. For example, the correlation coefficients for the biological replicates of three conditions in SRP018852 study were all above 0.93 (Figure 3). In addition, our reanalyzed RNA-Seq results of RPKM values for the transcripts had high

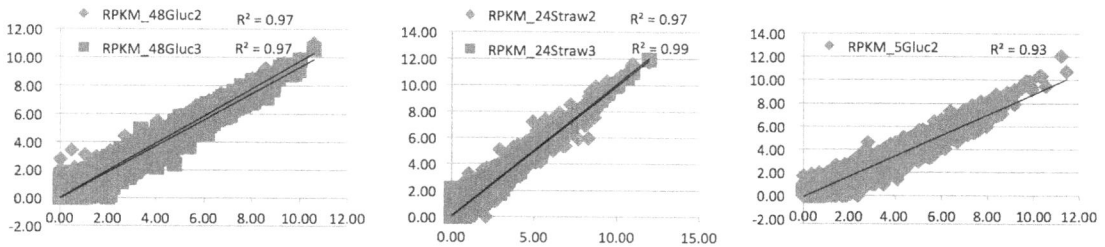

Figure 3

Correlation coefficients for biological replicates of 3 conditions in a RNA-Seq study SRP018852. Three conditions are 48 h growth in glucose-based minimal media, 24 h after transfer of mycelia from glucose-based media into media containing wheat straw as a sole carbon source, and 5 h after addition of glucose to straw cultures from left to right respectively.

correlations with published RPKM values. For example, our reanalyzed $\log_2$-based RPKM values for one sample of 48-h growth in glucose (SRR764963) had a correlation coefficient of 0.95 with the published $\log_2$-based RPKM values.

The expression values of these three RNA-Seq studies were then combined, and the average transcription abundances were calculated for each condition. The distributions of $\log_2$-transformed RPKM values for each condition were then calculated. In general, the average expression values of two studies using Illumina HiSeq2000 are similar, but greater than the first one using ABI SOLiD 4 System (SRP018852), and the average expression values in different conditions using same NGS platform were similar (Table 1). The gene expression abundance in each condition was compared to the statistical quartile values, with three artificial levels of abundant expression (the ones with the top 2.5%, 10%, or 25% abundance are highlighted by red, salmon, or cantaloupe color), and three artificial levels of scarce expression (the ones with the bottom 2.5%, 10%, or 25% abundance are highlighted by blue color from darkest to lightest), which were used for gene expression visualization on metabolic pathways (Figure 2).

The $\log_2$-based RPKM values for genes involved in glycolysis/fermentation and terpenoid biosynthesis pathways (Figure 2) were extracted, and the ratios were calculated (Table 2). Clearly, genes involved in glycolysis and the fermentation pathway had strong expression in different conditions. Nearly all of the genes were among the top 25%, most of them among the top 10%, and some of them among the top 2.5% abundant ones. The abundance of glycolysis/fermentation gene expression has been reported in other microorganisms, such as *Z. mobilis* and *Clostridium thermocellum*.[62,63] However, except for enzymes involved in five reactions, most transcripts encoding the enzymes involved in the terpenoid biosynthesis pathway did not express abundantly (Figure 2, Table 2), which is consistent with the role of glycolysis/fermentation on essential cellular metabolism and terpenoid biosynthesis for secondary metabolites.

Table 1: Distributions of three RNA-Seq studies (SRP018852, SRP034709, SRP024316) for each condition

RNA-Seq Study	SRP018852			SRP034709			SRP024316			
Conditions	48Gluc	24Straw	Straw5Gluc	QM6a_Gluc	QM6a_Sph	QM6a_Cel	TU6_Avicel	Step1_Avicel	STU6_Gluc	TU6_No_Carbon
100% (maximum)	10.71	11.88	11.63	17.88	17.16	17.52	16.11	15.78	15.11	15.44
Top 0.5%	7.70	8.22	7.75	10.34	10.11	10.24	10.64	10.65	11.05	11.50
Top 2.5%	5.98	6.27	5.68	8.69	8.46	8.53	9.06	8.84	9.32	9.08
Top 10%	4.29	4.75	3.89	7.08	7.02	6.98	6.91	6.64	7.06	6.66
Top 25%	3.01	3.68	2.57	5.89	5.94	5.85	5.53	5.22	5.59	4.75
50% (median)	1.67	2.55	1.43	4.52	4.67	4.66	4.20	3.84	4.16	3.08
Bottom 25%	0.40	1.10	0.36	2.30	2.87	2.85	2.09	1.89	1.86	1.24
Bottom 10%	0.02	0.09	0	0.33	0.61	0.62	0	0	0	0
Bottom 2.5%	0	0	0	0	0	0.06	0	0	0	0
0.50%	0	0	0	0	0	0	0	0	0	0
0% (minimum)	0	0	0	0	0	0	0	0	0	0
Mean	1.95	2.54	1.72	4.18	4.35	4.34	3.95	3.70	3.93	3.28
Std Dev	1.72	1.77	1.60	2.45	2.31	2.29	2.47	2.41	2.58	2.51
Std Err Mean	0.02	0.02	0.02	0.03	0.02	0.02	0.03	0.03	0.03	0.03
Transcript number	9126	9126	9126	9129	9129	9129	9129	9129	9129	9129

Gluc, glucose; Sph, sphorose; Cel, cellulose.

Table 2: The transcription strength and ratios for transcripts involved in glycolysis/fermentation and terpenoid biosynthesis pathway

Gluc, glucose; Sph, sphorose; Cel, cellulose.

Interestingly, when other carbon sources were used and compared to glucose, the gene expression patterns for glycolysis/fermentation and terpenoid biosynthesis were different. Except for the fact that several fermentation pathway genes from acetaldehyde to both ethanol and acetate were upregulated, nearly all glycolysis pathway genes and fermentation pathway genes from pyruvate to both acetaldehyde and acetyl-CoA were downregulated. Instead, the differentially expressed terpenoid biosynthesis pathway genes were upregulated when carbon sources other than glucose were used (Figure 2, Table 2).

The integration of transcriptomic data into the metabolic pathway thus provides additional information to identify genetic targets for strain improvement. For example, four genes (Trire2|122,653, Trire2|109,078, Trire2|124,222, and Trire2|22,093) encoding three enzymes involved in three metabolic reactions were upregulated, and two of them had more abundant transcripts than other terpenoid biosynthesis pathway enzymes (Figure 2, Table 2). Because these reactions are competing for the same substrate of farnesyl-PP for farnesene production, it is a reasonable strategy to divert the reactions from these reactions to increase farnesene production. Because the products of these two reactions (squalene and S-farnesyl protein) may be involved in essential metabolism, a potential knockdown instead of knockout of these four genes may help channel the farnesyl-PP to farnesene production while maintaining viable cellular growth. Comparing the RNA-Seq results with microarray data that contain more conditions, Trire2|122,653 and Trire2|109,078 were also highly expressed with strong intensity, whereas Trire2|22,093 was less abundant than Trire2|122,653 and Trire2|109,078 (Figure 4). Therefore, Trire2|122,653 and Trire2|109,078 could be the first two targets for knockdown.

Moreover, because genes involved in the terpenoid biosynthesis pathway generally had low transcript abundance, overexpression of several genes channeling acetyl-CoA to farnesyl-PP with less transcripts (e.g., Trire2|75,589, Trire2|59,887, and Trire2|120,106) may also help drive the key metabolite acetyl-CoA to farnesene production instead of other metabolic reactions, such as fatty acid biosynthesis (Figures 2 and 4, Table 2).

Investigate Transcriptional Regulators

Cellulase production in *T. reesei* is a secondary metabolism controlled by a complex regulatory network responding to various environmental stimuli, such as carbon and nitrogen sources, temperature, light, and pH.[64] Several global regulators involved in cellulase production have been identified and characterized, including the global carbon regulator CRE1 (Trire2|120,117) mediating carbon catabolite repression,[25–27] nitrogen regulator AreA (Trire2|76,817)[28] cellulase, and hemicellulase regulators of ACE1 (Trire2|75,418) and XYR1 (Trire2|122,208)[29,30] cellulase gene regulator ACE2 (Trire2|78,445)[31] and a $Zn(II)^2$ Cys^6-type beta-glucosidase regulator BglR (Trire2|52,368).[32]

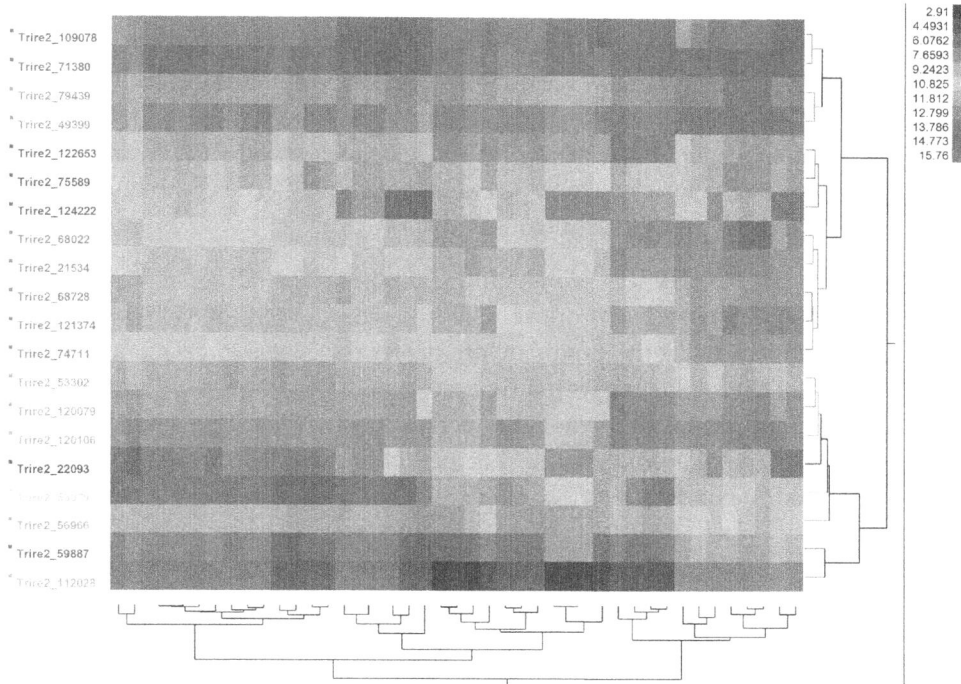

Figure 4

Hierarchical clustering the expression intensity of terpenoid biosynthesis pathway genes in different microarray experiments.

Because other carbon sources (e.g., straw biomass, cellulose, Avicel) were used to induce cellulase production in the recent two RNA-Seq studies,[56,57] the induction of terpenoid biosynthesis pathway genes under these conditions may indicate the competition between cellulase production and terpenoid biosynthesis. The expression profiles of these global regulators were investigated, and the results indicated that these regulators were induced when switching from glucose to biomass, and XYR1 was induced when either biomass or Avicel was used as carbon source compared to glucose, which is consistent with a previous report.[65] Further investigation is needed to understand the regulation of secondary metabolism to keep both excellent cellulolytic capability and high secondary metabolite productivity.

Identify Promoters with Different Strength for Metabolic Engineering

Except for several promoters related to cellobiohydrolase genes (e.g., *cbh1* and *cbh2*) and glycolysis pathway (e.g., GPD, PDC),[66-70] few works have been carried out to identify and characterize the *T. reesei* promoters systematically. The gene expression abundance data can also be used to help identify inducible promoters, or constitutive promoters with strong, medium, or weak intensity. The expression abundance in terms of log_2-based values for all microarray or RNA-Seq experiments were combined, and hierarchical clustering analyses were carried out for RNA-Seq datasets or array datasets separately (Figure 5). To exclude the discrepancy between sequencing platforms (Table 1), the log_2-based RPKM values were normalized for hierarchical clustering analyses (data not shown). The common ones between two sets of datasets were chosen as candidates.

It will be straightforward to identify the inducible promoters depending on the conditions of the available transcriptomic datasets. For example, we can select the ones highly expressed in Avicel or even biomass to select the candidate promoters for gene expression under these conditions. The promoter candidates with different strength of strong, medium, or weak were selected from the common ones between both microarray and RNA-Seq data hierarchical clustering results. Forty-nine promoters could be potential candidates as strong promoters

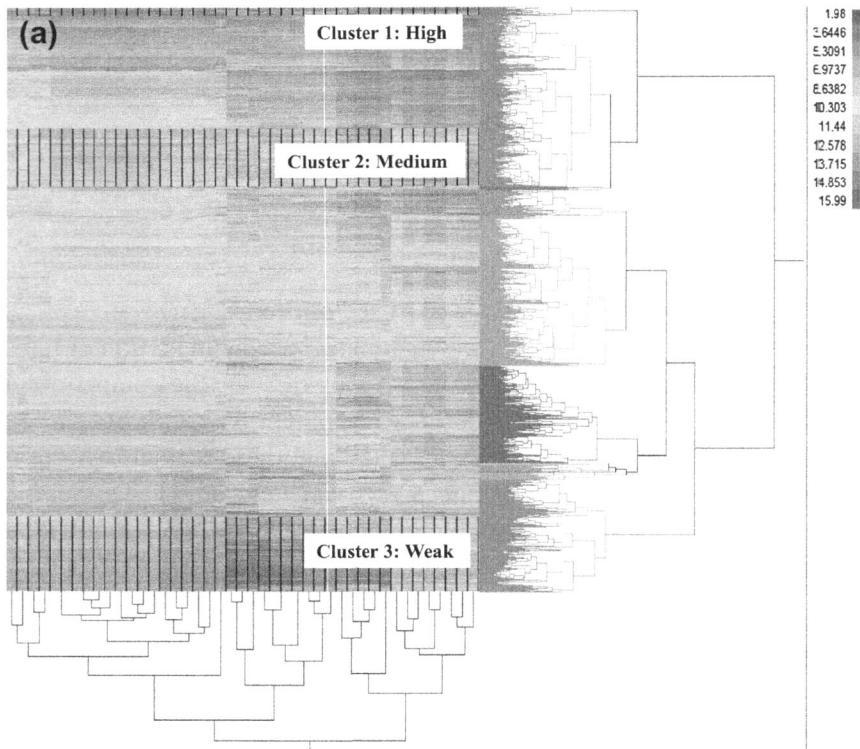

Figure 5
Hierarchical clustering the expression intensity of all *T. reesei* genes in different microarray (**a**) and RNA-Seq (**b**) experiments.

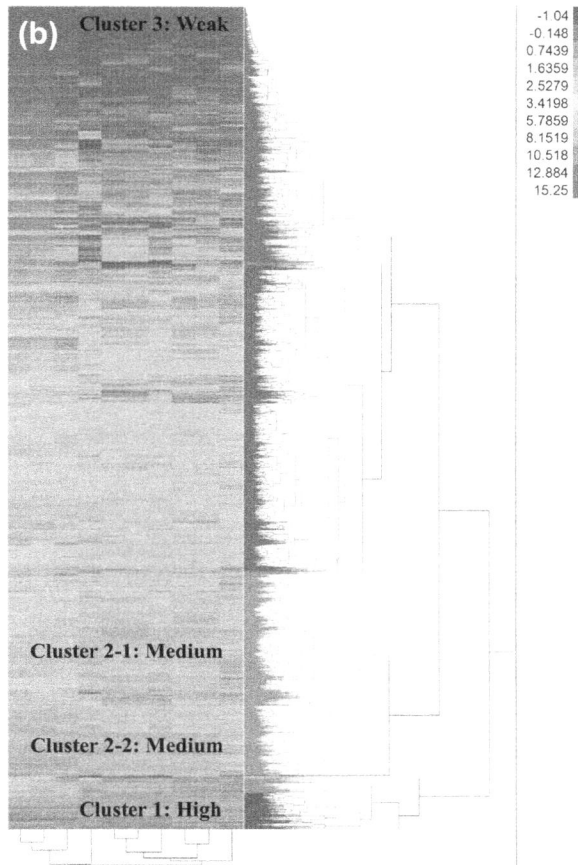

Figure 5—cont'd

(Figure 6), and actually some of these are consistent with previous literature report (e.g., the promoters of Trire2l119,735: gpd, glyceraldehyde-3-phosphate dehydrogenase and Trire2l23,200: fructose-bisphosphate aldolase) using qPCR to confirm the constitutive strong promoter for xylanase overexpression in *T. reesei*.[70] Compared to strong promoter candidates, there are many more candidates for weak promoters (Figure 6). The promoters with artificial medium strength are less obvious, among which potentially two clusters consistently have medium transcript abundance in different conditions, which make the numbers of promoter candidates with medium strength to be either 64 or 139 (Figure 6).

Conclusions and Perspectives

Although public genomic and systems biology datasets have heterogonous characteristics in terms of strain background, carbon source, culture method, growth condition, and sequencing

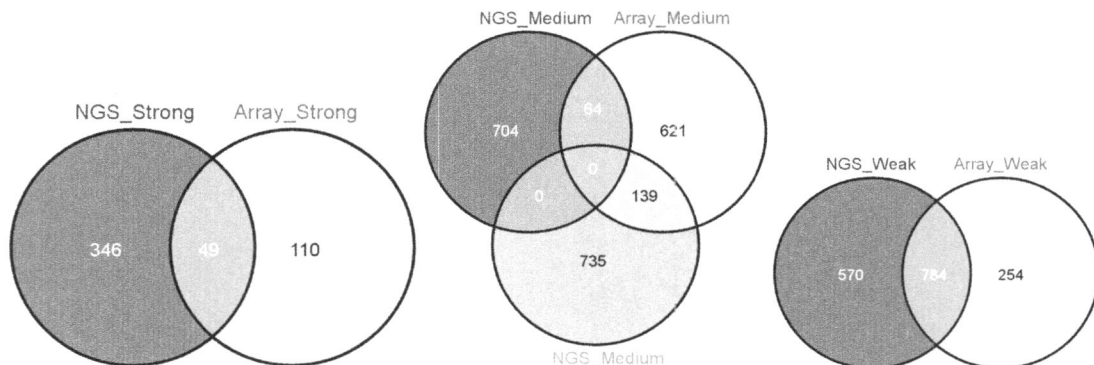

Figure 6
Venn diagrams to identify promoters with strong (**left**), medium (**center**), or weak (**right**) strength. The common ones identified from both microarray and RNA-Seq experiments are highly potential the candidate promoters for further investigation.

platform that may compromise accurate investigation, they do provide valuable information for the research community, especially for datasets retrieved from community-recognized data repositories. In this work, we clearly demonstrated the potential to identify genetic targets for farnesene production improvement in *T. reesei* using public genomic and transcriptomic datasets. The results indicate that *T. reesei* likely uses a mevalonate pathway for terpenoid biosynthesis, and heterogonous farnesene synthase(s) are needed for farnesene production. Several genetic targets were proposed to improve farnesene production based on genomic and transcriptomic pathway gene analyses. In addition, promoters with different strength were identified, while cautions for metabolic engineering based on genomic and transcriptomic analyses were also raised to deal with enzymes with multiple subunits or with multiple paralogous genes. Finally, the understanding of the regulation network controlling the biosynthesis of the secondary metabolites of terpenoids could help future efforts to increase the farnesene productivity, which needs further experimental work to compare the transcriptional profiles of farnesene synthase gene expression in different conditions.

Acknowledgment

Funding for this work was provided by the DOE EERE Bioenergy Technologies Office (BETO).

References

1. Peralta-Yahya PP, Ouellet M, Chan R, Mukhopadhyay A, Keasling JD, Lee TS. Identification and microbial production of a terpene-based advanced biofuel. *Nat Commun* 2011;**2**:483.
2. Peralta-Yahya PP, Keasling JD. Advanced biofuel production in microbes. *Biotechnol J* 2010;**5**:147–62.
3. Connor MR, Liao JC. Microbial production of advanced transportation fuels in non-natural hosts. *Curr Opin Biotechnol* 2009;**20**:307–15.

4. Atsumi S, Hanai T, Liao JC. Non-fermentative pathways for synthesis of branched-chain higher alcohols as biofuels. *Nature* 2008;**451**:86–9.

5. Hahn-Hagerdal B, Galbe M, Gorwa-Grauslund MF, Liden G, Zacchi G. Bio-ethanol–the fuel of tomorrow from the residues of today. *Trends Biotechnol* 2006;**24**:549–56.

6. Dien BS, Cotta MA, Jeffries TW. Bacteria engineered for fuel ethanol production: current status. *Appl Microbiol Biotechnol* 2003;**63**:258–66.

7. Alper H, Stephanopoulos G. Engineering for biofuels: exploiting innate microbial capacity or importing biosynthetic potential? *Nat Rev Micro* 2009;**7**:715–23.

8. Blombach B, Eikmanns BJ. Current knowledge on isobutanol production with *Escherichia coli, Bacillus subtilis* and *Corynebacterium* glutamicum. *Bioeng Bugs* 2011;**2**.

9. Blombach B, Riester T, Wieschalka S, Ziert C, Youn JW, Wendisch VF, et al. *Corynebacterium glutamicum* tailored for efficient isobutanol production. *Appl Envir Microbiol* 2011;**77**:3300–10.

10. Smith KM, Cho KM, Liao JC. Engineering *Corynebacterium glutamicum* for isobutanol production. *Appl Microbiol Biotechnol* 2010;**87**:1045–55.

11. Salimi F, Mahadevan R. Characterizing metabolic interactions in a clostridial co-culture for consolidated bioprocessing. *BMC Biotechnol* 2013;**13**.

12. Zuroff TR, Xiques SB, Curtis WR. Consortia-mediated bioprocessing of cellulose to ethanol with a symbiotic *Clostridium phytofermentans*/yeast co-culture. *Biotechnol Biofuels* 2013;**6**:59.

13. Minty JJ, Singer ME, Scholz SA, Bae CH, Ahn JH, Foster CE, et al. Design and characterization of synthetic fungal-bacterial consortia for direct production of isobutanol from cellulosic biomass. *Proc Natl Acad Sci USA* 2013;**110**:14592–7.

14. Hasunuma T, Okazaki F, Okai N, Hara KY, Ishii J, Kondo A. A review of enzymes and microbes for lignocellulosic biorefinery and the possibility of their application to consolidated bioprocessing technology. *Bioresour Technol* 2013;**135**:513–22.

15. Lynd LR, van Zyl WH, McBride JE, Laser M. Consolidated bioprocessing of cellulosic biomass: an update. *Curr Opin Biotechnol* 2005;**16**:577–83.

16. Hasunuma T, Rondo A. Development of yeast cell factories for consolidated bioprocessing of lignocellulose to bioethanol through cell surface engineering. *Biotechnol Adv* 2012;**30**:1207–18.

17. Yamada R, Hasunuma T, Kondo A. Endowing non-cellulolytic microorganisms with cellulolytic activity aiming for consolidated bioprocessing. *Biotechnol Adv* 2013;**31**:754–63.

18. la Grange DC, den Haan R, van Zyl WH. Engineering cellulolytic ability into bioprocessing organisms. *Appl Microbiol Biotechnol* 2010;**87**:1195–208.

19. Olson DG, McBride JE, Shaw AJ, Lynd LR. Recent progress in consolidated bioprocessing. *Curr Opin Biotechnol* 2012;**23**:396–405.

20. Xu Q, Singh A, Himmel ME. Perspectives and new directions for the production of bioethanol using consolidated bioprocessing of lignocellulose. *Curr Opin Biotechnol* 2009;**20**:364–71.

21. Schuster A, Schmoll M. Biology and biotechnology of *Trichoderma. Appl Microbiol Biotechnol* 2010;**87**:787–99.

22. Hakkinen M, Arvas M, Oja M, Aro N, Penttila M, Saloheimo M, et al. Re-annotation of the CAZy genes of *Trichoderma reesei* and transcription in the presence of lignocellulosic substrates. *Microb Cell Fact* 2012;**11**:134.

23. Verbeke J, Coutinho P, Mathis H, Quenot A, Record E, Asther M, et al. Transcriptional profiling of cellulase and expansin-related genes in a hypercellulolytic *Trichoderma reesei. Biotechnol Lett* 2009;**31**:1399–405.

24. Martinez D, Berka RM, Henrissat B, Saloheimo M, Arvas M, Baker SE, et al. Genome sequencing and analysis of the biomass-degrading fungus *Trichoderma reesei* (syn. *Hypocrea jecorina*). *Nat Biotechnol* 2008;**26**:553–60.

25. Dowzer CE, Kelly JM. Cloning of the creA gene from Aspergillus nidulans: a gene involved in carbon catabolite repression. *Curr Genet* 1989;**15**:457–9.

26. Ilmen M, Thrane C, Penttilä M. The glucose repressor genecre1 of *Trichoderma*: isolation and expression of a full-length and a truncated mutant form. *Mol Gen Genet MGG* 1996;**251**:451–60.

27. Portnoy T, Margeot A, Linke R, Atanasova L, Fekete E, Sandor E. et al. The CRE1 carbon catabolite repressor of the fungus *Trichoderma reesei*: a master regulator of carbon assimilation. *BMC Genomics* 2011;**12**:269.

28. Hynes M. Studies on the role of the are A gene in the regulation of nitrogen catabolism in *Aspergillus nidulans. Aust J Biol Sci* 1975;**28**:301–14.

29. Stricker AR, Grosstessner-Hain K, Würleitner E, Mach RL. Xyr1 (xylanase regulator 1) regulates both the hydrolytic enzyme system and D-xylose metabolism in *Hypocrea jecorina. Eukaryot Cell* 2006;**5**:2128–37.

30. Aro N, Ilmén M, Saloheimo A, Penttilä M. ACEI of *Trichoderma reesei* is a repressor of cellulase and xylanase expression. *Appl Envir Microbiol* 2003;**69**:56–65.

31. Aro N, Saloheimo A, Ilmén M, Penttilä M. ACEII, a novel transcriptional activator involved in regulation of cellulase and xylanase genes of *Trichoderma reesei. J Biol Chem* 2001;**276**:24309–14.

32. Nitta M, Furukawa T, Shida Y, Mori K, Kuhara S, Morikawa Y, et al. A new Zn(II)(2)Cys(6)-type transcription factor BglR regulates beta-glucosidase expression in *Trichoderma reesei. Fungal Genet Biol* 2012;**49**:388–97.

33. Schmoll M, Kubicek CP. Regulation of *Trichoderma* cellulase formation: lessons in molecular biology from an industrial fungus. A review. *Acta Microbiol Immunol Hung* 2003;**50**:125–45.

34. Nogawa M, Goto M, Okada H, Morikawa Y. L-Sorbose induces cellulase gene transcription in the cellulolytic fungus *Trichoderma reesei. Curr Genet* 2001;**38**:329–34.

35. Gibson DG, Glass JI, Lartigue C, Noskov VN, Chuang RY, Algire MA, et al. Creation of a bacterial cell controlled by a chemically synthesized genome. *Science* 2010;**329**:52–6.

36. Klein-Marcuschamer D, Yadav VG, Ghaderi A, Stephanopoulos GN. De novo metabolic engineering and the promise of synthetic DNA. *Adv Biochem Eng Biotechnol* 2010;**120**:101–31.

37. Salis HM, Mirsky EA, Voigt CA. Automated design of synthetic ribosome binding sites to control protein expression. *Nat Biotechnol* 2009;**27**:946–50.

38. Picataggio S. Potential impact of synthetic biology on the development of microbial systems for the production of renewable fuels and chemicals. *Curr Opin Biotechnol* 2009;**20**:325–9.

39. Tyo KE, Kocharin K, Nielsen J. Toward design-based engineering of industrial microbes. *Curr Opin Microbiol* 2010;**13**:255–62.

40. Stafford DE, Stephanopoulos G. Metabolic engineering as an integrating platform for strain development. *Curr Opin Microbiol* 2001;**4**:336–40.

41. Jarboe LR, Zhang X, Wang X, Moore JC, Shanmugam KT, Ingram LO. Metabolic engineering for production of biorenewable fuels and chemicals: contributions of synthetic biology. *J Biomed Biotechnol* 2010; **2010**:761042.

42. Clomburg JM, Gonzalez R. Biofuel production in *Escherichia coli*: the role of metabolic engineering and synthetic biology. *Appl Microbiol Biotechnol* 2010;**86**:419–34.

43. Lee JW, Kim HU, Choi S, Yi J, Lee SY. Microbial production of building block chemicals and polymers. *Curr Opin Biotechnol* 2011;**22**(6):758–67.

44. Yang S, Land ML, Klingeman DM, Pelletier DA, Lu T-YS, Martin SL, et al. Paradigm for industrial strain improvement identifies sodium acetate tolerance loci in *Zymomonas mobilis* and *Saccharomyces cerevisiae. Proc Natl Acad Sci USA* 2010;**107**:10395–400.

45. Brown SD, Guss AM, Karpinets TV, Parks JM, Smolin N, Yang S, et al. Mutant alcohol dehydrogenase leads to improved ethanol tolerance in *Clostridium thermocellum. Proc Natl Acad Sci USA* 2011;**108**:13752–7.

46. Na D, Kim TY, Lee SY. Construction and optimization of synthetic pathways in metabolic engineering. *Curr Opin Microbiol* 2010;**13**:363–70.

47. McArthur GHt, Fong SS. Toward engineering synthetic microbial metabolism. *J Biomed Biotechnol* 2010;**2010**:459760.

48. Lee SY. Systems biotechnology. *Genome Inf* 2009;**23**:214–6.

49. Kim TY, Sohn SB, Kim HU, Lee SY. Strategies for systems-level metabolic engineering. *Biotechnol J* 2008;**3**:612–23.

50. Prather KL, Martin CH. De novo biosynthetic pathways: rational design of microbial chemical factories. *Curr Opin Biotechnol* 2008;**19**:468–74.

51. Seidl V, Gamauf C, Druzhinina IS, Seiboth B, Hartl L, Kubicek CP. The *Hypocrea jecorina (Trichoderma reesei)* hypercellulolytic mutant RUT C30 lacks a 85 kb (29 gene-encoding) region of the wild-type genome. *BMC Genomics* 2008;**9**:327.

52. Vitikainen M, Arvas M, Pakula T, Oja M, Penttila M, Saloheimo M. Array comparative genomic hybridization analysis of *Trichoderma reesei* strains with enhanced cellulase production properties. *BMC Genomics* 2010;**11**:441.

53. Herpoel-Gimbert I, Margeot A, Dolla A, Jan G, Molle D, Lignon S, et al. Comparative secretome analyses of two *Trichoderma reesei* RUT-C30 and CL847 hypersecretory strains. *Biotechnol Biofuels* 2008;**1**:18.

54. Zuroff TR, Curtis WR. Developing symbiotic consortia for lignocellulosic biofuel production. *Appl Microbiol Biot* 2012;**93**:1423–35.

55. Chambergo FS, Bonaccorsi ED, Ferreira AJ, Ramos AS, Ferreira Junior JR, Abrahao-Neto J, et al. Elucidation of the metabolic fate of glucose in the filamentous fungus *Trichoderma reesei* using expressed sequence tag (EST) analysis and cDNA microarrays. *J Biol Chem* 2002;**277**:13983–8.

56. dos Santos Castro L, Pedersoli W, Antonieto AC, Steindorff A, Silva-Rocha R, Martinez-Rossi N, et al. Comparative metabolism of cellulose, sophorose and glucose in *Trichoderma reesei* using high-throughput genomic and proteomic analyses. *Biotechnol Biofuels* 2014;**7**:41.

57. Ries L, Pullan ST, Delmas S, Malla S, Blythe MJ, Archer DB. Genome-wide transcriptional response of *Trichoderma reesei* to lignocellulose using RNA sequencing and comparison with Aspergillus niger. *BMC Genomics* 2013;**14**:541.

58. Arvas M, Pakula T, Smit B, Rautio J, Koivistoinen H, Jouhten P, et al. Correlation of gene expression and protein production rate – a system wide study. *BMC Genomics* 2011;**12**:616.

59. Gotz S, Garcia-Gomez JM, Terol J, Williams TD, Nagaraj SH, Nueda MJ, et al. High-throughput functional annotation and data mining with the Blast2GO suite. *Nucleic Acids Res* 2008;**36**:3420–35.

60. Park BH, Karpinets TV, Syed MH, Leuze MR, Uberbacher EC. CAZymes Analysis Toolkit (CAT): web service for searching and analyzing carbohydrate-active enzymes in a newly sequenced organism using CAZy database. *Glycobiol* 2010;**20**:1574–84.

61. Yang S, Guarnieri MT, Smolinski S, Ghirardi M, Pienkos PT. De novo transcriptomic analysis of hydrogen production in the green alga *Chlamydomonas moewusii* through RNA-Seq. *Biotechnol Biofuels* 2013;**6**:118.

62. Yang S, Pan C, Tschaplinski TJ, Hurst GB, Engle NL, Zhou W, et al. Systems biology analysis of *Zymomonas mobilis* ZM4 ethanol stress responses. *PloS One* 2013;**8**:e68886.

63. Yang S, Giannone RJ, Dice L, Yang ZK, Engle NL, Tschaplinski TJ, et al. *Clostridium thermocellum* ATCC27405 transcriptomic, metabolomic and proteomic profiles after ethanol stress. *BMC Genomics* 2012;**13**:336.

64. Brakhage AA. Regulation of fungal secondary metabolism. *Nat Rev Microbiol* 2013;**11**:21–32.

65. Mach-Aigner AR, Pucher ME, Steiger MG, Bauer GE, Preis SJ, Mach RL. Transcriptional regulation of xyr1, encoding the main regulator of the xylanolytic and cellulolytic enzyme system in *Hypocrea jecorina*. *Appl Envir Microbiol* 2008;**74**:6554–62.

66. Meng F, Wei D, Wang W. Heterologous protein expression in *Trichoderma reesei* using the cbhII promoter. *Plasmid* 2013;**70**:272–6.

67. Liu T, Wang T, Li X, Liu X. Improved heterologous gene expression in *Trichoderma reesei* by cellobiohydrolase I gene (cbh1) promoter optimization. *Acta Biochim Biophys Sin* 2008;**40**:158–65.

68. Ilmen M, Onnela ML, Klemsdal S, Keranen S, Penttila M. Functional analysis of the cellobiohydrolase I promoter of the filamentous fungus *Trichoderma reesei*. *Mol Gen Genet MGG* 1996;**253**:303–14.

69. Stangl H, Gruber F, Kubicek CP. Characterization of the *Trichoderma reesei* cbh2 promoter. *Curr Genet* 1993;**23**:115–22.

70. Li JX, Wang J, Wang SW, Xing M, Yu SW, Liu G. Achieving efficient protein expression in *Trichoderma reesei* by using strong constitutive promoters. *Microb Cell Fact* 2012;**11**:84.

CHAPTER 11

Production of Ethanol from Engineered Trichoderma reesei

Qi Xu[1], Michael E. Himmel[1], Arjun Singh[2]

[1]Biosciences Center, National Renewable Energy Laboratory (NREL), Golden, CO, USA;
[2]National Bioenergy Center, National Renewable Energy Laboratory (NREL), Golden, CO, USA

Introduction

The traditional biochemical pathway for conversion lignocellulosic biomass includes pretreatment,[1,2] saccharification to convert the sugar polymers in the biomass into fermentable monomeric sugars, and finally fermentation of the sugars to ethanol using various organisms, such as *Saccharomyces cerevisiae*,[3–6] *Zymomonas mobilis*,[7,8] or *Escherichia coli*.[9,10] Currently, *Trichoderma reesei* is the source for all the hydrolytic enzymes used for the biomass saccharification process. Clearly, the economics of ethanol production would improve if some of the steps of the process could be eliminated or combined with other steps of the process. Significant efforts have been made in developing organisms for consolidated bioprocessing (CBP) in which the saccharification and fermentation step are performed by one organism.[11,12] Such an organism would produce cellulases and hemicellulases to produce monomeric sugars from pretreated biomass, but would then also ferment the resulting sugars into ethanol.

There are two technology paths discussed today for the development suitable CBP organisms. One is the convert an organism, such as *Clostridium thermocellum* or *T. reesei*, which possesses powerful cellulolytic systems, into an ethanologen. Conversely, the other path is to convert an outstanding ethanologen, such as *S. cerevisiae*, into an effective cellulolytic organism. To convert a cellulolytic organism into a CBP organism, the metabolic pathway for ethanol synthesis must be introduced and/or enhanced if it exists in the wild type strain. In addition, the ethanol tolerance of these organisms usually needs to be improved. To convert an ethanologen into a CBP organism, expression systems for various cellulases and hemicellulases must be introduced, such that active forms of these enzymes are produced in sufficient amounts and ratios to convert the polysaccharides in biomass to monomeric sugars efficiently.

Considerable effort has been made to develop the industrial ethanologen, yeast, into a commercially viable CBP organism. Expression systems for several cellulase-encoding genes, for

Direct Microbial Conversion of Biomass to Advanced Biofuels. http://dx.doi.org/10.1016/B978-0-444-59592-8.00011-7

example *T. reesei cbh*1 and *cbh*2, have been introduced into yeast. Although engineered yeast strains show some ability to produce glycoside hydrolases, these activities have not been sufficient for efficient hydrolysis of lignocellulosic biomass for use in commercial bioethanol production processes.[13–15] However, recent progress in engineering yeast into a better CBP organism remains promising.[16] To convert the already cellulolytic *C. thermocellum* into a commerical CBP organism, its ethanol yield and tolerance must be improved. Recent identification of a mutations that enhanced ethanol tolerance in this organism is a promising development.[17] Another cellulolytic bacterium, *Thermoanaerobacterium saccharolyticum* also offers promise as a possible CBP organism.[18,19]

In this chapter, we explore the prospect of developing the cellulolytic fungus, *T. reesei* into a suitable CBP organism. It is noteworthy that almost all hydrolytic enzymes currently being used for the saccharification of biomass for ethanol production are produced by *T. reesei*. This organism uses all primary biomass sugars (glucose, xylose, arabinose, mannose, and galactose) for growth; it has the basic pathway to produce ethanol from these sugars and, data presented here strongly suggest that the native ethanologenic fermentation pathway can be augmented and improved.

Trichoderma reesei *Produce Ethanol from Biomass Sugars*

We tested the wild-type strains of *T. reesei*, QM6a, for ethanol production from various biomass sugars. The QM6a strain was grown under aerobic condition for 3 days. The mycelia were collected and tested for fermentation of various sugars under anaerobic conditions for 12 days. As shown in Table 1, this strain is capable of fermenting glucose, galactose, mannose, and cellobiose and, to a lesser extent, also xylose and arabinose.

Fermentations were done in Vogel's medium.[20] Mycelia from 100 mL cultures, started from approximately 10^6 spores, were collected, washed with the Vogel's medium, and then anaerobically incubated in the same medium with various sugars. Two percent xylose and arabinose were used for growth and fermentation; all other sugars were used at 3%. Following the fermentation, the samples were analyzed for various products by HPLC using Aminex

Table 1: Fermentation of biomass sugars by *Trichoderma reesei* QM6a

Sugar	Products (g/L)			
	Acetic Acid	Glycerol	Xylitol	Ethanol
Glucose	0.31	0.10	0	4.25
Mannose	0.14	0	0	4.22
Galactose	0.28	0.07	0	3.33
Xylose	0.55	0.05	0.01	0.49
Arabinose	0.23	0	0	0.16
Cellobiose	0.17	0.06	0	3.17

HPX-87 column (Bio-Rad, Richmond, Calif). The values reported are average of two determinations.

As can be seen from the data in Table 1, the major byproduct made during this fermentation is acetic acid. We measured the kinetics of conversion of glucose to ethanol during a 16-day fermentation using the sugar at starting concentration of 5%. As seen in Figure 1, there is a steady increase in ethanol and a corresponding decrease in glucose concentrations during the course of the experiment. Maximum conversion efficiency achieved was 0.42 g ethanol/g of consumed glucose.

Under these fermentation conditions, we did not observe significant effects from the glucose concentrations used on ethanol production (Figure 2); glucose was varied from 3% to 15%.

The pH during Fermentation Affects Ethanol Yield

We conducted fermentation experiments at six different pH values varying from 2.5 to 7.0. It appears that very low pH conditions are detrimental to ethanol fermentation, and the optimum pH is 4.5–5.0 (Figure 3). The pH of the fermentation broth decreased in most cases due to the generation of acetic acid (Table 1). It is possible that other acids, not detected in our analysis, are also produced during the fermentation process.

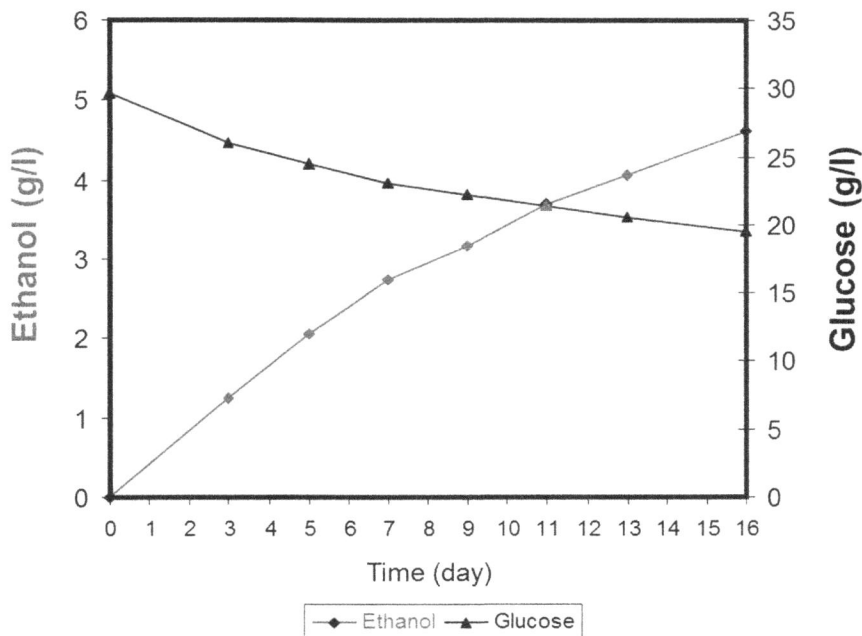

Figure 1
Time course of glucose consumption and ethanol production.

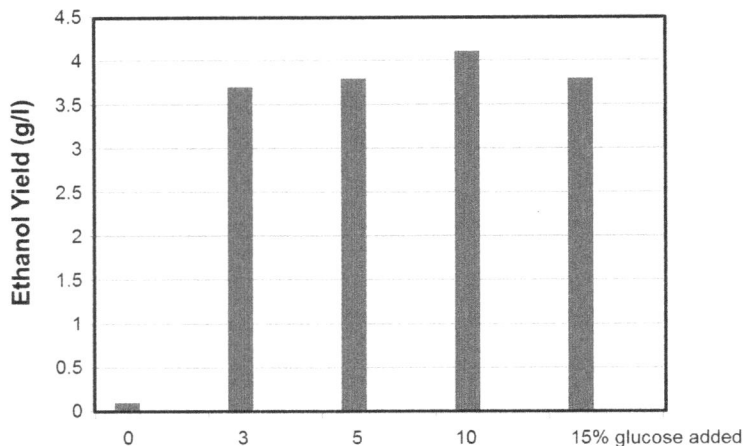

Figure 2
Ethanol yields from various amounts of glucose.

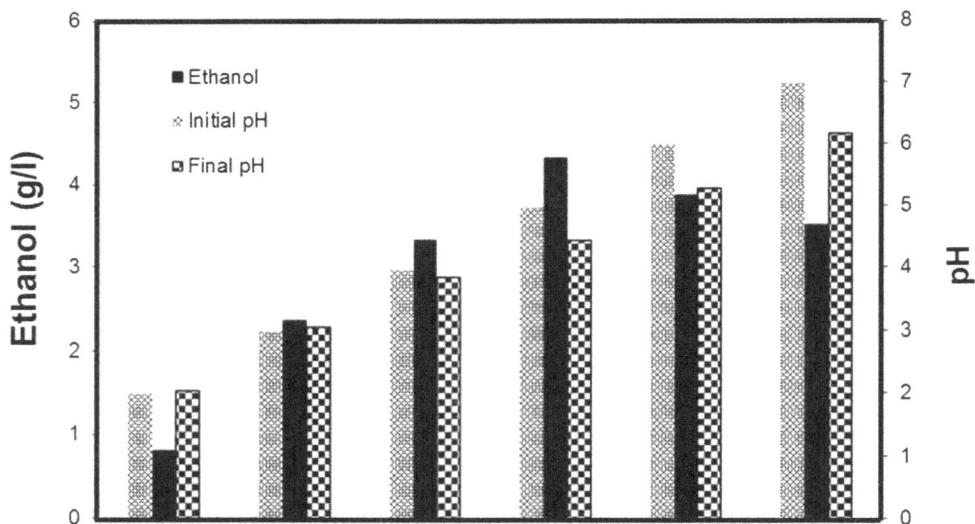

Figure 3
Effect of pH on ethanol production.

Sugar Used during Growth Phase Affects Xylose Fermentation

We determined whether or not the carbon source used during growth of mycelia, and subsequently used for fermentation experiments, affected ethanol yields. The parental strain, QM6a, was aerobically cultivated on D-glucose, D-galactose, D-mannose, D-xylose, L-arabinose, and glycerol. Equivalent amounts of mycelia (5 g) were obtained from each carbon

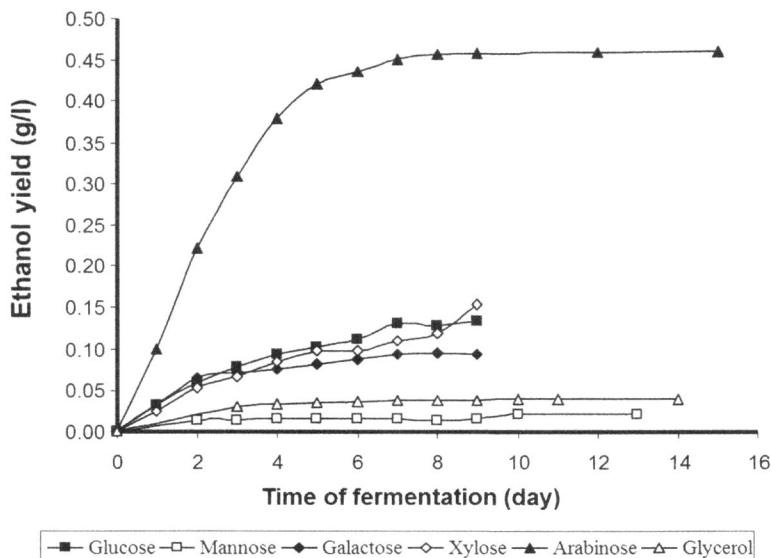

Figure 4
The effect of carbon source during aerobic mycelia growth on D-xylose fermentation.

source and used for the fermentation of 2% D-xylose under anaerobic conditions. The mycelial growth was collected by centrifugation. We found that the mycelia grown on D-glucose, D-galactose, D-mannose, D-xylose, and glycerol produced very little ethanol (Figure 4). However, the mycelia that were obtained on arabinose cultures produced comparatively higher levels of ethanol.

Although absolute amounts of ethanol from all carbon sources are low, the mycelia from arabinose cultures are clearly different in ethanol fermentation from all others. Surprisingly, xylose-grown mycelia performed similar to mycelia grown on glucose, mannose, galactose, and glycerol for fermentation of xylose.

Direct Conversion of Cellulose to Ethanol

The eventual goal of our experiments was to evaluate whether or not *T. reesei* could be developed into a CBP organism that would not only produce the hydrolytic enzymes to produce monomeric sugars, but would also further metabolize those sugars to ethanol or other advanced "drop-in" biofuels. We tested whether the *T. reesei* could produce ethanol directly from cellulose. In these experiments, conducted using the *T. reesei* strain QM9414, we also monitored cellulase activity both during the aerobic growth phase and during the anaerobic fermentation phase.

Cellulase induction during growth phase was done as described by Juhasz et al.[21] In brief, approximately 4×10^7 *T. reesei* QM9414 spores were inoculated into 20 mL of Mandel's medium supplemented with 0.7% Solka Floc 200 and grew for 3 days at 30 °C and 200 rpm. The culture was then mixed with 100 mL of Mendel's containing 1.5% Solka Floc 200. The medium also contained 0.1 M maleic acid to stabilize pH at 6.0. The cultures were grown aerobically for 6 days and then transferred to anaerobic condition for fermentation at 30 °C for 12 additional days. Periodic samples were taken for cellulase activity and ethanol measurements.

As shown in Figure 5, and as expected, the cellulase activity steadily increased during the growth phase, and there was no detectable ethanol during this phase. During the anaerobic fermentation phase, the ethanol levels steadily increased during the 10-day fermentation period, and there was barely a detectible amount of ethanol toward the end of this phase. In fact, the cellulase activity slightly declined during the fermentation phase. It is possible that this modest decrease in cellulase activity is due to inactivation of the protein on biomass itself and the result of protease action in the culture broth. It is in any case not surprising that cellulase levels did not increase during fermentation, because there is little or no cell growth during the anaerobic fermentation phase. It should be noted here that we envision that *T. reesei* would grow and produce the cellulolytic enzymes needed during the aerobic growth phase, and these levels would be sufficient for subsequent hydrolysis of the biomass polymers to produce sugars for ethanol or other biofuel production.

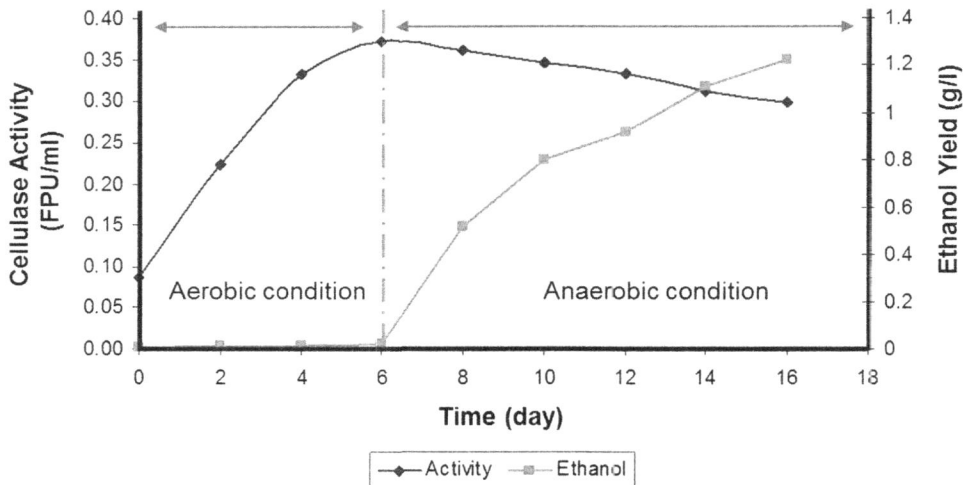

Figure 5
Direct conversion of cellulose to ethanol.

Enhancing Ethanol Synthesis by Metabolic Engineering

It is obvious that the current yield and rate of ethanol production by the wild-type strain of *T. reesei* are not sufficient for its use as a CBP organism, and thus improvements in yields and rates are required. However, these and other studies[22] have established that this organism has the essential pathways to use biomass sugars and to ferment them to ethanol. There is no obvious reason that the fermentation capacities of *T. reesei* cannot be increased. It follows that because the organism uses pentose sugars for growth, it has a complete and functioning pentose metabolic pathway. *T. reesei* also produces ethanol from various C_6 sugars, and it thus possesses the complete and functioning glycolytic pathway. We have conducted experiments to explore whether or not we can improve ethanol fermentation in *T. reesei* by introducing genes from other established and robust fermentative organisms.

We next introduced two key ethanol fermentation pathway genes from *S. cerevisiae* into *T. reesei* and tested the resulting transformants for ethanol fermentation. For this purpose, we chose the primary pyruvate decarboxylase gene, *ScPDC*1 (accession number: NP_013145), and the primary alcohol dehydrogenase gene, *ScADH*1 (accession number: DAA10699), from *S. cerevisiae*. We synthesized codon-optimized versions of these genes for expression in *T. reesei*. We designed and constructed an expression plasmid, pTR*ScADH*1-*ScPDC*1 (Figure 6), in which the expression of the *S. cerevisiae* genes is driven by the *T. reesei pgk*1 promoter.

The plasmid also has an expression system for *E. coli hph* gene coding for hygromycin B phosphotransferase, which confers resistance to the antibiotic, hygromycin B. The *T. reesei* strain, QM6a, was then transformed with this plasmid, and hygromycin B resistant transformants were selected. The transformants were purified by two rounds of sporulation and single colony isolation. Six independent colonies from one transformant and the parent strain QM6a were evaluated for their ethanol fermentation performance from glucose, as described earlier. Two important points may be made from the data summarized in Table 2.

Trichoderma reesei spheroplasts were prepared and transformed with slight modifications of various published methods.[23–25] Spheroplasts were generated by 3–4 h of incubation of fresh mycelia in 0.6 M KCl containing 4 mg/mL Glucanex (Sigma, St Louis, Mo.) and 2 mg/mL lysozyme (Sigma, St Louis, Mo.). After 7 h of recovery period in rich medium without hygromycin B, the transformants resistant to hygromycin B were selected by mixing the transformed spheroplasts with 10 mL of potato dextrose agar containing 1.0 M sorbitol and 100 mg/L hygromycin B and pouring over petri dishes containing the same medium. Transformants were purified by at least two rounds of sporulation and testing of single spores.

All transformants tested were found to produce more (11–38%) ethanol compared to the parent strain. Also, all transformants tested produced less (21–65%) of the byproduct, acetic acid, than the parent strain.

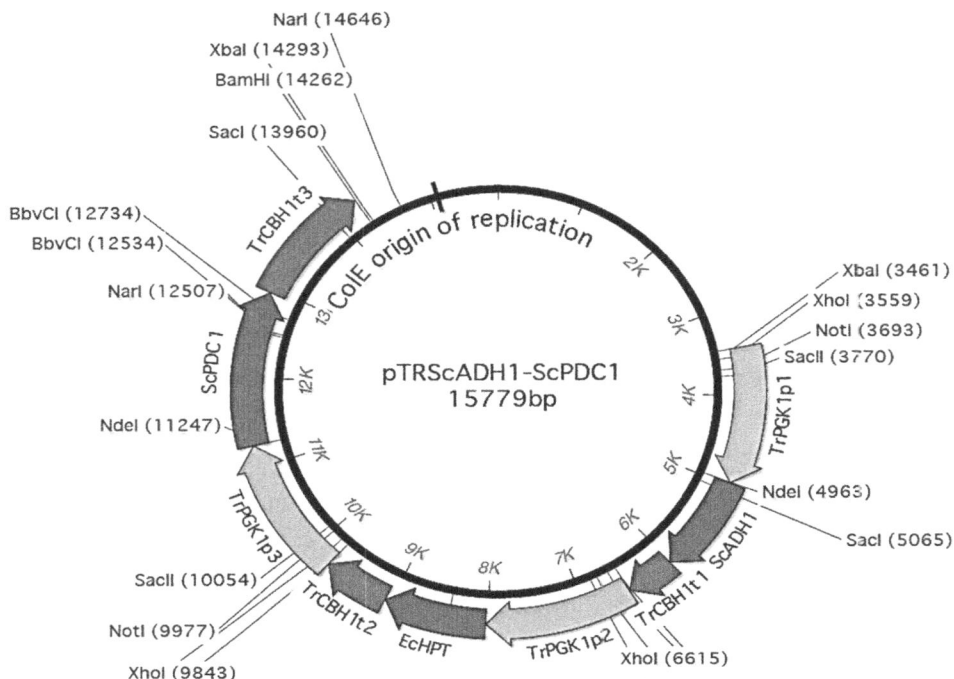

Figure 6

Plasmid for expression of *Saccharomyces cerevisiae* genes in *Trichoderma reesei*.

Table 2: Engineered strains produce higher levels of ethanol and lower levels of acetic acid

	Acetic Acid		Ethanol	
Strain	Yield (g/L)	Percent Decrease	Yield (g/L)	Percent Increase
Parent (QM6a)	1.56	0	3.87	0
TRf5-1-1	0.80	48.5	4.50	16.2
TRf5-4-1	0.66	57.4	4.66	20.4
TRf5-9-1	1.24	20.8	4.87	25.9
TRf5-10-1	0.54	65.6	4.30	11.1
TRf5-13-1	0.95	39.0	4.97	28.5
TRf5-14-1	0.58	62.5	5.34	38.0

Discussion

The preliminary studies described in the article show that *T. reesei*, which naturally possesses a potent cellulolytic system, also has the biochemical pathway needed to produce ethanol from biomass sugars. Furthermore, the ethanol-producing capacity may be enhanced by expression of appropriate heterologous genes. *Trichoderma harzianum* was

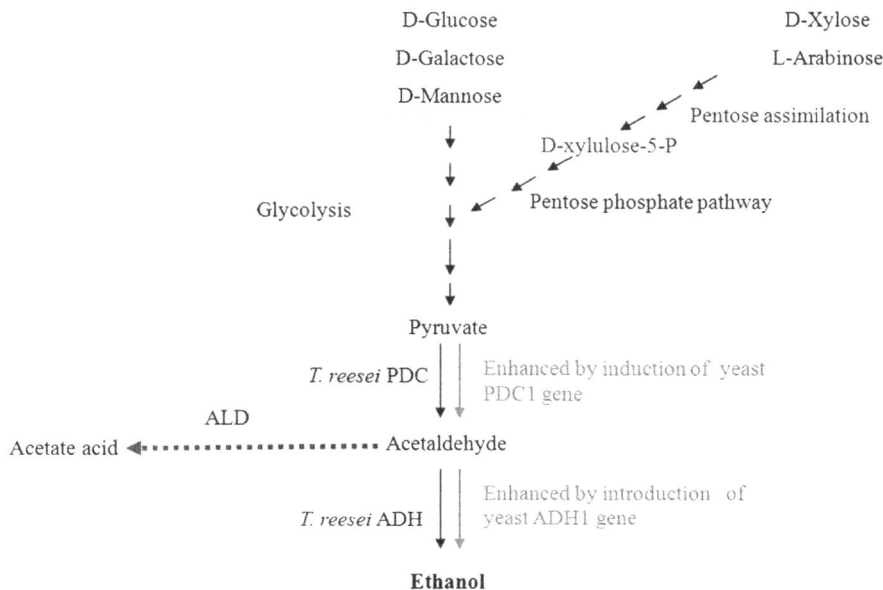

Figure 7
Possible targets for metabolic engineering to enhance ethanol fermentation.

previously shown to produce ethanol from cellulose at low levels.[26] The ethanol production in *T. harzianum* may also be increased by appropriate genetic engineering, as we have shown here, for *T. reesei*. However, we feel that *T. reesei* is more promising for such engineering to produce a CBP organism, because it possesses a much more effective cellulase system than does *T. harzianum*.[22,26] Other reasons for considering *T. reesei* for development into a CBP organism include: (1) it is already grown at large scale for cellulase production and therefore more likely to be more readily accepted and adapted by industry; and (2) useful systems for gene manipulation and expression have been developed for *T. reesei*.[27-29]

Figure 7 identifies some of targets for metabolic engineering to enhance ethanol yields.

We demonstrated that providing additional pyruvate decarboxylase and alcohol dehydrogenase activities improved ethanol formation relative to the wild type state. These strains also produced less acetic acid than the wild type strain, which suggests that the activity of alcohol dehydrogenase is not sufficient and leads to accumulation of acetaldehyde, which is subsequently converted to acetic acid. If we determine that acetic acid production, because of the build-up of acetaldehyde, remains a contributing factor in the observed stalling of ethanol fermentations, additional targets for improvement include deletion of one or more aldehyde dehydrogenase genes to prevent accumulation of the this substrate.

Although our work suggests that developing *T. reesei* as a CBP organism is promising, there are some issues that must be addressed before its use as an ethaologen becomes practical. *T. reesei* is an obligate aerobe, and fermentation conditions need to be determined such that, after the production of cellulases during the aerobic growth stage, the organism has sufficient enzyme activities for glycolysis and ethanolic fermentation. It may be necessary to alter or eliminate oxygen-dependent transcriptional control of the glycolytic pathway[30,31] by appropriate genetic engineering. Improvement of ethanol tolerance of the organism needs to be improved to use it to produce ethanol at economical and commercially feasible concentrations. It is hoped that the mechanism of ethanol tolerance in this ascomycete is similar to that found in efficient ascomycete, *S. cerevisiae*,[32–34] and therefore approaches similar to those used to enhance ethanol tolerance of *S. cerevisiae* may also be used for *T. reesei*. It is likely that partial cause of low ethanol yields in *T. reesei* compared to other established ethanologens such as *Z. mobilis* and *S. cerevisiae* is redox imbalance.[35] This limitation may be overcome by appropriate modifications of appropriate metabolic steps.[36,37]

Acknowledgment

This work was supported by the Laboratory Directed Research and Development (LDRD) Program at the National Renewable Energy Laboratory.

References

1. Weil J, Westgate P, Kohlmann K, Ladisch MR. Cellulose pretreatments of lignocellulosic substrates. *Enzyme Microbiol Technol* 1994;**16**:1002–4.
2. Jacobsen SE, Wyman CE. Cellulose and hemicellulose hydrolysis models for application to current and novel pretreatment processes. *Appl Biochem Biotechnol* 2000;**84–86**:81–96.
3. Karhumaa K, Wiedemann B, Hahn-Hagerdal B, Boles E, Gorwa-Grauslund MF. Co-utilization of L-arabinose and D-xylose by laboratory and industrial *Saccharomyces cerevisiae* strains. *Microbiol Cell Fact* 2006;**5**:18.
4. Katahira S, Fujita Y, Mizuike A, Fukuda H, Kondo A. Construction of a xylan-fermenting yeast strain through codisplay of xylanolytic enzymes on the surface of xylose-utilizing *Saccharomyces cerevisiae* cells. *Appl Environ Microbiol* 2004;**70**:5407–14.
5. Becker J, Boles E. A modified *Saccharomyces cerevisiae* strain that consumes L-arabinose and produces ethanol. *Appl Environ Microbiol* 2003;**69**:4144–50.
6. Katahira S, Mizuike A, Fukuda H, Kondo A. Ethanol fermentation from lignocellulosic hydrolysate by a recombinant xylose- and cellooligosaccharide-assimilating yeast strain. *Appl Microbiol Biotechnol* 2006;**72**:1136–43.
7. Lawford HG, Rousseau JD. Performance testing of *Zymomonas mobilis* metabolically engineered for cofermentation of glucose, xylose, and arabinose. *Appl Biochem Biotechnol* 2002;**98**:429–48.
8. Mohagheghi A, Evans K, Chou YC, Zhang M. Cofermentation of glucose, xylose, and arabinose by genomic DNA-integrated xylose/arabinose fermenting strain of *Zymomonas mobilis* AX101. *Appl Biochem Biotechnol* 2002;**98**:885–98.
9. Tao H, Gonzalez R, Martinez A, Rodriguez M, Ingram LO, Preston JF, et al. Engineering a homo-ethanol pathway in *Escherichia coli*: increased glycolytic flux and levels of expression of glycolytic genes during xylose fermentation. *J Bacteriol* 2001;**183**:2979–88.
10. Dien BS, Cotta MA, Jeffries TW. Bacteria engineered for fuel ethanol production: current status. *Appl Microbiol Biotechnol* 2003;**63**:258–66.

11. Lynd LR, van Zyl WH, McBride JE, Laser M. Consolidated bioprocessing of cellulosic biomass: an update. *Curr Opin Biotechnol* 2005;**16**:577–83.

12. Lynd LR, Weimer PJ, van Zyl WH, Pretorius IS. Microbial cellulose utilization: fundamentals and biotechnology. *Microbiol Mol Biol Res* 2002;**66**:739.

13. Den Haan R, Mcbride JE, La Grange DC, Lynd LR, Van Zyl WH. Functional expression of cellobiohydrolases in *Saccharomyces cerevisiae* towards one-step conversion of cellulose to ethanol. *Enzyme Microbiol Technol* 2007;**40**:1291–9.

14. Hong J, Tamaki H, Yamamoto K, Kumagai H. Cloning of a gene encoding a thermo-stable endo-beta-1,4-glucanase from *Thermoascus aurantiacus* and its expression in yeast. *Biotechnol Lett* 2003;**25**:657–61.

15. Takada G, Kawaguchi T, Sumitani J, Arai M. Expression of *Aspergillus aculeatus* no. F-50 cellobiohydrolase I (cbhI) and beta-glucosidase 1 (bgl1) genes by *Saccharomyces cerevisiae*. *Biosci Biotechnol Biochem* 1998;**62**:1615–8.

16. Van Zyl WH, Den Haan R, La Grange DC. Developing cellulolytic organisms for consolidated bioprocessing of lignocellulosics. In: Gupta VK, Tuohy MG, editors. *Biofuel technologies: recent developments*. New York: Springer Verlag; 2013. p. 189–220.

17. Brown SD, Guss AM, Karpinets TV, Parks JM, Smolin N, Yang SH, et al. Mutant alcohol dehydrogenase leads to improved ethanol tolerance in *Clostridium thermocellum*. *Proc Natl Acad Sci USA* 2011;**108**:13752–7.

18. Shaw AJ, Podkaminer KK, Desai SG, Bardsley JS, Rogers SR, Thorne PG, et al. Metabolic engineering of a thermophilic bacterium to produce ethanol at high yield. *Int Sugar J* 2009;**111**:164–71.

19. Currie DH, Herring CD, Guss AM, Olson DG, Hogsett DA, Lynd LR. Functional heterologous expression of an engineered full length CipA from *Clostridium thermocellum* in *Thermoanaerobacterium saccharolyticum*. *Biotechnol Biofuels* 2013;**6**:32.

20. Vogel HJ. A convenient growth medium for *Neurospora* (medium N). *Microb Genet Bull* 1956;**13**:42–3.

21. Juhasz T, Szengyel Z, Szijarto N, Reczey K. Effect of pH on cellulase production of *Trichoderma reesei* RUT C30. *Appl Biochem Biotechnol* 2004;**113**:201–11.

22. Xu Q, Singh A, Himmel ME. Perspectives and new directions for the production of bioethanol using consolidated bioprocessing of lignocellulose. *Curr Opin Biotechnol* 2009;**20**:364–71.

23. Deane EE, Whipps JM, Lynch JM, Peberdy JF. Transformation of *Trichoderma reesei* with a constitutively expressed heterologous fungal chitinase gene. *Enzyme Microbiol Technol* 1999;**24**:419–24.

24. Berges T, Barreau C. Isolation of uridine auxotrophs from *Trichoderma reesei* and efficient transformation with the cloned Ura3 and Ura5 genes. *Curr Genet* 1991;**19**:359–65.

25. Penttila M, Nevalainen H, Ratto M, Salminen E, Knowles J. A versatile transformation system for the cellulolytic filamentous fungus *Trichoderma reesei*. *Gene* 1987;**61**:155–64.

26. Stevenson DM, Weimer PJ. Isolation and characterization of a *Trichoderma* strain capable of fermenting cellulose to ethanol. *Appl Microbiol Biotechnol* 2002;**59**:721–6.

27. Guangtao Z, Hartl L, Schuster A, Polak S, Schmoll M, Wang TH, et al. Gene targeting in a nonhomologous end joining deficient *Hypocrea jecorina*. *J Biotechnol* 2009;**139**:146–51.

28. Hartl L, Seiboth B. Sequential gene deletions in *Hypocrea jecorina* using a single blaster cassette. *Curr Genet* 2005;**48**:204–11.

29. Steiger MG, Vitikainen M, Uskonen P, Brunner K, Adam G, Pakula T, et al. Transformation system for *Hypocrea jecorina* (*Trichoderma reesei*) that favors homologous integration and employs reusable bidirectionally selectable markers. *Appl Environ Microbiol* 2011;**77**:114–21.

30. Rautio JJ, Smit BA, Wiebe M, Penttila M, Saloheimo M. Transcriptional monitoring of steady state and effects of anaerobic phases in chemostat cultures of the filamentous fungus *Trichoderma reesei*. *BMC Genomics* 2006;**7**:247.

31. Bonaccorsi ED, Ferreira AJ, Chambergo FS, Ramos AS, Mantovani MC, Farah JP, et al. Transcriptional response of the obligatory aerobe *Trichoderma reesei* to hypoxia and transient anoxia: implications for energy production and survival in the absence of oxygen. *Biochemistry* 2006;**45**:3912–24.

32. Chi Z, Arneborg N. Relationship between lipid composition, frequency of ethanol-induced respiratory deficient mutants, and ethanol tolerance in *Saccharomyces cerevisiae*. *J Appl Microbiol* 1999;**86**:1047–52.

33. Susumu K. Improved ethanol tolerance and fermentation of *Saccharomyces cerevisiae* by alteration of fatty acid content in membrane lipids via metabolic engineering. *Foods Food Ingred J Jpn* 2003;**208**:276–82.

34. You KM, Rosenfield CL, Knipple DC. Ethanol tolerance in the yeast *Saccharomyces cerevisiae* is dependent on cellular oleic acid content. *Appl Environ Microbiol* 2003;**69**:1499–503.

35. Bruinenberg PM, Debot PHM, Vandijken JP, Scheffers WA. The role of redox balances in the anaerobic fermentation of xylose by yeasts. *Eur J Appl Microbiol* 1983;**18**:287–92.

36. Verho R, Londesborough J, Penttila M, Richard P. Engineering redox cofactor regeneration for improved pentose fermentation in *Saccharomyces cerevisiae*. *Appl Environ Microbiol* 2003;**69**:5892–7.

37. Jeffries TW, Jin YS. Metabolic engineering for improved fermentation of pentoses by yeasts. *Appl Microbiol Biotechnol* 2004;**63**:495–509.

CHAPTER 12

Remaining Challenges in the Metabolic Engineering of Yeasts for Biofuels

Sun-Mi Lee[1,2], Eric M. Young[1], Hal S. Alper[1,3]

[1]*Department of Chemical Engineering, The University of Texas at Austin, Austin, Texas, USA;* [2]*Clean Energy Research Center, Korea Institute of Science and Technology, Seongbuk-gu, Korea;* [3]*Institute for Cellular and Molecular Biology, The University of Texas at Austin, Austin, Texas, USA*

Introduction—Yeasts as the Catalyst for Biomass Consumption and Biofuel Production

Yeasts naturally possess many metabolic and physiological traits that make them advantageous organisms of choice for economical production of biofuels from biomass. These unicellular fungi are often found on live and decomposing plants, thus many species have already solved key challenges inherent in metabolizing lignocellulosic biomass. Additionally, many yeast species naturally produce metabolic byproducts that can be used as potential biofuels, such as alcohols or lipids.[1,2] Furthermore, yeasts possess several attractive industrial characteristics such as nearly ambient optimal growth temperatures, a dearth of lethal viruses like phage, facile separation due to large size and ability to flocculate, and tolerance to low pH values.[3] As eukaryotes, yeasts are complex—their cellular functions are compartmentalized into organelles[4]—and are more robust than prokaryotes like *Escherichia coli* due to their ability to mate and create more stable diploids.[5] As unicellular organisms, they are simpler than most higher eukaryotes, grow relatively quickly, and are easier to manipulate genetically.[6] This collection of qualities makes fungi and yeasts some of the most attractive living catalysts for the conversion of biomass to biofuels.

However, the immediate use of these organisms is limited without engineering. Production of energy-poor biofuels, low titers, and a mismatch between the natural timescale of digestion and desired biotechnological batch time are some of the factors that reduce the economics of the process. For example, with current technology, the yeast *Saccharomyces cerevisiae* is perhaps the most attractive and amenable host organism among the fungi clade. *S. cerevisiae* readily ferments glucose to carbon dioxide and ethanol. As such, *S. cerevisiae* is the catalyst for industrial production of such commodities as bread, wine, and beer. This fermentative ability also makes *S. cerevisiae* an attractive host for corn, sugar cane, and sugar beet ethanol production, because these feedstocks are primarily glucose. While glucose is also the primary

Direct Microbial Conversion of Biomass to Advanced Biofuels. http://dx.doi.org/10.1016/B978-0-444-59592-8.00012-9

constituent of lignocellulosic hydrolysates, many pentoses, other hexoses, and complex organic molecules are present in significant fractions. *S. cerevisiae* does not natively consume the pentose fraction due to the lack of robust pentose sugar metabolism,[7,8] and many of the complex organics are toxic. Furthermore, ethanol, the major byproduct of fermentation in *S. cerevisiae*, is not ideal for a transportation fuel.[9] Ethanol is not energy dense, it evaporates easily, and is hygroscopic—it readily absorbs water. Advanced biofuels, superior to ethanol, are not natively produced by *S. cerevisiae*. Other organisms in the fungi clade have been explored for use instead of *S. cerevisiae*; yet, to date, all have certain suboptimal qualities. For example, the yeast *Scheffersomyces stipitis* can efficiently convert pentoses, but does not produce much ethanol. The lack of a fungal host with the ideal combination of desirable characteristics limits the profitable industrial production of biofuels from lignocellulosic biomass.

Metabolic engineering has the potential to deliver an ideal strain. By rewiring the reactions that occur in a cell, the optimal living catalyst can be constructed. Beneficial characteristics from multiple organisms can be combined into a single host, overcoming the limitations of any individual yeast. As a result of years of research, it is possible to engineer *S. cerevisiae* more extensively than other organisms, making it an attractive host in which to import beneficial properties.[6] In this regard, many advances have already been made. *S. cerevisiae* has been engineered to consume xylose and arabinose and to be more tolerant to toxins present in lignocellulosic hydrolysates. Additionally, it should be noted that significant strides are enabling the engineering of many other yeasts that may prove to be useful hosts. By building on these advances, metabolic engineering has the potential to deliver a yeast strain that produces high-value biofuels from biomass in a profitable industrial fermentation process.

To summarize, *S. cerevisiae* seems a highly attractive host for economical lignocellulosic biomass conversion to fuels, but it must be altered and adapted from its wild type state. As a model organism for studying biology and metabolic engineering, many powerful biological tools already exist for altering and adapting *S. cerevisiae*. To date, metabolic engineering has improved *S. cerevisiae* by importing necessary traits for lignocellulosic biofuel production. However, the state of the art still remains suboptimal, preventing economical biofuel production from lignocellulosic biomass.

Therefore, this chapter will start by discussing the potential of metabolic engineering and then outline the major remaining challenges in fungi for going beyond glucose and beyond ethanol in a quest for an engineered organism that produces next generation biofuels from lignocellulosic biomass. The main focus of this chapter will be metabolic engineering of *S. cerevisiae*, because it currently seems to be the most attractive host for biofuel production with the longest industrial track record. Yet, as mentioned previously, other yeasts possess unique features and also serve as possible hosts for future biofuel production, and we will discuss those briefly.

Metabolic Engineering—An Overview

Metabolic engineering confers necessary traits to an organism by reconstructing metabolism. The success of metabolic engineering is dependent on how we study, design, and optimize metabolism in an organism. Advances in microbiology and biotechnology have built a strong foundation for understanding cellular processes and metabolism. Application of design and optimization principles, long the core of engineering disciplines, has helped provide a framework for developing tools to understand and alter metabolism.[10]

The basic premise of biological function, and thus biological engineering, is the central dogma of biology (Figure 1). The central dogma of biology states that DNA is the source of the biological information necessary to make proteins and more DNA. The DNA sequences that encode for proteins are referred to as genes. Proteins are made by the processes of transcription into RNA, followed by translation into the encoded protein (Figure 1). In transcription, regulatory sequences in a region of DNA called a promoter interact with

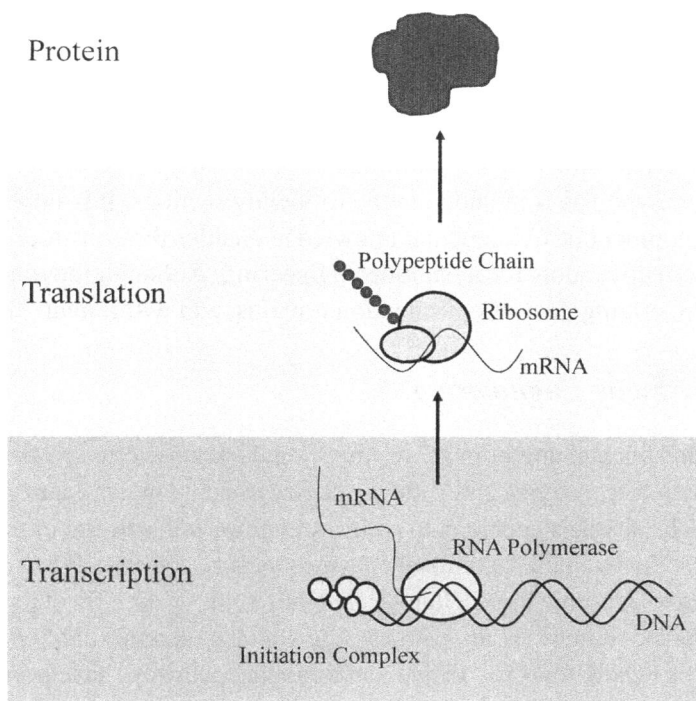

Figure 1
The central dogma of biology. Simplified diagram of gene expression in yeast. Transcription: transcription factors, a promoter, and RNA polymerase all interact to initiate synthesis of mRNA. Translation: The mRNA is read by the ribosome, which produces a polypeptide chain of amino acids. These amino acids then fold into the finished protein.

regulatory proteins, called transcription factors, to recruit RNA polymerase to produce messenger RNA (mRNA). In translation, mRNA is read by another set of interacting proteins and RNA, called the ribosome, which creates the encoded protein. Collectively, these processes give rise to cellular function and metabolism. Individually, each of these steps is a point of intervention to improve cellular traits, discussed further in the section on gene expression engineering.

Metabolism is, at its core, the series of reactions by which an organism extracts chemical energy from molecules (catabolism) to construct molecules necessary for its survival and growth (anabolism). Catabolism and anabolism are organized into pathways and cycles, which consist of specific enzymes catalyzing individual reaction steps. For central pathways, like the TCA cycle, the enzymes that catalyze the component reactions are always present—they are constitutively expressed. For other pathways, like galactose catabolism or antibiotic production, cells only express the genes in response to certain environmental or cellular conditions. These genes are inducible, in that a set of conditions will cause or cease their expression. The ability of the cell to alter the composition of expressed metabolic enzymes gives rise to a variety of possible metabolic states. Taken as a whole, the hundreds of possible interacting pathways in a cell form a complex and regulated network system.

Fundamental understanding of individual proteins, the reactions that enzymes catalyze, and the regulatory network controlling gene expression are critical for metabolic engineering. In light of this fundamental knowledge, a multitude of tools can be used to alter and optimize metabolism. A short synopsis is included in the following sections. It is important to note that the continued development of fundamental knowledge enables the construction of ever more sophisticated and effective tools for metabolic engineering. Although the type of tools used may change, the underlying goal of optimization remains, and will remain, the same.

Enzyme and Pathway Engineering

Enzymes are the fundamental unit of metabolism. Several enzymes often work in sequence to form a pathway. Therefore, enzyme and pathway engineering are major components of metabolic engineering. The simplest approach to pathway engineering is to add or remove and enzyme or enzymes. For example, additional enzymes must be expressed in *S. cerevisiae* to enable the organism to consume lignocellulosic derivatives like xylose. In other cases, an organism may have unwanted or unnecessary pathways, in which case removal of enzymes would be advantageous. In this regard, tools for adding and removing pathways have been developed to enable metabolic rewiring. As the central dogma dictates, one must be able to alter DNA to add or remove an enzyme or an entire pathway to an organism. We discuss some of these tools here.

To add genes, extrachromosomal DNA in the form of plasmid vectors, cosmids, or yeast artificial chromosomes (YACs) can be transformed into the nucleus of the yeast. In the nucleus, these DNA constructs express genes and replicate independently of the host genome. Vectors are circular loops of DNA that have been designed for foreign, or heterologous, gene expression in a host organism.[11] Vectors were some of the first tools to be developed in the biotechnological revolution.[12,13] As a result, there are many such vectors available for use in *S. cerevisiae*,[14,15] making *S. cerevisiae* an amenable host for addition of genes. Cosmids and YACs are constructs that enable large sections of DNA to be expressed in yeast. However, these constructs, especially those expressing nonessential genes, must be selected for at all times to ensure propagation as a culture grows. This can be done using antibiotics or restoration of auxotrophy, but these selections limit application in industrial contexts because of the additional costs in the form of expensive additives or formulations.

Genomic integration does not require continuous selection and is therefore more stable than extrachromosomal DNA addition techniques. Therefore, genomic integration is preferable in industrial contexts, especially for final strains. In *S. cerevisiae*, integration takes place via homologous recombination, in which flanking sequences identical to a location in the genome are placed around the gene expression cassette to be integrated.[16] When this cassette contains a selectable marker, *S. cerevisiae* will readily integrate the foreign DNA construct into its genome at the location of the homology. Recently, homologous recombination has been used not only to express an individual gene, but also to assemble complete metabolic pathways.[17,18]

Pathway construction may be relatively straightforward, if the enzymes that catalyze the desired reaction are known. However, if the desired enzyme is unknown, assembling the pathway is a challenge. In this case, inverse metabolic engineering can be used to search for the best candidate enzymes performing the specific function in the target pathway.[19,20] In this approach, a library of potential genes is expressed in a host organism with an incomplete metabolic pathway. Strains that possess a functioning metabolic pathway conferred by a gene from the library are then selected, and the gene that is responsible for the required function is identified. This approach can also be used to improve existing pathways by providing additional genes that resolve imbalances in metabolic flux or co-factors.[21] Pathway construction may also be difficult when multiple homologs of an enzyme are present in many different organisms. Combinatorial methods must be used to sample this diversity and choose the best enzyme.

Beyond gene addition, restructuring metabolism through pathway engineering sometimes requires the deletion of genes or pathways. Gene deletions operate on the same principle as genomic integration, but will not include an additional gene in the knockout construct. In some cases, gene knockouts can be intuitive, such as a branched pathway in which a reactant can be converted by different enzymes into either a desired or undesired product. The enzyme that catalyzes the branching reaction can be knocked out, thereby eliminating the undesired product and likely increasing the titer of the desired product. However, because of the complexity of the metabolic network, not all gene knockouts may be intuitive. In addition, some genes are essential for cell survival and should at least be minimally expressed instead of being deleted. Therefore, tools to measure and model metabolic fluxes have been devised to help identify gene knockout targets.[22] Each of these tools provides a method for altering the composition of pathways within cells.

Gene Expression Engineering

Even if the organism possesses all of the necessary pathways, the rate at which these reactions occur and the balance within the overall network may not be optimal for maximizing production of a desired molecule. In this case, metabolic engineering tools are needed that will change the rate at which a reaction occurs. This can be done in several ways. One may change an enzyme itself so that it is more efficient, or one may change the amount of the enzyme present, which affects overall rate of conversion.

Changing an enzyme itself requires changing the gene sequence. A gene sequence may be altered for a variety of reasons. In some cases, the enzyme that is encoded by the gene does not have the desired stability, reaction rate, or function. Protein engineering has provided a variety of tools to improve these characteristics. Directed evolution, a method of mutating a gene sequence and then selecting for improved mutants, has proved one of the most powerful tools in this regard.[23–26] There are a variety of methods of mutation. A gene sequence can be randomly mutated at any point,[27–29] switched by targeted mutagenesis based on understanding of essential amino acids,[30,31] altered by using a variety of techniques to limit library size,[32–34] or manipulated by using gene shuffling that recombines gene fragments into hybrids.[35–43] Often, these gene fragments are derived from homologs or similar genes from different organisms that have undergone moderate to extensive evolutionary change. Whether the method of mutation is random or rational, a screen must be applied to select for improved enzymes.[44] This is often the most difficult part of improving an enzyme, for a screen may not select for exactly the characteristics that are desired. However, with an effective screen, large improvements in protein structure and function can be achieved to offer more functionally efficient enzymes.

Aside from optimizing the enzyme itself, one may optimize the amount of enzyme present. This can be done by tuning the processes of gene expression, transcription, and translation. Perhaps the most robust tool for modulating transcription is promoter engineering.[45–49] A promoter is a sequence of DNA that attracts proteins for the initiation of transcription. By changing promoters and transcription factor binding sites within promoters, one can change how much mRNA, and therefore how much protein, is made. Tuning gene expression in this manner can overexpress proteins that catalyze desired reactions and decrease expression of genes that encode undesirable but essential reactions.

In terms of tuning translation, the best developed tool is codon optimization.[50–54] A gene sequence is altered to improve enzyme expression. While most organisms use the same genetic code for amino acids, which are building blocks of an enzyme, the redundant nature of the code makes certain codons more frequent than others, and this frequency can change from organism to organism. Optimizing this frequency for a specific host organism can ensure faster translation, and therefore better enzyme expression. Codon optimization is especially necessary if the organism from which the gene is derived and the host organism use different genetic codes. However, the codon optimization algorithms in use today are not entirely accurate, because they are based on broad averages of codon usage across the genome. It may be that these algorithms are refined to ensure consistent, predictable levels of gene expression in the future.

Engineering the Metabolic Network—Classical Strain Engineering and Systems Biology

Outside of pathway engineering, there are several metabolic engineering tools that take a broad view of metabolic networks. These tools are important for developing characteristics for which the genes are unknown or are the result of complex interactions among many genes.

One of the most effective tools for engineering multiple or complex trait is evolutionary engineering. Specifically, organisms are adapted toward a desired phenotype either through natural drift or forced mutagenesis. Evolutionary engineering is perhaps the most ancient metabolic engineering tool; it has been used in agriculture for eons, with farmers choosing the best animals and crops for breeding. In much the same way, the first industrial evolutionary engineering success was penicillin production from *Penicillium chrysogenum*. Over time, strains of *P. chrysogenum* that produced higher amounts of penicillin were selected, until yield was increased several orders of magnitude. In fact, *S. cerevisiae* is itself a product of long-term evolutionary engineering due to selection over time in ethanol fermentations of brewing and winemaking. This evolutionary process made *S. cerevisiae* the most attractive host for ethanol fermentation at industrial scale. This age-old selection method can be applied in almost any context. In fact, evolutionary engineering has been used to improve growth of *S. cerevisiae* on pentoses, as well as to build tolerance to toxins in lignocellulosic biomass derivatives.[55–60]

Another tool centered on normal biological processes is mating. As eukaryotes, yeasts are capable of sexual reproduction. This can be used for trait selection and combination among parental strains of yeast. In this regard, hybrid yeast can be more robust and can evolve more rapidly due to sexual exchange of genetic information.[5] This property of mating in yeasts is a chief advantage over simpler organisms such as bacteria in industrial contexts.

In addition to the above classical approaches, advances in basic science and computing have led to more rational and targeted tools to restructure the regulatory network. One of these advances is the development of the field of systems biology. Systems biology is a discipline that is intent on elucidating the regulatory network of many organisms and is capable of constructing large-scale interactome maps to discover key gene regulators.[22,61–64] By modifying these gene regulators, all the genes that the regulator affects are modified. To do this, one may use a technique such as global transcription machinery engineering (gTME). This tool can change the expression of multiple genes in an organism by mutating individual transcription factors, causing improvements in such complex phenotypes as tolerance to toxins.[65,66] Furthermore, the field of synthetic biology is endeavoring to construct ever more complicated biological circuits that could prove useful for designing de novo metabolic pathways and regulatory structures.[67–75] The increasing number of complex phenotypes made available by these modern techniques could greatly aid efficient biofuel production from lignocellulosic biomass.[76]

Computational Tools—Predictive Models for Metabolic Engineering

Beyond experimental tools, computational modeling and information databases can inform and design metabolic engineering solutions. Systems biology and metabolic network modeling, mentioned previously, rely heavily on large-scale data collection and computation. Maps of gene regulatory and signaling networks enable design of advanced gene control circuits. Models of metabolism enable determination of desired gene additions and gene knockouts by giving a complete picture of the metabolic network. However, large datasets are not confined to these two fields. Databases such as the KEGG (Kyoto Encyclopedia of Genes and Genomes) database catalog all of the known metabolic reactions to occur in nature. Simple cataloguing of genome sequences enables facile discovery of homologs via BLAST (Basic Local Alignment Search Tool).[77] Analysis of such homologs can be performed by engines such as Clustal, a multiple sequence alignment.[78,79] Furthermore, predictive models can prove to be a powerful tool from thermodynamically modeling mRNA stability, to creating a cell in silico. All of these tools aid in the design of experimental approaches for metabolic engineering.

In summary, metabolic engineering, with the above tools, has the potential to design and optimize yeasts such as *S. cerevisiae* for the purpose of constructing the ideal lignocellulosic biofuel producing organism.

Beyond Glucose

As mentioned previously, *S. cerevisiae* fermentation is currently the industry standard for corn, sugar cane, and sugar beet to ethanol processes. Very little design of metabolism is necessary, since *S. cerevisiae* is already well adapted to high glucose feedstock and high titer ethanol production. With high yields (>90%) and titers (up to 160 g/L) in ethanol production on starch and simple sugars, industrial bioethanol production is more than 15 billion gallons per year.[80] However, *S. cerevisiae* remains suboptimal for lignocellulosic biomass fermentation.

While glucose is the most common sugar in lignocellulosic biomass (32–50%), large quantities of other hexose sugars such as galactose and mannose (2–14%) and pentose sugars such as xylose and L-arabinose (11–25%) exist in varying fractions depending on the sources of lignocellulosic biomass.[1,7,81,82] Additionally, lignocellulosic biomass derivatives contain other molecules besides monomeric sugars generated from prior processes of pretreatment and hydrolysis for fermentation. Phenolic lignin derivatives are present in large fractions (10–20%), partial depolymerization during hydrolysis produces small carbohydrate polymers such as cellobiose, and harsh reaction conditions can degrade sugars to furfurals. Traditional metabolic engineering has relied on the conversion of readily consumed carbon sources. However, the composition of lignocellulosic biomass demands the development of pathways to consume non-native, or exogenous, carbon sources in *S. cerevisiae.*

Naturally, organisms are well adapted to take up and metabolize a wide variety of carbon sources, but few are adapted to consuming the above combinations of carbohydrates all at once. The simplest explanation is that these combinations of degradation products do not exist in nature at high titers, therefore no natural selection could bring about this characteristic. In addition, many organisms in nature are specialists that have evolved and adapted to a particular niche environment. Therefore, the challenge for metabolic engineers is to implement broad-sugar specificity and metabolic capability into a host organism and optimize the organism for the unique environment of an industrial process in a bioreactor.

As the first step, building and optimizing pentose metabolic pathways into *S. cerevisiae* has been the focus. Pentose sugars such as xylose and arabinose constitute significant portion of lignocellulosic biomass; especially xylose is the second most abundant sugar in lignocellulosic biomass constituting up to 22% of dry mass.[82] For the last few decades, two types of pentose pathways have been constructed in yeast, the oxidoreductase pathway and the isomerase pathway (Figure 2). Both xylose and arabinose can be metabolized through each of these pathways, although arabinose assimilation involves additional steps in both cases.[83] All four possible pathway variants have been previously constructed,[83–86] and all feed into native *S. cerevisiae* metabolism via D-xylulose or D-xylulose-5-phosphate. Once converted to xylulose-5-P, these sugars are further metabolized through the native pentose phosphate pathway (PPP).

The pentose oxidoreductase pathways are conserved between certain species of native fungi, and use common enzymes and redox cofactors to catalyze substrate conversion. The xylose oxidoreductase pathway was the first heterologous pentose pathway constructed in baker's yeast.[84] In this pathway, xylose is reduced to xylitol by an aldose reductase (AR), then xylitol is oxidized to xylulose by xylitol dehydrogenase (XDH). The AR most commonly used is encoded by *S. stipitis* (formerly *Pichia stipitis*) XYL1 (xylose reductase, XR), which prefers the cofactor NADPH over NADH. The XDH is encoded by *S. stipitis* XYL2, which is NAD$^+$ dependent.[87]

The L-arabinose oxidoreductase pathway was also the first variant constructed in baker's yeast for arabinose conversion.[88] The two enzymes from the xylose oxidoreductase pathway AR and XDH serve as catalysts of the first and last reactions, respectively. The remaining two steps have recently been constructed using two genes from the fungus *Trichoderma reesei*

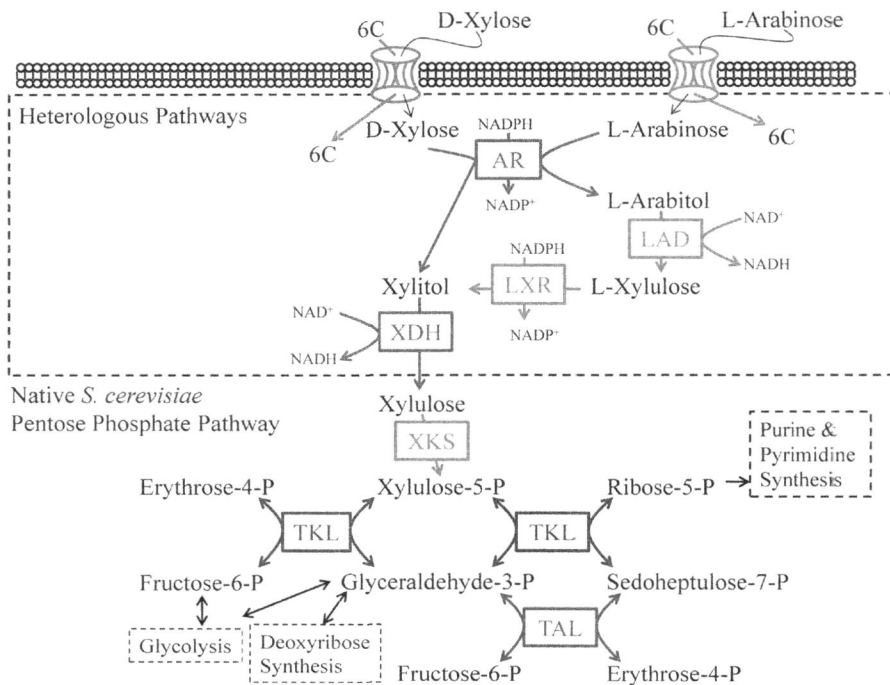

Figure 2

Pentose metabolism in *Saccharomyces cerevisiae*. Diagram depicting the initial assimilat on of pentoses in *S. cerevisiae*. First, xylose and arabinose leak in through hexose transporters at the cell membrane. Then a heterologous pathway (the oxidoreductase pathways are shown here) convert the pentoses to xylulose. Xylulose is then converted by the pentose phosphate pathway to make all of the necessary compounds for growth, maintenance, and, potentially, biofuel production. Abbreviations: AR, aldose reductase; LAD, L-arabinitol dehydrogenase; LXR, L-xylulose reductase; XDH, xylitol dehydrogenase; XKS, xylulokinase; TKL, transketolase; TAL, transaldolase.

(scientific name of *Hypocrea jecorina*). The first gene, LAD1, encodes arabinitol 4-dehydrogenase (ADH), which reduces arabitol to L-xylulose using NADPH as a co-factor.[89] The second is an L-xylulose reductase (XR) encoded by LXR1, which reduces L-xylulose to D-xylulose using NAD+ as a cofactor.[88] The D-xylulose from both the xylose and arabinose variants is then used by native metabolic steps, culminating in the production of biomass, carbon dioxide, and ethanol.

The main challenges in xylose and arabinose oxidoreductase pathways are cofactor imbalances limiting theoretical and actual pentose conversion by yeast.[90,91] Recent studies have revealed AR and XDH genes from other organisms, such as *Rhodototurula mucilaginosa*, may have more favorable co-factor usage, relieving co-factor imbalance.[92] Additional homologs have also been reported.[93–98] The pathways constructed from these homologous genes showed the potential to function better than the pathway constructed from *S. stipitis* AR and

XDH. Additionally, discovery of more efficient LAD and LXR homologs may enable faster rates of L-arabinose consumption. Significant work has gone into discovering more LAD and LXR[99–105]; however, this pathway remains slow and inefficient. Furthermore, there has been some disagreement over whether some proteins are in fact LXR.[106] This inability to replicate the arabinose growth rate of native organisms by constructing the arabinose oxidoreductase pathway could indicate that the enzymes are not functioning well in *S. cerevisiae* or that the pathway has not been fully elucidated. Thus, more work is required.

Alternatively, isomerase pathways may be constructed to facilitate the consumption of xylose and arabinose. In contrast to the pentose oxidoreductase pathway, the isomerase pathway variants require no co-factors. The pathway is native to bacterial species and to rare yeasts. The heterologous xylose isomerase pathway minimally consists of one enzyme, xylose isomerase (XI), which directly converts xylose to xylulose. Because most XIs are native to bacteria, difficulties for heterologous expression in yeast exist,[107] yet work has demonstrated functional bacterial XI pathways[85,108] and yielded functional heterologous XIs isolated from rare fungi.[109,110] Recently, pathway function and ethanol yields were shown to be improved via the directed evolution of xylose isomerase.[111] This represents a step forward in xylose pathway engineering, especially for the more attractive isomerase-based pathway. Furthermore, as with the oxidoreductase pathway, the complementation of a xylulokinase can further improve yields and assimilation rates. This pathway has improved ethanol conversion yields over the oxidoreductase pathway; however, strains show lower growth and sugar uptake rates, as discussed later. Nevertheless, this pathway is attractive because of its lack of cofactor imbalance.

Whereas the xylose isomerase pathway involves one step catalyzed by xylose isomerase, the arabinose isomerase pathway consists of three steps requiring three enzymes of arabinose isomerase, ribulosekinase, and ribulose-5-P-4-epimerase encoded by *araA, araB*, and *araD*, respectively. Two variations of this pathway have been constructed in yeast using distinct sets of heterologous genes from *Bacillus subtilis* (*araA*) and *Escherichia coli* (*araB* and *araD*)[86] and from *Lactobacillus plantarum* (*araA, araB,* and *araD*).[112] However, unlike the other pathways described above, evolutionary engineering was needed in both cases to isolate a yeast strain with an active arabinose isomerase pathway. Thus, only after mutations is a functional arabinose isomerase pathway in yeast possible. As a result, more work is necessary to describe a fully stand-alone arabinose catabolic pathway.

Whereas the enzymes discussed above alone are sufficient to enable xylose or arabinose metabolism in yeast, pentose assimilation remains suboptimal. This has led researchers to investigate downstream enzymatic steps in the PPP. The xylulokinase gene from *S. stipitis* (XYL3) is often complemented to further reduce xylitol production.[113–118] Transaldolase (TAL) and transketolase (TKL) alterations have also been shown to improve pathway function.[21,119,120] Another step to improve pentose assimilation is to optimize a step not typically included in metabolic pathways, transport across the cellular membrane. *S. cerevisiae* possesses many monosaccharide transporters, but these are almost exclusively hexose transporters.[121–125] Due to the broad substrate specificity of many of these transporters, xylose and arabinose leak in through several hexose transporters.[126] This is not optimal because these transporters are not adapted to efficiently uptake pentoses, and glucose can competitively inhibit pentose uptake limiting pentose assimilation. Heterologous expression of sugar transporters has produced some encouraging results.[127–136] However, the most potential likely lies in the directed evolution of transporters for pentose efficiency and specificity,[29] and there is still much improvement to be made for pentose uptake to rival that of glucose uptake.

Additionally, constituents such as lignins and furfurals have gone largely unaddressed in metabolic engineering of yeasts, but they represent a potential source of carbon. These constituents can be toxic to *S. cerevisiae* under industrially relevant conditions,[137] so most research in this area has consisted of engineering tolerance to these compounds. However, if these compounds could be metabolized by yeast, an additional source of carbon could be made accessible.

If the metabolism of pentose sugars and perhaps other compounds could be increased, then faster rates of production could be achieved. As it stands, growth rates of *S. cerevisiae* on glucose are roughly twice as fast as growth rates on xylose, and growth rates on arabinose are much slower.[76] However, growth rates of wild type yeasts such as *S. stipitis* on these pentoses are much faster than growth rates achieved by metabolic engineering of S. cerevisiae. Even so, there are reports of engineered *S. cerevisiae* that can grow on glucose and the two major pentoses.[57,138,139] With further improvements, simultaneous and rapid consumption of all constituent sugars may yet be achieved.

Suboptimal growth rates on pentoses for recombinant strains of *S. cerevisiae* is a clear indicator that additional metabolic engineering of the pathways and the larger metabolic network is necessary to achieve the optimal pathway. Incorporation of these new carbon sources causes perturbations in the metabolic network of *S. cerevisiae*. This is especially true for the pentoses, because catabolism creates a large flux of carbon through the PPP, which has not been adapted for that purpose. Allowing this network to evolve and adjust to the perturbations is necessary. Modeling of the metabolic network for *S. cerevisiae* strains that have been engineered to consume pentoses has recently been done.[58,140–142] If more work is done in this area, a greater understanding of the state of the metabolic network could be uncovered. This work may reveal other pathways or regulatory networks that may speed the rate of pentose assimilation. Alternatively, serial subculturing in mixed sugar media could select for improved mutants without a priori knowledge of what changes need to take place in the metabolic network. This technique, termed evolutionary engineering, is an effective way to improve recombinant strains. However, analysis of the changes that took place in evolutionary engineering experiments is not often discussed. Uncovering what is being altered could offer clues for even more powerful rational strain engineering methods.

Current metabolic engineering approaches have not yet achieved industrially relevant consumption rates of non-glucose carbon sources, especially pentoses. The continued study of pentose metabolism in other organisms, as well as the continued application of directed evolution and evolutionary engineering, is necessary to fully develop a hexose and pentose consuming strain of *S. cerevisiae*. While traditional metabolic engineering has successfully constructed pentose utilization pathways, steps remain to be taken to optimize them. While many of the tools available to metabolic engineering have been used in this context; an optimal solution has not been discovered despite decades of work on yeast pentose metabolism. In summary, the major remaining challenge for the metabolic engineering of yeast to consume lignocellulosic derivatives is improving the uptake and assimilation of the non-glucose carbon sources.

Beyond Bioethanol

Bioethanol is currently the most common but not the most attractive biofuel, as briefly mentioned previously. The low energy content, which is 70% of the energy content of gasoline, is a major drawback of bioethanol. The hygroscopicity, which is the tendency to attract water molecules, also causes problems with storage and transportation of bioethanol with widely distributed conventional infrastructure. In addition, the high miscibility and formation of an azeotrope with water makes bioethanol difficult to separate in distillation processes. As a result, there has been increasing interest in producing advanced biofuels that are more similar to petroleum-derived compounds with high energy content and drop-in capability for current infrastructures.

Despite the advantages of using advanced biofuels, production in microorganisms has been difficult. Unlike bioethanol, a native byproduct of fermentation for some yeasts including *S. cerevisiae*, most advanced biofuels are not naturally produced in microorganisms, especially at the industrial scale. This is not surprising considering the formation of petroleum over eons, under high temperature and high pressure. In addition, microorganisms have not been forced by natural selection to produce large amounts of petroleum-like products in nature, so the possession of a complete metabolic pathway for advanced biofuels is rare or does not exist. Though some organisms possess a native pathway for certain advanced biofuels such as short-chain alcohols, these natural biofuels still have less attractive features as transportation fuels than other advanced biofuels that are not produced natively in microorganisms. Therefore, metabolic engineering for advanced biofuel production commonly requires novel pathway construction.

To construct a novel pathway for advanced biofuel production, metabolic engineers often use bioethanol production pathways as benchmarks. Through metabolic engineering, advanced biofuel production is possible in *S. cerevisiae* (Figure 3). The inherent ability of ethanol

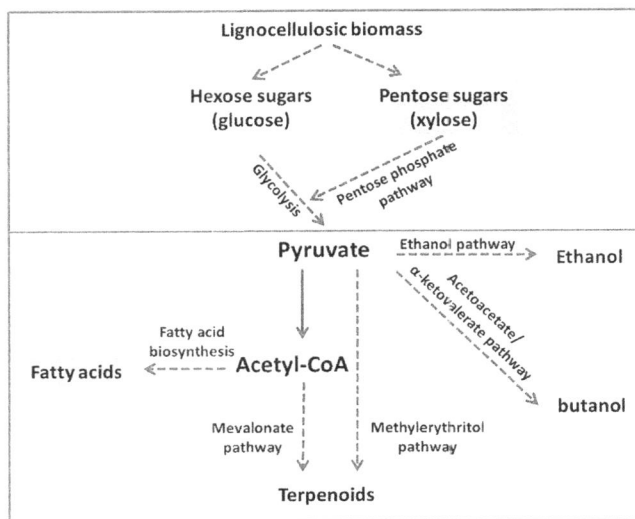

Figure 3

Advanced biofuel production in engineered *S. cerevisiae*. Pentose sugar utilization (blue (gray in print versions)), butanol, and terpenoid production (yellow (light gray in print versions)) have been demonstrated, but require further improvement for industrial scale applications. Fatty acids production (red (dark gray in print versions)) has not been successfully demonstrated in *S. cerevisiae*.

production and high tolerance to alcohol makes *S. cerevisiae* a suitable host for alcohol-based biofuels. Butanol production in *S. cerevisiae* has been demonstrated by rerouting acetyl-coA flux away from ethanol and toward butanol. Butanol is a short-chain alcohol with higher energy content and lower hygroscopicity than ethanol. By expressing genes from butanol producing *Clostridium* bacteria, *S. cerevisiae* was able to produce 2.5 mg/L of butanol.[143] Later, rerouting the valine pathway for butanol production improved titers to 143 mg/L,[144] and re-locating valine synthesis enzymes further increased butanol titers in *S. cerevisiae* to 630 mg/L.[145] While this is one of the leading achievements for advanced biofuels production, titer remains a limitation. Therefore, further metabolic engineering is required for pathway and metabolic network optimization.

With plentiful precursors, *S. cerevisiae* seems to be a desirable host for terpenoid (isoprenoid)-derived biofuels. Isoprenoid-derived biofuels are regarded as potential alternatives to both gasoline and diesels due to high octane number and low freezing point. The approaches for engineering high isoprenoid production in yeast commonly import metabolic pathways from plants, the major producers of isoprenoids. Plant-derived proteins are core parts for constructing metabolic pathways for isoprenoid-derived biofuel production, but are often reported to be toxic to when expressed heterologously in bacteria. However, *S. cerevisiae* is able to withstand the expression of these enzymes and thus is a potent host for isoprenoid biosynthesis. For example, amorphadiene, one of the precursors of a higher chain isoprenoid biofuel, was produced through engineered mevalonate pathway in *S. cerevisiae* at the high titer of 40 g/L.[146] Bisabolene, another isoprenoid biofuel precursor, recently has been produced in yeast with a titer of 5.2 g/L.[147] Bisabolene can be converted to bisabolane, a high-performance advanced drop-in biofuel, via a simple additional hydrogenation and reduction step.

Despite the possibility of producing these molecules, most alternative biofuels in yeast are produced at inferior yields, titers, and rates when compared with bioethanol production. At present, these values are not high enough to meet the demand for liquid fuels in the transportation sector or even to be comparable to those from metabolically engineered *E. coli*. For example, an engineered *E. coli* strain expressing heterologous genes from *Clostridium acetobutylicum*, the most well-known butanol producing bacteria, produced 30 g/L of butanol.[148] The butanol titer of engineered *E. coli* is 200-fold higher than that of engineered *S. cerevisiae* (0.143 g/L).[144] However, the robustness of yeast makes them promising hosts for advanced biofuels, especially when we consider using lignocellulosic biomass as feedstocks. Yeast strains are commonly more tolerant to products and inhibitors from lignocellulosic biomass hydrolysates.[80] Therefore, once the yields, titers, and production rates are improved, advanced biofuel production will be more feasible, with yeasts as a producing host. Here, metabolic engineering could play a role to make yeasts more promising host in producing advanced biofuels.

Next-generation biofuel production is necessary to overcome the challenges associated with ethanol production. With continued rewiring of carbon metabolism to produce energy-dense, drop-in biofuels, this remaining challenge may be overcome. Products such as isobutanol and terpenoids have great promise for production in *S. cerevisiae*. However, the optimal solution may not be reached with current capabilities of pathway assembly and common microbial hosts.

Beyond Current Capability

The final goal of metabolic engineering is to develop strains with the necessary traits for economical biofuel production from lignocellulosic biomass. In this endeavor, the high diversity of natural organisms may possess most of the components needed to reach this goal. The challenge remains on the efficiency in the process of searching and evaluating these components, and finding the optimal configuration, the best pathway. While metabolic engineering has proven proficient at assembling functional pathways, there are fewer technologies available to determine the optimal pathway. Additionally, with increasing efforts in engineering other fungal hosts besides *S. cerevisiae*, it may be necessary to go beyond the accepted host for industrial biofuel production. Therefore, the optimal host and pathway combination may require new tools to be discovered.

To uncover the optimal enzyme for a pathway step, a large diversity of homologs or a large library of mutants must be searched. In terms of homologs, the ever increasing genomic sequence information and search algorithms provide an enormous candidate pool for finding the necessary components, and help narrow down promising candidates without extensive experimentation. In terms of libraries, the advance of protein engineering, especially directed evolution, has produced more powerful techniques for sampling sequence space to find global maximums in enzyme function. However, predictive models for protein behavior do not yet exist, making screening of large libraries necessary. Furthermore, it may never be possible to predict synergistic effects when combining multiple homologs for multiple enzymes in a pathway. Systems biology may help to search and evaluate the possible candidates by predicting possible consequences of introducing certain enzymes and required missing components.[64]

Even now, a blend of computational search and experimentation can now be effective. For example, a functional survey of sugar transporters based on BLAST search offered efficient xylose transporter candidates.[149] There is also still use for nonbiased bioprospecting, as it is effective in discovering required components to construct the best pathway if a priori predictive capability is not available.[150] For example, bioprospecting of the xylose fermenting fungi of *S. stipitis* found genes necessary for improved xylose use.[151] However, these tools need to be generalized to much larger libraries and entire pathways to be more effective. In the future, more sophisticated search algorithms and better computer annotation of predicted enzyme function should yield better and better candidate pools for pathway construction. Improved modeling and library creation techniques should provide more effective and efficient optimization strategies. In this way, the optimal enzyme may be discovered.

Even with the use of search tools that sample the available biodiversity, the best pathway may not be discovered. The number of possible enzymatic pathways is not limited to those that already exist in nature. In this regard, the methods for finding and constructing new synthetic pathways are underdeveloped. Synthetic pathways are commonly constructed based on existing pathways from other organisms. Following this model, a microorganism that produces compounds present in petroleum, like alkanes, would need to be found in nature. Alternatively, a synthetic metabolic pathway can be constructed that incorporates disparate enzymes from several organisms. This has yet to be achieved with any great success. However, the de novo assembly of a pathway that performs novel organic chemistry reactions is within reach and would represent a major step forward for biological manufacture of chemicals. In this regard, finding and assembling the optimal pathways remains a major challenge for advanced biofuel synthesis in yeasts and is not possible with current bioprospecting approaches.

Contrary to traditional strategies, a metabolic engineering strategy proven to be promising in *E. coli* may not be a good starting point for engineering yeast. As metabolic pathways for advanced biofuels commonly benchmark bioethanol production pathways, metabolic engineering of yeasts for advanced biofuel production often use the same strategies that have been applied for *E. coli*. However, metabolic engineering of yeasts may require different strategies from those for *E. coli*. As eukaryotes, yeasts have a different cellular environment than *E. coli*, and this may require different properties of enzymes to be functionally expressed in yeasts.[80,152] As mentioned earlier, introducing a butanol production pathway from *Clostridium* sp. has proven to be a good strategy for metabolic engineering of *E. coli*. However, the same approach was found to be unsuccessful in *S. cerevisiae*, resulting in less than 1 g/L of butanol production, which is about 200-fold less than the titer from *E. coli*.[143] In *S. cerevisiae*, an alternative strategy has derived higher butanol production by targeting the gene overexpressions in the valine pathway.[152] Another example includes the pathways for fatty acid-based biofuel production. Although *S. cerevisiae* has the advantage of high tolerance to low pH over *E. coli*, most fatty-acid based biofuels are produced in *E. coli* rather than *S. cerevisiae* due to the lack of an optimal pathway in yeast.[153]

Rerouting to nonconventional pathways also offers better design strategies for successful production of advanced biofuels. Recent reports on higher alcohol production via amino acid pathways in engineered *E. coli* and *S. cerevisiae* are good examples of rerouting to nonconventional pathway.[154,155] In a strain with a conventional metabolic pathway, sugars are converted into alcohols through fermentative pathways. When the butanol pathway was rerouted from the fermentative pathway to the valine synthesis pathway, *S. cerevisiae* was able to produce more than 70-fold higher butanol.[155] In addition to nonconventional pathways, harnessing other strategies of nature would confer desired traits to the host strain in a more efficient manner and broaden available approaches beyond the conventional metabolic engineering tools. Surface display of cellulosomes[156] and hemicellulosome[157] on *S. cerevisiae*, which was adapted from a multiple enzyme complex in nature, significantly improved ethanol production from cellulose and hemicellulose, 1.4 g/L and 8.2 g/L, respectively, that could not be achieved by a single enzyme expression.

Whole pathway and whole cell techniques, aided by advanced computing and experimental tools, are necessary to go beyond the current stepwise optimization strategy of metabolic engineering. More fully sampling natural diversity and library diversity in a way that selects for synergistic effects between enzymes is necessary, as well as thinking outside the box of potentially limiting traditional pathways. Whole cell redesign of metabolism has the potential to blend both the ability to consume lignocellulosic derivatives and produce next generation biofuels with the optimal living catalyst. Metabolic engineering tools need to advance before this challenge is overcome.

Beyond Saccharomyces cerevisiae

S. cerevisiae is the most well-studied model yeast for biofuel production. However, there are many other yeasts that also hold promise. Not surprisingly, there is ongoing research to engineer strains of yeast isolated from lignocellulose-degrading ecosystems, or yeasts that produce superior precursors to fuels. Some yeasts with attractive bioprocessing traits include (but are not limited to) *S. stipitis* (formerly *Pichia stipitis*[158]) with its pentose sugar fermentation, *Yarrowia lipolytica* with its lipid production, and *Trichoderma reesei* with its cellulosic biomass utilization. Of these, the progress in research with filamentous fungi *T. reesei* has been covered in earlier chapters in this book (Chapters 10 and 11). Therefore, this chapter will briefly discuss the recent studies with other representative yeast strains.

S. stipitis is a native xylose fermenting yeast. The capability of xylose fermentation in *S. cerevisiae* was originally conferred by introducing the genes involved in xylose metabolic pathway of *S. stipitis*.[76] In addition to xylose fermentation, cellulosic ethanol fermentation has been reported by native *S. stipitis* with high titer of 41 g/L.[159] Recently, improved ethanol production from cellulolytic hydrolysate has been demonstrated via genome shuffling of *S. stipitis* and achieved titers of up to 140 g/L.[160] Although the genome sequence of *S. stipitis* has been available since 2007,[161] the lack of metabolic engineering tools for *S. stipitis* has slowed progress in this host. Most biofuel production studies have been reported by using either native or randomly modified strains of *S. stipitis* rather than rationally engineered strains.

Y. lipolytica, one of oleaginous yeasts that produce lipids in wild type state, has also been demonstrated as a promising microbial host for advanced biofuels, specifically biodiesel alternative fuels. Used extensively in industrial applications such as citric acid, protease, and lipase production, the whole genome of *Y. lipolytica* is fully sequenced and metabolic engineering tools for *Y. lipolytica* are available.[2] However, the recent development of hybrid promoters[162] and a gene overexpression platform[163] enable high expression of native and heterologous genes in *Y. lipolytica*, and thus enable metabolic engineering tools that were previously unavailable in this host. *Y. lipolytica* is of interest chiefly for its free fatty acid, precursor to biodiesel, accumulation properties. As an example, a recent study reports improved lipid production with a lipid content of up to 61.7% of dry cell weight.[163] Furthermore, the specific composition of lipid produced in *Y. lipolytica* can be adjusted for more energy-efficient biofuel production. So far, the composition of lipid produced in *Y. lipolytica* has been predominately modified by changing culture conditions rather than by engineering metabolic pathways[2] such as desaturases. However, lipid extraction issues remain, because the lipids accumulate in liposomes, requiring cell lysis to enable extraction. Additionally, *Y. lipolytica* is not suited for consuming many of the sugars present in lignocellulosic biomass. Even so, the ability to accumulate long-chain lipids makes this organism an attractive host.

Beyond these two strains, there are a great number of xylose-fermenting yeasts. Significant organisms include *Neurospora crassa*,[164] *Hansenula polymorpha*,[165,166] *Pachysolen tannophilus*,[167] *Candida arabinofermentans*,[168] and *Pichia guillermondii*.[168] A recent report describes the progress of and interest in the yeast *Spathaspora passalidarum* as a unique host for fuels production.[169] This yeast could potentially serve as a host for lignocellulosic ethanol production, because it has been recently shown to be tolerant to toxins and can conferment several components of lignocellulosic derivatives.[170,171]

It is unclear whether these yeasts will be better hosts for advanced biofuels production from lignocellulosic biomass over the model organism of *S. cerevisiae*. The lack of metabolic engineering tools for these yeasts prevents replicating approaches taken in *S. cerevisiae* to compensate for its shortcomings. By extending current metabolic engineering tools to these organisms, and potentially developing novel tools, more complex rewiring will become possible. Eventually, the ability to evaluate optimized strains of many species of yeast will be within reach, and the optimal strain for fuel production selected. In the end, it may be that a different organism is suited for production of each different biofuel. However, creating new synthetic tools for rapid testing in nonconventional organisms remains a challenge in the field.

Beyond Current Yield, Titers, and Production Rates

The advances and challenges in metabolic engineering described above all aim at improving the capacity to produce advanced biofuels. However, the yields, titers, and production rates of biofuel production are far less than desirable for industrial scale production. Additionally, the strains that are used for demonstrating improved carbon source utilization are not the strains in which advanced biofuel production has been demonstrated. A challenge remains in combining the ideal carbon source utilizing strain with the ideal biofuel producing strain, once both are achieved.

As discussed previously, the suboptimal performance of yeasts is a result of the inefficiency in biomass utilization, low carbon flux through an introduced heterologous pathway, cofactor and energy imbalance in the cell, and intolerance of the cell to the final product and other inhibitors. These difficulties are raised in the entire metabolic network of the host cell. To overcome these challenges, classical metabolic engineering tools could be applied to eliminate bottlenecks step-by-step, or new tools targeted to the entire network may be adapted. These techniques could be as simple as evolutionary engineering, described previously, or as complex as gTME. In the following paragraphs, the remaining challenges for engineering the different types of inefficiencies in the metabolic network are discussed.

The abilities to use or to produce biofuels are often conferred by new metabolic pathways that are commonly based on the expression of foreign enzymes adapted from organisms with desired features. However, integrating new metabolic capability into a whole cellular network can prove challenging due to the complexity of microbial metabolism. Often, the difficulty rests in managing carbon flux through the new pathway that can cause low titers and low production rates. The insufficient carbon flux through the new pathway is associated with the inefficient enzymatic step. The different gene expression machinery and metabolic network of a host strain often results in the failure of proper expression or folding of foreign proteins, and both can lead to an inefficient enzymatic step and low metabolic flux. For example, numerous efforts had failed to express bacterial xylose isomerases in *S. cerevisiae* over the last few decades due to the misfolding of this enzyme in yeast cell system.[76]

The functional expression of foreign enzymes in yeasts can be achieved by choosing promising candidate genes that encode enzymes.[149] The success in the functional expression of enzymes determines the maximum rate and yield. Even with the importance of choosing the proper version of the enzyme, however, the alterative choice for the different versions of enzyme is hardly investigated to date. For example, researchers have only begun looking into the alternative xylose reductases, a key enzyme in the initial metabolism of xylose.[97] Sampling this diversity of enzyme variants could be complimentary to directed evolution approaches, because directed evolution approaches typically only climb to local maxima, and increasing the number of potential starting points could increase the chance of finding the global maximum efficiency for that enzymatic step. Direct modifications of foreign genes by codon-optimization[50] and directed evolution[172] also improve the fitness of the foreign enzymes in a host cell system leading to an efficient enzymatic step and high metabolic flux toward desired product, thus finally improving titers and production rates. Gene knockouts, gene overexpression, and changes in metabolic regulation are often necessary to increase the metabolic flux toward the target product. Balancing enzyme expression levels by using optimal promoters and terminators is also important.[173] All these methods to force the metabolic flux toward products will be necessary for implementing efficient heterologous pathways for next generation biofuel production.

In addition, redox cofactor and the conservation of free energy (ATP) should be considered to better operate heterologous pathway in a whole cell system. The newly added metabolic pathway commonly causes imbalanced redox cofactors or ATP conservation in a host cell system. This causes compensation mechanisms to divert metabolic flux toward byproduct generation. Though xylose fermentation is possible with *S. cerevisiae* by successfully expressing heterologous xylose catabolic pathway from xylose-utilizing organisms of *S. stipitis*, the co-factor imbalance in this pathway limits the yield of ethanol production preventing economical xylose fermentation.[174] Changing the co-factor affinities, ATP stoichiometry, and precursor supply of a target product pathway will ease the disproportionate redox and energy balances. Recently, rerouting an original pathway to an alternative pathway has shown to resolve cofactor imbalance supporting successful production of isobutanol production in *E. coli* and *S. cerevisiae*.[155,175]

Economical biofuel production requires high titers, but the high concentration of product can be toxic to the cells. Although *S. cerevisiae* is known to be relatively tolerant to alcohols, cell growth is inhibited by high concentrations of alcohol, typically around 6% for ethanol[176] and 2% for butanol.[177] In addition, the environmental tolerance has more importance when lignocellulosic biomass is used as a substrate for biofuel production due to the high content of inhibitory compounds preventing cell growth and fermentation performance.[80] Therefore, improving tolerances should be considered in developing engineered strains for biofuel production.

The mechanisms of environmental tolerance are complex reactions involving numerous genes. Until recently, therefore, strain improvements have only been done by adaptive evolution or random mutagenesis. Without detailed a priori knowledge of cellular networks, adaptive evolution can be applied to improve tolerances to products in biofuel production.[60,178] Recent advances in genomics, transcriptomics, and proteomics and their applications in adaptive evolution studies offer clues that explain the adaptation processes taking place during the adaptive evolution.[59] However, more powerful tools or strategies to control and predict an adaptive evolution process are required for the use of adaptive evolution as a standard procedure for strain development, due to the lengthy time over which evolution occurs.

Another approach to improve tolerance is to engineer regulatory systems in *S. cerevisiae*. Unlike *E. coli*, eukaryotic yeast *S. cerevisiae* has a well-developed regulatory system that can be used to empower the overall control of cell systems for higher tolerance to final product or toxic intermediates. gTME is a representative approach for modifying regulatory systems in *S. cerevisiae* in this purpose.[179] The focus of regulatory control will be on the main pathway toward the target products, the transport systems that import and export substrates and final product, and feedback inhibition systems caused by final product or toxic intermediates.

With all these metabolic engineering strategies, designing and constructing a new pathway for successful biofuel production will be feasible and economical. Eventually, optimization of catabolism and anabolism in an optimal host will result in drop-in biofuel production from lignocellulosic biomass. However, challenges must be overcome to reach this goal. Challenges in pathway discovery, optimization, titer, tolerance, and integration into the greater metabolic network remain.

Conclusion

The yeast *S. cerevisiae*, along with other yeasts, has the potential to serve as an industrial host organism for converting lignocellulosic biomass into biofuels. Through pathway construction, optimization, and integration into the larger metabolic network, the metabolism of *S. cerevisiae* can be rewired to accomplish this goal. However, remaining challenges in pentose utilization, tolerance to compounds found in biomass derivatives, production of high energy density drop-in fuels, and integrating advances in these areas with advances in the others limit economic viability. Continually, the tools of metabolic engineering are being applied to each of these challenges. In the coming years, advances in computational and biological tools could make the application of metabolic engineering faster, more efficient, and more predictable. These advances will also help enable the engineering of nonconventional yeast strains. We have only begun to harness the power of metabolic engineering. As tools continue to advance and develop, the biological production of fuels from lignocellulosic biomass will become ever more feasible and efficient. In light of the eventual depletion of petroleum, let us hope that these remaining challenges do not remain for long.

References

1. Olsson L, HahnHagerdal B. Fermentation of lignocellulosic hydrolysates for ethanol production. *Enz Microb Technol* 1996;**18**:312–31.
2. Beopoulos A, Cescut J, Haddouche R, Uribelarrea J-L, Molina-Jouve C, Nicaud J-M. *Yarrowia lipolytica* as a model for bio-oil production. *Progr Lipid Res* 2009;**48**:375–87.
3. Jeffries TW. Engineering yeasts for xylose metabolism. *Curr Opin Biotechnol* 2006;**17**:320–6.
4. Farhi M, Marhevka E, Masci T, Marcos E, Eyal Y, Ovadis M, et al. Harnessing yeast subcellular compartments for the production of plant terpenoids. *Metabol Eng* 2011;**13**:474–81.
5. Sellis D, Callahan BJ, Petrov DA, Messer PW. Heterozygote advantage as a natural consequence of adaptation in diploids. *Proc Natl Acad Sci USA* 2011;**108**:20666–71.
6. Sherman F. Getting started with yeast. *Methods Enzymol* 2002;**350**:3–41.
7. Hahn-Hagerdal B, Linden T, Senac T, Skoog K. Ethanolic fermentation of pentoses in lignocellulose hydrolysates. *Appl Biochem Biotechnol* 1991;**28-9**:131–44.
8. Zaldivar J, Nielsen J, Olsson L. Fuel ethanol production from lignocellulose: a challenge for metabolic engineering and process integration. *Appl Microbiol Biotechnol* 2001;**56**:17–34.
9. Fortman JL, Chhabra S, Mukhopadhyay A, Chou H, Lee TS, Steen E, et al. Biofuel alternatives to ethanol: pumping the microbial well. *Trends Biotechnol* 2008;**26**:375–81.
10. Crook N, Alper HS. Model-based design of synthetic, biological systems. *Chem Eng Sci* 2013;**103**:2–11.
11. Sambrook J. In: Russell, D.W. editor. 3rd ed. Cold Spring Harbor Laboratory Press, Cold Spring Harbor, NY 2000.
12. Goeddel DV, Kleid DG, Bolivar F, Heyneker HL, Yansura DG, Crea R, et al. Expression in *Escherichia coli* of chemically synthesized genes for human insulin. *Proc Natl Acad Sci USA* 1979;**76**:106–10.
13. Nagata S, Taira H, Hall A, Johnsrud L, Streuli M, Ecsodi J, et al. Synthesis in *Escherichia coli* of a polypeptide with human-leukocyte interferon activity. *Nature* 1980;**284**:316–20.
14. Mumberg D, Muller R, Funk M. Yeast vectors for the controlled expression of heterologous proteins in different genetic backgrounds. *Gene* 1995;**156**:119–22.
15. Crook NC, Freeman ES, Alper HS. Re-engineering multicloning sites for function and convenience. *Nucleic Acids Res* 2011;**39**.
16. Hegemann JH, Heick SB. In: Williams JA, editor. *Strain engineering: methods and protocols*, vol. 765. Springer Science and Business Media; 2011. p. 189–206.
17. Nair NU, Zhao HM. Mutagenic inverted repeat assisted genome engineering (MIRAGE). *Nucleic Acids Res* 2009;**37**(1):e9.
18. Shao ZY, Zhao H, Zhao HM. DNA assembler: an in vivo genetic method for rapid construction of biochemical pathways. *Nucleic Acids Res* 2009;**37**(2):e16.
19. Gill RT. Enabling inverse metabolic engineering through genomics. *Curr Opin Biotechnol* 2003;**14**:484–90.
20. Alper H, Stephanopoulos G. Metabolic engineering challenges in the post-genomic era. *Chem Eng Sci* 2004;**59**:5009–17.
21. Jin YS, Alper H, Yang YT, Stephanopoulos G. Improvement of xylose uptake and ethanol production in recombinant *Saccharomyces cerevisiae* through an inverse metabolic engineering approach. *Appl Envir Microbiol* 2005;**71**:8249–56.
22. Blazeck J, Alper H. Systems metabolic engineering: genome-scale models and beyond. *Biotechnol J* 2010;**5**:647–59.

23. Moore JC, Arnold FH. Directed evolution of a para-nitrobenzyl esterase for aqueous-organic solvents. *Nat Biotechnol* 1996;**14**:458–67.
24. Shao Z, Arnold FH. Engineering new functions and altering existing functions. *Curr Opin Struct Biol* 1996;**6**:513–8.
25. You L, Arnold FH. Directed evolution of subtilisin E in *Bacillus subtilis* to enhance total activity in aqueous dimethylformamide. *Protein Eng* 1996;**9**:77–83.
26. Zhao H, Li Y, Arnold FH. Strategy for the directed evolution of a peptide ligase. *Ann N Y Acad Sci* 1996;**799**:1–5.
27. Matsumura I, Rowe LA. Whole plasmid mutagenic PCR for directed protein evolution. *Biomol Eng* 2005;**22**:73–80.
28. Emond S, Mondon P, Pizzut-Serin S, Douchy L, Crozet F, Bouayadi K, et al. A novel random mutagenesis approach using human mutagenic DNA polymerases to generate enzyme variant libraries. *Protein Eng Des Sel* 2008;**21**:267–74.
29. Young EM, Comer AD, Huang HS, Alper HS. A molecular transporter engineering approach to improving xylose catabolism in *Saccharomyces cerevisiae*. *Metabol Eng* 2012;**14**:401–11.
30. Persson B, Roepe PD, Patel L, Lee J, Kaback HR. Site-directed mutagenesis of lysine-319 in the lactose permease of *Escherichia coli*. *Biochem* 1992;**31**:8892–7.
31. Lu JM, Bush DR. His-65 in the proton-sucrose symporter is an essential amino acid whose modification with site-directed mutagenesis increases transport activity. *Proc Natl Acad Sci USA* 1998;**95**:9025–30.
32. Voigt CA, Mayo SL, Arnold FH, Wang ZG. Computational method to reduce the search space for directed protein evolution. *Proc Natl Acad Sci USA* 2001;**98**:3778–83.
33. Xiong AS, Peng RH, Zhuang J, Liu JG, Gao F, Xu F, et al. A semi-rational design strategy of directed evolution combined with chemical synthesis of DNA sequences. *Biol Chem* 2007;**388**:1291–300.
34. Ho BK, Agard DA. Probing the flexibility of large conformational changes in protein structures through local perturbations. *PLoS Comput Biol* 2009;**5**:13.
35. Moore JC, Jin HM, Kuchner O, Arnold FH. Strategies for the in vitro evolution of protein function: enzyme evolution by random recombination of improved sequences. *J Mol Biol* 1997;**272**:336–47.
36. Zhao HM, Arnold FH. Optimization of DNA shuffling for high fidelity recombination. *Nucleic Acids Res* 1997;**25**:1307–8.
37. Zhao HM, Giver L, Shao ZX, Affholter JA, Arnold FH. Molecular evolution by staggered extension process (StEP) in vitro recombination. *Nat Biotechnol* 1998;**16**:258–61.
38. Volkov AA, Arnold FH. *Applications of chimeric genes and hybrid proteins* 2000;**328**(Pt C):447–56.
39. Gibbs MD, Nevalainen KMH, Bergquist PL. Degenerate oligonucleotide gene shuffling (DOGS): a method for enhancing the frequency of recombination with family shuffling. *Gene* 2001;**271**:13–20.
40. Meyer MM, Silberg JJ, Voigt CA, Endelman JB, Mayo SL, Wang ZG, et al. Library analysis of SCHEMA-guided protein recombination. *Protein Sci* 2003;**12**:1686–93.
41. Silberg JJ, Endelman JB, Arnold FH. *Protein Eng* 2004;**388**:35–42.
42. Meyer MM, Hochrein L, Arnold FH. Structure-guided SCHEMA recombination of distantly related beta-lactamases. *Protein Eng Des Sel* 2006;**19**:563–70.
43. Heinzelman P, Snow CD, Wu I, Nguyen C, Villalobos A, Govindarajan S, et al. A family of thermostable fungal cellulases created by structure-guided recombination. *Proc Natl Acad Sci USA* 2009;**106**:5610–5.
44. Forsburg SL. The art and design of genetic screens: yeast. *Nat Rev Genet* 2001;**2**:659–68.
45. Nevoigt E, Fischer C, Mucha O, Matthäus F, Stahl U, Stephanopoulos G. Engineering promoter regulation. *Biotechnol Bioeng* 2007;**96**:550–8.
46. Alper H, Fischer C, Nevoigt E, Stephanopoulos G. Tuning genetic control through promoter engineering. *Proc Natl Acad Sci USA* 2006;**103**:3006.
47. Nevoigt E, Kohnke J, Fischer CR, Alper H, Stahl U, Stephanopoulos G. Engineering of promoter replacement cassettes for fine-tuning of gene expression in *Saccharomyces cerevisiae*. *Appl Envir Microbiol* 2006;**72**:5266–73.
48. Venter M. Synthetic promoters: genetic control through *cis* engineering. *Trends Plant Sci* 2007;**12**:118–24.
49. Blazeck J, Garg R, Reed B, Alper HS. Controlling promoter strength and regulation in *Saccharomyces cerevisiae* using synthetic hybrid promoters. *Biotechnol Bioeng* 2012;**109**:2884–95.

50. Gustafsson C, Govindarajan S, Minshull J. Codon bias and heterologous protein expression. *Trends Biotechnol* 2004;**22**:346–53.

51. Burgess-Brown NA, Sharma S, Sobott F, Loenarz C, Oppermann U, Gileadi O. Codon optimization can improve expression of human genes in *Escherichia coli*: a multi-gene study. *Protein Expr Purif* 2008;**59**:94–102.

52. Sasaguri S, Maruyama J, Moriya S, Kudo T, Kitamoto K, Arioka M. Codon optimization prevents premature polyadenylation of heterologously-expressed cellulases from termite-gut symbionts in *Aspergillus oryzae*. *J Gen Appl Microbiol* 2008;**54**:343–51.

53. Wiedemann B, Boles E. Codon-optimized bacterial genes improve L-arabinose fermentation in recombinant *Saccharomyces cerevisiae*. *Appl Envir Microbiol* 2008;**74**:2043–50.

54. Wu ZL, Qiao J, Zhang ZG, Guengerich FP, Liu Y, Pei XQ. Enhanced bacterial expression of several mammalian cytochrome P450s by codon optimization and chaperone coexpression. *Biotechnol Lett* 2009;**10**:1589–93.

55. Wahlbom CF, Otero RRC, van Zyl WH, Hahn-Hagerdal B, Jonsson LJ. Molecular analysis of a *Saccharomyces cerevisiae* mutant with improved ability to utilize xylose shows enhanced expression of proteins involved in transport, initial xylose metabolism, and the pentose phosphate pathway. *Appl Envir Microbiol* 2003;**69**:740–6.

56. Kuyper M, Toirkens MJ, Diderich JA, Winkler AA, van Dijken JP, Pronk JT. Evolutionary engineering of mixed-sugar utilization by a xylose-fermenting *Saccharomyces cerevisiae* strain. *FEMS Yeast Res* 2005;**5**:925–34.

57. Wisselink HW, Toirkens MJ, Wu Q, Pronk JT, van Maris AJA. Novel evolutionary engineering approach for accelerated utilization of glucose, xylose, and arabinose mixtures by engineered *Saccharomyces cerevisiae* strains. *Appl Envir Microbiol* 2009;**75**:907–14.

58. Cadiere A, Ortiz-Julien A, Camarasa C, Dequin S. Evolutionary engineered *Saccharomyces cerevisiae* wine yeast strains with increased in vivo flux through the pentose phosphate pathway. *Metabol Eng* 2011;**13**:263–71.

59. Hong K-K, Vongsangnak W, Vemuri GN, Nielsen J. Unravelling evolutionary strategies of yeast for improving galactose utilization through integrated systems level analysis. *Proc Natl Acad Sci USA* 2011;**108**:12179–84.

60. Çakar ZP, Turanlı-Yıldız B, Alkım C, Yılmaz Ü. Evolutionary engineering of *Saccharomyces cerevisiae* for improved industrially important properties. *FEMS Yeast Res* 2012;**12**:171–82.

61. Westergaard SL, Oliveira AP, Bro C, Olsson L, Nielsen J. A systems biology approach to study glucose repression in the yeast *Saccharomyces cerevisiae*. *Biotechnol Bioeng* 2007;**96**:134–45.

62. Cantone I, Marucci L, Iorio F, Ricci MA, Belcastro V, Bansal M, et al. A yeast synthetic network for *in vivo* assessment of reverse-engineering and modeling approaches. *Cell* 2009;**137**:172–81.

63. Pryciak PM. Designing new cellular signaling pathways. *Chem Biol* 2009;**16**:249–54.

64. Curran KA, Alper HS. Expanding the chemical palate of cells by combining systems biology and metabolic engineering. *Metabol Eng* 2012;**14**:289–97.

65. Alper H, Moxley J, Nevoigt E, Fink GR, Stephanopoulos G. Engineering yeast transcription machinery for improved ethanol tolerance and production. *Science* 2006;**314**:1565–8.

66. Alper H, Stephanopoulos G. Global transcription machinery engineering: a new approach for improving cellular phenotype. *Metab Eng* 2007;**9**:258–67.

67. Andrianantoandro E, Basu S, Karig DK, Weiss R. Synthetic biology: new engineering rules for an emerging discipline. *Mol Syst Biol* 2006;**2**: 2006 0028.

68. Isaacs FJ, Dwyer DJ, Collins JJ. RNA synthetic biology. *Nat Biotechnol* 2006;**24**:545–54.

69. Greber D, Fussenegger M. Mammalian synthetic biology: engineering of sophisticated gene networks. *J Biotechnol* 2007;**130**:329–45.

70. Tyo KE, Alper HS, Stephanopoulos GN. Expanding the metabolic engineering toolbox: more options to engineer cells. *Trends Biotechnol* 2007;**25**:132–7.

71. Sayut DJ, Kambam PKR, Sun L. Engineering and applications of genetic circuits. *Mol Biosyst* 2007;**3**:835–40.

72. Grilly C, Stricker J, Pang WL, Bennett MR, Hasty J. A synthetic gene network for tuning protein degradation in *Saccharomyces cerevisiae*. *Mol Syst Biol* 2007;**3**:127.

73. Young E, Alper H. Synthetic biology: tools to design, build, and optimize cellular processes. *J Biomed Biotechnol* 2010;**2010**. http://dx.doi.org/10.1155/2010/130781. Article ID 130781.

74. Krivoruchko A, Siewers V, Nielsen J. Opportunities for yeast metabolic engineering: lessons from synthetic biology. *Biotechnol J* 2011;**6**(3):262–76.

75. Lanza AM, Crook NC, Alper HS. Innovation at the intersection of synthetic and systems biology. *Curr Opin Biotechnol* 2012;**23**:712–7.

76. Young E, Lee S-M, Alper H. Optimizing pentose utilization in yeast: the need for novel tools and approaches. *Biotechnol Biofuels* 2010;**3**:24.

77. Altschul SF, Gish W, Miller W, Myers EW, Lipman DJ. Basic local alignment search tool. *J Mol Biol* 1990;**215**:403–10.

78. Thompson JD, Higgins DG, Gibson TJ. CLUSTAL W: improving the sensitivity of progressive multiple sequence alignment through sequence weighting, position-specific gap penalties and weight matrix choice. *Nucleic Acids Res* 1994;**22**:4673–80.

79. Larkin MA, Blackshields G, Brown NP, Chenna R, McGettigan PA, McWilliam H, et al. Clustal W and Clustal X version 2.0. *Bioinformatics* 2007;**23**:2947–8.

80. Fischer CR, Klein-Marcuschamer D, Stephanopoulos G. Selection and optimization of microbial hosts for biofuels production. *Metabol Eng* 2008;**10**:295–304.

81. Lee J. Biological conversion of lignocellulosic biomass to ethanol. *J Biotechnol* 1997;**56**:1–24.

82. Hamelinck CN, Hooijdonk Gv, Faaij APC. Ethanol from lignocellulosic biomass: techno-economic performance in short-, middle- and long-term. *Biomass Bioenergy* 2005;**28**:384–410.

83. Richard P, Verho R, Putkonen M, Londesborough J, Penttila M. Production of ethanol from L-arabinose by *Saccharomyces cerevisiae* containing a fungal L-arabinose pathway. *FEMS Yeast Res* 2003;**3**:185–9.

84. Kotter P, Amore R, Hollenberg CP, Ciriacy M. Isolation and characterization of the *Pichia stipitis* xylitol dehydrogenase gene, *XYL2*, and construction of a xylose-utilizing *Saccharomyces cerevisiae* transformant. *Curr Genet* 1990;**18**:493–500.

85. Walfridsson M, Bao XM, Anderlund M, Lilius G, Bulow L, HahnHagerdal B. Ethanolic fermentation of xylose with *Saccharomyces cerevisiae* harboring the *Thermus thermophilus* xylA gene, which expresses an active xylose (glucose) isomerase. *Appl Envir Microbiol* 1996;**62**:4648–51.

86. Becker J, Boles E. A modified *Saccharomyces cerevisiae* strain that consumes L-arabinose and produces ethanol. *Appl Envir Microbiol* 2003;**69**:4144–50.

87. Nevoigt E. Progress in metabolic engineering of *Saccharomyces cerevisiae*. *Microbiol Mol Biol Rev* 2008;**72**:379–412.

88. Richard P, Putkonen M, Vaananen R, Londesborough J, Penttila M. The missing link in the fungal L-arabinose catabolic pathway, identification of the L-xylulose reductase gene. *Biochem* 2002;**41**:6432–7.

89. Richard P, Londesborough J, Putkonen M, Kalkkinen N, Penttila M. Cloning and expression of a fungal L-arabinitol 4-dehydrogenase gene. *J Biol Chem* 2001;**276**:40631–7.

90. Van Vleet JH, Jeffries TW. Yeast metabolic engineering for hemicellulosic ethanol production. *Curr Opin Biotechnol* 2009;**20**:300–6.

91. van Maris AJ, Abbott DA, Bellissimi E, van den Brink J, Kuyper M, Luttik MA, et al. Alcoholic fermentation of carbon sources in biomass hydrolysates by *Saccharomyces cerevisiae*: current status. *Ant Van Leeuwenhoek* 2006;**90**:391–418.

92. Xu P, Bura R, Doty SL. Genetic analysis of D-xylose metabolism by endophytic yeast strains of *Rhodotorula graminis* and *Rhodotorula mucilaginosa*. *Genet Mol Biol* 2011;**34**:471–8.

93. Ho NW, Lin FP, Huang S, Andrews PC, Tsao GT. Purification, characterization, and amino terminal sequence of xylose reductase from *Candida shehatae*. *Enz Microb Technol* 1990;**12**:33–9.

94. Zhao X, Gao P, Wang Z. The production and properties of a new xylose reductase from fungus *Neurospora crassa*. *Appl Biochem Biotechnol* 1998;**70-72**:405–14.

95. Woodyer R, Simurdiak M, van der Donk WA, Zhao H. Heterologous expression, purification, and characterization of a highly active xylose reductase from *Neurospora crassa*. *Appl Envir Microbiol* 2005;**71**:1642–7.

96. Berghall S, Hilditch S, Penttila M, Richard P. Identification in the mould Hypocrea jecorina of a gene encoding an NADP(+): D-xylose dehydrogenase. *FEMS Microbiol Lett* 2007;**277**:249–53.

97. Zhang F, Qiao D, Xu H, Liao C, Li S, Cao Y. Cloning, expression, and characterization of xylose reductase with higher activity from *Candida tropicalis*. *J Microbiol* 2009;**47**:351–7.

98. Tran LH, Kitamoto N, Kawai K, Takamizawa K, Suzuki T. Cloning and expression of a NAD(+)-dependent xylitol dehydrogenase gene (xdhA) of *Aspergillus oryzae*. *J Biosci Bioeng* 2004;**97**:419–22.

99. Verho R, Putkonen M, Londesborough J, Penttila M, Richard P. A novel NADH-linked L-xylulose reductase in the L-arabinose catabolic pathway of yeast. *J Biol Chem* 2004;**279**:14746–51.

100. Nair N, Zhao H. Biochemical characterization of an L-Xylulose reductase from *Neurospora crassa*. *Appl Environ Microbiol* 2007;**73**:2001–4.

101. Sullivan R, Zhao H. Cloning, characterization, and mutational analysis of a highly active and stable L-arabinitol 4-dehydrogenase from *Neurospora crassa*. *Appl Microbiol Biotechnol* 2007;**77**:845–52.

102. Bae B, Sullivan RP, Zhao HM, Nair SK. Structure and engineering of l-arabinitol 4-Dehydrogenase from *Neurospora crassa*. *J Mol Biol* 2010;**402**:230–40.

103. Bera AK, Sedlak M, Khan A, Ho NWY. Establishment of l-arabinose fermentation in glucose/xylose co-fermenting recombinant *Saccharomyces cerevisiae* 424A(LNH-ST) by genetic engineering. *Appl Microbiol Biotechnol* 2010;**87**:1803–11.

104. Suzuki T, Tran LH, Yogo M, Idota O, Kitamoto N, Kawai K, et al. Cloning and expression of NAD+-dependent L-arabinitol 4-dehydrogenase gene (ladA) of *Aspergillus oryzae*. *J Biosci Bioeng* 2005;**100**:472–4.

105. Mojzita D, Penttila M, Richard P. Identification of an L-arabinose reductase gene in *Aspergillus niger* and its role in L-Arabinose catabolism. *J Biol Chem* 2010;**285**:23622–8

106. Mojzita D, Vuoristo K, Koivistoinen OM, Penttila M, Richard P. The "true" L-xylulose reductase of filamentous fungi identified in *Aspergillus niger*. *FEBS Lett* 2010;**584**:3540–4.

107. Gardonyi M, Hahn-Hagerdal B. The *Streptomyces rubiginosus* xylose isomerase is misfolded when expressed in *Saccharomyces cerevisiae*. *Enz Microb Technol* 2003;**32**:252–9.

108. Brat D, Boles E, Wiedemann B. Functional expression of a bacterial xylose isomerase in *Saccharomyces cerevisiae*. *Appl Envir Microbiol* 2009;**75**:2304–11.

109. Kuyper M, Harhangi HR, Stave AK, Winkler AA, Jetten MSM, de Laat W, et al. High-level functional expression of a fungal xylose isomerase: the key to efficient ethanolic fermentation of xylose by *Saccharomyces cerevisiae*? *FEMS Yeast Res* 2003;**4**:69–78.

110. Madhavan A, Tamalampudi S, Ushida K, Kanai D, Katahira S, Srivastava A, et al. Xylose isomerase from polycentric fungus *Orpinomyces*: gene sequencing, cloning, and expression in *Saccharomyces cerevisiae* for bioconversion of xylose to ethanol. *Appl Microbiol Biotechnol* 2009;**82**:1067–78.

111. Lee SM, Jellison T, Alper HS. Directed evolution of xylose isomerase for improved xylose catabolism and fermentation in the yeast *Saccharomyces cerevisiae*. *Appl Envir Microbiol* 2012;**78**:5708–16.

112. Wisselink HW, Toirkens MJ, Berriel MDF, Winkler AA, van Dijken JP, Pronk JT, et al. Engineering of *Saccharomyces cerevisiae* for efficient anaerobic alcoholic fermentation of L-arabinose. *Appl Envir Microbiol* 2007;**73**:4881–91.

113. Ho NW, Chen Z, Brainard AP. Genetically engineered *Saccharomyces* yeast capable of effective cofermentation of glucose and xylose. *Appl Envir Microbiol* 1998;**64**:1852–9.

114. Ho NW, Chen Z, Brainard AP, Sedlak M. Successful design and development of genetically engineered *Saccharomyces* yeasts for effective cofermentation of glucose and xylose from cellulosic biomass to fuel ethanol. *Adv Biochem Eng Biotechnol* 1999;**65**:163–92.

115. Eliasson A, Christensson C, Wahlbom CF, Hahn-Hagerdal B. Anaerobic xylose fermentation by recombinant *Saccharomyces cerevisiae* carrying *XYL1*, *XYL2*, and *XKS1* in mineral medium chemostat cultures. *Appl Envir Microbiol* 2000;**66**:3381–6.

116. Richard P, Toivari MH, Penttila M. The role of xylulokinase in *Saccharomyces cerevisiae* xylulose catabolism. *FEMS Microbiol Lett* 2000;**190**:39–43.

117. Jin YS, Ni H, Laplaza JM, Jeffries TW. Optimal growth and ethanol production from xylose by recombinant *Saccharomyces cerevisiae* require moderate D-xylulokinase activity. *Appl Envir Microbiol* 2003;**69**:495–503.

118. Matsushika A, Sawayama S. Efficient bioethanol production from xylose by recombinant *Saccharomyces cerevisiae* requires high activity of xylose reductase and moderate xylulokinase activity. *J Biosci Bioeng* 2008;**106**:306–9.

119. Metzger MH, Hollenberg CP. Isolation and characterization of the *Pichia stipitis* transketolase gene and expression in a xylose utilizing *Saccharomyces-cerevisiae* transformant. *Appl Microbiol Biotechnol* 1994;**42**:319–25.

120. Walfridsson M, Hallborn J, Penttila M, Keranen S, Hahnhagerdal B. Xylose-metabolizing *Saccharomyces cerevisiae* strains overexpressing the TKL1 and TAL1 genes encoding the pentose-phosphate pathway enzymes transketolase and transaldolase. *Appl Envir Microbiol* 1995;**61**:4184–90.

121. Lagunas R. Sugar transport in *Saccharomyces cerevisiae*. *FEMS Microbiol Rev* 1993;**10**:229–42.

122. Andre B. An overview of membrane transport proteins in S*accharomyces cerevisiae*. *Yeast* 1995;**11**:1575–611.

123. Boles E, Hollenberg CP. The molecular genetics of hexose transport in yeasts. *FEMS Microbiol Rev* 1997; **21**:85–111.

124. Pao SS, Paulsen IT, Saier MH. Major facilitator superfamily. *Microbiol Mol Biol Rev* 1998;**62**:1–34.

125. Ozcan S, Johnston M. Function and regulation of yeast hexose transporters. *Microbiol Mol Biol Rev* 1999;**63**:554–69.

126. Hamacher T, Becker J, Gardonyi M, Hahn-Hagerdal B, Boles E. Characterization of the xylose-transporting properties of yeast hexose transporters and their influence on xylose utilization. *Microbiol* 2002;**148**:2783–8.

127. Weierstall T, Hollenberg CP, Boles E. Cloning and characterization of three genes (*SUT1-3*) encoding glucose transporters of the yeast *Pichia stipitis*. *Mol Microbiol* 1999;**31**:871–83.

128. Leandro MJ, Goncalves P, Spencer-Martins I. Two glucose/xylose transporter genes from the yeast *Candida intermedia*: first molecular characterization of a yeast xylose-H⁺ symporter. *Biochem J* 2006;**395**:543–9.

129. Saloheimo A, Rauta J, Stasyk OV, Sibirny AA, Penttila M, Ruohonen L. Xylose transport studies with xylose-utilizing *Saccharomyces cerevisiae* strains expressing heterologous and homologous permeases. *Appl Microbiol Biotechnol* 2007;**74**:1041–52.

130. Hector RE, Qureshi N, Hughes SR, Cotta MA. Expression of a heterologous xylose transporter in a *Saccharomyces cerevisiae* strain engineered to utilize xylose improves aerobic xylose consumption. *Appl Microbiol Biotechnol* 2008;**80**:675–84.

131. Du J, Li SJ, Zhao HM. Discovery and characterization of novel D-xylose-specific transporters from *Neurospora crassa* and *Pichia stipitis*. *Mol Biosyst* 2010;**6**:2150–6.

132. Galazka JM, Tian CG, Beeson WT, Martinez B, Glass NL, Cate JHD. Cellodextrin transport in yeast for improved biofuel production. *Science* 2010;**330**:84–6.

133. Runquist D, Hahn-Hagerdal B, Radstrom P. Comparison of heterologous xylose transporters in recombinant *Saccharomyces cerevisiae*. *Biotechnol Biofuels* 2010;**3**:5.

134. Subtil T, Boles E. Improving L-arabinose utilization of pentose fermenting *Saccharomyces cerevisiae* cells by heterologous expression of L-arabinose transporting sugar transporters. *Biotechnol Biofuels* 2011;**4**:38.

135. Verho R, Penttila M, Richard P. Cloning of two genes (LAT1,2) encoding specific L-arabinose transporters of the L-arabinose fermenting yeast *Ambrosiozyma monospora*. *Appl Biochem Biotechnol* 2011;**164**:604–11.

136. Young E, Poucher A, Comer A, Bailey A, Alper H. Functional survey for heterologous sugar transport proteins, using *Saccharomyces cerevisiae* as a host. *Appl Envir Microbiol* 2011;**77**:3311–9.

137. Palmqvist E, Grage H, Meinander NQ, Hahn-Hagerdal B. Main and interaction effects of acetic acid, furfural, and p-hydroxybenzoic acid on growth and ethanol productivity of yeasts. *Biotechnol Bioeng* 1999;**63**:46–55.

138. Kim SR, Ha SJ, Wei N, Oh EJ, Jin YS. Simultaneous co-fermentation of mixed sugars: a promising strategy for producing cellulosic ethanol. *Trends Biotechnol* 2012;**30**:274–82.

139. Karhumaa K, Wiedemann B, Hahn-Hagerdal B, Boles E, Gorwa-Grauslund MF. Co-utilization of L-arabinose and D-xylose by laboratory and industrial *Saccharomyces cerevisiae* strains. *Microb Cell Fac* 2006;**5**:18.

140. Parachin NS, Bergdahl B, van Niel EWJ, Gorwa-Grauslund MF. Kinetic modelling reveals current limitations in the production of ethanol from xylose by recombinant S*accharomyces cerevisiae*. *Metabol Eng* 2011;**13**:508–17.

141. Wisselink HW, Cipollina C, Oud B, Crimi B, Heijnen JJ, Pronk JT, et al. Metabolome, transcriptome and metabolic flux analysis of arabinose fermentation by engineered *Saccharomyces cerevisiae*. *Metabol Eng* 2010;**12**:537–51.

142. Pitkanen JP, Aristidou A, Salusjarvi L, Ruohonen L, Penttila M. Metabolic flux analysis of xylose metabolism in recombinant *Saccharomyces cerevisiae* using continuous culture. *Metab Eng* 2003;**5**:16–31.
143. Steen EJ, Chan R, Prasad N, Myers S, Petzold CJ, Redding A, et al. Metabolic engineering of *Saccharomyces cerevisiae* for the production of n-butanol. *Microb Cell Fac* 2008;**7**:36.
144. Kondo T, Tezuka H, Ishii J, Matsuda F, Ogino C, Kondo A. Genetic engineering to enhance the Ehrlich pathway and alter carbon flux for increased isobutanol production from glucose by *Saccharomyces cerevisiae*. *J Biotechnol* 2012;**159**:32.
145. Brat D, Weber C, Lorenzen W, Bode H, Boles E. Cytosolic re-localization and optimization of valine synthesis and catabolism enables inseased isobutanol production with the yeast *Saccharomyces cerevisiae*. *Biotechnol Biofuels* 2012;**5**:65.
146. Westfall PJ, Pitera DJ, Lenihan JR, Eng D, Woolard FX, Regentin R, et al. Production of amorphadiene in yeast, and its conversion to dihydroartemisinic acid, precursor to the antimalarial agent artemisinin. *Proc Natl Acad Sci USA* 2012;**109**:E111–8.
147. Özaydın B, Burd H, Lee TS, Keasling JD. Carotenoid-based phenotypic screen of the yeast deletion collection reveals new genes with roles in isoprenoid production. *Metab Eng* 2012;**15**:174–83.
148. Shen CR, Lan EI, Dekishima Y, Baez A, Cho KM, Liao JC. Driving forces enable high-titer anaerobic 1-Butanol synthesis in *Escherichia coli*. *Appl Envir Microbiol* 2011;**77**:2905–15.
149. Young E, Poucher A, Comer A, Bailey A, Alper H. Functional survey for heterologous sugar transport proteins, using *Saccharomyces cerevisiae* as a host. *Appl Environ Microbiol* 2011;**77**:3311–9.
150. Nevoigt E. Progress in metabolic engineering of *Saccharomyces cerevisiae*. *Microbiol Mol Biol Rev* 2008;**72**:379–412.
151. Jin Y-S, Alper H, Yang Y-T, Stephanopoulos G. Improvement of xylose uptake and ethanol production in recombinant *Saccharomyces cerevisiae* through an inverse metabolic engineering approach. *Appl Environ Microbiol* 2005;**71**:8249–56.
152. Zhang F, Rodriguez S, Keasling JD. Metabolic engineering of microbial pathways for advanced biofuels production. *Curr Opin Biotechnol* 2011;**22**:775–83.
153. Peralta-Yahya PP, Zhang F, del Cardayre SB, Keasling JD. Microbial engineering for the production of advanced biofuels. *Nature* 2012;**488**:320–8.
154. Atsumi S, Hanai T. Non-fermentative pathways for synthesis of branched-chain higher alcohols as biofuels. *Nature* 2008;**451**:86–9.
155. Chen X, Nielsen KF, Borodina I, Kielland-Brandt MC, Karhumaa K. Increased isobutanol production in *Saccharomyces cerevisiae* by overexpression of genes in valine metabolism. *Biotechnol Biofuels* 2011;**4**:21.
156. Fan L-H, Zhang Z-J, Yu X-Y, Xue Y-X, Tan T-W. Self-surface assembly of cellulosomes with two miniscaffoldins on *Saccharomyces cerevisiae* for cellulosic ethanol production. *Proc Natl Acad Sci USA* 2012;**109**:13260–5.
157. Sakamoto T, Hasunuma T, Hori Y, Yamada R, Kondo A. Direct ethanol production from hemicellulosic materials of rice straw by use of an engineered yeast strain codisplaying three types of hemicellulolytic enzymes on the surface of xylose-utilizing *Saccharomyces cerevisiae* cells. *J Biotechnol* 2012;**158**:203–10.
158. Kurtzman CP, Suzuki M. Phylogenetic analysis of ascomycete yeasts that form coenzyme Q-9 and the proposal of the new genera *Babjeviella, Meyerozyma, Millerozyma, Priceomyces*, and *Scheffersomyces*. *Mycosci* 2010;**51**:2–14.
159. Agbogbo FK, Coward-Kelly G. Cellulosic ethanol production using the naturally occurring xylose-fermenting yeast, *Pichia stipitis*. *Biotechnol Lett* 2008;**30**:1515–24.
160. Bajwa PK, Phaenark C, Grant N, Zhang X, Paice M, Martin VJJ, et al. Ethanol production from selected lignocellulosic hydrolysates by genome shuffled strains of *Scheffersomyces stipitis*. *Bioresour Technol* 2011;**102**:9965–9.
161. Jeffries TW, Grigoriev IV, Grimwood J, Laplaza JM, Aerts A, Salamov A, et al. Genome sequence of the lignocellulose-bioconverting and xylose-fermenting yeast *Pichia stipitis*. *Nat Biotechnol* 2007;**25**:319–26.
162. Blazeck J, Liu LQ, Redden H, Alper H. Tuning gene expression in *Yarrowia lipolytica* by a hybrid promoter approach. *Appl Envir Microbiol* 2011;**77**:7905–14.

163. Tai M, Stephanopoulos G. Engineering the push and pull of lipid biosynthesis in oleaginous yeast *Yarrowia lipolytica* for biofuel production. *Metab Eng* 2013;**15**:1–9.
164. Zhao X, Gao P, Wang Z. The production and properties of a new xylose reductase from fungus. *Neurospora crassa. Appl Biochem Biotechnol* 1998;**70-72**:405–14.
165. Grabek-Lejko D, Ryabova OB, Oklejewicz B, Voronovsky AY, Sibirny AA. Plate ethanol-screening assay for selection of the *Pichia stipitis* and *Hansenula polymorpha* yeast mutants with altered capability for xylose alcoholic fermentation. *J Ind Microbiol Biotechnol* 2006;**33**:934–40.
166. Ryabova EB, Chmil OM, Sibirny AA. Xylose and cellobiose fermentation to ethanol by the thermotolerant methylotrophic yeast *Hansenula polymorpha. FEMS Yeast Res* 2003;**4**:157–64.
167. Xu J, Taylor KB. In: Himmel ME, Baker JO, Overend RP, editors. *Enzymatic conversion of biomass for fuels production*, vol. 566. 1994. p. 468–81.
168. Fonseca C, Spencer-Martins I, Hahn-Hagerdal B. L-Arabinose metabolism in *Candida arabinofermentans* PYCC 5603T and *Pichia guilliermondii* PYCC 3012: influence of sugar and oxygen on product formation. *Appl Microbiol Biotechnol* 2007;**75**:303–10.
169. Hou X. Anaerobic xylose fermentation by *Spathaspora passalidarum. Appl Microbiol Biotechnol* 2012; **94**:205–14.
170. Long TM, Su YK, Headman J, Higbee A, Willis LB, Jeffries TW. Cofermentation of glucose, xylose, and cellobiose by the beetle-associated yeast *Spathaspora passalidarum. Appl Envir Microbiol* 2012; **78**:5492–500.
171. Hou XR, Yao S. Improved inhibitor tolerance in xylose-fermenting yeast *Spathaspora passalidarum* by mutagenesis and protoplast fusion. *Appl Microbiol Biotechnol* 2012;**93**:2591–601.
172. Bloom JD, Arnold FH. In the light of directed evolution: pathways of adaptive protein evolution. *Proc Natl Acad Sci USA* 2009;**106**:9995–10000.
173. Lu C, Jeffries T. Shuffling of promoters for multiple genes to optimize xylose fermentation in an engineered *Saccharomyces cerevisiae* strain. *Appl Envir Microbiol* 2007;**73**:6072–7.
174. Van Vleet JH, Jeffries TW. Yeast metabolic engineering for hemicellulosic ethanol production. *Curr Opin Biotechnol* 2009;**20**:300–6.
175. Bastian S, Liu X, Meyerowitz JT, Snow CD, Chen MMY, Arnold FH. Engineered ketol-acid reductoisomerase and alcohol dehydrogenase enable anaerobic 2-methylpropan-1-ol production at theoretical yield in *Escherichia coli. Metab Eng* 2011;**13**:345–52.
176. Teixeira MC, Raposo LR, Mira NP, Lourenco AB, Sa-Correia I. Genome-wide identification of *Saccharomyces cerevisiae* genes required for maximal tolerance to ethanol. *Appl Envir Microbiol* 2009;**75**:5761–72.
177. Knoshaug EP, Zhang M. Butanol tolerance in a selection of microorganisms. *Appl Biochem Biotechnol* 2009;**153**:13–20.
178. Almeida JR, Modig T, Petersson A, Hähn-Hägerdal B, Lidén G, Gorwa-Grauslund MF. Increased tolerance and conversion of inhibitors in lignocellulosic hydrolysates by *Saccharomyces cerevisiae. J Chem Technol Biotechnol* 2007;**82**:340–9.
179. Alper H, Moxley J, Nevoigt E, Fink GR, Stephanopoulos G. Engineering yeast transcription machinery for improved ethanol tolerance and production. *Science* 2006;**314**:1565–8.

PART 4

Fuels from Bacteria

CHAPTER 13

New Tools for the Genetic Modification of Industrial Clostridia

Katrin Schwarz, Ying Zhang, Wouter Kuit, Muhammad Ehsaan, Katalin Kovács, Klaus Winzer, Nigel P. Minton

Clostridia Research Group, Centre for Biomolecular Sciences, BBSRC Sustainable Bioenergy Centre, Nottingham Digestive Diseases Centre NIHR Biomedical Research Unit, School of Life Sciences, University of Nottingham, University Park, Nottingham, UK

Introduction

The genus *Clostridium* comprises Gram-positive, anaerobic, endospore-forming, rod-shaped bacteria of both medical and industrial importance. *Clostridium tetani*, *Clostridium perfringens*, *Clostridium botulinum*, and *Clostridium difficile* are of medical significance due to their pathogenicity to humans and animals,[1–4] whereas the importance of *Clostridium sporogenes* and *Clostridium novyi* resides with their potential use as anticancer drug-delivery vehicles.[5,6] Industrially important strains for the production of biofuels and commodity chemicals are the solvent producers *Clostridium acetobutylicum*, *Clostridium beijerinckii*, and *Clostridium pasteurianum*,[7–10] the $CO/CO_2/H_2$ using acetogens *Clostridium ljungdahlii* and *Clostridium carboxidovorans*,[11–13] and the lignocellulose-degrading microorganisms *Clostridium thermocellum*, *Clostridium cellulolyticum*, and *Clostridium phytofermentans*.[14–16]

The type strain of the genus, *C. acetobutylicum*, was isolated in the 1910s by Chaim Weizmann,[17] just a few years before the advent of the molecular biological era in the 1930s.[18,19] However, despite the longstanding economical importance of the type strain since the first half of the twentieth century,[20] the genus has been relatively little-studied. This is largely a consequence of its challenging genetic accessibility.[21–25] Barriers to gene transfer are high nuclease activities, thick outer layers, and strong restriction-modification (R-M) systems.[26–29] Early milestones in the field of clostridial molecular biology were the transfer of exogenous genetic material into a number of *Clostridium* strains by either transformation or conjugation[30–37] and the development of *Escherichia coli/Clostridium* shuttle vectors.[30,32]

In 2001, the first clostridial genome, that of the type strain *C. acetobutylicum*, was published.[7] However, targeted genetic modifications remained challenging,[38,39] despite the ever-increasing rise of the medical and industrial profile of the genus. *C. difficile* had emerged as a major

Direct Microbial Conversion of Biomass to Advanced Biofuels. http://dx.doi.org/10.1016/B978-0-444-59592-8.00013-0

worldwide cause of healthcare-associated disease,[40,41] whereas *C. perfringens* had evolved into a major problem in the poultry industry as the causative agent of necrotic enteritis.[42] Last, but not least, dwindling fuel resources, high oil prices, and the need to move toward renewable and sustainable energy and fuel resources had revived the interest in *Clostridium* strains for the industrial production of biofuels and/or commodity chemicals.[25] Vital for the development of robust, efficient, and reliable clostridial genetic tools was the introduction of next generation sequencing systems in the early/mid 2000s, allowing the sequencing of whole genomes in a cost-effective high-throughput manner.[43] Since its implementation, it has resulted in the release of genome sequences of all major *Clostridium* species.[44]

Accompanied by the renewed interest and research efforts in the genus, substantial progress was made, leading to the development of gene knock-down methods based on antisense RNA (asRNA),[45,46] gene disruption methods on the basis of the bacterial mobile group II intron from the *ltrB* gene of *Lactococcus lactis*,[22,24,47,48] and gene knockout procedures based on homologous recombination and the use of replicative and non-replicative vectors.[27,38,39,49–54] *Clostridium* shuttle plasmids were improved and standardized,[55] attempts made to increase homologous recombination frequencies through the use of the *Bacillus subtilis* resolvase gene *recU*,[56–59] protocols for the transformation of exogenous genetic material revised for both higher efficiencies and application in previously genetically recalcitrant *Clostridium* strains,[27,29,60] counter (negative) selection methods developed to improve gene deletion efficiencies and to select for double-crossover events,[38,39,61–64] and inducible gene expression (IGE) systems established to regulate gene expression.[21,65–68] Progress was also made in the field of forward genetics. In addition to the well-known and well-characterized conjugative transposons Tn*916* and Tn*5397*, new mutagens, EZ-Tn*5* and the *mariner*-transposable element *Himar*1, were established to ensure a more effective and random mutagenesis.[69–72] This chapter will summarize recent achievements and give an overview of the new genetic tools developed for the genus *Clostridium*. The main focus will be on the industrially valuable strains. However, new genetic tools were developed to be applicable in a broad range of *Clostridium* species, making it often impossible to distinguish between tools for pathogenic and non-pathogenic clostridia.

Transfer of Exogenous Genetic Material

The pivotal first step in the genetic modification of an organism is the physical transfer of the exogenous genetic material into the target cell. In *Clostridium*, the presence of strong R-M systems, high nuclease activities, and thick outer layers can present a formidable barrier to DNA transfer[27] and have to be examined on a case-by-case basis.[26,29,60] DNA transfer is most often achieved using electroporation or through conjugation.[30] One or both procedures have been established for all major *Clostridium* species. However, for some industrially important clostridia, such as *Clostridium saccharoacetobutylicum*, *Clostridium saccharobutylacetonicum*, *Clostridium saccharoperbutylicum*, and *Clostridium klyveri*, the successful application of either procedure has yet to be reported.

In the past, transformation was in some cases performed by PEG-induced DNA uptake into protoplasts. It resulted in varying transformation efficiencies of 28 to 10^4–10^5 transformants per µg DNA.[33,73,74]

Electroporation

Due to varying transformation efficiencies, often complex and time-consuming protocols for the generation of protoplasts and the need for protoplast regeneration in specially formulated, species-specific regeneration media, protoplast transformation became unpopular, especially with the manifestation of electroporation technology.[30,31,33,37] Electroporation increases the cell membrane permeability to ions and macromolecules by exposing the cells to an external and short high electric field pulse.[75,76] Once set up for a specific organism, electroporation offers a rapid and easy to apply protocol, small-scale operation (DNA concentration and cells), and high efficiencies.[77] Studies on *E. coli* showed that 80% of the cells received the exogenous DNA.[78] However, if not set up appropriately, electroporation can cause cell damage and rupture. The transport of material into and out of the cell during the pore formation is unspecific and can cause ion imbalance and cell death.[79,80] The development of a successful and efficient electroporation procedure is reliant on a number of factors, including cell growth (medium, growth phase), composition of the electroporation buffer (ion strength, osmolarity), temperature during the procedure, number of washes, pulse parameters (time, voltage, field strength, resistance), cell:exogenous DNA ratio, quality and characteristics of the exogenous DNA (e.g., origin of replicon, circularity, linearity, size), recovery conditions (time, medium, temperature), and quality of the water and cuvette.[78,79,81–83]

Electroporation procedures have been reported for many of the major *Clostridium* species, such as *C. acetobutylicum*,[84,85] *C. beijerinckii*[86] (formerly *C. acetobutylicum*), *C. pasteurianum*,[29] *Clostridium tyrobutyricum*,[87] *C. ljungdahlii*,[27] *C. thermocellum*,[60] *C. cellulolyticum*,[26] *C. botulinum*,[88,89] *C. perfringens*,[90,91] and *C. sporogenes*.[92] Transformation efficiencies vary. For *C. ljungdahlii*, transformation efficiencies of up to 10^4 transformants per µg DNA were reported,[27] for *C. thermocellum* up to 10^5,[60] and for *C. acetobutylicum* up to 10^2,[55] 10^5,[84] and 10^6.[85] High-efficiency transformations are desirable for effective suicide (conditional, nonreplicative; Clostridial Vector Systems section) vector-based allelic exchange or transposition procedures (Random Mutagenesis By Biological Mutagens and Recombination-Based Methods (Allelic Exchange) Sections). Because the suicide plasmid is unable to replicate in the transformed cell, but will on transformation (rarely) insert into the chromosome by homologous recombination or directly deliver the transposon, mutagenesis rates are highly dependent on transformation efficiency.[93] Consequently, even established electroporation protocols are constantly reviewed for higher efficiencies.

Conjugation

Conjugation, the transfer of genetic material from one bacterial cell (donor) to another (recipient) by direct cell-to-cell contact[94,95] has been reported for a number of *Clostridium* species. These include *C. beijerinckii*[35] (formerly *C. acetobutylicum*), *C. cellulolyticum*,[26] *Clostridium phytofermentas*,[96] *C. sporogenes*,[97] *C. novyi*,[97] *C. difficile*,[98,99] *C. botulinum*,[100] and *C. perfringens*.[101] For some of these species, conjugation is either the only documented way of plasmid transfer to date (e.g., *C. difficile*) or the most efficient (e.g., some *C. botulinum* strains). In the past, *B. subtilis*, *Lactococcus lactis* (*L. lactis subsp. lactis*, formerly *Streptococcus lactis*), or *Enterococcus faecalis*[102,103] were used as donor strains. Today, the preferred donor is *E. coli* (for review see references above). However, independent of the donor strain used, conjugation requires not only close cell-to-cell contact, but also a *cis*-acting nick side (*oriT*, origin of transfer) and a number of *trans*-acting functions (Tra-functions). No mating pair formation, DNA processing, and plasmid transfer to the recipient cell will occur without these elements.[104] In the case of an *E. coli* donor and a *Clostridium* recipient, the *cis*-acting *oriT* of the broad host range IncP family of plasmids (e.g., RP4, RK2) is incorporated into the *E. coli/Clostridium* shuttle vector and used to mobilize it.[35] The Tra-functions are provided by the *E. coli* donor (Tra+) and are either plasmid-encoded (e.g., IncP type helper plasmid R702 in *E. coli* CA434)[35] or integrated into the chromosome (e.g., *E. coli* SM10).[105]

The advantages of plasmid transfer by conjugation are the simplicity of the method, the needlessness of specialized and expensive equipment, the minimal disruption of the cell membrane of the target organism, the prevention of extracellular nuclease activities due to the close cell-to-cell contact,[106] and potentially higher plasmid transfer frequencies.[55,88,89] However, certain *Clostridium* species/strains are unable to receive DNA via conjugation from an *E. coli* donor. One instance is *C. acetobutylicum*. For this organism, a successful conjugational plasmid transfer with *E. coli* as a donor has never been demonstrated. The reason for this is unclear, especially given the fact that the organism accepts DNA via conjugation from *B. subtilis*, *L. lactis*, and *E. faecalis* donors.[102,103] Conjugation of other clostridia, such as *C. cellulolyticum*, proved to be problematic due to the different growth requirements of the donor and recipient strain.[26] Furthermore, conjugation can also be affected by the strong R-M systems of the recipient strain.[99] To account for these barriers, more labor-intensive measures are compulsory to equip the *E. coli* donor with the necessary methylase gene(s) or to develop different conjugation procedures (e.g., the use of two donor strains and one recipient).

Restriction-Modification Systems

The successful transfer of exogenous DNA into *Clostridium* can depend considerably on the presence of one or more endogenous R-M systems.[26,28,29,107] R-M systems are commonly used by bacteria to protect themselves against foreign DNA, most obviously phage DNA, and consist typically of a restriction endonuclease (R) and a modification (M) enzyme of identical specificity. Generally, the restriction endonuclease cleaves DNA at a specific recognition sequence, whereas the modification enzyme, usually a methyltransferase, prevents restriction at this site.[108] Based on sequence analysis, four types of endonucleases (REases, type I, II, III and IV) are recognized. Type II REases are subdivided (A, B, C, E, E, G, H, M, S, and T) due to the characteristics of their recognition sequence.[109] With the advent of cost-effective high-throughput sequencing procedures and, hence, the availability of an increasing number of *Clostridium* genomes, numerous clostridial R-M systems have been identified in silico. An overview of predicated and confirmed R-M systems is given in the database REBASE®[110] (http://rebase.neb.com/rebase/rebase.html). It is noteworthy that all four types of endonucleases (I, II, III, and IV), most prominently type II systems, are described or postulated for *Clostridium*. A good in silico prediction is a powerful tool to achieve DNA transfer into *Clostridium*. The predicted methyltransferase can be cloned and expressed in *E. coli* and used to equip the exogenous genetic material with the right methylation pattern.

Experimentally, appropriate methylation of the vector DNA was shown to be vital for successful DNA transfer into *C. acetobutylicum* ATCC 824,[28,107] *C. pasteurianum* ATCC 6013,[29] *C. cellulolyticum* ATCC 35319,[26] *C. ljungdahlii* DSM 13528,[27] *C. botulinum* ATCC 25765,[88] and *C. difficile* CD3 and CD6.[99] Interestingly, except for *C. ljungdahlii*, in which the applied in vivo methylation proved to be disadvantageous, identified R-M systems were class II systems. In other instances, restriction has not been a problem. Examples include *C. beijerinckii* NCIMB 8052,[28] *C. perfringens* strain 13,[111] *C. difficile* strains CD37[99] and CD630,[112] and *C. botulinum* ATCC 3502.[31] Although endonucleases and/or methyltransferases are annotated in the genomes of these strains,[110] they, especially the endonucleases, might not be functional or lack their R-M system partner (orphan methyltransferases). The potential effect of type I R-M systems has not been analyzed yet in detail for *Clostridium*.

Clostridial Vector Systems

Several *Clostridium* species have been described to possess native plasmids.[99,113–117] This section will, however, describe vectors/plasmids that were developed to genetically manipulate clostridial species. Their development for forward and reverse genetics was a necessity because *Clostridium* cannot be transformed with linear DNA. In the past three decades, many of these manipulation plasmids were generated, shown to be transferable into *Clostridium* cells and be maintainable by antibiotic selection.[30,32,55] Because many other plasmids designated to modify Gram-positive organisms, the clostridial vectors are constructed in the genetically more amenable organism *E. coli* and are subsequently (ready-to-use), with or without prior methylation, transferred into the clostridial target strain (Transfer Of Exogenous Genetic Material Section). The replicative *E. coli/Clostridium* shuttle vectors developed to date minimally comprise a selectable marker, a Gram-positive and a Gram-negative replicon, and, if needed for conjugational plasmid transfer, a *cis*-acting nick side (*oriT*).[30,32,55] The selectable markers used almost exclusively encode positively selectable antibiotic resistances ideally functional in both, *E. coli* and *Clostridium*.[30] Genes encoding resistances against erythromycin and/or lincomycin (*erm*) or chloramphenicol/thiamphenicol (*cat*) proved to be of greatest value.[22,26,55,118] Spectinomycin (*aad*) and tetracycline (*tet*) resistance genes are useful in a selected number of clostridial species.[55] Beside these, choices for antibiotic selection markers are limited which can complicate manipulation strategies or even restrain the development of new genetic tools.

Gram-positive replicons of the shuttle plasmids are either of clostridial origin or are derived from the plasmids of other Gram-positive bacteria. Clostridial origins of replication are usually the replicons of the *C. perfringens* plasmid pIP404,[119] the *Clostridium butyricum* plasmids pCB102 and pCB101,[120,121] the *C. botulinum* plasmid pBP1,[122] or the *C. difficile* plasmid pCD6[99]. Nonclostridial but Gram-positive replicons are typically the replicons of the *B. subtilis* plasmid pIM13,[123–125] the *L. lactis* pWV01[126] (*L. lactis* subsp. *cremoris* formerly *Streptococcus cremoris*), the *Staphylococcus aureus* plasmid replicon pUB110[127] or the *E. faecalis* plasmid pAMβ1.[26] Most of these replicons enable plasmid replication via the rolling circle replication (RCR) mechanism (RCR; Table 1). This mechanism involves highly recombinogenic single stranded DNA (ssDNA) and is thought to account for structural and segregational plasmid instability.[32,140] Interestingly, plasmids harboring the replicons of pUB110 and pIM13 were found to be thermo-stable and applicable to *C. thermocellum*.[60,127] The plasmids pCB102, pCD6, pIP404, and pAMβ1 do not replicate via the RCR. pCB102 replicates via an unknown mechanism. The small, single open reading frame (ORF) that potentially encodes the replication protein does not exhibit any homologies to known genes of this function.[55] pIP404 and pCD6 replicate via a similar mechanism involving a large polypeptide (RepA) and repeat elements (iterons) located down-stream of the RepA.[99,128] Replication of pAMβ1 occurs via a unidirectional theta

Table 1: Gram-positive replicons typically used in *Clostridium*

Progenitor Plasmid/Replicon	Host	Application In[a]	Replication Mechanism	References
pIP404	*Clostridium perfringens*	*C. perfringens*[b]	RepA, iterons	128
pCD6	*Clostridium difficile*	*C. difficile*[b]	RepA, iterons	99
pCB101	*Clostridium butyricum*	*Clostridium*	RCR	113,120,121
pCB102	*C. butyricum*	*Clostridium beijerinckii*[b]	Unknown	55,120,121
pBP1	*Clostridium botulinum*	*Clostridium*	Unknown	55
pAMβ1	*Enterococcus faecalis*	*C. beijerinckii, Clostridium cellulolyticum*	Unidirectional theta	129,130
pIM13	*Bacillus subtilis*	*Clostridium acetobutylicum* [b]	RCR	131
pUB110	*Staphylococcus aureus*	*C. acetobutylicum*[c] *Clostridium thermocellum*[d]	RCR	132,133
pWV01	*Lactococcus lactis*	Broad host range [d]	RCR	134–136

127.
[a]Application-specific reference for target strain.
[b]137.
[c]73.
[d]138,139.

mechanism,[129,130] which is thought to account for its high segregational stability.[134] All clostridial shuttle plasmids constructed to date exhibit varying degrees of segregational instability and are lost from the cell and population in the absence of selective pressure.[55] A list of the replicons, their origin, their mechanism of replication and their usual target strain is shown in Table 1. As for the Gram-negative replicon of the generated shuttle plasmids, requirements are modest. Widely used are the high-copy number replicon ColEI of the plasmid pMTL20[141] and the low-copy number replicon p15a.[142] Both replicons are compatible and can be maintained in the same cell if necessary.[55] In the case of required conjugational transfer, the conjugational transfer function (*oriT*) of the plasmid RK2 is most widely used[35] (Conjugation Section).

Until recently, generally available *E. coli*/*Clostridium* shuttle plasmids did not share a common structure. They were typically designed ad-hoc with little or no emphasis on a standard format. Standardization, however, would allow the modification of plasmids for application-specific needs in a straightforward, targeted, and time-saving way and facilitate the direct comparison of the functional properties of various plasmid components. Hence, a more rational approach to vector design was undertaken and the pMTL80000 modular shuttle plasmid series was created.[55] Specifications for the design and implementation were: (1) a modular component format to provide combinatorial freedom in the construction of new plasmids; (2) a fast and facile realization of any changes to a given combination of modules; (3) reversibility of executed changes; and (4) the possibility to extend the system for new modules or applications. To realize these specifications, a standard arrangement was defined, in which every plasmid contains exactly one of each of four distinct modules which were always arranged in the same order and always flanked by the same four rare 8-bp-type II restriction enzyme (RE) recognition sites (*Asc*I, *Fse*I, *Pme*I, *Sbf*I; Figure 1). The four modules comprise a Gram-negative replicon with or without the *oriT*, a Gram-positive replicon, a selectable marker, and an application-specific module. They are numbered and, thereby, in number and position, determine the precise nomenclature of the modular pMTL80000 plasmid (Table 2). The modules were chosen to provide different properties, be relevant in a broad range of *Clostridium* species and to be useful for various applications[55] (Table 2). For example, the Gram-positive replicons supplied with the system exhibit variable mechanisms of replication (Table 1) and, therefore, different segregational stabilities in different clostridial species[55,134] (Table 3). Furthermore, by choosing different application-specific modules or further modifying them, the modular vector series is more than just a series of shuttle plasmid. It can be used (1) as a reporter gene system to test the strengths of different promoters, (2) to overexpress genes heterologously, and (3) to modify the clostridial genome via homologous recombination[55] (Table 2). As a reporter gene, the versatile chloramphenicol/thiamphenicol resistance gene (*catP*) of *C. perfringens* was chosen.[145–147] For the heterologous expression of genes, the strong promoter and ribosome binding site of the *C. acetobutylicum* thiolase gene (*thl*)[66,148] or the *C. sporogenes* ferredoxin gene (*fdx*)[144] were used. Shuttle plasmids have been used in the past as reporter systems[66,145,146,148–150] or to heterologously

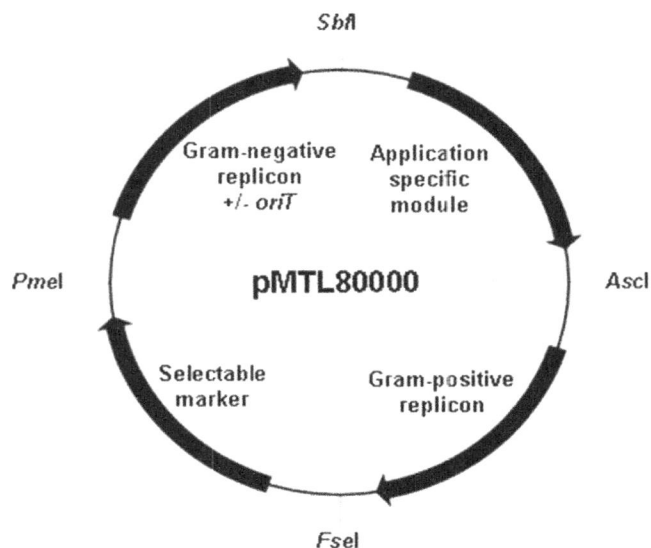

Figure 1
Schematic of the pMTL80000 modular vector series[55] modified. In the standard arrangement, the four plasmid modules (Gram-negative replicon +/− *oriT*, Gram-positive replicon, selection marker, application-specific module) are always arranged in the same order and flanked by the same four rare type II restriction enzyme (RE) recognition sites (8 bp, *Asc*I, *Fse*I, *Pme*I, *Sbf*I). The modules are numbered and determined with their position and combination the nomenclature of the particular plasmid (Table 2).

Table 2: Numeric module assignment and nomenclature of the pMTL80000 vector series[137,557]

Modular Plasmid[a]	Gram-Positive Replicon	Marker	Gram-Negative Replicon	Application-Specifc Module
pMTL80110	0. spacer	1. *catP*	1. p15a	0. spacer
pMTL82254	2. pBP1	2. *ermB*	5. ColE1 + *oriT*	4. *catP* Reporter
pMTL83353	3. pCB102	3. *aad9*	5. ColE1 + *oriT*	3. P$_{fdx}$ + MCS
pMTL84422	4. pCD6	4. *tetA*	2. p15a + *oriT*	2. P$_{thl}$ + MCS
pMTL85141	5. pIM13	1. *catP*	4. ColE1	1. MCS

[a]The shown five plasmids represent the core set and include all 18 standard modules available to date. These plasmids facilitate the easy and quick construction of all possible 395 plasmid variations in just a few cloning steps by the use of the REs *Asc*I, *Fse*I, *Pme*I, and *Sbf*I and, if needed, for the provision of the *oriT*, *Apa*I. The position and combination of the modules determine the nomenclature of the modular plasmids, e.g., plasmid pMTL82353 carries the Gram-positive replicon of the plasmid pBP1, the spectinomycin adenyltransferase gene (*aad9*) from *Enterococcus faecalis*,[143] the Gram-negative replicon ColE1[141] with the pRK2 derived *oriT* region,[35] and the promoter and ribosome binding site of the *Clostridium sporogenes* ferredoxin gene (*fdx*)[144] followed by a multiple cloning site (MCS). Annotatec sequences of any modular pMTL80000 plasmid can be downloaded from www.clostron.com. Requests for the purchase of the core set are handled via the same website.

Table 3: Characteristics of selected modular plasmids[137,55]

Plasmid	Replicon	*Clostridium acetobutylicum* ATCC 824		*Clostridium botulinum* ATCC 3502		*Clostridium difficile* CD630	
		Transfer Frequency[a]	Segregational Stability (%)[b]	Transfer Frequency[a]	Segregational Stability (%)[b]	Transfer Frequency[a]	Segregational Stability (%)[b]
pMTL82151	pBP1	1.38×10^2	99.4 ± 0.9	1.01×10^{-3}	99.9 ± 0.3	3.36×10^{-6}	87.3 ± 1.3
pMTL83151	pCB102	2.45×10^2	76.5 ± 6.0	2.90×10^{-4}	99.4 ± 0.3	2.23×10^{-6}	76.2 ± 0.5
pMTL84151	pCD6	8.47×10^1	82.4 ± 9.6	5.71×10^{-6}	81.6 ± 3.7	7.00×10^{-6}	77.4 ± 2.1
pMTL85151	pIM13	2.92×10^2	81.6 ± 8.6	7.80×10^{-6}	89.6 ± 0.7	4.18×10^{-7}	69.0 ± 1.1

[a]Values shown are for *C. acetobutylicum* ATCC 824 transformant colonies per µg plasmid DNA, for *C. botulinum* ATCC 3502, and *C. difficile* 630 transconjugant colonies per *Escherichia coli* donor colony forming units.[55]

[b]Segregational stability per generation has been calculated using the formula $\sqrt[n]{R}$. Thereby, R equals the portion of cells that feature the plasmid at the latest point its existence that could be verified; n are the generations of growth without selection at that particular time.[55]

express genes[30,151] or asRNA.[46,148,152] However, the modular vector series, offers a systematically designed and standardized approach to these applications.[55]

Recently, modular pMTL80000 plasmids have been used as pseudosuicide vectors to modify the clostridial chromosome by transposon-mediated random mutagenesis[69] (Random Mutagenesis by Biological Mutagens Section) or double-crossover homologous recombination (allelic exchange, ACE; Recombination-Based Methods (Allelic Exchange) Section).[39,61,63] Due to their (if applicable) replicative but segregationally highly unstable nature, they are most suitable to deliver a transposon or allelic exchange cassette into a vast number of chromosomes, but are readily lost in the absence of selective pressure when no longer required.[69] A further powerful application of the modular vectors series is their potential use as suicide (conditional, nonreplicative) plasmids, i.e, plasmids that cannot replicate in the target strain to which they have been transferred.[93] These plasmids either lack the necessary replicon[27,153] or require specific conditions for the replicon to be functional.[154] Typically, they possess sequences (homologous sequence[s], transposon) that facilitate their integration into the bacterial chromosome during or just after the transformation. By omitting the Gram-positive replicon between the *Acs*I and *Fse*I sites (Figure 1) and the integration of a transposon or homologous sequences into the application-specific module[39,63,69] or the plasmid backbone,[61] modular plasmids can be easily converted into suicide plasmids. The fact that primary transformants are exclusively integrants is the great advantage of their use in transposon-mediated random mutagenesis or allelic exchange. It facilitates and quickens the isolation of the desired mutant.[93] Beyond that, no efforts have to be undertaken to determine plasmids with pseudosuicide function.[69,154] The broad application of suicide plasmids, especially in allelic exchange, is, however, impeded by the low frequency of directly obtainable double-crossover mutants. This has only been reported for a limited number of organisms.[155,156] To be applicable for genetically intractable organisms, such as *Clostridium*, counter (negative) selection markers (Counter (Negative) Selection Markers Section) and high transformation efficiencies (Electroporation Section) are essential.[93] The identification of suitable counter selection markers including appropriate selection strategies and conditions and the development and improvement of electroporation protocols is, however, challenging.[93] Nonreplicative (nonstandardized, nonmodular) plasmids have been used in the past to genetically modify clostridial genomes. However, these plasmids lacked appropriate counter selection markers and only resulted in the isolation of unstable single-crossover mutants[49,50,56,153,157–159] (Recombination-Based Methods (Allelic Exchange) Section).

Forward Genetics by Random Mutagenesis

Forward genetics addresses the identification of a genotype responsible for a particular phenotype. Random and (if needed) extensive mutagenesis is one of several means to achieve forward genetics. Combined with selective breeding, it is a powerful tool to generate mutants randomly, select for the desired phenotype, and finally determine the genotype related to the observed phenotype. In addition, random mutagenesis allows the engineering of complex and highly regulated metabolic or regulatory pathways in which rational design fails due to inadequate knowledge of the pathway and consequently, a lack of well-defined and well-understood targets.[160] Genetically intractable organisms with high medical or industrial value are, under certain conditions, approachable by random mutagenesis as well. Randomly acting chemical, physical, or biological agents are the key to an effective random mutagenesis approach because they increase the frequency of mutations significantly above the normal background level and, therefore, grant the subsequent selective isolation of the desired phenotype. Mutagens either act directly on the DNA or interfere with the replication machinery or chromosomal partitioning. In some cases the mutagen (promutagen) is not mutagenic itself, but its metabolites are. To date, much emphasis has been placed on the development of random mutagenesis methods, for example, to investigate the mechanisms of pathogenicity or to generate metabolically improved strains for industry. The following section will give an overview of the most common approaches.

Random Mutagenesis by Chemical and Physical Mutagens

In early days, physical mutagens such as ultraviolet radiation and ionizing radiation (e.g., X-rays, γ-rays, β-rays) were used to promote mutations in *C. acetobutylicum*.[161] However, physical agents were concluded to be poor mutagens in *Clostridium* due to the lack of error-prone repair mechanisms.[162] Alternatively, chemical mutagens, mostly alkylating agents, which act directly to induce base substitutions, deletions, or frameshift mutations in DNA, were reported to be significantly more effective in *Clostridium* species.[163–167] The mutagenesis of *C. acetobutylicum* and related saccharolytic clostridia was mainly accomplished by the use of the mutagen, *N*-methyl-*N'*-nitro-*N*-nitrosoguanidine (MNNG).[163,168,169] MNNG remains the mutagen of choice to date and has recently led to the isolation of the hypersolvent producing *C. acetobutylicum* strains EA2018[168] and BKM19[170] and the hyperbutanol producer *C. beijerinckii* BA101.[169] The strain *C. beijerinckii* BA101 was selected in the presence of the glucose analogue 2-deoxyglucose and showed increased amylolytic activity. It has been licensed for commercial use to TetraVitae Bioscience (http://www.tetravitae.com).[171] Similarly, the *C. acetobutylicum* strain EA2018, which produces higher butanol-to-solvent ratios (0.7) than its parental strain (0.6)[168] has been licensed to numerous commercial producers in China.[171] Despite these advances, it has to be noted that chemical mutagens should be used with caution, because it has been reported that they can lead to strain degeneration.[114,172]

Random mutagenesis as a result of genome shuffling[173,174] is another reported method used for the evolutionary engineering of solventogenic clostridia. Using genome shuffling and selection in the presence of high butanol concentrations, the *C. acetobutylicum* Rh8 mutant was obtained, which exhibited a higher butanol tolerance and produced more butanol compared to the parental strain *C. acetobutylicum* DSM 1731.[175]

Random Mutagenesis by Biological Mutagens

Biological mutagens that cause random mutagenesis are reactive oxygen species produced by the same or another organism, viruses, and transposons. Transposons are mobile/transposable genetic elements (TEs) that can "jump" from one location in the genome to another and thereby induce mutations.[176] They have been widely used in genetics to alter genome functions.[177,178] Based on their mechanism of transposition, transposons are assigned to one of two classes, the copy-and-paste (class I TEs) or the cut-and-paste (class II TEs).[179] To date, class II DNA transposons have been the mutagen of choice in clostridia.[69,71,72,180,181] These transposons never use RNA intermediates. They move on their own by inserting and excising (cut and paste) themselves from the genome. Typically, they consist of a transposase gene and terminal inverted repeats (TIRs) at each end of the sequence (Figure 2(a)). The TIRs are recognized by the transposase, which subsequently catalyses the movement of the transposons to another part of the genome. On integration, the TIRs are duplicated, resulting in target-site duplications, a unique hallmark of each DNA transposon.[178,179] Depending on their delivery vehicle, TEs are subcategorized into conjugative or nonconjugative transposons. Conjugative transposons encode their own cell-to-cell transfer machinery beside modules for regulation, integration, and excision.[182] Nonconjugative transposons rely on multicopy plasmids or bacteriophages and high electroporation efficiencies for their transfer.[183] Transposomes, another subcategory, are in vitro preassembled synaptic transposition complexes that

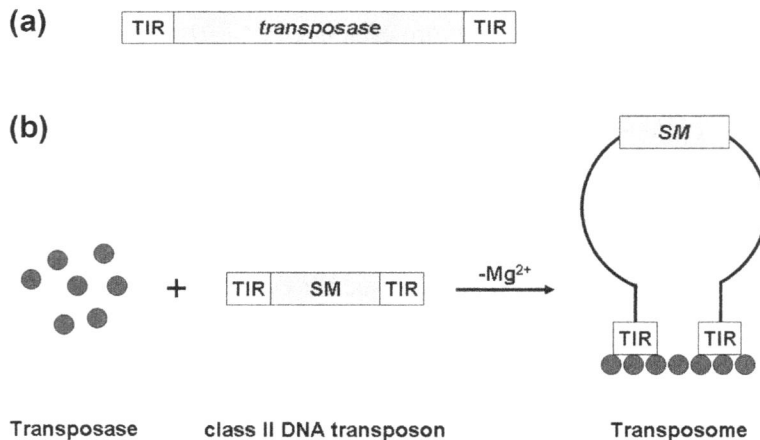

Figure 2
Schematic view of (a) a class II DNA transposon[178] modified and (a) a transposome[190] Epicentre®, modified. (a) The class II transposon consists minimally of a gene encoding the transposition-mediating transposase and terminal inverted repeats (TIRs) that act as recognition sites for the transposase. (b) Transposomes are in vitro assembled synaptic complexes of a transposase and transposon DNA in the absence of cations such as Mg^{2+}. A selectable marker (SM) located between the TIR sequences enables the rapid isolation of cells with transposon integrations.

consist of a transposase and the transposon DNA[184] (Figure 2(b)). In vivo, these complexes are formed transiently during transposition. But, in vitro, in the absence of divalent cations such as Mg^{2+}, they can be formed and maintained stably. After their transfer into the target strain by electropration, they become active due to the presence of divalent cations such as Mg^{2+}. This results in the integration of the transposon DNA into the genome. Typically, transposons contain a selectable marker for the rapid isolation of cells with transposon integrations.[184–189]

Conjugative transposons

Conjugative transposons or integrated conjugative elements are typically integrated into the chromosome or endogenous host plasmid. However, they are able to excise themselves from it and, subsequently, form a covalently closed circular transposable intermediate that can either reintegrate into the chromosome of the same cell (intracellular transposition) or be transferred by conjugation to another cell of the same or a different species. In the recipient, they integrate into the genome or endogenous host plasmid (intercellular transposition).[183,191] Conjugative transposons vary widely in size (18–500 kbp) and contribute as much as plasmids to the distribution of antibiotic resistance genes, for example in *C. difficile*.[3,192–195]

The best characterized conjugative transposons in clostridia are Tn*5397* and Tn*916*. Both are members of the large Tn*916*/Tn*1545* family of conjugative transposons, of which almost all members carry the tetracycline resistance gene *tetM*.[180,196] Despite high similarity, Tn*5397* and Tn*916* differ in their integration/excision system.[197,198] Particularly, Tn*916* has been widely used as a mutagenic tool in both pathogenic and nonpathogenic *Clostridium* species. The earliest demonstrations were reported in *C. botulinum* and *C. perfringens* and resulted in the generation of a pleiotrophic[199] or several auxotrophic mutants.[200,201] In both strains, an observed decrease in toxin production was caused by large Tn*916*-mediated deletion events. Analyses of the mutants confirmed that the transposition occurred at different sites[201,202] and randomly in single copy or multicopy.[200] Initial attempts at using Tn*916* in the nontoxinogenic *C. difficile* strain CD37 indicated the presence of a chromosomal "hot spot," at which the transposons inserted.[203,204] This strong target site preference was later concluded to be strain specific. When a different conjugative transposon, Tn*5397*, was tested in the same strain, it inserted at a "hot spot" (*attB_{Cd}*) as well, whereas it integrated at multiple sites in *B. subtilis*.[198] Subsequently, it was demonstrated that Tn*916*, and its derivative Tn*916*ΔE, integrate at random positions in the genome of other *C. difficile* strains.[3,205,206]

In nonpathogenic clostridia, the use of conjugative transposons has principally focused on acquiring mutants defective in solvent production or sporulation. The Tn*916* transposon was initially used in two model organisms, *C acetobutylicum* and *C. saccharobutylicum* P262 (formerly *C. acetobutylicum*).[207–209] However, the mutants obtained still remain largely uncharacterized to this day. In addition, Tn*916* and Tn*1514* transposons were tested in the solventogenic species *C. beijerinckii* NCIMB 8052 (formerly *C. acetobutylicum*).[103] Tn*916* was found to insert at a preferred "hot spot," whereas Tn*1545* (encoding antibiotic resistance genes *aphA*-3, *ermAM*, *tetM*) integrated at multiple sites in the chromosome. Of the *C. beijerinckii* mutants generated by Tn*1545*, one was deficient in "degeneration" (the loss of the ability to produce solvents in for example prolonged cultures),[210] and others exhibited increased butanol tolerance.[211] The phenotype of the antidegeneration mutant is thought to be a result of the Tn*1545* integration into the peptide deformylase encoding gene *fms*. This causes a reduction of the growth rate and eventually leads to enhanced stability in solvent production. The antidegeneration mechanism can be mimicked in the wildtype by reducing the growth rate.[212] The butanol-tolerant mutants were postulated to be a result of reduced glycerol dehydrogenase (GDH) activity due to the chromosomal insertion of Tn*1545* adjacent to *gldA* and the production of outwardly directed RNA. This RNA acts as *gldA* asRNA and inhibits the expression of GDH.[211]

In summary, the major drawbacks encountered by using conjugative transposons for random mutagenesis in *Clostridium* species are their: (1) large size,[191] (2) low efficiency of transposition, (3) limitation to bacterial strains that can be conjugated (Conjugation Section), (4) existence of (a) chromosomal "hot spot(s),"[103,203,204] and (5), most unfavorably, their predilection to insert in multiple copies (probably attributed to a high rate of vector retention), which significantly complicates the association of genotype with phenotype.[103,201]

Nonconjugative transposons and transposome mutagenesis systems

The first random nonconjugative transposon mutagenesis systems were developed for *C. perfringens* due to its routinely achieved high electroporation efficiencies.[111,181] The very first systems were based on the delivery of preassembled transposomes, namely Mu bacteriophage and Tn*5* transposon-based systems. The bacteriophage Mu-based transposition system, consisting of the MuA transposase and Mu transposon DNA with an erythromycin-resistance cassette,[190,213] resulted in 239 and 134 transposon insertions per μg of DNA in a laboratory strain (JIR325, a derivative of strain 13) and a field isolate (strain 56) of *C. perfringens*, respectively.[181] Although this system allowed it to generate single-insertion transposon mutants, it showed a high preference for the integration of the transposon at a few "hot spots," for example an rRNA gene (43%) or a certain intergenic region (12.5%). This hot spot preference, therefore, limits the practicability of the system. In another study, the EZ-Tn*5* random mutagenesis system (Epicentre®), comprising a hyperactive Tn*5* transposase and a Tn*5*-derived transposon,[184] was used in *C. perfringens* strain 13. Compared to the Mu phage-based system, this system exhibited two important advantages, a higher transposition frequency and the highest degree of randomness.[72] The use of EZ-Tn*5* led to the identification of a regulatory network (homologous to the accessory gene regulatory [*agr*] system in staphylococci) responsible for early production of both alpha toxin and perfringolysin O. All EZ-Tn*5* mutants were shown to possess a single, random transposon insertion, with a significantly lower insertion frequency into rRNA genes (18%) compared to the Mu phage-based system.

In comparison to conjugative transposons, both mutagenesis systems, Mu and EZ-Tn5, represent a considerable improvement because they allowed the generation of single-insertion mutants. However, both systems exhibit a preference toward the insertion into rRNA genes. To date, it is unknown why. Furthermore, both systems rely on high transformation efficiencies. For *C. perfringens*, such high frequencies are of no concern,[111] but they are for other *Clostridium* species (Electroporation Section).

Due to these limitations, other nonconjugative plasmid-based transposable mutagenesis systems are under investigation. In particular, the *mariner* transposable element *Himar1* has recently found broad application and has been shown to insert randomly into the genomes of many different bacterial species.[214–218] As a member of the Tc1/*mariner* superfamily, the most widely distributed family of TEs,[178] *Himar1* requires no other factors for its cut-and-paste transposition than its self-encoded transposase and the TIR sequences at either end of the transposon. A noticeable bias for insertion at "bent" or "bendable" DNA sequences and TA target sites has been reported.[219,220] The latter is of advantage if random mutagenesis libraries are to be generated in species with low GC content such as *Clostridium*. To date, *Himar1* has been used in the clostridial species *C. difficile*[69] and *C. perfringens*.[71] In both studies, a plasmid-based approach was taken to ensure transposition in every transformed cell. The transposase was placed under the control of either a native (P_{tcdB}, transcription reliant on sigma factor TcdR) or an inducible (P_{bglA}, lactose inducible) promoter (Inducible Gene Expression Section) to prevent transposase expression and activation of the system in the *E. coli* strain used to assemble the transposome encoding plasmid. A selection marker (*catP*, *ermBP*) was placed between the TIR sequences to select for transposition events after the transformation. In *C. difficile*, R20291 transposon insertions arose at a frequency of 4.5 (± 0.4) $\times 10^{-4}$ per cell.[69] Nucleotide sequencing of the insertion sites of 60 mutants demonstrated that insertions occurred randomly at a "TA" di-nucleotide target site in the plus (+) and minus (−) strand, with no evidence for a "hot spot." With one exception, all isolated mutants were single-insertion mutants. The utility of the system for forward genetic studies was demonstrated through the isolation of a mutant defective in sporulation/germination and an auxotrophic mutant. The only drawback of this study was the lack of a counter selection marker and, therefore, the need to use segregationally unstable pseudosuicide plasmids and a more laborious screening approach. To eradicate the plasmid backbone and the *Himar1* C9 transposase gene from the population, subculturing in the absence of the plasmid backbone-encoded selection was required. In case of high-throughput mutagenesis strategies subculturing would become time and labor intensive and considerably restrict the implementation of the system. This limitation was addressed in a follow-up study, and galactokinase (*galK*) was used as a counter selection marker (selection for loss of plasmid backbone and *Himar1* C9 transposase gene from the population after transposition occurred) in the *C. perfringens* strain HN13, a derivative of *C. perfringens* strain 13 with a deletion of the genes *galK* and Gal-1-phosphate uridylyltransferase (*galT*)[71] (Inducible Gene Expression Section). The study confirmed the ability of the *Himar1 mariner* transposon to generate single and random mutations in the genome. The suitability for forward genetics was demonstrated by screening for mutants with altered gliding motility. Of 10,000 mutants, 24 showed the desired phenotype, but only one showed the expected genotype. It was argued, that this might be due to the fact that the number of mutants screened was below the number necessary to reach saturation mutagenesis.[71,221] The drawback of this system is the use of an endogenous counter selection marker (Counter (Negative) Selection Markers Section), which required the use of a *galKT* deletion mutant. Deletion of genes involved in carbohydrate use might cause unnecessary pressure on cell survival or other unwanted effects, especially in combination with random mutagenesis.

However, both studies demonstrated the value of the *mariner*-based transposon system to generate random mutagenesis libraries. It is expected, with appropriate adaptations, such as higher electroporation efficiencies (Section 2.1), that this system will also be applicable in industrially important *Clostridium* species.

Reverse Genetics

In contrast to forward genetics, in which an observed phenotype is linked to a specific genotype, reverse genetics investigates the effect(s) a targeted modification of a genetic sequence has on the phenotype. After a gene or sequence of interest (target gene/sequence) has been identified, its nucleotide sequence is altered (disruption, deletion, insertion, base-exchange), and subsequently, phenotypic consequences are analyzed. Reverse genetics is a very important complement to forward genetics, but requires: (1) sufficient knowledge of the pathways and regulatory networks, which are under investigation, to pinpoint a well-defined target gene (rational design); (2) the availability of at least the nucleotide sequence of the gene/sequence of interest; and (3) the existence of reliable techniques to alter its sequence. This section will discuss two techniques currently available in clostridial reverse genetics, mobile group II introns and allelic exchange. Both techniques allow the alteration of chromosomal genetic sequences. However, group II introns are mainly used to insertionally disrupt genes, whereas allelic exchange is used to disrupt, delete, or insert genes or alter their nucleotide sequence by base substitution(s).

Recombination-Independent Methods (Group II Introns)

One method of gene inactivation that has proven to be highly successful is based on the use of group II introns. These introns, in their active form, are catalytic RNAs, found frequently in organelles of plants and lower eukaryotes, but also in bacteria. Bacterial group II introns differ significantly from their eukaryotic counterparts; they generally are exon-less (residing outside of structural genes) and are in most cases associated with mobile elements.[222] Mobile group II introns carry an ORF encoding an intron-encoded protein (IEP). The IEP is responsible for both RNA splicing and insertion. RNA splicing is accomplished by the reverse transcriptase activity of its N-terminal domain, integration by the endonuclease acitivty of its C-terminal domain that is involved in mobility. The *L. lactis* Ll.LtrB intron, found within the *ltrB* gene of conjugative plasmids, is used as a model system for these kinds of introns. The specificity for a certain insertion site is, for the largest part, determined by the intron RNA, which base-pairs with the DNA nucleotides of the target site. The sequences in the intron RNA that determine its insertion site have been elucidated.[223] These finding led to the formulation of an algorithm that allowed the reliable prediction of the necessary changes to the intron RNA sequence, so that it would target the intron to a new, desired, location instead of the native insertion site.[224,225] Because a few of the nucleotides around the insertion site are recognized by the IEP and not by the intron RNA, DNA sequences shorter than 400 bp may not contain an insertion site in which it can integrate with high efficiency. Another important finding was that the *ltrA* gene, which encodes the IEP of the intron and is natively present within the intron II sequence, can be provided in *cis* or in *trans*. This allows it to be positioned on the backbone of the intron delivery plasmid. Loss of this plasmid following successful integration of the intron II in the target site prevents a continuing mobility of the inserted intron due to the loss of the splicing-mediating LtrA. Re-targeted introns resulting from this work are called "TargeTrons," and the system has been commercialized under that name and is sold by Sigma Aldrich.[226] The commercially available kit supplies the uses with a template that can be used to generate a small DNA fragment (c. 350 bp) by PCR. The resulting product is cloned into the TargeTron plasmid in place of the native sequence that targets the L1.LtrB intron. The primers used in this amplification introduce the desired intron sequence changes "retargeting." Software for the prediction of the primer sequences is at hand online on a "pay-per-click" basis (http://www.sigma-genosys.com/targetron).

TargeTron-mediated gene inactivation

The first exemplification of the TargeTron technology within the clostridial genus was reported for *C. perfringens*.[47] The basic vector available at the time, pACD3, needed extensive modifications because most of the elements did not function in *Clostridium*. A new clostridial replicon was necessary, as was a selectable marker (both from plasmid pJIR750). The T7 promoter, responsible for the transcription of the Ll.LtrB group II intron, was replaced with a clostridial promoter region, namely of the beta-2 toxin gene (*cpb2*) from a *C. perfringens* type A isolate. The final vector (pJIR750ai) was, after retargeting, shown to be able to successfully insert into the alpha toxin targeted gene *plc*. PCR screening showed that two of the 38 individual colonies screened contained a mixture of wildtype and mutant. Restreaks from these colonies for phenotypical screening revealed that approximately 10% of the colonies formed were pure mutants. Since this demonstration, TargeTrons have been used to generate mutants in four nonpathogenic clostridia, *C. acetobutylicum*, *C. beijerinckii*, *C. cellulolyticum*, and *C. phytofermentans*, affecting solvent production, substrate use, and restriction enzymes.[24,107,227–232]

ClosTron-mediated gene inactivation

Although TargeTron-based mutagenesis in clostridia is possible, there are certain drawbacks. The rational alteration of intron sequences is not an exact science, and integration frequencies vary widely between target sites. As a consequence, if insertional inactivation of a target gene does not result in a readily detectable phenotype, substantial screening efforts may be needed to isolate the desired mutant. To circumvent this deficiency, a retrotransposition-activated marker (RAM) was developed,[226,233] but it needed a modification of the marker to be suitable for the use in a wide range of clostridia. In the clostridial system, the RAM consists of an *ermB* gene, which confers resistance to erythromycin, interrupted by a group I intron.[22] Essential for the functioning of a RAM is the orientation the group I intron and the *ermB* gene relative to the group II intron in which they are present. Group I introns are capable of self-catalytic splicing from messenger RNA (mRNA) transcripts, but only when transcribed in the correct orientation. When the group II intron is transcribed, the group I intron is orientated correctly and will remove itself, thereby restoring a functional *ermB* gene, once it is integrated and retrotranscribed in the target site. Because the *ermB* gene with its promoter is activated when the group II intron is inserted in the DNA target site, resistance to erythromycin is acquired by the insertional mutant. The plasmid that introduced the group II intron also gives rise to *ermB* transcripts, but as the orientation of the group I intron present in there is incorrect, it cannot splice itself out, and no resistance is conferred by the plasmid-derived transcript.

A further modification of the TargeTron plasmid was the use of a segregational unstable Gram-positive replicon (pCB102; pseudosuicide plasmid[69]; Clostridial Vector Systems Section). The advantage of using such a replicon is that after obtaining the integrant, which is identified though the obtained erythromycin resistance, the plasmid is rapidly lost on omission of the backbone-encoded chloramphenicol/thiamphenicol (*catP*) resistance and thus avoids extensive rounds of restreaking. The constructed plasmid, pMTL007, was applicable in a wide variety of pathogenic and nonpathogenic *Clostridium* species and transferable into the clostridial cell by electroporation or conjugation.[22] An improved version of the ClosTron, plasmid pMTL007C-E2, became available recently and features, besides its: (1) conformity to the standards of the pMTL80000 modular vector series[55] (Clostridial Vector Systems Section); (2) easier intron retargeting through blue white screening; (3) improved group II intron expression; and (4) marker recycling by flanking the *ermB* RAM with short yeast FLP recombinase target sites (FRT).[54,234] After isolation of the desired disruption mutant, the selection marker can be removed from the genome by introduction of an FLP recombinase expressing plasmid into the cell. The FLP recombinase recognizes its FRT target sites and excises any DNA flanked by these sites.[235,236] Consequently, the markerless disruption mutant can be subjected to another round of ClosTron-mediated mutagenesis, allowing multiple mutations in one strain.[237] Further improvements of the ClostTron technology included: (1) the incorporation of alternative Gram-positive replicons (e.g., pCD6), which allowed the application of the pseudosuicide vector approach[69,238] and other selection markers (e.g., *aad9*, *ermB*)[239] to extend the number of *Clostridium* species the system can be used in; (2) the provision of an easily accessible intron design tool; and (3) the outsourcing of the construction of the retargeted plasmid.[237] The new free-of-charge intron design tool was based on the algorithm and data described earlier[225] and is available via a publicly accessible website (www.clostron.com). To reduce the labor and costs associated with the retargeting of the plasmid, the website also provides instructions on how to have the retargeted region synthesized and cloned into a modular ClosTron plasmid such as pMTL007C-E2 by a DNA synthesis company. Compared to the conventional in-house preparation, this commercial approach is favorable in terms of costs, labor-intensity, failure rate, and time from start-to-finish. Company-based plasmid retargeting can be carried out within two weeks (order to deliver time). If one still opts for the more classical SOE PCR approach, two plasmids are available (pMTL20IT1 and pMTL20IT2) that can be mixed together and used as the template for the SOE PCR, negating the need for the template supplied with the Sigma kit.

Besides its default use for insertional gene disruption, the group II intron mutagenesis system has been tested for its feasibility to insert additional exogenous DNA sequences (other than the RAM and the flanking native intron sequences) at a chosen chromosomal target site. In *C. perfringens*, such an application of the Ll.LtrB intron was demonstrated.[240,241] Sequences to be integrated in the genome should be placed in domain IV of the intron, a region which in the ClosTron is already occupied by the *ermB* RAM, and in the native Ll.LtrB intron contains the *ltrA* gene. The amount of additional sequences that could be delivered by the system, besides the *ermB* RAM, was evaluated in a systematic study using fragments of phage lambda DNA.[237] Phage lambda fragments of 1.0, 2.0, and 2.3 kb were tested using *C. sporogenes* NCIMB 10696 as the host organism. Each intron variant containing the additional fragment was targeted to a previously validated target site in the *pyrF* gene. No integrant colonies were observed using introns containing the larger two fragments, but integrants were reproducibly obtained that contained the 1.0 kb fragment. An intriguing approach for targeted deletions was the use of the cargo region of the ClosTron to introduce a sequence into the chromosome of *C. acetobutylicum* DSM1731 homologous to a neighboring chromosomal region.[53] The idea behind the approach was to screen for deletion mutants resulting from a homologous recombination event that would occur between the native chromosomal sequence and the one introduced by the intron. The method was exemplified by deleting the chromosomal operon *cac1493-cac1494* encoding for polypeptides of unknown function and the megaplasmid-based *ctfAB* operon involved in acetone formation. Although deletion mutants were found for both targets (two of 648 colonies for *cac1493/1494* and one of 1998 colonies for *ctfAB*), extensive screening was required to identify the desired mutants. Other drawbacks of the approach are the dependency on the intron insertion site and the remainder of a scar upstream of the cargo domain.

The advantages of the ClosTron technology are the speed and the ease with which mutants can be generated. At least for *C. acetobutylicum*, it has been demonstrated that the resistance marker can be removed by the FLP recombinase-based system.[237] However, mutations are insertional gene disruption events which are not in-frame and can cause polar effects on adjacent genes upstream or downstream of the target site.[242,243] If, as is usually the case, the marker cannot be removed from the genome, its presence might cause so-far unknown effects on, for example, metabolism or growth. Furthermore, additional modifications in the same genome are complicated due to the limited availability of clostridial selection markers (Clostridial Vector Systems Section). The size of exogenous DNA that can be introduced into the chromosome besides the RAM and the intron sequences is limited. The mobility of the Ll.LtrB intron is severely impeded if additional DNA sequences of >1 kbp are inserted.[237,241] Taken together, TargeTron and ClosTron technologies are less suitable to create multiple, markerless, scarless, in-frame deletions in the clostridial genome or to integrate whole pathways or more complex genetic arrangements. Here, recombination-based methods such as allelic exchange are more promising.

Recombination-Based Methods (Allelic Exchange)

In applied genetics, allelic exchange, the movement of a mutated gene (allele) typically from a plasmid onto the chromosome by double-crossover homologous recombination, is a very powerful tool. It enables the alteration (deletion, disruption, insertion, base exchange) of any genetic material (genes, intergenic regions, gene parts) located on a chromosome or plasmid without leaving scars behind or exhibiting polar effects (insertion/deletion frameshift, non-sense/antisense disruption) on adjacent genetic material.[93] For organisms that can be efficiently transformed with linear DNA, such as yeast,[244] *Mycobacterium tuberculosis*,[245] *Bordetella pertussis*,[246] or naturally competent *B. subtilis*,[247] allelic exchange can be achieved in one step. However, most bacteria are not transformable with linear DNA. For allelic exchange, the mutated gene and/or further region(s) of homology have to be placed onto a circular DNA molecule, a plasmid. Commonly used plasmids for allelic exchange are: (1) pseudosuicide,[69] (2) suicide (conditional, nonreplicative), or (3) incompatible plasmids[93,248] (Clostridial Vector Systems Section). However, due to their circularity, a single homologous recombination event (single-crossover) by a Campbell-like mechanism will lead to the integration of entire plasmid between the target sequence of the chromosome and create direct repeats of the homologous sequence on either end of the inserted plasmid.[249–252] This arrangement makes the chromosomal locus an active substrate for a second intramolecular recombination event (double-crossover) and, therefore, extremely unstable. Without a selective pressure, the occurring double-crossover will result in the excision of the plasmid and the reversion to the wildtype genotype. If, however, a second homologous sequence and an appropriate second selectable marker are provided by the plasmid, the second recombination event can occur at a different site, resulting in the desired stable double-crossover mutant genotype (allelic exchange). The latter recombination event is rare, and the selection is laborious if the desired allelic exchange does not cause a selectable phenotype. Selection can be facilitated by the use of the second selectable maker, preferably a counter (negative) selection marker (Counter

(Negative) Selection Markers Section). In fact, most effective allelic exchange methods rely on robust counter selection markers and were only possible after they had been developed for the particular organism.[93] Combined with counter selection, allelic exchange can be carried out not only scarless, in-frame, and without any polar effects, but also efficiently and markerless. If double-crossover frequencies are low and the desired allelic exchange is not directly selectable, most commonly a double-selection strategy is used[93] (Figure 3). In brief, in the first step transformants that have integrated the entire plasmid into the chromosome by single-crossover are selected due to the presence of a (positive) selection marker provided by the plasmid. Subsequently, cells are grown to allow the desired double-crossover (allelic exchange) to occur. Cells that have excised and lost the plasmid and therefore, the counter selection marker encoded on the plasmid backbone are identified by plating on counter selection medium.[93] A summary of recently developed counter-selection markers for the use in *Clostridium* is given in Counter (Negative) Selection Markers Section.

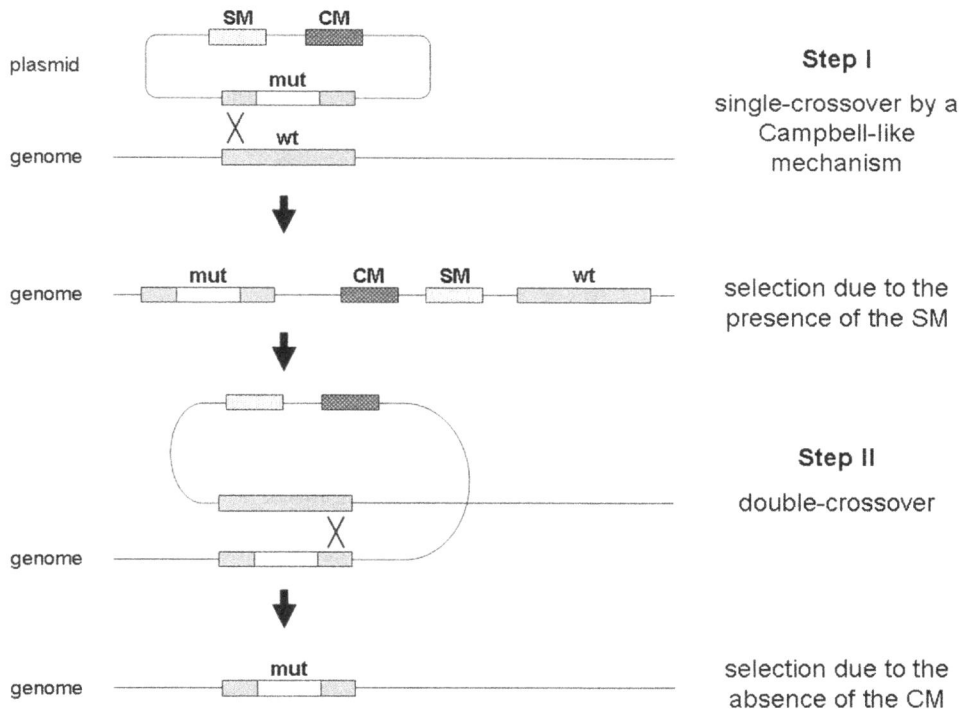

Figure 3

Double-selection strategy using a selectable and a counter selectable marker[93] modified. The strategy is most commonly used for the isolation of allelic exchange mutants that cannot be directly selected because the mutated gene does not exhibit a selectable phenotype, for example auxotrophy. Abbreviations: SM, selection marker; CM, counter selection marker, wt, wildtype allele; mut, mutant allele.

For *Clostridium*, early reports on the generation of mutants by homologous recombination described the use of nonreplicative (suicide; Clostridial Vector Systems Section) plasmids to insertionally inactivate chromosomal genes by single-crossover.[49,50,157,159] Organisms used were the solvent producers *C. acetobutylicum* and *C. beijerinckii* and genes that encoded enzymes that participate in acid or solvent formation or sporulation. The generated single-crossover mutants were segregationally unstable due to the duplication of the homologous sequence and could only be maintained in the presence of an appropriate antibiotic. The presence of an antibiotic can have unknown effects on the metabolism and growth. The industrial application of such a strain is most unlikely due to costs and the release of high quantities of (an) antibiotic(s). Furthermore, generated strains are difficult to engineer further, because the number of available selection markers is limited for *Clostridium* (Clostridial Vector Systems Section). The first reports on the generation of clostridial double-crossover mutants were published for *C. perfringens*[253] and *C. acetobutylicum*.[51] In *C. perfringens*, a suicide plasmid was used to inactivate the genes coding for the Θ- or α-toxin by the insertion of a gene-encoding erythromycin resistance. In *C. acetobutylicum*, *spo0A* was insertionally inactivated by a macrolide, lincosamide, and streptogramin B (MLS) resistance encoding gene and the use of a replicative plasmid. Although the generated mutants were the result of a double-crossover, they were not markerless and therefore featured the same disadvantages as described above. In the case of *C. acetobutylicum*, the double-crossover was not as intended and arose as the result of an unexpected recombination event.[51] Recently, several efficient allelic exchange methods were developed in both pathogenic[61–63] and industrially important clostridial species.[27,38,39,54,64,254] These methods use replicative, pseudosuicide or suicide plasmids and make use of positive (*cat*, *erm*) and exogenous or endogenous counter selection markers (*codA*, *mazF*, *galK*, *hpt/tdk*, *pyrF*, *pyrE*, *upp*) or couple a promoterless antibiotic resistance (*erm*) or orotate phosphoribosyltransferase (*pyrE*) gene with a constitutively expressed chromosomal promoter.[39] An overview of the methods is given in Table 4; details on the counter selection markers are described in Counter (Negative) Selection Markers Section. The following section discusses significant features, such as multiple use (marker recycling), markerless, scarless, and in-frame of selected methods.

The first advanced method for the generation of stable markerless double-crossover mutants was described for *C. acetobutylicum* and based on the use of a replicative plasmid carrying two homologous sequences and two selection markers (*catP*, *mlsR*), of which one (*mlsR*) was placed between the homology arms and flanked on either site by short *Saccharomyces cerevisiae* FRT.[54] The desired double-crossover was selected due to its erythromycin resistance (*mlsR*) and thiamphenicol sensitivity (*catP*). The remaining *mlsR* marker was, if needed, removed from the genome by transforming an FLP recombinase-expressing[234–236] plasmid into the cells. Expressed FLP recombinase recognized its FRT target sites and removed the *mlsR* marker from the genome. Loss of the FLP-expressing plasmid was stimulated by two successive subcultures without selective pressure. The resulting double-crossover mutants were markerless, but retained a scar, because removal of any DNA by flanking FRT sites and FLP recombinase typically causes scars of 82–85 nucleotides.[255] Hence, double-crossover mutants generated by this method either retain the *mlsR* marker in the genome or exhibit scars. The retention of an antibiotic marker compromises the use of the engineered strain in industrial applications and complicates the introduction of further mutations into the same strain, because the number of selection markers available for *Clostridium* is limited. The generation of a scar impedes most likely with the generation of in-frame modifications. A variation of the method describes for the first time the use of an (endogenous) counter selection marker in a clostridial species (uracil phosphoribosyltransferase, *upp*; Counter (Negative) Selection Markers Section). The two-antibiotic resistance strategy was successfully used to determine the role of PerR in the oxygen tolerance of *C. acetobutylicum*.[256] Recently, two other allelic exchange procedures have been developed for *C. acetobutylicum*. The method established by Al-Hinai et al. (2012)[38] features the use of the wildtype strain and uses a thiamphenicol resistance (Th[r]) marker for positive selection and an (exogenous) toxin (*mazF*) marker for counter selection (Counter (Negative) Selection Markers Section) in a one-step screening procedure. Briefly, Th[r] was placed onto the plasmid between the homology arms, while the expression of *mazF* on the plasmid backbone was controlled by a lactose-inducible promoter.[67] Double-crossover events were selected directly by plating cells onto medium containing thiamphenicol and lactose. Only cells that had integrated the allelic exchange cassette (Th[r] ± exogenous genetic material flanked by the homology arms) into the genome and excised and lost the plasmid survived. The results were disruption, deletion, or integration mutants still containing the Th[r] marker. However, if required, the marker was removed by the FLP recombinase as described previously,[54] with one modification, loss of the FLP-expressing plasmid was stimulated by the lactose-mediated expression of an asRNA, which targeted the plasmid's Gram-positive origin of replication.[38] Despite the novelty of the use of an exogenous counter selection marker, this method features the same drawbacks as the method previously used.[54] In-frame deletions or insertions typically contain the Th[r] marker. Removal of Th[r] by FRT/FLP generates scars.[38] In addition, the method used the *B. subtilis* resolvase or recombinase RecU by positioning its gene onto the plasmid backbone.[257] RecU is thought to increase homologous recombination frequencies in *C. acetobutylicum*.[56,58,59] However, no direct evidence of its superior function over the *Clostridium* host recombination machinery is available.

Another method for the genetic chromosomal modification of *C. acetobutylicum* was described by Heap et al. (2012).[39] The method, named allele-coupled exchange (ACE), differs from all other methods mentioned so far because it primarily focuses on the integration of exogenous genetic material into the chromosome rather than the deletion or disruption of genes. Three variations of the method have been established. They use either the chromosomal thiolase (*thl*) or *pyrE* locus and the wildtype or a *pyrE* truncation strain. In all cases, a two-step selection strategy was applied. Single-crossover mutants were also selected due to their gained thiamphenicol resistance. The screening for the double-crossover varied according to the strategy applied. Using the wildtype, the *thl* locus, and a promoterless erythromycin resistance (*ermB*) gene located between the homology arms, the double-crossover event placed the *ermB* marker under the control of the chromosomal *thl* promoter and provided the mutant strain with erythromycin resistance. Although an effective approach, it is disadvantaged by the retention of an antibiotic marker in the genome. Therefore, a variation of the method used the *thl* locus of a Δ*pyrE* strain and consequently replaced the *ermB* gene with the *pyrE* gene. Screening for double-crossover involved the isolation of uracil prototrophs. A third variation of the method used either the wildtype or the Δ*pyrE* strain and replicative integration plasmids, which truncated or repaired the chromosomal *pyrE*-enabling screening for uracil auxotrophy or prototrophy. By alternating the presence of *pyrE* with *ermB* (variation two) or its state on the chromosome (variation three) and, if needed, the simultaneous adjustment of the homology arms, the method enabled it to alternate between uracil prototrophy and auxotrophy and, thereby, insert (stepwise) large sequences of exogenous DNA such as the lambda phage genome (52.5 kbp) into the clostridial chromosome.[39] However, due to the use of a promoterless *ermB* or *pyrE* marker, which relies on a strong chromosomal promoter or the truncation or repair of the chromosomal *pyrE*, the method is restricted to a limited number of chromosomal loci. Besides the chosen loci (*thl*, *pyrE*), there is little information available about the strength and continuity of other chromosomal promoters that could substitute the used *thl* promoter. Furthermore, the identification of new suitable counter and auxotrophy markers such as *pyrE* can be very challenging. Nonetheless, by combining ACE with allelic exchange targeted, markerless, scarless, in-frame gene deletions can be combined with the insertion of (extensive) exogenous and endogenous DNA sequences into the chromosomal.[39,63]

In terms of other industrially important *Clostridium* species, the allelic exchange method described by Tripathi et al. (2010)[64] and later used by Olson et al. (2010)[258] led to the successful deletion of a phosphotransacetylase (*pta*) and a cellulase gene (*cel48S*) in *C. thermocellum* Δ*pyrF*. Within the two-step selection procedure, the first screening targeted transformants with restored uracil prototrophy due to the *pyrF* (orotidine 5′-phosphate decarboxylase) marker provided by the plasmid; the second screening targeted clones that reverted back to uracil auxotrophy, but gained thiamphenicol resistance. The remainder of the antibiotic marker in the genome as well as the growth defects observed for the Δ*pyrF* strain made this method less favorable.[254]

A more advanced allelic exchange approach for *C. thermocellum* is the method described by Argyros et al. (2011).[254] It involves a *C. thermocellum* Δ*hpt* strain, one selection (*cat*) and two counter selection markers, hypoxanthinephosphoribosyl transferase (*hpt*) and thymidine kinase (*tdk*), in a three-step selection procedure (Tables 4 and 5; Counter (Negative) Selection Markers Section). The genes *cat* and *hpt* were joined in an operon driven by the *C. thermocellum gapDH* promoter (*gapDHp*) and located between the two homology arms. The *tdk* gene was positioned on the plasmid backbone. In the first step of the three-step selection procedure, transformants were selected that took up the plasmid and/or inserted it

Table 4: Allelic exchange methods developed for *Clostridium*

Strain	Plasmid	Selection Marker	Counter Selection	Exemplification	Comments	References
Clostridium acetobutylicum ΔCac15Δupp	Replicative	*catP*, *mlsR*	*upp*	Insertional inactivation of *perR*	Transformation, two-step selection (Second step by *mlsR* and if desired *upp*), removal of *mlsR* by FLP recombinase target sites (FRT) and FLR recombinase (marker-less, but contains scar, not in-frame)	54[e] 256
C. acetobutylicum ATCC 824 *C. acetobutylicum* M5	Replicative	Th[r]	*mazF*	Deletion of *ca_p0167*, *sigF* or *sigK*, disruption of *ca_p0167* or *sigF*, integration of *fdh*	Transformation, one-step selection, removal of Th[r] by FRT and FLR recombinase (marker-less, but contains scar, not in-frame), in-frame possible if FRT/FLR not used, use of RecU	38
C. acetobutylicum ATCC 824 *C. acetobutylicum* ATCC 824 ΔpyrE	Pseudosui-cide	*catP*[a]	*ermB*[b]	Insertion of *ermB* and *adh*	Transformation, two-step selection, *ermB* markers remains in genome, in-frame	39
C. acetobutylicum ATCC 824			*pyrE*[b]	Insertion of *pyrE* into *thl* locus, insertion of lambda phage genome	Transformation, two-step selection, *pyrE* used as a selection marker (easing uracil auxotro-phy on integration into the *thl* locus), in-frame, markerless, multiple modifications in one genome	
			pyrE[c]	Truncation and repair of chromosomal *pyrE*	Transformation, two-step selection, achievement of second selection by alternat-ing between truncated and repaired chromo-somal *pyrE*, in-frame, markerless	
Clostridium ljungdahlii DSM 13528	Suicide	*ermC*	none	Replacement of *fliA* for *ermC* and *adhE1* and/ or *adhe2* for *ermC*	transformation, singlestep selection, *ermC* remains in the genome, in-frame described for *adhE1*	27

Continued

Table 4: Allelic exchange methods developed for *Clostridium*—cont'd

Strain	Plasmid	Selection Marker	Counter Selection	Exemplification	Comments	References
Clostridium thermocellum DSM 1313 *ΔpyrF*	Replicative	*cat*	*pyrF*[d]	Exchange of *pta* for *cat*	Transformation, two-step selection, after double-crossover *cat* stays in chromosome and replaces *pta*	64
C. thermocellum DSM 1313 *Δhpt*	Replicative	*cat*	*hpt/tdk*	Deletion of *ldh* and *pta*	Transformation, three-step selection (*cat, tdk, hpt*), multiple gene deletions in one strain, in-frame, markerless	254
Clostridium difficile R20291 *C. difficile* 630	Pseudosuicide	*catP*[a]	*codA*	Restoration and in-frame deletion of *tcdC*	Conjugation, double selection strategy, in-frame, markerless	61
C. difficile R20291 *ΔpyrE* *C. difficile* Δerm *ΔpyrE*	Pseudosuicide	*catP*[a]	*pyrE*[c]	In-frame deletion of *spo0A, cwp84* or *mtlD*	Conjugation, double selection strategy, allelic exchange and ACE, in-frame, markerless	63
C. difficile M68 *C. difficile* 630	Pseudosuicide	*catP*[a]	None	deletion of *ermB* and *fliC*	Conjugation, two-step selection (second step without any plasmid-based marker, but use of selectable phenotype), markerless	259
Clostridium perfringens strain13 *ΔgalKT*	Suicide	*catP*[a]	*galK*	Disruption of α-, θ and κ-toxins genes or *virRS*-operon	Transformation, double-selection strategy, multiple gene deletions in one strain, in-frame, markerless	62

[a]*catP* chloramphenicol/thiamphenicol resistance genes of *C. perfringens*.[145–147]
[b]Used as a positive and counter (negative) selection marker.[39,63]
[c]Used as a (positive) selection marker.[39]
[d]Used as a (positive) selection and counter (negative) selection marker.[64]
[e]Preferred embodiment of the patent.

into the chromosome (single-crossover) and, therefore, exhibit thiamphenicol resistance. The second selection step screened for clones that had inserted the allelic exchange cassette (homology arms with *gapDHp-cat-hpt*) into the chromosome and excised and lost the plasmid (double-crossover). Mutants with these characteristics exhibited thiamphenicol and 5-fluorodeoxyuridine (FUDR) resistance due to the presence of *cat*, but the absence of *tdk*. In a third step, cells were selected that had lost the *gapDHp-cat-hpt* cassette through recombination (third-crossover) between an internal sequence and a chromosomal one and

Table 5: Counter selection markers used for *Clostridium*

Type	Marker	Marker Gene	Selection Criterion	Origin of the Marker[a]	Application	Host Requirements	Medium Requirements	References
Exogenous	Pyrimidine metabolism	*codA*	5-fluorouracil resistance	*E. coli*	*Clostridium difficile*	Wildtype	FC	61
	Programmed cell death	*mazF*	Cell survival	*E. coli*	*Clostridium acetobutylicum*	Wildtype	Lactose	38
Endogenous	Pyrimidine metabolism	*pyrE*	5-fluoroorotic acid (FOA) resistance, uracil phototrophy	*C. acetobutylicum*[b]	*C. acetobutylicum*	Δ*pyrE*	FOA, uracil	39
			FOA resistance	*Clostridium sporogenes*	*C. difficile*	Δ*pyrE*	FOA, uracil	63
	Pyrimidine metabolism	*pyrF*	FOA resistance	*Clostridium thermocellum*	*C. thermocellum*	Δ*pyrF*	FOA, uracil	64
	Purine/pyrimidine metabolism	*tdk/hpt*[c]	5-fluorodeoxyuridine (FUDR) and 8-aza-hypoxanthine (AZH) resistance	*T. saccharolyticum/C. thermocellum*	*C. thermocellum*	Δ*hpt*	FUDR AZH	254
	Galactose metabolism	*galK*	D-galactose analog 2-deoxy-D-galactose (DOG) resistance	*C. acetobutylicum*	*Clostridium perfringens*	Δ*galKT*	DOG	62

[a]For endogenous types of markers, the marker in the plasmid backbone complements the chromosomal deletion.
[b]The sequence of the plasmid-based homology arms allows it to switch the state of *pyre*.[39]
[c]*tdk* is an exogenous marker, *hpt* an endogenous one.

therefore, exhibited 8-aza-hypoxanthine (AZH) resistance.[254] Although this three-step selection procedure is more laborious than a two-step one, it has been shown to generate multiple markerless in-frame deletions in one strain, namely of the lactate dehydrogenase (*ldh*) and the phosphotransacetylase (*pta*).

Of the procedures described for pathogenic clostridia, each applies the selection and counter selection marker in the classical double-selection strategy (Figure 3). The method established by Nariya et al. (2011)[62] was shown to facilitate multiple markerless in-frame deletions in one strain. The procedure developed by Ng et al. (2013)[63] combines two approaches. By using a *C. difficile pyrE* deletion mutant, a *C. sporogenes pyrE* gene can be used as a plasmid-based counter selection marker for allelic exchange. On the other hand, the chromosomal *C. difficile pyrE* mutant allele can be restored by ACE,[39] allowing the complementation of previously performed allelic exchange deletions at the chromosomal level or the concomitant integration of (extensive) exogenous genetic material into the chromosome.

Of all the recently developed allelic exchange methods, the ones described by Faulds-Pain and Wren (2013)[259] and Leang et al. (2013)[27] for *C. difficile* and *C. ljungdahlii* get by with the use of just one selection marker (*catP*, *ermC*) and no counter selection marker. In *C. difficle*, generation of an *ermB* deletion mutant by alleleic exchange was described to be facilitated by longer homology arms.[259] The longer the homology arms (300 bp, 600 bp, 1200 bp), the higher the frequency of double-crossover events inclusive of the desired one and the less the need to apply counter selection markers. However, the generated *C. difficile* Δ*ermB* mutant exhibited an easy selectable phenotype and therefore, did not call for the use of a second (counter) selection marker. Consequently, the generation of a mutant with a less easy-to-select phenotype, such as the flagellin deletion mutant Δ*fliC*, required an increased number of nonselective subcultures and a more extensive screening approach.[259] In addition, the application of long homology arms might hamper the integration of exogenous DNA into the chromosome of the target strain. In *C. ljungdahlii*, transfer of an allelic exchange suicide plasmid into the clostridial host resulted in 30% of all cases in the desired disruption of *fliA* by double-crossover.[27] The gene *fliA* encodes a sigma factor that is involved in the regulation of flagellar biosynthesis and motility. The disadvantage of the method is the retention of the antibiotic-resistance marker in the genome. The marker cannot be easily removed and therefore, complicates future modifications.

The development of allelic exchange methods and their related counter selection markers within the last couple of years is one of the major breakthroughs in the molecular biology of *Clostridium*. It allows the targeted generation of stable markerless and scarless in-frame double-crossover mutants. It is expected that the methods described here will, in the near future, lead to the generation of *Clostridium* strains with superior biofuel and/or chemical commodity production.

Other Advanced Genetic Tools

Counter (Negative) Selection Markers

In contrast to (positive) selection markers (Clostridial Vector Systems Section), counter (negative) selection markers are used to remove unwanted plasmids or cells from a given population. They are particularly of value if only a limited number of selection markers are available and multiple markerless mutations in a gene or chromosome are required. In reverse genetics, especially of genetically more intractable organisms, they greatly facilitate the identification of rare recombination events and the generation of double-crossover homologous recombination (allelic exchange) mutants.[93] Commonly used counter selection markers provoke, for example, sensitivity to otherwise tolerable compounds (e.g., *tetAR*,[260] *sacB*,[261] *ura3*,[262] *codA*,[263,264] *hpt*,[265] incorporation of toxic analogs into cell components (e.g., *pheS*,[266] sensitivity to antibiotics (e.g., *rpsL*[267]) or are part of toxin-antitoxin systems (e.g., *ccdB*,[268] *mazF*[269]). Counter selection can also be obtained by using auxotrophic mutants (e.g., *ura3*,[262] *upp*,[270] *trp1*[271]). The latter markers allow both positive and negative selection (see below, *pyrE/pyrF*). In general, counter selection markers can be classified into two groups, exogenous and endogenous markers.[61] Exogenous markers are derived from another species and do not have chromosomal counterparts in the host strain. Endogenous markers are closely related orthologs or copies of chromosomal genes. Their function is well known in the target strain. Consequently, they can only be used in a mutant background in which the original gene has been inactivated or deleted. Due to the variety of counter selection markers and their application for allelic exchange, various selection strategies have been described (Recombination-Based Methods (Allelic Exchange) Section). The most frequently used selection strategy follows a two-step protocol and is always used when the mutated gene cannot be directly selected (i.e., does not cause a selectable phenotype)[93] (Figure 3). The main challenge for a broad application of counter selection markers is their identification and the development of appropriate selection strategies and conditions.[93]

For *Clostridium*, the development of such markers, especially for allelic exchange, has only recently been reported (Table 5). Exogenous markers used to date are the *E. coli* genes *codA*[61] and *mazF*.[38] *E. coli codA* encodes for cytosine deaminase (CodA), an enzyme involved in the pyrimidine metabolism by converting cytosine into uracil. Its substrate specificity is reasonably relaxed to allow the conversion of the harmless fluorinated pyrimidine analogue 5-fluorocytosine (FC) into the uracil analog 5-fluorouracil (FU).[272] Subsequently, FU is metabolized into several active highly toxic compounds, which lead eventually to: (1) the inactivation of the thymidylate synthase (ThyA or TS), a key enzyme of the nucleotide biosynthesis; (2) the misincorporation of fluoronucleotides into RNA and DNA; and (3) cell death[273,274] (Figure 4). In the case of *C. difficile*, FU can be metabolized, but not generated. The provision of *codA* on a plasmid backbone will lead to cell death in the presence of FC. Applied in allelic exchange, double-crossover mutants can be identified due to their ability to grow in the presence of FC. These clones have excised and lost the plasmid including *codA* in the second recombination event.[61]

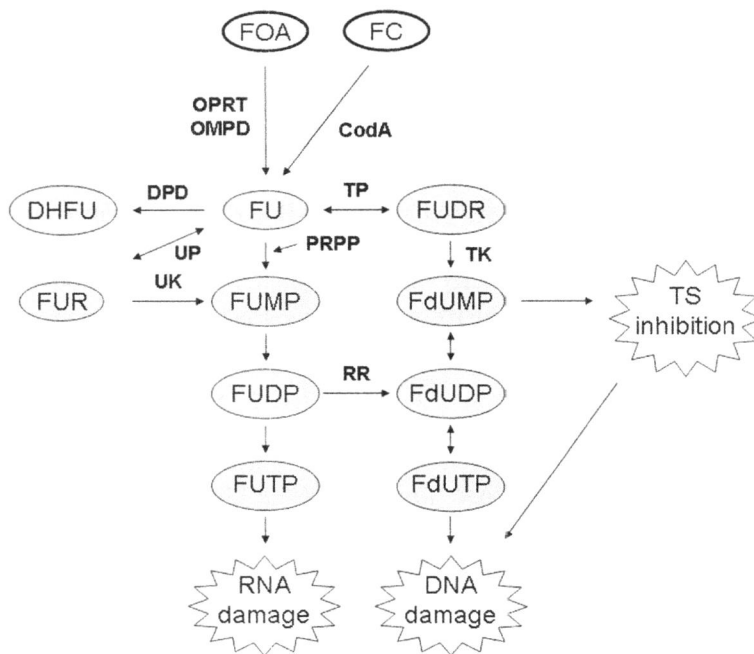

Figure 4

Metabolism of 5-Fluorouracil (FU)[274] modified. Fluorinated pyrimidine analogs 5-fluoroorotic acid (FOA) and FC are converted by the endogenous markers orotate phosphoribosyltransferase (OPRT, *pyrE*) or orotidine 5′-phosphate decarboxylase (OMPD, *pyrF*) or the exogenous marker cytosine deaminase (CodA) to FU. FOA, FC and FC are not metabolically active. They exhibit their cytotoxic effect through their active highly toxic fluorinated metabolites fluorodeoxyuridine monophosphate (F-dUMP), fluorodeoxyuridine triphosphate (FdUTP) and fluorouridine triphosphate (FUTP). These compounds cause the inactivation of thymidylate synthase (ThyA or TS; key enzyme of the nucleotide biosynthesis), miscomposition of RNA and DNA and cell death.[273,274] In detail, activation of FU occurs via three pathways: (1) the direct conversion to 5-fluorouridine monophosphate (FUMP) by the action of orotate phosphoribosyltransferase (OPRT) and the cofactor phosphoribosylpyrophosphate (PRPP); (2) the indirect conversion to FUMP via 5-fluorouridine (FUR) and the activity of uridine phosphorylase (UP) and uridine kinase (UK); and (3) the catalysation to 5-fluorodeoxyuridine (FUDR) by thymidine phosphorylase (TP). FUMP is phosphorylated to 5-fluorouridine diphosphate (FUDP) which is either further phosphorylated to the cytotoxic compound FUTP or metabolised to 5-fluorodeoxyuridine diphosphate (FdUDP) by ribonucleotidereductase (RR). FdUDP is phosphorylated or dephosphorylated to the cytotoxic compounds FdUTP or FdUMP. Conversion of FU to FUDR by TS is followed by the phosphorylation of FUDR to FdUMP by thymidine kinase (TDK or TK). Conversion of FU to dihydrofluorouracil (DHFU) is mediated by dihydropyrimidine dehydrogenase (DPD).

The *E. coli mazE-mazF* genes encode a toxin-antitoxin system involved in the programmed cell death under nutrient deprivation. The toxin MazF, an mRNA interferase with ACA-specific endoribonuclease activity, is stable; the antitoxin MazE is not. Inactivation or omission of MazE will leave active MazF behind and cause cell death.[269,275,276] To facilitate counter selection in *C. acetobutylicum*, *mazF* was placed onto the plasmid backbone under the control of a lactose inducible promoter.[38,67] Addition of lactose to the medium permitted a controlled induction of *mazF* and caused death of cells that have not yet excised and lost the plasmid in the second recombination event.[38]

Endogenous markers developed to date for *Clostridium* are *upp*,[54] *pyrE*,[39,63] *pyrF*,[64] *hpt/tdk*,[254] and *galKT*[62] (Table 5). Of these, most are involved in pyrimidine biosynthesis, *pyrE* and *pyrF* in the de novo pathway, *upp* in the salvage pathway.[277] Deletion or inactivation of *upp* (encoding uracil phosphoribosyltransferase) has been shown to create strains, which cannot convert FU into one of its highly toxic metabolites[273,274] (Figure 4) and are, therefore, able to grow in the presence of FU. If incorporated into the plasmid used for allelic exchange cells that have not undergone double-crossover and excised and lost, the plasmid can be counter selected in the presence of FU.[54,270] The use of *upp* was first been described by Fabret et al. (2002)[270] and later adapted for use in *Clostridium* by Soucaille et al. (2006).[54] Of the de novo biosynthesis pathway, deletions of the genes encoding the *pyrE* and *pyrF* have been shown to generate strains that are resistant to the bactericidal degradation products of the fluorinated pyrimidine analog 5-fluoroorotic acid (FOA) and auxotroph for uracil.[262] Beside their natural substrates, both enzymes convert FOA to FU. As mentioned above, the latter is subsequently metabolized into several highly toxic compounds, which eventually leads to cell death[273,274] (Figure 4).

The beauty of *pyrE* and *pyrF* is that they can be used for both positive and negative selection. Positive selection is achieved by complementation of the auxotrophic mutant with the phototrophic wildtype allele (*pyrE*, *pyrF*) and growth on unsupplemented medium, negative selection by the use of FOA to counter select the wildtype allele. If needed, the auxotrophic mutant can be selected against by growth on defined medium lacking uracil supplementation.[262] The ability of *pyrE* to serve as both a selection or counter selection marker for *Clostridium* has been extensively explored in the ACE and the allelic exchange method developed recently for *C. acetobutylicum*[39] and *C. difficile*[63] (Tables 4 and 5; Recombination-Based Methods (Allelic Exchange) Section). it is noteworthy that within two variations of ACE, *pyrE* was not used as a plasmid backbone-based (counter) selection marker as is usually the case. The state of *pyrE* and, therefore, uracil auxotrophy/prototrophy was determined either by the truncation or repair of its chromosomal gene or by the alternation of its presence in the allelic exchange cassette with another selection marker (*ermB*) in a Δ*pyrE* strain.

In contrast to *C. acetobutylicum*, a *C. thermocellum* Δ*pyrF* strain has been shown to exhibit a growth defect and not be optimal for future genetic manipulations.[64,254] Therefore, another counter selection system was developed for this organism. The system combines the endogenous *C. thermocellum* hypoxanthinephosphoribosyl transferase (*hpt*) gene and the exogenous *Thermoanaerobacterium saccharolyticum* thymidine kinase (*tdk*) gene in a three-step selection procedure[254] (Recombination-Based Methods (Allelic Exchange) Section). Both enzymes catalyze reactions of the nucleic acid metabolism. The HPT or HPRT is involved in the recycling of purines such as adenine, guanine, xanthine, and hypoxanthine.[278] Purine analogs like

8-aza-2,6-diaminopurine (ADP), 8-aza-guanine (AZG), and AZH are also converted by the enzyme, but lead to cell death.[254,265] The TDK (or TK) metabolizes deoxythymidine to deoxythymidine 5'-phosphate, but can also convert FUDR, an analog of FU, to fluoro-dUMP (F-dUMP), an inhibitor of the ThyA or TS[274] (Figure 4). Another endogenous counter selection system, the GalK/galactose (Gal)-based system developed for *C. perfringens* used the *galK* gene and the nonmetabolizable D-galactose analog 2-deoxy-D-galactose (DOG).[62] GalK catalyzes the phosphorylation of galactose to galactose-1-phosphate. Due to its relaxed substrate specificity, it can also phosphorylate DOG to 2-deoxy-galactose-1-phosphate (DOGP). DOGP cannot be further metabolized and accumulates to toxic levels and causes cell death. Hence, cells expressing GalK exhibit sensitivity to DOG,[279–281] and a *galKT* deletion mutant had to be used in this approach as *C. perfringens* possesses an endogenous *galKT* operon.[62]

Inducible Gene Expression

Systems for the targeted, regulated, and effective control of gene expression by the use of chemical or physical inducers are another powerful tool in genetics.[282] They facilitate: (1) the investigation of the function(s) of proteins in vitro and/or in vivo; (2) the production of high levels of industrially and/or medically valuable proteins; and (3) the genetic modification of organisms. Functional in vitro analyses are associated with the heterologous expression and subsequent purification and characterization of the protein of interest[282]; in vivo analysis is associated with the complementation of mutant phenotype[63,283] or the overexpression of the target protein in the wildtype strain.[151] Genetic manipulation by IGE systems can, for example, be achieved by the expression of (key) regulatory or metabolic enzymes in the target strain[25,284] or the controlled induction of counter selection markers.[38] Efficient IGE systems: (1) work in a variety of organisms; (2) are easy to use and do not require temperature shifts, specific media, or expensive inducers; (3) allow a dose-dependent control of the gene expression over a broad range of inducer concentrations; (4) react rapidly to the presence of the inducer; and (5) have a low basal gene expression in the absence of the inducers.[21,67]

The development of reliable IGE systems for use in *Clostridium* has been pursued for many years. Early systems equipped the otherwise strong promoters of the *thl* gene[66,148,285] or the *C. pasteurianum* ferredoxin (*fdx*) gene with *lac* operator sequences of the *E. coli lacZ* gene[22] to endow inducibility by IPTG. However, particularly for the latter promoter (P_{fac}), low levels of induction on addition of the inducer and high levels of basal gene expression in the absence of the inducer were reported.[237] Two other systems focused on the use of the *C. acetobutylicum recA* promoter, which is induced by UV-radiation[286] or a weakened version of the *thl* promoter.[287] However, only low levels of induction were achieved using *recA*. The use of the weakened *thl* promoter represents a general downregulation of the gene expression, rather than a controlled expression.

Table 6: Inducible gene expression systems for *Clostridium*

Inducer	Applied Components	Origin	Application	Location	Comments	References
Xylose	xylR-P$_{xylA}$	S. xylosus	Clostridium acetobutylicum	Plasmid	17-fold induction, potentially effected by glucose-mediated catabolite repression	66
Xylose	xylR-P$_{xylB}$	Clostridium difficile	Clostridium perfringens	Plasmid	10-fold induction, not effected by glucose, divergent transcription system	68
Lactose	bgaR-P$_{bgaL}$	C. perfringens	C. perfringens	Plasmid	Up to 80-fold induction (strain-dependent), low basal expression in absence of inducer, not effected by glucose or IPTG, divergent transcription system	67
AHT	P$_{tetR}$-tetR; P$_{tet}$-gusA	Escherichia coli Bacillus subtilis	C. difficile	Plasmid	Approximately 200-fold induction, divergent transcription, dose-dependent, low basal expression, up to 500 ng/mL aTc	65
	P$_{thl}$-tetR; P$_{cm}$-$_{tetO1}$-gusA	C. acetobutylicum, C. perfringens, E.coli	C. acetobutylicum	Plasmid	119-fold induction, dose-dependent, low basal expression, aTc toxicity above 100 ng/mL	21

AHT = anhydrotetracycline hydrochloride.

The first more advanced IGE system was reported for *C. acetobutylicum* and based on the xylose operon promoter-repressor regulatory system of *Staphylococcus xylosus*[66] (Table 6). In brief, the *S. xylosus* xylose use operon comprises the genes for a xylose isomerase (*xylA*) and a xylulokinase (*xylB*).[288,289] The repressor of the operon, XylR, is located in direct proximity to the *xylA-xylB* operon and transcribed in the same direction. In the absence of xylose, transcription of the *xylA-xylB* operon is repressed due to the binding of XylR to the *xylA* operator palindrome downstream of its promoter. In the presence of the inducer xylose, XylR is inactivated, and transcription of the *xylA-xylB* operon enabled.[290] To apply this system for a regulated and targeted control of gene expression in *C. acetobutylicum*, *xylR* and the *xylA* promoter-operator sequence were localized to a plasmid.[66] A 17-fold induction was achieved in the presence of xylose as the sole carbon source. However, use of this IGE system is limited because it is potentially affected

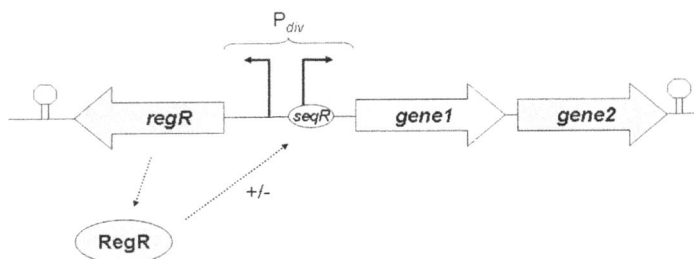

Figure 5

Schematics of divergent transcription. The gene of a regulator R (RegR) and its target genes (genes 1 and 2) are located in direct proximity but are transcribed in different directions by the use of the same or directly adjacent promoters. The RegR controls the transcription of the genes 1 and 2 either: (1) negatively by binding to the sequence *seqR* in the absence of the inducer (*seqR* acts as an operator sequence)[68] or (2) positively in the presence of the inducer.[67] Divergent transcription is a known feature in carbohydrate metabolism.[67,68,292] Abbreviations: *regR*/RegR, regulator R; P_{div}, divergent acting promoter, *seqR*, regulatory sequence and binding site of RegR.

by glucose-mediated catabolite repression.[66] Using a medium with a carbon source other than glucose can compromise growth rates and therefore, expression rates.

More recently developed IGE systems are based on divergently transcribed genes (Figure 5, Table 6). Of these, a xylose inducible system was developed for *C. perfringens* based on the directly adjacent divergently transcribed *C. difficile* genes *xylR* and *xylB-xylA*[68] (Table 5). The intergenic region of the genes contains a divergently transcribed promoter and the putative *xylO* operator sequence, the upstream of *xylBA* located binding site of the repressor XylR. It is assumed that the same regulation pattern occurs as described above for *S. xylosus*[290] or *Bacillus subtilis*.[291] To serve as an IGE system in *C. perfringens*, the chromosomal *C. difficile* *xylR*-P_{xylB} region was placed on a replicative plasmid. Using the *catP* of *C. perfringens* as a reporter gene, an almost 10-fold increase in gene expression and a tightly regulated control in response to the inducer concentration was demonstrated. Due to the fact that *C. perfringens* is unable to use xylose, the system was not affected by catabolite repression.[68]

Another IGE system developed for *C. perfringens* is inducible by lactose and uses elements of the divergently transcribed genes *bgaR* and *bgaL*[67] (Table 6). The gene *bgaL* encodes a β-galactosidase, the gene *bgaR* an activator for *bgaL* expression. For the generation of a plasmid-based IGE system, the entire *bgaR* gene, the intergenic divergent promoter region (P_{bgaL}), and an N-terminal region of *bgaL* were placed on an *E. coli*/*C. perfringens* shuttle vector. Using β-glucuronidase as a reporter, an 80-fold increase in gene expression on lactose induction was demonstrated for a *C. perfringens* strain with low endogenous β-glucuronidase activity. In the absence of the inducer, expression levels were only two-fold increased, indicating a low basal expression. The IGE system was lactose dose-dependent and unaffected by glucose or IPTG. The same system has been applied in *C. acetobutylicum* to regulate the expression of the plasmid-based counter selection marker *mazF*.[38] Here, reporter assays showed a 10- to 15-fold increase of β-glucuronidase activity on lactose addition.

Recently, anhydrotetracycline (aTc) IGE systems have been developed for *C. difficile*[65] and *C. acetobutylicum*.[21] Discovered in the *E. coli* transposon Tn*10*, the tetracycline (Tc) responsive expression system consists of a *tetA* gene encoding a membrane spanning Tc-exporting protein and a divergently orientated *tetR* gene coding for a Tc-responsive repressor (TetR).[293] The transcription of *tetA* is controlled by TetR due to its binding to operator sequences (*tetO1*, *tetO2*) in the promoter region of *tetA*. In the absence of Tc, TetR binds tightly to *tetO* disabling *tetA* transcription. The presence of Tc leads to conformational changes of TetR, rendering it unable to bind to *tetO* and therefore, releases the repression of *tetA* transcription.[294] In IGE systems, the Tc analog anhydrotetracycline (aTc) is usually used as an inducer because it exhibits a lower toxicity and higher binding efficiency to TetR.[295] The Tc (aTc) IGE system has been shown to be one of the most effective systems for a controlled and regulated gene expression and has been used in many organisms.[296,297] In *C. difficile* and *C. acetobutylicum*, the β-glucuronidase (*gusA*) gene was used as a reporter gene to evaluate the strength and inducibility of the tetracycline inducible promoter. Expression of *gusA* was controlled by the *tet* promoter (P_{tet}) in *C. difficile* and by the $P_{cm-tetO1}$ promoter in *C. acetobutylicum*. P_{tet} consists of the strong *B. subtilis* xylose operon (*xyl*) promoter equipped with Tn*10 tet* regulatory elements[298,299]; $P_{cm-tetO1}$ comprises the promoter of the *C. perfringens* chloramphenicol acetyltransferase (*catP*) gene equipped with *tet* operator (*tetO*) sequences.[21]

A further difference between the two studies was the expression of the *tetR* gene. In *C. difficile*, *tetR* was divergently expressed using its native promoter, whereas in *C. acetobutylicum*, the expression was driven by the constitutive *thl* gene promoter (P_{thl}) in the same direction as the *gusA* gene. In *C. difficile*, it was demonstrated that the plasmid-based IGE system was aTc dose-dependent with a maximum almost 200-fold increased gene expression at 500 ng/mL aTc. The P_{tet} promoter proved to be tightly controlled in the absence of aTc.[65] For *C. acetobutylicum*, a maximum 119-fold increased gene expression could be achieved.[21] Binding of TetR to *tetO* in the absence of the inducer was stringent.[21] The drawback of the *C. acetobutylicum* system is, however, the inhibition of growth at aTc concentrations higher than 100 ng/mL.

Conclusion

The clostridial molecular biology has seen major changes over the last decade. The number of sequenced genomes increased drastically. For reverse genetics, new methods for the targeted disruption and deletion of chromosomal genes by either mobile group II introns or allelic exchange have been described. Homologous recombination also permitted the integration of large exogenous DNA sequences into the chromosome. It has been shown that multiple, markerless, in-frame, and scarless mutations can be made in one strain using allelic exchange. Besides these major achievements, standardized plasmid systems, counter selection markers, and inducible gene expression systems have been developed. For forward genetics, the use of the mariner transposon enables effective random mutagenesis for the discovery of new genetic functions. The clostridial molecular biology has stepped into a new era, allowing not only the more thorough investigation of pathogens and measures to counteract them, but also the targeted modification of industrially valuable strains to generate superior biofuel and/or chemical commodities producing strains.

Acknowledgments

This work was supported by the BBSRC (grant reference numbers: BB/G016224/1, BB/L000105/1, BB/LOO4356/1) and TMO Renewables.

References

1. Brüggemann H, Bäumer S, Fricke WF, Wiezer A, Liesegang H, Decker I, et al. The genome sequence of *Clostridium tetani*, the causative agent of tetanus disease. *Proc Natl Acad Sci USA* 2003;**100**:1316–21.
2. Sakaguchi Y, Hayashi T, Kurokawa K, Nakayama K, Oshima K, Fujinaga Y, et al. The genome sequence of *Clostridium botulinum* type C neurotoxin-converting phage and the molecular mechanisms of unstable lysogeny. *Proc Natl Acad Sci USA* 2005;**102**:17472–7.
3. Sebaihia M, Wren BW, Mullany P, Fairweather NF, Minton NP, Stabler R, et al. The multidrug-resistant human pathogen *Clostridium difficile* has a highly mobile, mosaic genome. *Nat Genet* 2006;**38**:779–86.
4. Shimizu T, Ohtani K, Hirakawa H, Ohshima K, Yamashita A, Shiba T, et al. Complete genome sequence of *Clostridium perfringens*, an anaerobic flesh-eater. *Proc Natl Acad Sci USA* 2002;**99**:996–1001.
5. Minton NP. Clostridia in cancer therapy. *Nat Rev Microbiol* 2003;**1**:237–42.
6. Minton NP, Mauchline ML, Lemmon MJ, Brehm JK, Fox M, Michael NP, et al. Chemotherapeutic tumour targeting using clostridial spores. *FEMS Microbiol Rev* 1995;**17**:357–64.
7. Nölling J, Breton G, Omelchenko MV, Makarova KS, Zeng Q, Gibson R, et al. Genome sequence and comparative analysis of the solvent-producing bacterium *Clostridium acetobutylicum*. *J Bacteriol* 2001;**183**:4823–38.
8. Wang Y, Li X, Mao Y, Blaschek H. Single-nucleotide resolution analysis of the transcriptome structure of *Clostridium beijerinckii* NCIMB 8052 using RNA-Seq. *BMC Genomics* 2011;**12**:479.
9. Wang Y, Li X, Mao Y, Blaschek H. Genome-wide dynamic transcriptional profiling in *Clostridium beijerinckii* NCIMB 8052 using single-nucleotide resolution RNA-Seq. *BMC Genomics* 2012;**13**:102.
10. Winogradsky S. Recherches sur l'assimilation de l'azote libre de l'atmosphère par les microbes. *Arch Sci Biol* 1895;**3**:297–352.
11. Bruant G, Lévesque M-J, Peter C, Guiot SR, Masson L. Genomic analysis of carbon monoxide utilization and butanol production by *Clostridium carboxidivorans* strain P7[T]. *PLoS One* 2010;**5**:e13033.
12. Debarati P, Austin FW, Arick T, Bridges SM, Burgess SC, Dandass YS, et al. Genome sequence of the solvent-producing bacterium *Clostridium carboxidivorans* strain P7[T]. *J Bacteriol* 2010;**192**:5554–5.
13. Köpke M, Held C, Hujer S, Liesegang H, Wiezer A, Wollherr A, et al. *Clostridium ljungdahlii* represents a microbial production platform based on syngas. *Proc Natl Acad Sci USA* 2010;**107**:13087–92.
14. Feinberg L, Foden J, Barrett T, Davenport KW, Bruce D, Detter C, et al. Complete genome sequence of the cellulolytic thermophile *Clostridium thermocellum* DSM1313. *J Bacteriol* 2011;**193**:2906–7.
15. Petitdemange E, Caillet F, Giallo J, Gaudin C. *Clostridium cellulolyticum* sp. nov., a cellulolytic mesophilic species from decayed grass. *Int J Syst Bacteriol* 1984;**34**:155–9.
16. Warnick TA, Methe BA, Leschine SB. *Clostridium phytofermentans* sp. nov., a cellulolytic mesophile from forest soil. *Int J Syst Evol Microbiol* 2002;**52**:1155–60.
17. Gabriel CL. Butanol fermentation process. *Ing Eng Chem* 1928;**20**:1063–7.

18. Darden L, Tabery J. Molecular biology. In: Zalta EN, editor. *The Stanford encyclopedia of philosophy*. Fall 2010 ed 2010.

19. Rheinberger H-J. A short history of molecular biology. In: Lorenzano P, Rheinberger H-J, Ortiz E, Galles CD, editors. *History and philosophy of science and technology. Encyclopedia of life support systems (EOLSS)*, vol. 2. 2007, ISBN: 978-1-84826-324-6. p. 1–31. ebook.

20. Dürre P. Formation of solvents in clostridia. In: Dürre P, editor. *Handbook on clostridia*. Boca Raton: CRC Press; 2005. p. 671–98.

21. Dong H, Tao W, Zhang Y, Li Y. Development of an anhydrotetracycline-inducible gene expression system for solvent-producing *Clostridium acetobutylicum*: a useful tool for strain engineering. *Metab Eng* 2012;**14**:59–67.

22. Heap JT, Pennington OJ, Cartman ST, Carter GP, Minton NP. The ClosTron: a universal gene knock-out system for the genus *Clostridium*. *J Microbiol Methods* 2007;**70**:452–64.

23. Papoutsakis ET. Engineering solventogenic clostridia. *Curr Opin Biotechnol* 2008;**19**:420–9.

24. Shao L, Hu S, Yang Y, Gu Y, Chen J, Yang Y, et al. Targeted gene disruption by use of a group II intron (targetron) vector in *Clostridium acetobutylicum*. *Cell Res* 2007;**17**:963–5.

25. Tracy BP, Jones SW, Fast AG, Indurthi DC, Papoutsakis ET. Clostridia: the importance of their exceptional substrate and metabolite diversity for biofuel and biorefinery applications. *Curr Opin Biotechnol* 2012;**23**:364–81.

26. Jennert KCB, Tardif C, Young DI, Young M. Gene transfer to *Clostridium cellulolyticum* ATCC 35319. *Microbiology* 2000;**146**:3071–80.

27. Leang C, Ueki T, Nevin KP, Lovley DR. A genetic system for *Clostridium ljungdahlii*: a chassis for autotrophic production of biocommodities and a model homoacetogen. *Appl Environ Microbiol* 2013;**79**:1102–9.

28. Mermelstein LD, Papoutsakis ET. In vivo methylation in *Escherichia coli* by the *Bacillus subtilis* phage phi 3T I methyltransferase to protect plasmids from restriction upon transformation of *Clostridium acetobutylicum* ATCC 824. *Appl Environ Microbiol* 1993;**59**:1077–81.

29. Pyne M, Moo-Young M, Chung D, Chou C. Development of an electrotransformation protocol for genetic manipulation of *Clostridium pasteurianum*. *Biotechnol Biofuels* 2013;**6**:50.

30. Davis IJ, Carter G, Young M, Minton NP. Gene cloning in clostridia. In: Dürre P, editor. *Handbook on clostridia*. Boca Raton: CRC Press; 2005. p. 37–52.

31. Mauchline ML, Davis TO, Minton NP. Genetics of clostridia. In: Demain AL, Davies JE, editors. *Manual of industrial microbiology and biotechnology*. 2nd ed. Washington: ASM Press; 1999. p. 475–90.

32. Minton NP, Brehm JK, Swinfield T-J, Whelan SM, Mauchline ML, Bodsworth N, et al. Clostridial cloning vectors. In: Woods DR, editor. *The clostridia and biotechnology*. Stoneham: Butterworth-Heinemann; 1993. p. 119–50.

33. Reysset G. Transformation and electrotransformation in clostridia. In: Sebald M, editor. *Genetics and molecular biology of anaerobic bacteria*. New York: Springer; 1993. p. 111–9.

34. Rood JI, Cole ST. Molecular genetics and pathogenesis of *Clostridium perfringens*. *Microbiol Rev* 1991;**55**:621–48.

35. Williams DR, Young DI, Young M. Conjugative plasmid transfer from *Escherichia coli* to *Clostridium acetobutylicum*. *J Gen Microbiol* 1990;**136**:819–26.

36. Young DI, Evans VJ, Jefferies JR, Jennert KCB, Phillips ZEV, Ravagnani A, et al. Genetic methods in clostridia. *Method Microbiol* 1999;**29**:191–207. Academic Press.

37. Young M, Minton NP, Staudenbauer WL. Recent advances in the genetics of the clostridia. *FEMS Microbiol Lett* 1989;**63**:301–25.

38. Al-Hinai MA, Fast AG, Papoutsakis ET. Novel system for efficient isolation of clostridium double-crossover allelic exchange mutants enabling markerless chromosomal gene deletions and DNA integration. *Appl Environ Microbiol* 2012;**78**:8112–21.

39. Heap JT, Ehsaan M, Cooksley CM, Ng Y-K, Cartman ST, Winzer K, et al. Integration of DNA into bacterial chromosomes from plasmids without a counter-selection marker. *Nucleic Acids Res* 2012;**40**:e59.

40. Redelings MD, Sorvillo F, Mascola L. Increase in *Clostridium difficile*-related mortality rates, United States, 1999–2004. *Emerg Infect Dis* 2007;**13**:1417–9.
41. Office for National Statistics. *Deaths involving Clostridium difficile: England and Wales, 2011.* Mortality Analysis Team, Health and Life Events Division; 2012.
42. Bannam TL, Yan X-X, Harrison PF, Seemann T, Keyburn AL, Stubenrauch C, et al. Necrotic enteritis-derived *Clostridium perfringens* strain with three closely related independently conjugative toxin and antibiotic resistance plasmids. *mBio* 2011;**2**:e00190–11.
43. Schuster SC. Next-generation sequencing transforms today's biology. *Nat Methods* 2008;**5**:16–8.
44. Pruitt KD, Tatusova T, Maglott DR. NCBI reference sequence (RefSeq): a curated non-redundant sequence database of genomes, transcripts and proteins. *Nucleic Acids Res* 2005;**33**:D501–4.
45. Desai RP, Papoutsakis ET. Antisense RNA strategies for metabolic engineering of *Clostridium acetobutylicum*. *Appl Environ Microbiol* 1999;**65**:936–45.
46. Tummala SB, Junne SG, Papoutsakis ET. Antisense RNA downregulation of coenzyme A transferase combined with alcohol-aldehyde dehydrogenase overexpression leads to predominantly acohologenic *Clostridium acetobutylicum* fermentations. *J Bacteriol* 2003;**185**:3644–53.
47. Chen Y, McClane BA, Fisher DJ, Rood JI, Gupta P. Construction of an alpha toxin gene knockout mutant of *Clostridium perfringens* type A by use of a mobile group II intron. *Appl Environ Microbiol* 2005;**71**:7542–7.
48. Heap JT, Cartman ST, Pennington OJ, Cooksley CM, Scott JC, Blount B, et al. Development of genetic knock-out systems for clostridia. In: Brüggemann H, Gottschalk G, editors. *Clostridia: molecular biology in the post-genomic era.* Caister Academic Press; 2009. p. 179–98.
49. Green EM, Bennett GN. Inactivation of an aldehyde/alcohol dehydrogenase gene from *Clostridium acetobutylicum* ATCC 824. *Appl Biochem. Biotechnol* 1996;**57–58**:213–21.
50. Green EM, Boynton ZL, Harris LM, Rudolph FB, Papoutsakis ET, Bennett GN. Genetic manipulation of acid formation pathways by gene inactivation in *Clostridium acetobutylicum* ATCC 824. *Microbiology* 1996;**142**:2079–86.
51. Harris LM, Welker NE, Papoutsakis ET. Northern, morphological, and fermentation analysis of spo0A inactivation and overexpression in *Clostridium acetobutylicum* ATCC 824. *J Bacteriol* 2002;**184**:3586–97.
52. Heap JT, Minton NP. 2009. Patent PCT/GB2009/000380.
53. Jia K, Zhu Y, Zhang Y, Li Y. Group II intron-anchored gene deletion in *Clostridium*. *PLoS One* 2011;**6**:e16693.
54. Soucaille P, Figge R, Croux C. Process for chromosomal integration and DNA sequence replacement in clostridia. 2006. Patent PCT/EP2006/066997.
55. Heap JT, Pennington OJ, Cartman ST, Minton NP. A modular system for *Clostridium* shuttle plasmids. *J Microbiol Methods* 2009;**78**:79–85.
56. Jones SW, Tracy BP, Gaida SM, Papoutsakis ET. Inactivation of sigmaF in *Clostridium acetobutylicum* ATCC 824 blocks sporulation prior to asymmetric division and abolishes sigmaE and sigmaG protein expression but does not block solvent formation. *J Bacteriol* 2011;**193**:2429–40.
57. Papoutsakis ET, Tracy BP. Methods and compositions for genetically engineering clostridia species. 2012. Patent US20120301964 A1.
58. Tracy BP, Jones SW, Papoutsakis ET. Inactivation of σE and σG in *Clostridium acetobutylicum* illuminates their roles in clostridial-cel-fom biogenesis, granulose synthesis, solventogenesis, and spore morphogenesis. *J Bacteriol* 2011;**193**:1414–26.
59. Tracy BP, Papoutsakis ET. Methods and compositions for genetically engineering Clostridia species. 2010. Patent US2010/0075424.
60. Tyurin MV, Desai SG, Lynd LR. Electrotransformation of *Clostridium thermocellum*. *Appl Environ Microbiol* 2004;**70**:883–90.
61. Cartman ST, Kelly ML, Heeg D, Heap JT, Minton NP. Precise manipulation of the *Clostridium difficile* chromosome reveals a lack of association between the *tcdC* genotype and toxin production. *Appl Environ Microbiol* 2012;**78**:4683–90.
62. Nariya H, Miyata S, Suzuki M, Tamai E, Okabe A. Development and application of a method for counterselectable in-frame deletion in *Clostridium perfringens*. *Appl Environ Microbiol* 2011;**77**:1375–82.

63. Ng YK, Ehsaan M, Philip S, Collery MM, Janoir C, Collignon A, et al. Expanding the repertoire of gene tools for precise manipulation of the *Clostridium difficile* genome: allelic exchange using *pyrE* alleles. *PLoS One* 2013;**8**:e56051.

64. Tripathi SA, Olson DG, Argyros DA, Miller BB, Barrett TF, Murphy DM, et al. Development of *pyrF*-based genetic system for targeted gene deletion in *Clostridium thermocellum* and creation of a *pta* Mutant. *Appl Environ Microbiol* 2010;**76**:6591–9.

65. Fagan RP, Fairweather NF. *Clostridium difficile* has two parallel and essential Sec secretion systems. *J Biol Chem* 2011;**286**:27483–93.

66. Girbal L, Mortier-Barriere I, Raynaud F, Rouanet C, Croux C, Soucaille P. Development of a sensitive gene expression reporter system and an inducible promoter-repressor system for *Clostridium acetobutylicum*. *Appl Environ Microbiol* 2003;**69**:4985–8.

67. Hartman AH, Liu H, Melville SB. Construction and characterization of a lactose-inducible promoter system for controlled gene expression in *Clostridium perfringens*. *Appl Environ Microbiol* 2011;**77**:471–8.

68. Nariya H, Miyata S, Kuwahara T, Okabe A. Development and characterization of a xylose-inducible gene expression system for *Clostridium perfringens*. *Appl Environ Microbiol* 2011;**77**:8439–41.

69. Cartman ST, Minton NP. A mariner-based transposon system for *in vivo* random mutagenesis of *Clostridium difficile*. *Appl Environ Microbiol* 2010;**76**:1103–9.

70. Hussain HA, Roberts AP, Whalan R, Mullany P. Transposon mutagenesis in *Clostridium difficile*. In: Mullany P, Roberts AP, editors. *Clostridium difficile—methods and protocols*, vol. XII. Humana Press; 2010. p. 203–11.

71. Liu H, Bouillaut L, Sonenshein AL, Melville SB. Use of a mariner-based transposon mutagenesis system to isolate *Clostridium perfringens* mutants deficient in gliding motility. *J Bacteriol* 2012;**195**:629–36.

72. Vidal JE, Chen J, Li J, McClane BA. Use of an EZ-Tn5-based random mutagenesis system to identify a novel toxin regulatory locus in *Clostridium perfringens* strain 13. *PLoS One* 2009;**4**:e6232.

73. Lin Y-L, Blaschek HP. Transformation of heat-treated *Clostridium acetobutylicum* protoplasts with pUB110 plasmid DNA. *Appl Environ Microbiol* 1984;**48**:737–42.

74. Reysset G, Hubert J, Podvin L, Sebald M. Transfection and transformation of *Clostridium acetobutylicum* strain N1-4081 protoplasts. *Biotechnol Tech* 1988;**2**:199–204.

75. Chen C, Smye SW, Robinson MP, Evans JA. Membrane electroporation theories: a review. *Med Biol Eng Comput* 2006;**44**:5–14.

76. Weaver JC, Chizmadzhev YA. Theory of electroporation: a review. *Bioelectrochem Bioenerg* 1996;**41**:135–60.

77. Andreason GL. Electroporation as a technique for the transfer of macromolecules into mammalian cell lines. *J Tissue Cult Methods* 1993;**15**:56–62.

78. Dower WJ, Miller JF, Ragsdale CW. High efficiency transformation of *E.coli* by high voltage electroporation. *Nucleic Acids Res* 1988;**16**:6127–45.

79. Ho SY, Mittal GS. Electroporation of cell membranes: a review. *Crit Rev Biotechnol* 1996;**16**:349–62.

80. Weaver JC. Electroporation: a general phenomenon for manipulating cells and tissues. *J Cell Biochem* 1993;**51**:426–35.

81. McIntyre DA, Harlander SK. Improved electroporation efficiency of intact *Lactococcus lactis* subsp. lactis cells grown in defined media. *Appl Environ Microbiol* 1989;**55**:2621–6.

82. Trevors JT, Chassy BM, Dower WJ, Blaschek HP. Electrotransformation of bacteria by plasmid DNA. In: Chang DC, Chassy BM, Saunders JA, Sowers AE, editors. *Guide to electroporation and electrofusion*. Academic Press; 1991. p. 265–90.

83. Wu N, Matand K, Kebede B, Acquaah G, Williams S. Enhancing DNA electrotransformation efficiency in *Escherichia coli* DH10B electrocompetent cells. *Electron J Biotechnol* 2010;**13**. http://dx.doi.org/10.2225/vol13-issue5-fulltext-11.

84. Mermelstein LD, Welker NE, Bennett GN, Papoutsakis ET. Expression of cloned homologous fermentative genes in *Clostridium acetobutylicum* ATCC 824. *Nat Biotechnol* 1992;**10**:190–5.

85. Tyurin M, Padda R, Huang KX, Wardwell S, Caprette D, Bennett GN. Electrotransformation of *Clostridium acetobutylicum* ATCC 824 using high-voltage radio frequency modulated square pulses. *J Appl Microbiol* 2000;**88**:220–7.

86. Oultram JD, Loughlin M, Swinfield TJ, Brehm JK, Thompson DE, Minton NP. Introduction of plasmids into whole cells of *Clostridium acetobutylicum* by electroporation. *FEMS Microbiol Lett* 1988;**56**:83–8.

87. Zhu Y, Liu X, Yang S-T. Construction and characterization of *pta* gene-deleted mutant of *Clostridium tyrobutyricum* for enhanced butyric acid fermentation. *Biotechnol Bioeng* 2005;**90**:154–66.

88. Davis TO, Henderson I, Brehm JK, Minton NP. Development of a transformation and gene reporter system for group II, non-proteolytic *Clostridium botulinum* type B strains. *J Mol Microbiol Biotechnol* 2000;**2**:59–69.

89. Zhou Y, Johnson E. Genetic transformation of *Clostridium botulinum* hall a by electroporation. *Biotechnol Lett* 1993;**15**:121–6.

90. Allen SP, Blaschek HP. Factors involved in the electroporation-induced transformation of *Clostridium perfringens*. *FEMS Microbiol Lett* 1990;**70**:217–20.

91. Phillips-Jones MK. Plasmid transformation of *Clostridium perfringens* by electroporation methods. *FEMS Microbiol Lett* 1990;**66**:221–6.

92. Liu S-C, Minton NP, Giaccia AJ, Brown JM. Anticancer efficacy of systemically delivered anaerobic bacteria as gene therapy vectors targeting tumor hypoxia/necrosis. *Gene Ther* 2002;**9**:291–6.

93. Reyrat J-M, Pelicic V, Gicquel B, Rappuoli R. Counterselectable markers: untapped tools for bacterial genetics and pathogenesis. *Infect Immun* 1998;**66**:4011–7.

94. Holmes RK, Jobling MG. Genetics. In: Baron S, editor. *Medical microbiology*. 4 ed. Galveston (TX): University of Texas Medical Branch; 1996.

95. Tatum EL, Lederberg J. Gene recombination in *Escherichia coli*. *Nat* 1946;**158**:558.

96. Tolonen AC, Chilaka AC, Church GM. Targeted gene inactivation in *Clostridium phytofermentans* shows that cellulose degradation requires the family 9 hydrolase Cphy3367. *Mol Microbiol* 2009;**74**:1300–13.

97. Theys J, Pennington O, Dubois L, Anlezark G, Vaughan T, Mengesha A, et al. Repeated cycles of *Clostridium*-directed enzyme prodrug therapy result in sustained antitumour effects *in vivo*. *Br J Cancer* 2006;**95**:1212–9.

98. Mani N, Lyras D, Barroso L, Howarth P, Wilkins T, Rood JI, et al. Environmental response and autoregulation of *Clostridium difficile* TxeR, a sigma factor for toxin gene expression. *J Bacteriol* 2002;**184**:5971–8.

99. Purdy D, O'Keeffe TAT, Elmore M, Herbert M, McLeod A, Bokori-Brown M, et al. Conjugative transfer of clostridial shuttle vectors from *Escherichia coli* to *Clostridium difficile* through circumvention of the restriction barrier. *Mol Microbiol* 2002;**46**:439–52.

100. Bradshaw M, Goodnough MC, Johnson EA. Conjugative transfer of the *Escherichia coli*-*Clostridium perfringens* shuttle vector pJIR1457 to *Clostridium botulinum* type A strains. *Plasmid* 1998;**40**:233–7.

101. Lyras D, Rood JI. Conjugative transfer of RP4-oriT shuttle vectors from *Escherichia coli* to *Clostridium perfringens*. *Plasmid* 1998;**39**:160–4.

102. Bertram J, Dürre P. Conjugal transfer and expression of streptococcal transposons in *Clostridium acetobutylicum*. *Arch Microbiol* 1989;**151**:551–7.

103. Woolley RC, Pennock A, Ashton RJ, Davies A, Young M. Transfer of Tn1545 and Tn916 to *Clostridium acetobutylicum*. *Plasmid* 1989;**22**:169–74.

104. Trieu-Cuot P, Carlier C, Martin P, Courvalin P. Plasmid transfer by conjugation from *Escherichia coli* to Gram-positive bacteria. *FEMS Microbiol Lett* 1987;**48**:289–94.

105. Simon R, Priefer U, Puhler A. A broad host range mobilization system for in vivo genetic engineering: transposon mutagenesis in Gram negative bacteria. *Nat Biotech* 1983;**1**:784–91.

106. Mermelstein LD, Welker NE, Petersen DJ, Bennett GN, Papoutsakis ET. Genetic and metabolic engineering of *Clostridium acetobutylicum* ATCC 824. *Annu. NY Acad Sci* 1994;**721**:54–68.

107. Dong H, Zhang Y, Dai Z, Li Y. Engineering *Clostridium* strain to accept unmethylated DNA. *PLoS One* 2010;**5**:e9038.

108. Wilson GG. Organization of restriction-modification systems. *Nucleic Acids Res* 1991;**19**:2539–66.

109. Roberts RJ, Belfort M, Bestor T, Bhagwat AS, Bickle TA, Bitinaite J, et al. A nomenclature for restriction enzymes, DNA methyltransferases, homing endonucleases and their genes. *Nucleic Acids Res* 2003;**31**:1805–12.

110. Roberts RJ, Vincze T, Posfai J, Macelis D. REBASE—a database for DNA restriction and modification: enzymes, genes and genomes. *Nucleic Acids Res* 2010;**38**:D234–6.

111. Scott PT, Rood JI. Electroporation-mediated transformation of lysostaphin-treated *Clostridium perfringens*. *Gene* 1989;**82**:327–33.

112. Herbert M, O'Keeffe TA, Purdy D, Elmore M, Minton NP. Gene transfer into *Clostridium difficile* CD630 and characterisation of its methylase genes. *FEMS Microbiol Lett* 2003;**229**:103–10.

113. Brehm JK, Pennock A, Bullman HMS, Young M, Oultram JD, Minton NP. Physical characterization of the replication origin of the cryptic plasmid pCB101 isolated from *Clostridium butyricum* NCIB 7423. *Plasmid* 1992;**28**:1–13.

114. Cornillot E, Nair RV, Papoutsakis ET, Soucaille P. The genes for butanol and acetone formation in *Clostridium acetobutylicum* ATCC 824 reside on a large plasmid whose loss leads to degeneration of the strain. *J Bacteriol* 1997;**179**:5442–7.

115. Eklund MW, Poysky FT, Mseitif LM, Strom MS. Evidence for plasmid-mediated toxin and bacteriocin production in *Clostridium botulinum* type G. *Appl Environ Microbiol* 1988;**54**:1405–8.

116. Finn CWJ, Silver RP, Habig WH, Hardegree MC, Zon G, Garon CF. The structural gene for tetanus neurotoxin is on a plasmid. *Science* 1984;**224**:881–4.

117. Katayama S, Dupuy B, Daube G, China B, Cole ST. Genome mapping of *Clostridium perfringens* strains with I-CeuI shows many virulence genes to be plasmid-borne. *Mol Gen Genet* 1996;**251**: 720–6.

118. Sloan J, Warner TA, Scott PT, Bannam TL, Berryman DL, Rood JI. Construction of a sequenced *Clostridium perfringens-Escherichia coli* shuttle plasmid. *Plasmid* 1992;**27**:201–19.

119. Brefort G, Magot M, Ionesco H, Sebald M. Characterization and transferability of *Clostridium perfringens* plasmids. *Plasmid* 1977;**1**:52–66.

120. Collins ME, Oultram JD, Young M. Identification of restriction fragments from two cryptic *Clostridium butyricum* plasmids that promote the establishment of a replication-defective plasmid in *Bacillus subtilis*. *J Gen Microbiol* 1985;**131**:2097–105.

121. Minton N, Morris JG. Isolation and partial characterization of three cryptic plasmids from strains of *Clostridium butyricum*. *J Gen Microbiol* 1981;**127**:325–31.

122. Davis TO. Regulation of botulinum toxin complex formation in Clostridium botulinum type A NCTC 2916 [Ph.D. thesis]. Open University; 1998.

123. Azeddoug H, Hubert J, Reysset G. Stable inheritance of shuttle vectors based on plasmid pIM13 in a mutant strain of *Clostridium acetobutylicum*. *J Gen Microbiol* 1992;**138**:1371–8.

124. Monod M, Denoya C, Dubnau D. Sequence and properties of pIM13, a macrolide-lincosamide-streptogramin B resistance plasmid from *Bacillus subtilis*. *J Bacteriol* 1986;**167**:138–47.

125. Truffaut N, Hubert J, Reysset G. Construction of shuttle vectors useful for transforming *Clostridium acetobutylicum*. *FEMS Microbiol Lett* 1989;**58**:15–9.

126. Vosman B, Venema G. Introduction of a *Streptococcus cremoris* plasmid in *Bacillus subtilis*. *J Bacteriol* 1983;**156**:920–1.

127. Soutschek-Bauer E, Hartl L, Staudenbauer WL. Transformation of *Clostridium thermohydrosulfuricum* DSM 568 with plasmid DNA. *Biotechnol Lett* 1985;**7**:705–10.

128. Garnier T, Cole ST. Identification and molecular genetic analysis of replication functions of the bacteriocinogenic plasmid pIP404 from *Clostridium perfringens*. *Plasmid* 1988;**19**:151–60.

129. Bruand C, Ehrlich SD. Transcription-driven DNA replication of plasmid pAMβ1 in *Bacillus subtilis*. *Mol Microbiol* 1998;**30**:135–45.

130. Jannière L, Gruss A, Ehrlich SD. Plasmids. In: Sonenshein AL, Hoch J, Losick R, editors. *Bacillus subtilis and other Gram-positive bacteria: biochemistry, physiology, and molecular genetics*. Washington, DC: American Society for Microbiology Press; 1993. p. 625–44.

131. Projan SJ, Monod M, Narayanan CS, Dubnau D. Replication properties of pIM13, a naturally occurring plasmid found in *Bacillus subtilis*, and of its close relative pE5, a plasmid native to *Staphylococcus aureus*. *J Bacteriol* 1987;**169**:5131–9.

132. Bidnenko VE, Gruss A, Ehrlich SD. Mutation in the plasmid pUB110 Rep protein affects termination of rolling circle replication. *J Bacteriol* 1993;**175**:5611–6.

133. Boe L, Gros MF, te Riele H, Ehrlich SD, Gruss A. Replication origins of single-stranded-DNA plasmid pUB110. *J Bacteriol* 1989;**171**:3366–72.

134. Kiewiet R, Kok J, Seegers JFML, Venema G, Bron S. The mode of replication is a major factor in segregational plasmid instability in *Lactococcus lactis. Appl Environ Microbiol* 1993;**59**:358–64.

135. Kok J, van der Vossen JM, Venema G. Construction of plasmid cloning vectors for lactic streptococci which also replicate in *Bacillus subtilis* and *Escherichia coli. Appl Environ Microbiol* 1984;**48**:726–31.

136. Seegers JFML, Meijer WJJ, Venema G, Bron S, Zhao AC, Khan SA. Structural and functional analysis of the single-strand origin of replication from the lactococcal plasmid pWV01. *Mol Gen Genet* 1995; **249**:43–50.

137. Cartman ST, Heap JT, Kuehne S, Cooksley CM, Ehsaan M, Winzer K, Minton NP. Clostridial gene tools. In Dürre P (ed.), *Systems biology of Clostridium.* London Imperial College Press (ICP) Press, 2014.

138. Leenhouts KJ, Tolner B, Bron S, Kok J, Venema G, Seegers JFML. Nucleotide sequence and characterization of the broad-host-range lactococcal plasmid pWVO1. *Plasmid* 1991;**26**:55–66.

139. Shareck J, Choi Y, Lee B, Miguez CB. Cloning vectors based on cryptic plasmids isolated from lactic acid bacteria: their characteristics and potential applications in biotechnology. *Crit Rev Biotechnol* 2004; **24**:155–208.

140. Gruss A, Ehrlich SD. The family of highly interrelated single-stranded deoxyribonucleic acid plasmids. *Microbiol Rev* 1989;**53**:231–41.

141. Chambers SP, Prior SE, Barstow DA, Minton NP. The pMTL nic⁻ cloning vectors. I. Improved pUC polylinker regions to facilitate the use of sonicated DNA for nucleotide sequencing. *Gene* 1988;**68**:139–49.

142. Chang AC, Cohen SN. Construction and characterization of amplifiable multicopy DNA cloning vehicles derived from the P15A cryptic miniplasmid. *J Bacteriol* 1978;**134**:1141–56.

143. LeBlanc DJ, Lee LN, Inamine JM. Cloning and nucleotide base sequence analysis of a spectinomycin adenyltransferase AAD(9) determinant from *Enterococcus faecalis. Antimicrob Agents Chemother* 1991;**35**:1804–10.

144. Pennington O. *The development of molecular tools for the expression of prodrug converting enzmes in Clostridium sporogenes* [PhD thesis]. The University of Nottingham; 2006.

145. Matsushita C, Matsushita O, Koyama M, Okabe A. A *Clostridium perfringens* vector for the selection of promoters. *Plasmid* 1994;**31**:317–9.

146. Scotcher MC, Huang K-X, Harrison ML, Rudolph FB, Bennett GN. Sequences affecting the regulation of solvent production in *Clostridium acetobutylicum. J Ind Microbiol Biotechnol* 2003;**30**:414–20.

147. Steffen C, Matzura H. Nucleotide sequence analysis and expression studies of a chloramphenicol-acetyltransferase-coding gene from *Clostridium perfringens. Gene* 1989;**75**:349–54.

148. Tummala SB, Welker NE, Papoutsakis ET. Development and characterization of a gene expression reporter system for *Clostridium acetobutylicum* ATCC 824. *Appl Environ Microbiol* 1999;**65**:3793–9.

149. Bullifent HL, Moir A, Titball RW. The construction of a reporter system and use for the investigation of *Clostridium perfringens* gene expression. *FEMS Microbiol Lett* 1995;**131**:99–105.

150. Feustel L, Nakotte S, Dürre P. Characterization and development of two reporter gene systems for *Clostridium acetobutylicum. Appl Environ Microbiol* 2004;**70**:798–803.

151. Tomas CA, Welker NE, Papoutsakis ET. Overexpression of *groESL* in *Clostridium acetobutylicum* results in increased solvent production and tolerance, prolonged metabolism, and changes in the cell's transcriptional program. *Appl Environ Microbiol* 2003;**69**:4951–65.

152. Tummala SB, Welker NE, Papoutsakis ET. Design of antisense RNA constructs for downregulation of the acetone formation pathway of *Clostridium acetobutylicum. J Bacteriol* 2003;**185**:1923–34.

153. Lee SY, Bennett GN, Papoutsakis E. Construction of *Escherichia coli-Clostridium acetobutylicum* shuttle vectors and transformation of *Clostridium acetobutylicum* strains. *Biotechnol Lett* 1992;**14**:427–32.

154. Philippe N, Alcaraz J-P, Coursange E, Geiselmann J, Schneider D. Improvement of pCVD442, a suicide plasmid for gene allele exchange in bacteria. *Plasmid* 2004;**51**:246–55.

155. Allam AB, Reyes L, Assad-Garcia N, Glass JI, Brown MB. Enhancement of targeted homologous recombination in *Mycoplasma mycoides* subsp. *capri* by inclusion of heterologous *recA*. *Appl Environ Microbiol* 2010;**76**:6951–4.

156. McFadden J. Recombination in mycobacteria. *Mol Microbiol* 1996;**21**:205–11.

157. Nair RV, Green EM, Watson DE, Bennett GN, Papoutsakis ET. Regulation of the *sol* locus genes for butanol and acetone formation in *Clostridium acetobutylicum* ATCC 824 by a putative transcriptional repressor. *J Bacteriol* 1999;**181**:319–30.

158. O'Connor JR, Lyras D, Farrow KA, Adams V, Powell DR, Hinds J, et al. Construction and analysis of chromosomal *Clostridium difficile* mutants. *Mol Microbiol* 2006;**61**:1335–51.

159. Wilkinson SR, Young M. Targeted integration of genes into the *Clostridium acetobutylicum* chromosome. *Microbiology* 1994;**140**:89–95.

160. Nielsen J. Metabolic engineering: techniques for analysis of targets for genetic manipulations. *Biotechnol Bioeng* 1998;**58**:125–32.

161. Bowring SN, Morris JG. Mutagenesis of *Clostridium acetobutylicum*. *J Appl Bacteriol* 1985;**58**:577–84.

162. Rogers P. Genetics and biochemistry of *Clostridium* relevant to development of fermentation processes. *Adv Appl Microbiol* 1986;**31**:1–60.

163. Annous BA, Blaschek HP. Isolation and characterization of *Clostridium acetobutylicum* mutants with enhanced amylolytic activity. *Appl Environ Microbiol* 1991;**57**:2544–8.

164. Elkanouni A, Junelles AM, Janatiidrissi R, Petitdemange H, Gay R. *Clostridium acetobutylicum* mutants isolated for resistance to the pyruvate halogen analogs. *Curr Microbiol* 1989;**18**:139–44.

165. Hermann M, Fayolle F, Marchal R, Podvin L, Sebald M, Vandecasteele JP. Isolation and characterization of butanol-resistant mutants of *Clostridium acetobutylicum*. *Appl Environ Microbiol* 1985;**50**:1238–43.

166. Lemmel SA. Mutagenesis in *Clostridium acetobutylicum*. *Biotechnol Lett* 1985;**7**:711–6.

167. Mattaelammouri G, Janatiidrissi R, Rambourg JM, Petitdemange H, Gay R. Acetone butanol fermentation by a *Clostridium acetobutylicum* mutant with high solvent productivity. *Biomass* 1986;**10**:109–19.

168. Hu SY, Zheng HJ, Gu Y, Zhao JB, Zhang WW, Yang YL, et al. Comparative genomic and transcriptomic analysis revealed genetic characteristics related to solvent formation and xylose utilization in *Clostridium acetobutylicum* EA 2018. *BMC Genomics* 2011;**12**:93.

169. Qureshi N, Blaschek HP. Recent advances in ABE fermentation: hyper-butanol producing *Clostridium beijerinckii* BA101. *J Ind Microbiol Biotechnol* 2001;**27**:287–91.

170. Jang YS, Malaviya A, Lee SY. Acetone-butanol-ethanol production with high productivity using *Clostridium acetobutylicum* BKM19. *Biotechnol Bioeng* 2013;**110**:1646–53.

171. Green EM. Fermentative production of butanol—the industrial perspective. *Curr Opin Biotechnol* 2011;**22**:337–43.

172. Clark SW, Bennett GN, Rudolph FB. Isolation and characterization of mutants of *Clostridium acetobutylicum* ATCC 824 deficient in acetoacetyl-coenzyme A:acetate/butyrate:coenzyme A-transferase (EC 2.8.3.9) and in other solvent pathway enzymes. *Appl Environ Microbiol* 1989;**55**:970–6.

173. Gong J, Zheng H, Wu Z, Chen T, Zhao X. Genome shuffling: progress and applications for phenotype improvement. *Biotechnol Adv* 2009;**27**:996–1005.

174. Zhang Y-X, Perry K, Vinci VA, Powell K, Stemmer WPC, del Cardayre SB. Genome shuffling leads to rapid phenotypic improvement in bacteria. *Nature* 2002;**415**:644–6.

175. Mao S, Luo Y, Zhang T, Li J, Bao G, Zhu Y, et al. Proteome reference map and comparative proteomic analysis between a wild type *Clostridium acetobutylicum* DSM 1731 and its mutant with enhanced butanol tolerance and butanol yield. *J Proteome Res* 2010;**9**:3046–61.

176. McClintock B. The origin and behavior of mutable loci in maize. *Proc Natl Acad Sci USA* 1950;**36**:344–55.

177. Choi KH, Kim KJ. Applications of transposon-based gene delivery system in bacteria. *J Microbiol Biotechnol* 2009;**19**:217–28.

178. Muñoz-López M, García-Pérez JL. DNA transposons: nature and applications in genomics. *Curr Genomics* 2010;**11**:115–28.

179. Wicker T, Sabot F, Hua-Van A, Bennetzen JL, Capy P, Chalhoub B, et al. A unified classification system for eukaryotic transposable elements. *Nat Rev Genet* 2007;**8**:973–82.

180. Ciric L, Jasni A, de Vries LE, Agersø Y, Mullany P, Roberts AP. The *Tn916/Tn*1545 family of conjugative transposons. In: Roberts AP, Mullany P, editors. *Bacterial integrative mobile genetic elements*. Landes Bioscience; 2013.

181. Lanckriet A, Timbermont L, Happonen LJ, Pajunen MI, Pasmans F, Haesebrouck F, et al. Generation of single-copy transposon insertions in *Clostridium perfringens* by electroporation of phage mu DNA transposition complexes. *Appl Environ Microbiol* 2009;**75**:2638–42.

182. Burrus V, Pavlovic G, Decaris B, Guédon G. Conjugative transposons: the tip of the iceberg. *Mol Microbiol* 2002;**46**:601–10.

183. Vizváryová M, Valková D. Transposons—the useful genetic tools. *Biol Brat* 2004;**59**:309–18.

184. Goryshin IY, Jendrisak J, Hoffman LM, Meis R, Reznikoff WS. Insertional transposon mutagenesis by electroporation of released *Tn*5 transposition complexes. *Nat Biotech* 2000;**18**:97–100.

185. Bhasin A, Goryshin IY, Steiniger-White M, York D, Reznikoff WS. Characterization of a *Tn*5 pre-cleavage synaptic complex. *J Mol Biol* 2000;**302**:49–63.

186. Chaconas G, Harshey RM. Transposition of phage Mu DNA. In: Craig NL, Craigie R, Gellert M, Lambowitz AM, editors. *Mobile DNA II*. Washington, DC: ASM; 2002. p. 384–402.

187. Goryshin IY, Reznikoff WS. Tn5 in vitro transposition. *J Biol Chem* 1998;**273**:7367–74.

188. Reznikoff WS, Bhasin A, Davies DR, Goryshin IY, Mahnke LA, Naumann T, et al. Tn5: a molecular window on transposition. *Biochem Biophys Res Commun* 1999;**266**:729–34.

189. Savilahti H, Rice PA, Mizuuchi K. The phage Mu transpososome core: DNA requirements for assembly and function. *EMBO J* 1995;**14**:4893–903.

190. Pajunen MI, Pulliainen AT, Finne J, Savilahti H. Generation of transposon insertion mutant libraries for Gram-positive bacteria by electroporation of phage Mu DNA transposition complexes. *Microbiology* 2005;**151**:1209–18.

191. Salyers AA, Shoemaker NB, Stevens AM, Li LY. Conjugative transposons: an unusual and diverse set of integrated gene transfer elements. *Microbiol Rev* 1995;**59**:579–90.

192. Ionesco H. Transferable tetracycline resistance in *Clostridium difficile*. *Annu. Microbiol (Paris) A* 1980;**131**:171–9.

193. Mullany P, Wilks M, Lamb I, Clayton C, Wren B, Tabaqchali S. Genetic-analysis of a tetracycline resistance element from *Clostridium difficile* and its conjugal transfer to and from *Bacillus subtilis*. *J Gen Microbiol* 1990;**136**:1343–9.

194. Smith CJ, Markowitz SM, Macrina FL. Transferable tetracycline resistance in *Clostridium difficile*. *Antimicrob Agents Chemother* 1981;**19**:997–1003.

195. Wust J, Hardegger U. Transferable resistance to clindamycin, erythromycin, and tetracycline in *Clostridium difficile*. *Antimicrob Agents Chemother* 1983;**23**:784–6.

196. Clewell DB, Flannagan SE, Jaworski DD, Clewell DB. Unconstrained bacterial promiscuity: the *Tn*916-*Tn*1545 family of conjugative transposons. *Trends Microbiol* 1995;**3**:229–36.

197. Roberts AP, Johanesen PA, Lyras D, Mullany P, Rood JI. Comparison of Tn5397 from *Clostridium difficile*, Tn916 from *Enterococcus faecalis* and the CW459tet(M) element from *Clostridium perfringens* shows that they have similar conjugation regions but different insertion and excision modules. *Microbiology* 2001;**147**:1243–51.

198. Wang H, Smith MC, Mullany P. The conjugative transposon Tn*5397* has a strong preference for integration into its *Clostridium difficile* target site. *J Bacteriol* 2006;**188**:4871–8.

199. Lyristis M, Bryant AE, Sloan J, Awad MM, Nisbet IT, Stevens DL, et al. Identification and molecular analysis of a locus that regulates extracellular toxin production in *Clostridium perfringens*. *Mol Microbiol* 1994;**12**:761–77.

200. Kaufmann P, Lehmann Y, Meile L. Conjugative transposition of Tn916 from *Enterococcus faecalis* and *Escherichia coli* into *Clostridium perfringens*. *Syst Appl Microbiol* 1996;**19**:35–9.

201. Lin WJ, Johnson EA. Transposon Tn916 mutagenesis in *Clostridium botulinum*. *Appl Environ Microbiol* 1991;**57**:2946–50.

202. Lin WJ, Johnson EA. Genome analysis of *Clostridium botulinum* type A by pulsed-field gel electrophoresis. *Appl Environ Microbiol* 1995;**61**:4441–7.

203. Mullany P, Wilks M, Tabaqchali S. Transfer of *Tn916* and *Tn916* delta E into *Clostridium difficile*: demonstration of a hot-spot for these elements in the *C. difficile* genome. *FEMS Microbiol Lett* 1991;**63**:191–4.

204. Wang H, Roberts AP, Mullany P. DNA sequence of the insertional hot spot of Tn916 in the *Clostridium difficile* genome and discovery of a Tn916-like element in an environmental isolate integrated in the same hot spot. *FEMS Microbiol Lett* 2000;**192**:15–20.

205. Hussain HA, Roberts AP, Mullany P. Generation of an erythromycin-sensitive derivative of *Clostridium difficile* strain 630 (630Delta*erm*) and demonstration that the conjugative transposon Tn*916*DeltaE enters the genome of this strain at multiple sites. *J Med Microbiol* 2005;**54**:137–41.

206. Roberts AP, Hennequin C, Elmore M, Collignon A, Karjalainen T, Minton N, et al. Development of an integrative vector for the expression of antisense RNA in *Clostridium difficile*. *J Microbiol Methods* 2003;**55**:617–24.

207. Babb BL, Collett HJ, Reid SJ, Woods DR. Transposon mutagenesis of *Clostridium acetobutylicum* P262-isolation and characterization of solvent deficient and metronidazole-resistant mutants. *FEMS Microbiol Lett* 1993;**114**:343–8.

208. Bertram J, Kuhn A, Durre P. Tn916-Induced mutants of *Clostridium acetobutylicum* defective in regulation of solvent formation. *Arch Microbiol* 1990;**153**:373–7.

209. Mattsson DM, Rogers P. Analysis of Tn916-Induced mutants of *Clostridium- acetobutylicum* altered in solventogenesis and sporulation. *J Ind Microbiol* 1994;**13**:258–68.

210. Kashket ER, Cao ZY. Isolation of a degeneration-resistant mutant of *Clostridium acetobutylicum* NCIMB 8052. *Appl Environ Microbiol* 1993;**59**:4198–202.

211. Liyanage H, Young M, Kashket ER. Butanol tolerance of *Clostridium beijerinckii* NCIMB 8052 associated with down-regulation of gldA by antisense RNA. *J Mol Microbiol Biotechnol* 2000;**2**:87–93.

212. Evans VJ, Liyanage H, Ravagnani A, Young M, Kashket ER. Truncation of peptide deformylase reduces the growth rate and stabilizes solvent production in *Clostridium beijerinckii* NCIMB 8052. *Appl Environ Microbiol* 1998;**64**:1780–5.

213. Haapa-Paananen S, Rita H, Savilahti H. DNA transposition of bacteriophage Mu: a quantitative analysis of target site selection *in vitro*. *J Biol Chem* 2002;**277**:2843–51.

214. Cao M, Bitar AP, Marquis H. A mariner-based transposition system for *Listeria monocytogenes*. *Appl Environ Microbiol* 2007;**73**:2758–61.

215. Gao LY, Groger R, Cox JS, Beverley SM, Lawson EH, Brown EJ. Transposon mutagenesis of *Mycobacterium marinum* identifies a locus linking pigmentation and intracellular survival. *Infect Immun* 2003; **71**:922–9.

216. Le Breton Y, Mohapatra NP, Haldenwang WG. In vivo random mutagenesis of *Bacillus subtilis* by use of TnYLB-1, a *mariner*-based transposon. *Appl Environ Microbiol* 2006;**72**:327–33.

217. Maier TM, Pechous R, Casey M, Zahrt TC, Frank DW. In vivo *Himar1*-based transposon mutagenesis of *Francisella tularensis*. *Appl Environ Microbiol* 2006;**72**:1878–85.

218. Wilson AC, Perego M, Hoch JA. New transposon delivery plasmids for insertional mutagenesis in *Bacillus anthracis*. *J Microbiol Methods* 2007;**71**:332–5.

219. Lampe DJ, Churchill ME, Robertson HM. A purified *mariner* transposase is sufficient to mediate transposition in vitro. *EMBO J* 1996;**15**:5470–9.

220. Lampe DJ, Grant TE, Robertson HM. Factors affecting transposition of the *Himar1 mariner* transposon in vitro. *Genetics* 1998;**149**:179–87.

221. Cameron DE, Urbach JM, Mekalanos JJ. A defined transposon mutant library and its use in identifying motility genes in *Vibrio cholerae*. *Proc Natl Acad Sci USA* 2008;**105**:8736–41.

222. Lambowitz AM, Zimmerly S. Mobile group II introns. *Annu Rev Genet* 2004;**38**:1–35.

223. Mohr G, Smith D, Belfort M, Lambowitz AM. Rules for DNA target-site recognition by a lactococcal group II intron enable retargeting of the intron to specific DNA sequences. *Genes Dev* 2000;**14**:559–73.

224. Cousineau B, Smith D, Lawrence-Cavanagh S, Mueller JE, Yang J, Mills D, et al. Retrohoming of a bacterial group II intron: mobility via complete reverse splicing, independent of homologous DNA recombination. *Cell* 1998;**94**:451–62.

225. Perutka J, Wang W, Goerlitz D, Lambowitz AM. Use of computer-designed group II introns to disrupt *Escherichia coli* DExH/D-box protein and DNA helicase genes. *J Mol Biol* 2004;**336**:421–39.
226. Zhong J, Karberg M, Lambowitz AM. Targeted and random bacterial gene disruption using a group II intron (targetron) vector containing a retrotransposition-activated selectable marker. *Nucleic Acids Res* 2003; **31**:1656–64.
227. Cui G-Z, Hong W, Zhang J, Li W-L, Feng Y, Liu Y-J, et al. Targeted gene engineering in *Clostridium cellulolyticum* H10 without methylation. *J Microbiol Methods* 2012;**89**:201–8.
228. Jiang Y, Xu C, Dong F, Yang Y, Jiang W, Yang S. Disruption of the acetoacetate decarboxylase gene in solvent-producing *Clostridium acetobutylicum* increases the butanol ratio. *Metab Eng* 2009;**11**:284–91.
229. Ren C, Gu Y, Hu S, Wu Y, Wang P, Yang Y, et al. Identification and inactivation of pleiotropic regulator CcpA to eliminate glucose repression of xylose utilization in *Clostridium acetobutylicum*. *Metab Eng* 2010;**12**:446–54.
230. Wilson DB. The first evidence that a single cellulase can be essential for cellulose degradation in a cellulolytic microorganism. *Mol Microbiol* 2009;**74**:1287–8.
231. Xiao H, Li Z, Jiang Y, Yang Y, Jiang W, Gu Y, et al. Metabolic engineering of d-xylose pathway in *Clostridium beijerinckii* to optimize solvent production from xylose mother liquid. *Metab Eng* 2012;**14**:569–78.
232. Zhang L, Leyn SA, Gu Y, Jiang W, Rodionov DA, Yang C. Ribulokinase and transcriptional regulation of arabinose metabolism in *Clostridium acetobutylicum*. *J Bacteriol* 2012;**194**:1055–64.
233. Karberg M, Guo H, Zhong J, Coon R, Perutka J, Lambowitz AM. Group II introns as controllable gene targeting vectors for genetic manipulation of bacteria. *Nat Biotech* 2001;**19**:1162–7.
234. Cherepanov PP, Wackernagel W. Gene disruption in *Escherichia coli*: TcR and KmR cassettes with the option of Flp-catalyzed excision of the antibiotic-resistance determinant. *Gene* 1995;**158**:9–14.
235. Dymecki SM. Flp recombinase promotes site-specific DNA recombination in embryonic stem cells and transgenic mice. *Proc Natl Acad Sci USA* 1996;**93**:6191–6.
236. Zhu X-D, Sadowski PD. Cleavage-dependent ligation by the FLP recombinase: characterization of a mutant FLP protein with an alteration in a catalytic amino acid. *J Biol Chem* 1995;**270**:23044–54.
237. Heap JT, Kuehne SA, Ehsaan M, Cartman ST, Cooksley CM, Scott JC, et al. The ClosTron: mutagenesis in *Clostridium*: refined and streamlined. *J Microbiol Methods* 2010;**80**:49–55.
238. Kuehne SA, Cartman ST, Heap JT, Kelly ML, Cockayne A, Minton NP. The role of toxin A and toxin B in *Clostridium difficile* infection. *Nature* 2010;**467**:711–3.
239. Kuehne SA, Minton NP. ClosTron-mediated engineering of *Clostridium*. *Bioengineered* 2012;**3**:247–54.
240. Chen Y, Caruso L, McClane B, Fisher D, Gupta P. Disruption of a toxin gene by introduction of a foreign gene into the chromosome of *Clostridium perfringens* using targetron-induced mutagenesis. *Plasmid* 2007;**58**:182–9.
241. Plante I, Cousineau B. Restriction for gene insertion within the *Lactococcus lactis* Ll.LtrB group II intron. *RNA* 2006;**12**:1980–92.
242. Cooksley CM, Zhang Y, Wang H, Redl S, Winzer K, Minton NP. Targeted mutagenesis of the *Clostridium acetobutylicum* acetone-butanol-ethanol fermentation pathway. *Metab Eng* 2012;**14**:630–41.
243. Steiner E, Dago AE, Young DI, Heap JT, Minton NP, Hoch JA, et al. Multiple orphan histidine kinases interact directly with Spo0A to control the initiation of endospore formation in *Clostridium acetobutylicum*. *Mol Microbiol* 2011;**80**:641–54.
244. Orr-Weaver TL, Szostak JW, Rothstein RJ. Yeast transformation: a model system for the study of recombination. *Proc Natl Acad Sci USA* 1981;**78**:6354–8.
245. Balasubramanian V, Pavelka MS, Bardarov SS, Martin J, Weisbrod TR, McAdam RA, et al. Allelic exchange in *Mycobacterium tuberculosis* with long linear recombination substrates. *J Bacteriol* 1996;**178**:273–9.
246. Zealey GR, Loosmore SM, Yacoob RK, Cockle SA, Boux LJ, Miller LD, et al. Gene replacement in *Bordetella pertussis* by transformation with linear DNA. *Nat Biotechnol* 1990;**8**:1025–9.
247. Itaya M, Tanaka T. Gene-directed mutagenesis on the chromosome of *Bacillus subtilis* 168. *Mol Gen Genet* 1990;**223**:268–72.
248. Maloy SR, Stewart VJ, Taylor RK. *Genetic analysis of pathogenic bacteria. A laboratory manual*. Cold Spring Harbor, NY: Cold Spring Harbor Laboratory Press; 1996.
249. Campbell AM. Episomes. *Adv Genet* 1962;**11**:101–46.

250. Dowds BCA, O'Kane C, Gormley E, McConnell DJ, Devine KM. Integrating plasmids in the genetic engineering of bacilli. In: Thomson JA, editor. *Recombinant DNA and bacterial fermentation.* Boca Raton, Florida: CRC Press; 1988. p. 137–56.

251. Perego M. Integrational vectors for genetic manipulation in *Bacillus subtilis.* In: Sonenshein AL, Hoch JA, Losick R, editors. *Bacillus subtilis and other Gram-positive bacteria: biochemistry, physiology and molecular genetics.* Washington DC: American Society for Microbiology; 1993. p. 615–24.

252. Young M, Hranueli D. Chromosomal gene amplification in Gram-positive bacteria. In: Thomson JA, editor. *Recombinant DNA and bacterial fermentation.* Boca Raton, Florida: CRC Press; 1988. p. 157–200.

253. Awad MM, Bryant AE, Stevens DL, Rood JI. Virulence studies on chromosomal alpha-toxin and theta-toxin mutants constructed by allelic exchange provide genetic evidence for the essential role of alpha-toxin in *Clostridium perfringens*-mediated gas gangrene. *Mol Microbiol* 1995;**15**:191–202.

254. Argyros DA, Tripathi SA, Barrett TF, Rogers SR, Feinberg LF, Olson DG, et al. High ethanol titers from cellulose by using metabolically engineered thermophilic, anaerobic microbes. *Appl Environ Microbiol* 2011;**77**:8288–94.

255. Datsenko KA, Wanner BL. One-step inactivation of chromosomal genes in *Escherichia coli* K-12 using PCR products. *Proc Natl Acad Sci USA* 2000;**97**:6640–5.

256. Hillmann F, Fischer R-J, Saint-Prix F, Girbal L, Bahl H. PerR acts as a switch for oxygen tolerance in the strict anaerobe *Clostridium acetobutylicum. Mol Microbiol* 2008;**68**:848–60.

257. Ayora S, Carrasco B, Doncel E, Lurz R, Alonso JC. *Bacillus subtilis* RecU protein cleaves Holliday junctions and anneals single-stranded DNA. *Proc Natl Acad Sci USA* 2004;**101**:452–7.

258. Olson DG, Tripathi SA, Giannone RJ, Lo J, Caiazza NC, Hogsett DA, et al. Deletion of the Cel48S cellulase from *Clostridium thermocellum. Proc Natl Acad Sci USA* 2010;**107**:17727–32.

259. Faulds-Pain A, Wren BW. Improved bacterial mutagenesis by high-frequency allele exchange, demonstrated in *Clostridium difficile* and *Streptococcus suis. Appl Environ Microbiol* 2013;**79**(15):4768–71.

260. Bochner BR, Huang HC, Schieven GL, Ames BN. Positive selection for loss of tetracycline resistance. *J Bacteriol* 1980;**143**:926–33.

261. Gay P, Le Coq D, Steinmetz M, Berkelman T, Kado CI. Positive selection procedure for entrapment of insertion sequence elements in gram-negative bacteria. *J Bacteriol* 1985;**164**:918–21.

262. Boeke JD, Croute F, Fink GR. A positive selection for mutants lacking orotidine-5'-phosphate decarboxylase activity in yeast: 5-fluoro-orotic acid resistance. *Mol Gen Genet* 1984;**197**:345–6.

263. Hartzog PE, Nicholson BP, McCusker JH. Cytosine deaminase MX cassettes as positive/negative selectable markers in *Saccharomyces cerevisiae. Yeast* 2005;**22**:789–98.

264. Mullen CA, Kilstrup M, Blaese RM. Transfer of the bacterial gene for cytosine deaminase to mammalian cells confers lethal sensitivity to 5-fluorocytosine: a negative selection system. *Proc Natl Acad Sci USA* 1992;**89**:33–7.

265. Pritchett MA, Zhang JK, Metcalf WW. Development of a markerless genetic exchange method for *Methanosarcina acetivorans* C2A and its use in construction of new genetic tools for methanogenic archaea. *Appl Environ Microbiol* 2004;**70**:1425–33.

266. Kast P. pKSS - a second-generation general purpose cloning vector for efficient positive selection of recombinant clones. *Gene* 1994;**138**:109–14.

267. Lederberg J. Streptomycin resistance: a genetically recessive mutation. *J Bacteriol* 1951;**61**:549–50.

268. Bernard P, Couturier M. Cell killing by the F plasmid CcdB protein involves poisoning of DNA-topoisomerase II complexes. *J Mol Biol* 1992;**226**:735–45.

269. Yamaguchi Y, Inouye M. mRNA interferases, sequence-specific endoribonucleases from the toxin-antitoxin systems. *Prog Mol Biol Transl Sci* 2009;**85**:467–500.

270. Fabret C, Dusko Ehrlich S, Noirot P. A new mutation delivery system for genome-scale approaches in *Bacillus subtilis. Mol Microbiol* 2002;**46**:25–36.

271. Toyn JH, Gunyuzlu PL, White WH, Thompson LA, Hollis GF. A counterselection for the tryptophan pathway in yeast: 5-fluoroanthranilic acid resistance. *Yeast* 2000;**16**:553–60.

272. Austin EA, Huber BE. A first step in the development of gene therapy for colorectal carcinoma: cloning, sequencing, and expression of *Escherichia coli* cytosine deaminase. *Mol Pharmacol* 1993;**43**:380–7.

273. Heidelberger C, Danenberg PV, Moran RG. Fluorinated pyrimidines and their nucleotides. *Adv Enzymol Relat Areas Mol Biol* 1983;**54**:58–119.

274. Longley DB, Harkin DP, Johnston PG. 5-fluorouracil: mechanisms of action and clinical strategies. *Nat Rev Cancer* 2003;**3**:330–8.

275. Aizenman E, Engelberg-Kulka H, Glaser G. An *Escherichia coli* chromosomal "addiction module" regulated by guanosine [corrected] 3',5'-bispyrophosphate: a model for programmed bacterial cell death. *Proc Natl Acad Sci USA* 1996;**93**:6059–63.

276. Marianovsky I, Aizenman E, Engelberg-Kulka H, Glaser G. The regulation of the *Escherichia coli mazEF* promoter involves an unusual alternating palindrome. *J Biol Chem* 2001;**276**:5975–84.

277. Moffatt BA, Ashihara H. Purine and pyrimidine nucleotide synthesis and metabolism. *The Arabidopsis Book*, vol. 1. American Society of Plant Biologists; 2002. http://dx.doi.org/10.1199/tab. 0018. e0018.

278. Stout JT, Caskey CT. HPRT: gene structure, expression, and mutation. *Annu Rev Genet* 1985;**19**:127–48.

279. Alper MD, Ames BN. Positive selection of mutants with deletions of the *gal-chl* region of the *Salmonella* chromosome as a screening procedure for mutagens that cause deletions. *J Bacteriol* 1975;**121**:259–66.

280. Ueki T, Inouye S, Inouye M. Positive-negative KG cassettes for construction of multi-gene deletions using a single drug marker. *Gene* 1996;**183**:153–7.

281. Warming S, Costantino N, Court DL, Jenkins NA, Copeland NG. Simple and highly efficient BAC recombineering using galK selection. *Nucleic Acids Res* 2005;**33**:e36.

282. Terpe K. Overview of bacterial expression systems for heterologous protein production: from molecular and biochemical fundamentals to commercial systems. *Appl Microbiol Biotechnol* 2006;**72**:211–22.

283. Olson DG, Giannone RJ, Hettich RL, Lynd LR. Role of the CipA scaffoldin protein in cellulose solubilization, as determined by targeted gene deletion and complementation in *Clostridium thermocellum*. *J Bacteriol* 2013;**195**:733–9.

284. Lütke-Eversloh T, Bahl H. Metabolic engineering of *Clostridium acetobutylicum*: recent advances to improve butanol production. *Curr Opin Biotechnol* 2011;**22**:634–47.

285. Perret S, Casalot L, Fierobe HP, Tardif C, Sabathe F, Belaich JP, et al. Production of heterologous and chimeric scaffoldins by *Clostridium acetobutylicum* ATCC 824. *J Bacteriol* 2004;**186**:253–7.

286. Nuyts S, Van Mellaert L, Theys J, Landuyt W, Lambin P, Anné J. The use of radiation-induced bacterial promoters in anaerobic conditions: a means to control gene expression in clostridium-mediated therapy for cancer. *Radiat Res* 2001;**155**:716–23.

287. Mingardon F, Perret S, Belaich A, Tardif C, Belaich J-P, Fierobe H-P. Heterologous production, assembly, and secretion of a minicellulosome by *Clostridium acetobutylicum* ATCC 824. *Appl Environ Microbiol* 2005;**71**:1215–22.

288. Ho NWY, Chen Z, Brainard AP, Sedlak M. Successful design and development of genetically engineered *Saccharomyces* Yeasts for effective cofermentation of glucose and xylose from cellulosic biomass to fuel ethanol. In: Tsao GT, editor. *Advances in biochemical engineering/biotechnology: recent progress in bioconversion of lignocellulosics*, vol. 65. Berlin, Heidelberg: Springer-Verlag; 1999. p. 163–92.

289. Sizemore C, Buchner E, Rygus T, Witke C, Götz F, Hillen W. Organization, promoter analysis and transcriptional regulation of the *Staphylococcus xylosus* xylose utilization operon. *Mol Gen Genet* 1991;**227**:377–84.

290. Sizemore C, Wieland B, Götz F, Hillen W. Regulation of *Staphylococcus xylosus* xylose utilization genes at the molecular level. *J Bacteriol* 1992;**174**:3042–8.

291. Bhavsar AP, Zhao X, Brown ED. Development and characterization of a xylose-dependent system for expression of cloned genes in *Bacillus subtilis*: conditional complementation of a teichoic acid mutant. *Appl Environ Microbiol* 2001;**67**:403–10.

292. Guzman LM, Belin D, Carson MJ, Beckwith J. Tight regulation, modulation, and high-level expression by vectors containing the arabinose PBAD promoter. *J Bacteriol* 1995;**177**:4121–30.

293. Chalmers R, Sewitz S, Lipkow K, Crellin P. Complete nucleotide sequence of Tn10. *J Bacteriol* 2000;**182**:2970–2.

294. Chopra I, Roberts M. Tetracycline antibiotics: mode of action, applications, molecular biology, and epidemiology of bacterial resistance. *Microbiol Mol Biol Rev* 2001;**65**:232–60.

295. Gossen M, Bujard H. Anhydrotetracycline, a novel effector for tetracycline controlled gene expression systems in eukaryotic cells. *Nucleic Acids Res* 1993;**21**:4411–2.

296. Bertram R, Hillen W. The application of Tet repressor in prokaryotic gene regulation and expression. *Microb Biotechnol* 2008;**1**:2–16.

297. Orth P, Schnappinger D, Hillen W, Saenger W, Hinrichs W. Structural basis of gene regulation by the tetracycline inducible Tet repressor-operator system. *Nat Struct Mol Biol* 2000;**7**:215–9.

298. Bateman BT, Donegan NP, Jarry TM, Palma M, Cheung AL. Evaluation of a tetracycline-inducible promoter in *Staphylococcus aureus* in vitro and in vivo and its application in demonstrating the role of *sigB* in microcolony formation. *Infect Immun* 2001;**69**:7851–7.

299. Corrigan RM, Foster TJ. An improved tetracycline-inducible expression vector for *Staphylococcus aureus*. *Plasmid* 2009;**61**:126–9.

CHAPTER 14

Outlook for the Production of Butanol from Cellulolytic Strains of Clostridia

Jennifer L. Takasumi, James C. Liao

Department of Chemical and Biomolecular Engineering, University of California, Los Angeles, California, USA

Introduction

Lignocellulosic biomass is widely accepted as a desirable feedstock for biofuel production because it is an abundant, renewable, nonfood carbon source that is an order of magnitude less expensive than simple sugars and starches.[1] However, industrialization of lignocellulose processing has been troubled by plant cell-wall recalcitrance, which necessitates expensive thermochemical pretreatment and the production of cellulolytic enzymes to release fermentable sugars.[2] Consolidated bioprocessing (CBP) offers an economical alternative to current multistep processing, in which the capacity for feedstock hydrolysis and fuel production are contained within a single microorganism. Cellulolytic *Clostridium* species are among the most promising organisms to serve as CBP hosts because they possess robust lignocellulose-degrading machinery. Recent efforts have focused on overproduction of ethanol and molecular hydrogen from cellulolytic hosts, such as *Clostridium thermocellum*.[3–5] Furthermore, microbial productions of non-native products such as *n*-butanol and isobutanol by CBP are also of interest.[6–8] These C4 alcohols have emerged as prominent advanced biofuels because of their favorable fuel properties, compatibility with current infrastructure, and ability to serve as chemical feedstocks. Microbial productions of such compounds from sugars have been demonstrated with reasonably encouraging yields and productivities.[9–11] However, direct production of these compounds is still in its infancy, despite the demonstration of feasibility.[12] Production of butanol from cellulosic materials will ultimately be the goal.

Cellulolytic Clostridia *and the Cellulosome*

Members of the genus *Clostridium* are strictly anaerobic, spore-forming bacteria. In particular, cellulolytic *Clostridia* are of interest for direct microbial conversion of biomass to liquid fuels because of their native, robust cellulolytic machinery—cellulosomes. Cellulosomes are lignocellulose-degrading, multienzymatic complexes that have been found in anaerobic

Direct Microbial Conversion of Biomass to Advanced Biofuels. http://dx.doi.org/10.1016/B978-0-444-59592-8.00014-2

microorganisms such as *Clostridia* and *Ruminococci*. They are composed of scaffoldins; cohesins; dockerins; carbohydrate-binding modules (CBMs); and catalytic, sugar-degrading subunits.[13] The flexible backbone of a cellulosome is formed by scaffoldin subunits, which are cohesin-containing moieties. Cohesins form highly specific, calcium-dependent bonds with dockerin domains of other subunits, such as catalytic domains and CBMs.[13–15] Catalytic domains can include cellulases, hemicellulases, and other polysaccharide-degrading enzymes and vary between species. Finally, a CBM allows for the attachment of the cellulosome to the biomass substrate it is degrading.

These structural and catalytic features of a cellulosome provide many benefits for efficient biomass degradation. First, cellulosomes are typically bound to the cell surface and the substrate, which provides close proximity between cells and released cellodextrins, thus minimizing losses due to diffusion.[14,15] In addition, catalytic components of the cellulosome are thought to redistribute under different conditions, providing an adaptive structure.[13,16] This concept of synergism among diverse hydrolytic enzymes has demonstrated improved efficiency of substrate utilization.[15,17] Furthermore, enzyme-microbe synergy has been observed, in which cellulosome-cell attachment was found to improve cellulose degradation in *Clostridium thermocellum* as well as a synthetic minicellulosome displayed in *Bacillus subtilis*.[18,19] Because the biomass degradation capacity of cellulosomes is natively robust, cellulolytic *Clostridia* provide a promising platform for direct microbial conversion of biomass to fuels. Below, the cellulolytic and metabolic features of CBP candidates will be discussed as well as the status of genetic techniques and examples of metabolic engineering.

Clostridium thermocellum

C. thermocellum is one of the most investigated cellulosome-expressing bacteria and is the model thermophilic cellulolytic *Clostridium*, growing optimally at approximately 60 °C.[3] It has one of the fastest growth rates on cellulose,[15] and its cellulosomes are more complex (based on cellulosome size) than mesophilic *Clostridia* such as *Clostridium cellulolyticum*.[20] Furthermore, the temperature for optimal cellulosomal activity corresponds to the host's growth temperature, and the cellulosome has demonstrated resistance to inhibitors and fermentative products.[21] Its thermophilic nature minimizes the chance of contamination and facilitates product recovery.[3] *C. thermocellum* produces hydrogen, ethanol, and acetate as major fermentative products, and the effects of end product accumulation on metabolism have been investigated.[22] Metabolic and cellulolytic features have been examined by microarray analysis to help elucidate some novel features of this organism.[23,24] For example, *C. thermocellum* uses an atypical pathway for synthesizing the central metabolite, pyruvate. Instead of pyruvate kinase, which produces pyruvate and ATP from phosphoenolpyruvate, it uses the malate shunt (i.e., the transhydrogenase-malate pathway), which directs flux to oxaloacetate and malate and is dependent on different cofactors.[25,26]

In addition to studies on *C. thermocellum* metabolism, development of genetic techniques and examples of metabolic engineering provide a background for introducing and optimizing butanol production pathways. Tyurin et al. demonstrated uptake of plasmid DNA by electro-transformation, and Guss et al. improved the transformation efficiency to strain DSM 1313 by preparing plasmid DNA with a dam+dcm - *Escherichia coli* strain.[27,28] Gene deletions and heterologous expression via chromosomal insertion in *C. thermocellum* have also been achieved.[26,29,30] Efforts in metabolic engineering have focused on improving bioethanol production. For example, Deng et al. overexpressed pyruvate kinase from *Thermoanaerobacterium saccharolyticum*, which, combined with a lactate dehydrogenase deletion, improved ethanol production by 3.25-fold.[26] Argyros et al. was also successful in redirecting carbon flux from biomass by deleting genes from competing fermentative pathways—lactate dehydrogenase and phosphotransacetylase. Evolving the strain over 2000 h resulted in improving ethanol titers and selectivity.[31] Genetic engineering tools have significantly aided progress toward *C. thermocellum* as a CBP host, but additional technologies will be necessary to achieve goals of industrialization. Because *C. thermocellum* is not a native butanol-producer, expression of heterologous genes will be necessary. Consequently, development of a dependable multigene expression system is one major challenge moving forward. In addition, strategies for achieving high titers and yields in this organism must be developed.

Clostridium cellulolyticum

C. cellulolyticum is a model mesophilic cellulolytic *Clostridium* with a growth temperature of 34 °C. In addition to its ability to degrade cellulose, the cellulosomes of *C. cellulolyticum* also contain components for degrading hemicelluloses and pectin.[20] Techniques for DNA transfer and gene deletions have been established[32,33] and have been utilized for biofuel production in this organism. It is interesting to note that it secretes the central metabolite, pyruvate, in nutrient-rich conditions, suggesting an imbalance carbon flow at this node.[34] Guedon et al. overexpressed two genes from the ethanol-producer, *Zymomonas mobilis*—pyruvate decarboxylase and alcohol dehydrogenase—to metabolize accumulated pyruvate. This improved cellulose consumption and ethanol production.[35] On the other hand, Li et al. deleted lactate dehydrogenase and malate dehydrogenase of competing pathways to improve the ethanol production 8.5-fold from crystalline cellulose.[33] Higashide et al. recently reported the first instance of cellulose to isobutanol by a CBP organism. A recombinant strain of *C. cellulolyticum* containing five heterologous genes produced 660 mg/L isobutanol from crystalline cellulose in 7–9 days.[12] This demonstrative work is encouraging for continuing research on the CBP of cellulose to non-native products.

Other Cellulolytic Hosts

In addition to *C. thermocellum* and *C. cellulolyticum*, other organisms merit consideration as native cellulolytic CBP hosts. For example, *Clostridium cellulovorans* is a mesophilic, cellulosome-expressing *Clostridium* that can degrade a broad range of substrates such as cellulose, xylan, and pectin. It also ferments acetate and butyrate, in addition to ethanol, lactate, hydrogen, formate, and CO_2.[36] *Clostridium phytofermentans* is another interesting mesophilic cellulolytic *Clostridium*. Unlike other species discussed here, there is no evidence of cellulosome expression because it lacks scaffolding and dockerin domains, but some catalytic enzymes do adhere to the substrate via CBMs.[37] *C. phytofermentans* is an attractive host because it contains the highest number of genes for lignocellulose degradation among sequenced *Clostridia*[38] and has a broad range of carbon substrates, which include diverse polysaccharides, oligosaccharides, and monosaccharides.[39]

The industrial *n*-butanol-producer, *Clostridium acetobutylicum*, has also been considered as a CBP host. Although it is unable to grow on cellulose, it contains 11 cellulosomal components[40] and secretes hemicellulose-degrading enzymes.[41] Modifying this inactive system to enable cellulolytic capacity presents another promising strategy for butanol CBP. Finally, thermophilic cellulolytic species of the genus *Caldicellulosiruptor*, such as *Caldicellulosiruptor bescii*, are promising hosts because they can efficiently degrade plant substrates that have not undergone chemical pretreatments.[5,42] Eliminating the biomass pretreatment step in lignocellulose processing is an economical benefit that has led to continued research in organisms of this genus. Their cellulolytic systems are composed of noncellulosomal, multidomain cellulases.[43] Furthermore, the recent ability to transform DNA to *C. bescii* may soon enable metabolic engineering for biofuel production.[44]

Microbial n-Butanol- and Isobutanol-Producing Pathways

Microbial production of higher alcohols has become of increasing interest in the past decade for use as transportation fuels and chemical precursors.[7] Native and synthetic pathways have been constructed in microorganisms to utilize central metabolites such as pyruvate and acetyl-coenzyme A (CoA) and direct flux to desired chemical products. *n*-Butanol and isobutanol are desirable substitutes for liquid transportation fuels because of their comparable octane number and heating value to gasoline.[6] In addition, their low hygroscopicity makes them compatible fuels for storage and distribution[9]. This section will introduce enzymatic pathways for producing either *n*-butanol or isobutanol, present metabolic engineering of these pathways into desirable hosts, and outline general and species-specific challenges leading to industrialization of such microbial processes.

Microbial n-Butanol Pathways

CoA-Dependent n-Butanol Production in Solventogenic Clostridia

n-Butanol production by solventogenic *Clostridia* was industrialized in the early 1900s and has recently regained attention.[45] This group of *Clostridia* natively produces acetone, *n*-butanol, and ethanol in what is known as ABE fermentation. Growth occurs in two phases: acidogenesis, in which organic acids such as acetate and butyrate are produced with ATP, and then solventogenesis, in which the acids are reassimilated to produce acetone, butanol, and ethanol at a ratio of 3:6:1, respectively.[46] Recent efforts in optimizing native butanol production have aimed at better understanding key gene regulation as well as improving titers and butanol selectivity against other fermentative products.[45]

The CoA-dependent butanol pathway proceeds in six steps from acetyl-CoA. First, two molecules of acetyl-CoA are converted to butyryl-CoA by *thiL*, *hbd*, *crt*, and *bcd/etfA/etfB* in a pathway analogous to fatty acid biosynthesis. Butyryl-CoA is then converted to butanol by an aldehyde dehydrogenase and alcohol dehydrogenase, such as the bifunctional aldehyde-CoA/alcohol dehydrogenases encoded by *adhE* or *adhE1*, which catalyze both reactions.[45,47] Alternatively, butyrate can be converted to *n*-butanol by CtfAB and AdhE1 while producing the acetone precursor, acetoacetate. The schematic in Figure 1 shows the pathways for Clostridial butanol production.[48]

Butanol titers exceeding 15 g/L have been achieved in *C. acetobutylicum*,[49,50] *Clostridium beijerinckii*,[11,51] and *Clostridium saccharoperbutylacetonicum*.[52] Genetic manipulations have focused on *C. acetobutylicum* with strategies such as knocking out competing pathways[49,53] and overexpressing butanol production genes.[48,49,53] In addition, Nair et al. demonstrated a successful regulatory strategy. They identified a repressor, SolR, which acts on the *sol* locus to downregulate expression of solventogenic genes (*aad*, *ctfA*, *ctfB*, and *adc*). Inactivating the *solR* gene led to significant improvement in butanol and acetone titers.[50] It is worth noting that all of these productions occurred in acidic conditions because it is a trigger for solventogenesis.[46] Although much has been revealed about the metabolism of solventogenic *Clostridia* and effective ways to manipulate it, research continues in improving genetic techniques, understanding biphasic regulation, and improving solvent tolerance.

n-*Butanol Production in Non-Native Hosts*

In the past 5 years, heterologous expression of clostridial CoA-dependent pathways has enabled butanol production in many non-native hosts. Strategies, such as deleting competing pathways and increasing expression of pathway genes, have been implemented, as well as investigating homologous enzymes and utilizing host-specific driving forces.

As a highly investigated bacterial host with well-developed genetic techniques, *E. coli* serves as an attractive host for heterologous butanol production. Atsumi et al. were the first to produce *n*-butanol in *E. coli* with this pathway. They expressed *C. acetobutylicum* butanol genes (*hbd*, *crt*, *bcd/etfA/etfB*, *adhE2*), but they replaced *thiL* with a native *E. coli* thiolase gene, *atoB*. To improve butanol production, they knocked out genes from competing fermentative pathways (Δ*adhE*, Δ*ldhA*, Δ*frdBC*, Δ*pta*) and the anaerobic regulator, Fnr. The best strain microaerobically produced 550 mg/L butanol from 20 g/L glucose.[54] Inui et al. also introduced the *C. acetobutylicum* butanol pathway, including *thiL*, to anaerobically produce

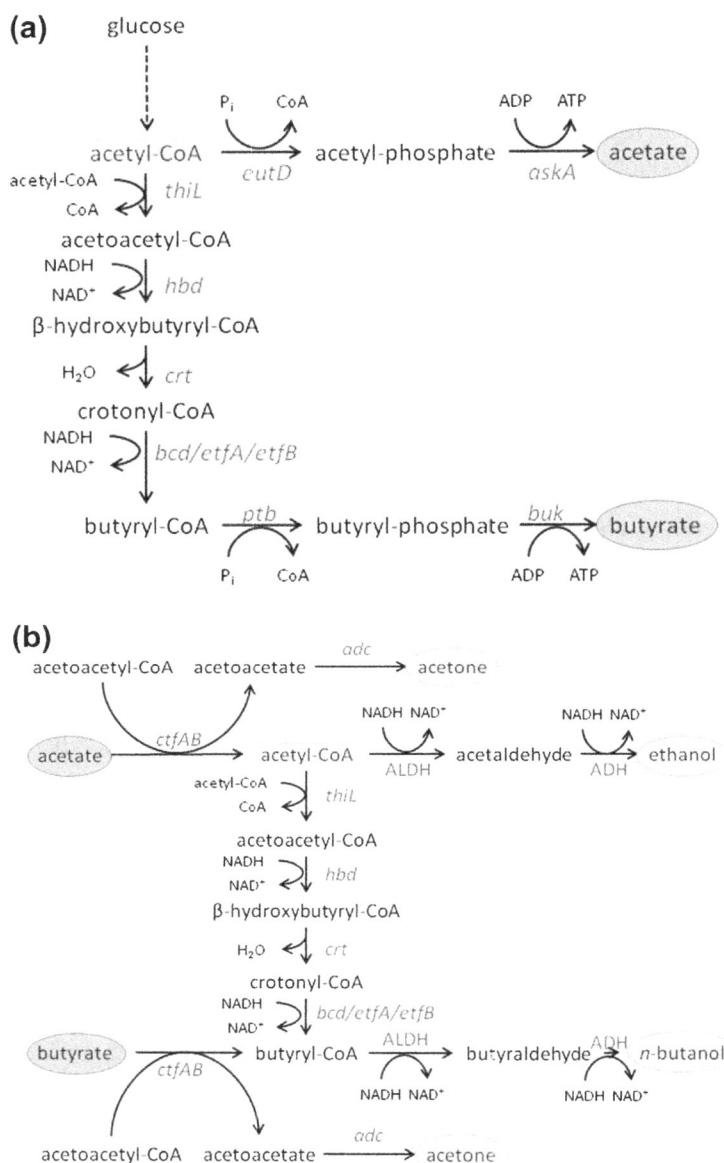

(a)

glucose

acetyl-CoA $\xrightarrow[\text{eutD}]{P_i \quad CoA}$ acetyl-phosphate $\xrightarrow[\text{askA}]{ADP \quad ATP}$ **acetate**

acetyl-CoA
CoA $\Bigg\downarrow$ thiL

acetoacetyl-CoA

NADH
NAD⁺ $\Bigg\downarrow$ hbd

β-hydroxybutyryl-CoA

H₂O $\Bigg\downarrow$ crt

crotonyl-CoA

NADH
NAD⁺ $\Bigg\downarrow$ bcd/etfA/etfB

butyryl-CoA $\xrightarrow[P_i \quad CoA]{ptb}$ butyryl-phosphate $\xrightarrow[ADP \quad ATP]{buk}$ **butyrate**

(b)

acetoacetyl-CoA   acetoacetate $\xrightarrow{adc}$ acetone

acetate $\xrightarrow[\text{ctfAB}]{}$ acetyl-CoA $\xrightarrow[\text{ALDH}]{NADH \quad NAD⁺}$ acetaldehyde $\xrightarrow[\text{ADH}]{NADH \quad NAD⁺}$ ethanol

acetyl-CoA
CoA $\Bigg\downarrow$ thiL

acetoacetyl-CoA

NADH
NAD⁺ $\Bigg\downarrow$ hbd

β-hydroxybutyryl-CoA

H₂O $\Bigg\downarrow$ crt

crotonyl-CoA

NADH
NAD⁺ $\Bigg\downarrow$ bcd/etfA/etfB

butyrate $\xrightarrow[\text{ctfAB}]{}$ butyryl-CoA $\xrightarrow[NADH \quad NAD⁺]{ALDH}$ butyraldehyde $\xrightarrow[NADH \quad NAD⁺]{ADH}$ n-butanol

acetoacetyl-CoA   acetoacetate $\xrightarrow{adc}$ acetone

Figure 1

Fermentative pathways of *Clostridium acetobutylicum*. Schematic of (a) acidogenic fermentative pathways and (b) solventogenic pathways. *thiL* = acetyl-CoA acetyltransferase, *hbd* = 3-hydroxybutyryl-CoA dehydrogenase, *crt* = 3-hydroxybutyryl-CoA dehydratase, *bcd* = butyryl-CoA dehydrogenase, *etfAB* = electron transfer flavoprotein, *eutD* = phosphotransacetylase, *askA* = acetate kinase, *ptb* = phosphate butyryltransferase, *buk* = butyrate kinase, *ctfAB* = butyrate-acetoacetate CoA-transferase, *adc* = acetoacetate decarboxylase, *ALDH* = aldehyde dehydrogenase, *ADH* = alcohol dehydrogenase.

1.2 g/L of butanol from 40 g/L glucose.[47] The most successful examples of high-titer, high-yield butanol productions were achieved by replacing *bcd/etfA/etfB* with *ter* from *Treponema denticola*[55,56] and increasing pathway flux with NADH and acetyl-CoA driving forces.[56] Shen et al. knocked out fermentative pathways that consume acetyl-CoA and NADH (Δpta, $\Delta adhE$, $\Delta ldhA$, $\Delta frdBC$) and overexpressed a formate dehydrogenase from *Candida boidinii* to direct carbon flux from acetyl-CoA through the butanol pathway. Anaerobic growth on glucose generates NADH, a product of glycolysis, which could not be recycled to NAD^+ because all of the native fermentative pathways in the host were deleted. The synthetic pathway, which requires four NADH-consuming reactions, allowed the strain to regenerate the NAD^+ necessary for continued glucose consumption. The deletion of the acetyl-CoA-consuming pathway mediated by *pta* further boosted the production. Utilization of these driving forces enabled anaerobic production of 15 g/L of butanol, or 30 g/L with continuous product removal, yielding 70–88% of the theoretical maximum (see Figure 2).[10]

The clostridial CoA-dependent butanol pathway has been introduced to other hosts including *Saccharomyces cerevisiae*,[57] *Pseudomonas putida*,[58] *B. subtilis*,[58] *Lactobacillus brevis*,[59] and *Synechococcus elongatus*, a cyanobacterium[60,61]; however, production in these organisms lags

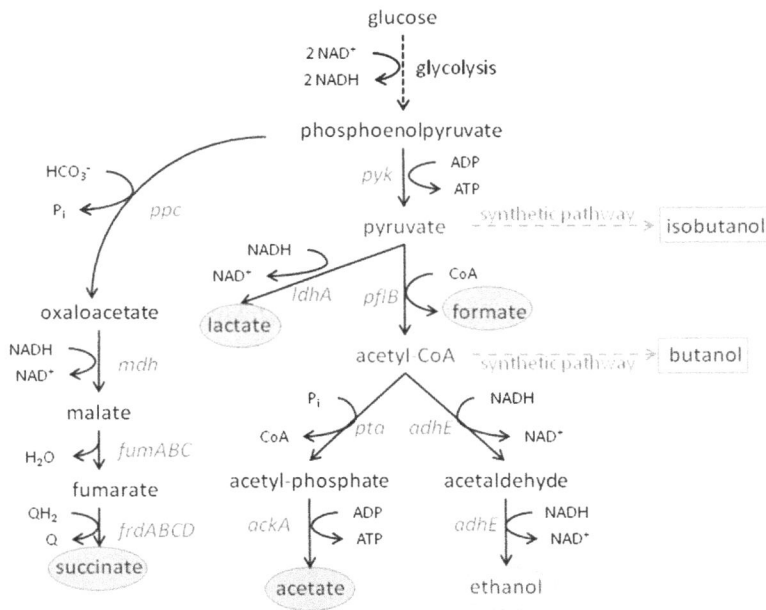

Figure 2

Fermentative pathways of *Escherichia coli*. Schematic of fermentative pathways. *ppc* = phosphoenol-pyruvate carboxylase, *mdh* = malate dehydrogenase, *fumABC* = fumarase, *frdABCD* = fumarate reductase, *pyk* = pyruvate kinase, *pflB* = pyruvate formate-lyase, *ldhA* = lactate dehydrogenase, *pta* = phosphate acetyltransferase, *ackA* = acetate kinase, *adhE* = acetaldehyde/alcohol dehydrogenase, *ldhA* = lactate dehydrogenase.

behind native *Clostridia* and *E. coli*. In many of these cases, metabolic engineering principles were applied to improve butanol production from initial pathway introduction, such as using host enzymes or other homologs to replace activities of the clostridial enzymes or by designing host-specific driving forces. Berezina et al. selected the lactic acid bacterium, *L. brevis*, for its butanol tolerance and native expression of a thiolase, aldehyde dehydrogenase, and alcohol dehydrogenase. Recombinant introduction of five clostridial genes (*hbd*, *crt*, and *bcd/etfA/etfB*) resulted in production of 300 mg/L butanol.[59] Lan and Liao demonstrated a 4-fold improvement in photosynthetic butanol production in *S. elongatus* by requiring irreversible ATP-consuming reactions and selecting enzymes that used the cofactor NADPH instead of NADH.[61] These instances of heterologous butanol pathways demonstrate promise for use in cellulolytic *Clostridia*, but they also illustrate the common challenge of increasing production to industrial levels.

In addition to CoA-dependent synthesis, *n*-butanol production has also been successful from 2-keto acid intermediates of amino acid biosynthesis. Atsumi et al.[9] demonstrated that these metabolites can be utilized for alcohol production via two steps: decarboxylation and reduction. The expression of a keto acid decarboxylase (Kdc) and alcohol dehydrogenase (Adh) in *E. coli* enabled microbial production of *n*-butanol as well as other alcohols such as *n*-propanol, 2-methyl-1-butanol, and isobutanol, which will be discussed later.

Butanol production by the keto acid pathway, (i.e., amino acid pathway) stems from synthesis of the unnatural amino acid, norvaline (see Figure 3). From threonine, a deaminase (*ilvA*) produces 2-ketobutryate, which is catalyzed by the leucine pathway (*leuABCD*) to form the keto acid precursor, 2-ketovalerate. 2-Ketovalerate is then catalyzed by exogenous Kdc and Adh to produce butanol. To improve the titer, *ilvA* and the leucine pathway were overexpressed and *ilvD* was deleted to minimize competitive substrates and flux toward leucine biosynthesis.[9] In another study, Shen and Liao co-produced *n*-propanol and butanol in *E. coli* at about 1 g/L each. Production of these alcohols was achieved by a combination of pathway overexpression (*ilvA*, *leuABCD*, a feedback-resistant *thrA* of leucine biosynthesis, and *kivd* and *ADH2* of the keto acid pathway) and competing pathway deletion (Δ*metA*, Δ*tdh*, Δ*ilvB*, Δ*ilvI*, Δ*adhE*).[56] Currently, butanol production via this pathway lags behind the clostridial pathway, but its promise remains.

Microbial Isobutanol Pathways

Production of isobutanol by the amino acid pathway uses the valine biosynthesis pathway to produce the intermediate, 2-ketoisovalerate (KIV). Two pyruvate molecules undergo condensation and decarboxylation to form 2-acetolactate by an acetohydroxy acid synthase (AHAS). Then, reduction and dehydration by IlvC and IlvD yields KIV, the substrate for isobutanol production by Kdc and Adh (see Figure 3).

Atsumi et al. overexpressed the valine pathway (*ilvIHCD*) along with *kivd* and *ADH2* in *E. coli* and removed competing pathways (Δ*adhE*, Δ*ldhA*, Δ*frdAB*, Δ*fnr*, Δ*pta*) to produce 2.3 g/L

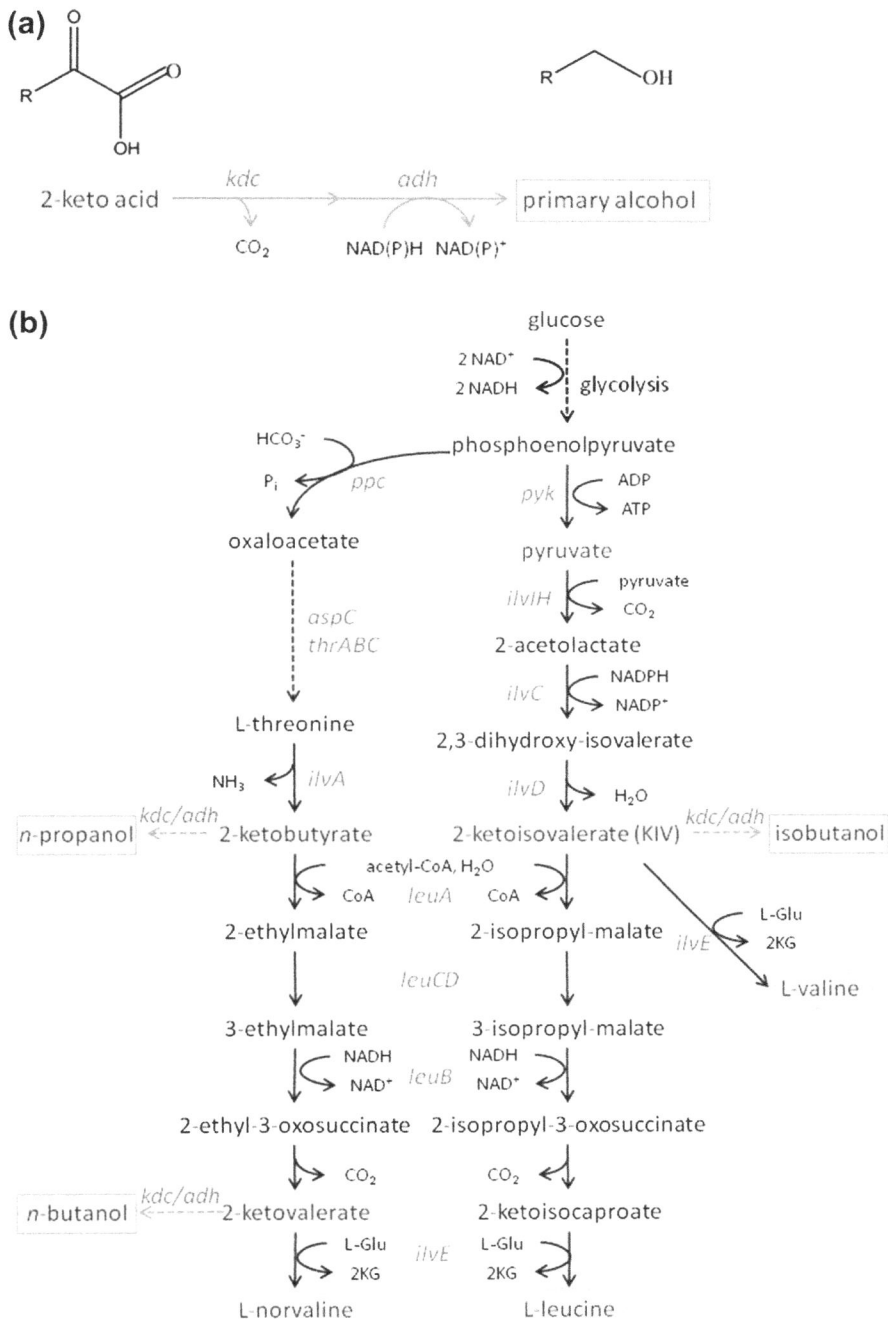

Figure 3

Keto acid pathways for *n*-butanol and isobutanol production. (a) Production of a primary alcohol from a 2-keto acid intermediate by *kdc* (keto acid decarboxylase) and *adh* (alcohol dehydrogenase). (b) Schematic of keto acid pathways for *n*-butanol and isobutanol production. *ppc* = phosphoenol-pyruvate carboxylase, *aspC* = aspartate aminotransferase, *thrA* = aspartate kinase, *thrB* = homoserine kinase, *thrC* = threonine synthase, *ilvA* = threonine deaminase, *leuA* = 2-isopropylmalate synthase, *leuCD* = isopropylmalate isomerase, *leuB* = 3-isopropylmalate dehydrogenase, *ilvE* = branched-chain amino acid aminotransferase, *pyk* = pyruvate kinase, *ilvIH* = acetolactate synthase I, *ilvC* = acetohy-droxy acid isomeroreductase, *ilvD* = dihydroxy acid dehydratase.

isobutanol.[9] Exchanging the AHAS, encoded by *ilvIH*, with a catabolic enzyme from *B. subtilis*, AlsS, and deleting *pflB* led to production of 22 g/L microaerobically.[9] AlsS has a significantly higher affinity for pyruvate than IlvIH, which enhanced flux from the pyruvate node through the isobutanol pathway. Moving production to a bioreactor with gas stripping enabled titers to exceed 50 g/L.[62] Alternatively, an evolutionary approach was used to develop an *E. coli* strain capable of similar isobutanol titers. Multiple rounds of random mutagenesis and selection were conducted to evolve strains resistant to norvaline, a branched chain amino acid analog that becomes toxic at high concentrations. Introduction of isobutanol pathway genes (*alsS*, *ilvC*, *ilvD*, *kivd*, *adhA*) to the evolved strain enabled production of 21.2 g/L isobutanol.[63]

Isobutanol has also been produced by the keto acid pathway in *Corynebacterium glutamicum*,[64] *B. subtilis*,[65,66] and *S. cerevisiae*.[67–70] Furthermore, using renewable resources such as CO_2 in *S. elongatus*,[71] electricity in *Ralstonia eutropha*,[72] and waste proteins in *E. coli*[73] highlights the potential for renewable advanced biofuels. Similar metabolic engineering strategies were used to improve isobutanol production after initial introduction of the pathway. For instance, expressing genes to increase pyruvate and KIV pools was used in multiple hosts, as was deleting competing pathways.

Improving enzymes is another strategy for enhancing microbial fuel production. Bastian et al. used in vitro enzyme evolution to improve isobutanol yields in *E. coli*.[74] They improved the balance of the cofactor, NADH, which is produced during glycolysis and oxidized during isobutanol production, by engineering IlvC to prefer NADH as an electron donor over NADPH, the preferred substrate of the wild-type enzyme. They also overexpressed a transhydrogenase to transfer electrons between NADH and $NADP^+$. Finally, they improved the affinity of AdhA for isobutryaldehyde to achieve 100% of the theoretical yield from glucose anaerobically.[74] Another approach was used by Matsuda et al., who engineered *S. cerevisiae* to express pathway enzymes in the cytosol instead of the mitochondrion, where it is normally expressed. This resulted in doubling production and demonstrates the role that compartmentalization may play in microbial chemical production.[70]

As is the case for clostridial pathway *n*-butanol production, engineering high-titer, high-yield isobutanol production has been most successful in *E. coli*. Although work in other hosts has demonstrated improvements in titer and yield, a more comprehensive understanding of how to manipulate metabolism will be necessary to reach industrial levels. Thus, continuing research on the level of basic metabolism will be necessary to understand relevant cell regulations and enable the design of host-specific driving forces.

Some thermophilic and hyperthermophilic archaea also possess enzymes that offer an alternative route to isobutanol from KIV. 2-Ketoisovalerate ferredoxin oxidoreductase (VOR) is a class of enzymes capable of CoA-dependent decarboxylation of 2-keto acids analogous to the reaction of the more common enzyme, pyruvate ferredoxin oxidoreductase. VORs have been identified and characterized in *Thermococcus litoralis* and *Pyrococcus* sp.,[75] *Methanobacterium thermoautotrophicum*,[76] and *Thermococcus profundus*.[77] These multimeric enzymes are composed of either three or four subunits and are often sensitive to oxygen. VORs are most relevant for isobutanol CBP in the thermophile *C. thermocellum* because of their activity at high temperatures. Biochemical assays have demonstrated their activity in vitro, but heterologous expression has not been reported in the literature.

Progress toward Butanol CBP in Cellulolytic Clostridia

Examples of native butanol production in *Clostridia* and heterologous C4 alcohol production in other microbes elucidate a general strategy for enabling and improving production of these chemicals. The first step is overexpressing pathway genes in the selected host. Alternative enzymes should also be considered for host compatibility, preferable cofactor usage,[74] or superior activity, such as the pyruvate specificity of *B. subtilis* AlsS[9] or the irreversibility of *T. denticola* Ter.[10] Once the pathway is functional, production can be improved by deleting competing pathways, overexpressing additional genes to improve pathway flux, or disrupting unproductive regulation.[50] Finally, optimizing production conditions, developing host-specific driving forces, and strain evolution have demonstrated success for achieving goals of high titer and yield. In solventogenic *Clostridia*, controlling pH and the switch from acidogenesis to solventogenesis was essential for improved titers.[11] Balancing cofactors and providing effective driving forces are essential for synthetic pathways. For example, anaerobic NADH and acetyl-CoA accumulation in the specific *E. coli* knockout strain[10] provided the necessary driving forces for CoA-dependent butanol production. In addition, continuous product removal may be used to reduce product toxicity.[62] Cellulolytic *Clostridia* are still in the early stages of strain development,[12] but this section will outline three promising strategies for developing *n*-butanol or isobutanol CBP strains.

Isobutanol CBP in Clostridium cellulolyticum

CBP of crystalline cellulose to isobutanol in *C. cellulolyticum* was the first demonstration of isobutanol CBP. Five genes from the *E. coli* pathway were expressed on a plasmid downstream of a ferredoxin promoter to produce 660 mg/L isobutanol. Challenges and unexpected results that occurred in engineering *C. cellulolyticum* will help inform future engineering of cellulolytic microbes for isobutanol CBP. For example, transformants containing wild-type *alsS* directly downstream of a constitutive promoter could not be obtained. Enzyme toxicity was believed to be the cause. However, moving the gene to the third position in the operon enabled successful transformation and isobutanol production thereafter. In addition, in vitro assays indicated that recombinant strains did not have statistically improved activities for IlvC, IlvD, or alcohol dehydrogenase. Thus, native enzymes may be sufficient to manage the pathway flux. On the other hand, negative controls demonstrated the absence of AlsS and Kivd activities, which implies that these are the most important steps for isobutanol CBP in cellulolytic *Clostridia*.[12]

Toward Isobutanol CBP in Clostridium thermocellum

C. thermocellum is one of the most promising hosts for cellulolytic CBP because of the microbe's rapid growth on cellulose and favorable properties of high-temperature anaerobic bioprocessing.[3,78] However, engineering a heterologous pathway in this thermophilic organism has its unique challenges. High temperatures are believed to increase chemical toxicity, which may include unnatural pathway products and intermediates.[79] Likewise, intermediate aldehydes have increased volatility. For instance, the boiling point of pure isobutyraldehyde is 63 °C. The effect of increased temperature on the viability of cells and preventing loss of volatile intermediates should be considered. In addition, the thermostability of heterologous enzymes becomes an issue. To our knowledge, C4 alcohol production has not been demonstrated in thermophiles; therefore, it is necessary to identify thermophilic enzymes. However, archaeal CoA-dependent VORs may be promising options because they are from thermophilic hosts. Finally, the unique pyruvate metabolism of *C. thermocellum*[25] may factor into the success of butanol and isobutanol pathways derived from pyruvate and acetyl-CoA. Developments in genetic techniques, increasing instances of metabolic engineering, and recent findings on native metabolism are encouraging for the future development of CBP strains.

Clostridium acetobutylicum *Cellulosome Development*

As mentioned earlier, *C. acetobutylicum* secretes hemicellulose-degradation enzymes and possesses cellulosomal genes, although it cannot grow on crystalline cellulose. Combined with native *n*-butanol fermentation, the development of a chimeric cellulosomal system in this organism offers another encouraging route to *n*-butanol CBP. Fierobe et al. demonstrated that functional chimeric cellulosomes could be produced by mixing dockerin and catalytic domains from *C. thermocellum* and *C. cellulolyticum*[17] with the previous knowledge that cohesin-dockerin interactions are specific to species.[80] Perret et al. then demonstrated that these heterologous miniscaffoldins could be expressed and secreted by *C. acetobutylicum* in mature and active forms.[81] Recently, minicellulosomes containing *C. cellulolyticum* catalytic subunits, including a mannose and some cellulases, were functionally expressed.[41,82] Unfortunately, expression of larger catalytic modules was unsuccessful because of secretion problems. The identification of an additional chaperone protein for proper secretion of larger proteins may be necessary for enabling CBP in *C. acetobutylicum*.[41] Although activity of the minicellulosomes was demonstrated in vitro, CBP capacity has yet to be reported.

Conclusions

Butanol CBP has emerged as a promising route for producing renewable advanced fuels, although the ultimate role that butanol CBP will play among diverse biofuel production strategies remains unclear. Several challenges, such as the development of heterologous expression systems, limited understanding of metabolic regulation, and toxicity of intermediates and products, still face this field. Nonetheless, the diversity of research and the significant progress that has been made over the past 20 years supports a bright future. The various strategies discussed here will not only contribute to developing a butanol CBP organism, but they will also help shape the general design principles used in the broader field of metabolic engineering.

Acknowledgment

The BioEnergy Science Center is a U.S. Department of Energy (DOE) Bioenergy Research Center supported by the Office of Biological and Environmental Research in the DOE Office of Science.

References

1. Chu S, Majumdar A. Opportunities and challenges for a sustainable energy future. *Nature* 2012;**488**:294–303.
2. Himmel ME, Ding SY, Johnson DK, Adney WS, Nimlos MR, Brady JW, et al. Biomass recalcitrance: engineering plants and enzymes for biofuels production. *Science* 2007;**315**:804–7.
3. Demain AL, Newcomb M, Wu JHD. Cellulase, clostridia, and ethanol. *Microbiol Mol Biol Rev* 2005; **69**:124–54.
4. Levin DB, Carere CR, Cicek N, Sparling R. Challenges for biohydrogen production via direct lignocellulose fermentation. *Int J Hydrogen Energy* 2009;**34**:7390–403.
5. Olson DG, McBride JE, Shaw AJ, Lynd LR. Recent progress in consolidated bioprocessing. *Curr Opin Biotechnol* 2012;**23**:396–405.
6. Li H, Cann AF, Liao JC. Biofuels: biomolecular engineering fundamentals and advances. *Annu Rev Chem Biomol Eng* 2010;**1**:19–36.
7. Mainguet SE, Liao JC. Bioengineering of microorganisms for C-3 to C-5 alcohols production. *Biotechnol J* 2010;**5**:1297–308.
8. Lan EI, Liao JC. Microbial synthesis of *n*-butanol, isobutanol, and other higher alcohols from diverse resources. *Bioresour Technol* 2013;**135**:339–49.
9. Atsumi S, Hanai T, Liao JC. Non-fermentative pathways for synthesis of branched-chain higher alcohols as biofuels. *Nature* 2008;**451**:86–U13.
10. Shen CR, Lan EI, Dekishima Y, Baez A, Cho KM, Liao JC. Driving forces enable high-titer anaerobic 1-Butanol synthesis in *Escherichia coli*. *Appl Environ Microbiol* 2011;**77**:2905–15.
11. Mutschlechner O, Swoboda H, Gapes JR. Continuous two-stage ABE-fermentation using *Clostridium beijerinckii* NRRL B592 operating with a growth rate in the first stage vessel close to its maximal value. *J Mol Microbiol Biotechnol* 2000;**2**:101–5.
12. Higashide W, Li YC, Yang YF, Liao JC. Metabolic engineering of *Clostridium cellulolyticum* for production of isobutanol from cellulose. *Appl Environ Microbiol* 2011;**77**:2727–33.
13. Bayer EA, Belaich JP, Shoham Y, Lamed R. The cellulosomes: multienzyme machines for degradation of plant cell wall polysaccharides. *Annu Rev Microbiol* 2004;**58**:521–54.
14. Schwarz WH. The cellulosome and cellulose degradation by anaerobic bacteria. *Appl Microbiol Biot* 2001;**56**:634–49.
15. Lynd LR, Weimer PJ, van Zyl WH, Pretorius IS. Microbial cellulose utilization: fundamentals and biotechnology. *Microbiol Mol Biol Rev* 2002;**66**:506–77.
16. Doi RH, Kosugi A. Cellulosomes: plant-cell-wall-degrading enzyme complexes. *Nat Rev Microbiol* 2004;**2**:541–51.
17. Fierobe HP, Mechaly A, Tardif C, Belaich A, Lamed R, Shoham Y, et al. Design and production of active cellulosome chimeras - selective incorporation of dockerin-containing enzymes into defined functional complexes. *J Biol Chem* 2001;**276**:21257–61.

18. Lu YP, Zhang YHP, Lynd LR. Enzyme-microbe synergy during cellulose hydrolysis by *Clostridium thermocellum*. *Proc Natl Acad Sci U S A* 2006;**103**:16165–9.

19. You C, Zhang XZ, Sathitsuksanoh N, Lynd LR, Zhang YH. Enhanced microbial utilization of recalcitrant cellulose by an ex vivo cellulosome-microbe complex. *Appl Environ Microbiol* 2012;**78**:1437–44.

20. Desvaux M. The cellulosome of *Clostridium cellulolyticum*. *Enzyme Microb Technol* 2005;**37**:373–85.

21. Xu CG, Qin Y, Li YD, Ji YT, Huang JZ, Song HH, et al. Factors influencing cellulosome activity in consolidated bioprocessing of cellulosic ethanol. *Bioresour Technol* 2010;**101**:9560–9.

22. Rydzak T, Levin DB, Cicek N, Sparling R. End-product induced metabolic shifts in Clostridium *thermocellum* ATCC 27405. *Appl Microbiol Biotechnol* 2011;**92**:199–209.

23. Raman B, McKeown CK, Rodriguez Jr M, Brown SD, Mielenz JR. Transcriptomic analysis of *Clostridium thermocellum* ATCC 27405 cellulose fermentation. *BMC Microbiol* 2011;**11**:134.

24. Riederer A, Takasuka TE, Makino S, Stevenson DM, Bukhman YV, Elsen NL, et al. Global gene expression patterns in *Clostridium thermocellum* as determined by microarray analysis of chemostat cultures on cellulose or cellobiose. *Appl Environ Microbiol* 2011;**77**:1243–53.

25. Burton E, Martin VJJ. Proteomic analysis of *Clostridium thermocellum* ATCC 27405 reveals the upregulation of an alternative transhydrogenase-malate pathway and nitrogen assimilation in cells grown on cellulose. *Can J Microbiol* 2012;**58**:1378–88.

26. Deng Y, Olson DG, Zhou J, Herring CD, Joe Shaw A, Lynd LR. Redirecting carbon flux through exogenous pyruvate kinase to achieve high ethanol yields in *Clostridium thermocellum*. *Metab Eng* 2013;**15**:151–8.

27. Tyurin MV, Desai SG, Lynd LR. Electrotransformation of *Clostridium thermocellum*. *Appl Environ Microb* 2004;**70**:883–90.

28. Guss AM, Olson DG, Caiazza NC, Lynd LR. Dcm methylation is detrimental to plasmid transformation in *Clostridium thermocellum*. *Biotechnol Biofuels* 2012;**5**:30.

29. Tripathi SA, Olson DG, Argyros DA, Miller BB, Barrett TF, Murphy DM, et al. Development of pyrF-based genetic system for targeted gene deletion in *Clostridium thermocellum* and creation of a pta mutant. *Appl Environ Microbiol* 2010;**76**:6591–9.

30. Olson DG, Lynd LR. Transformation of *Clostridium thermocellum* by electroporation. *Methods Enzymol* 2012;**510**:317–30.

31. Argyros DA, Tripathi SA, Barrett TF, Rogers SR, Feinberg LF, Olson DG, et al. High ethanol titers from cellulose by using metabolically engineered thermophilic, anaerobic microbes. *Appl Environ Microbiol* 2011;**77**:8288–94.

32. Jennert KCB, Tardif C, Young DI, Young M. Gene transfer to *Clostridium cellulolyticum* ATCC 35319. *Microbiology-UK* 2000;**146**:3071–80.

33. Li YC, Tschaplinski TJ, Engle NL, Hamilton CY, Rodriguez M, Liao JC, et al. Combined inactivation of the *Clostridium cellulolyticum* lactate and malate dehydrogenase genes substantially increases ethanol yield from cellulose and switchgrass fermentations. *Biotechnol Biofuels* 2012;**5**:2.

34. Guedon E, Desvaux M, Payot S, Petitdemange H. Growth inhibition of *Clostridium cellulolyticum* by an inefficiently regulated carbon flow. *Microbiology-UK* 1999;**145**:1831–8.

35. Guedon E, Desvaux M, Petitdemange H. Improvement of cellulolytic properties of *Clostridium cellulolyticum* by metabolic engineering. *Appl Environ Microbiol* 2002;**68**:53–8.

36. Tamaru Y, Miyake H, Kuroda K, Nakanishi A, Matsushima C, Doi RH, et al. Comparison of the mesophilic cellulosome-producing *Clostridium cellulovorans* genome with other cellulosome-related clostridial genomes. *Microb Biotechnol* 2011;**4**:64–73.

37. Tolonen AC, Haas W, Chilaka AC, Aach J, Gygi SP, Church GM. Proteome-wide systems analysis of a cellulosic biofuel-producing microbe. *Mol Syst Biol* 2011;**7**:461.

38. Jin MJ, Balan V, Gunawan C, Dale BE. Consolidated bioprocessing (CBP) Performance of *Clostridium phytofermentans* on AFEX-treated corn Stover for ethanol production. *Biotechnol Bioeng* 2011;**108**:1290–7.

39. Zhang XZ, Zhang ZM, Zhu ZG, Sathitsuksanoh N, Yang YF, Zhang YHP. The noncellulosomal family 48 cellobiohydrolase from *Clostridium phytofermentans* ISDg: heterologous expression, characterization, and processivity. *Appl Microbiol Biotechnol* 2010;**86**:525–33.

40. Nolling J, Breton G, Omelchenko MV, Makarova KS, Zeng QD, Gibson R, et al. Genome sequence and comparative analysis of the solvent-producing bacterium *Clostridium acetobutylicum. J Bacteriol* 2001;**183**:4823–38.

41. Mingardon F, Chanal A, Tardif C, Fierobe HP. The issue of secretion in heterologous expression of Clostridium cellulolyticum cellulase-encoding genes in *Clostridium acetobutylicum* ATCC 824. *Appl Environ Microbiol* 2011;**77**:2831–8.

42. Yang SJ, Kataeva I, Hamilton-Brehm SD, Engle NL, Tschaplinski TJ, Doeppke C, et al. Efficient degradation of lignocellulosic plant biomass, without pretreatment, by the thermophilic anaerobe *"Anaerocellum thermophilum"* DSM 6725. *Appl Environ Microbiol* 2009;**75**:4762–9.

43. Kanafusa-Shinkai S, Wakayama J, Tsukamoto K, Hayashi N, Miyazaki Y, Ohmori H, et al. Degradation of microcrystalline cellulose and non-pretreated plant biomass by a cell-free extracellular cellulase/hemicellulase system from the extreme thermophilic bacterium *Caldicellulosiruptor bescii. J Biosci Bioeng* 2013;**115**:64–70.

44. Chung D, Farkas J, Huddleston JR, Olivar E, Westpheling J. Methylation by a unique alpha-class N4-cytosine methyltransferase is required for DNA transformation of c*aldicellulosiruptor bescii* DSM6725. *PloS One* 2012;**7**. http://dx.doi.org/10.1371/journal.pone.0043844.

45. Jang YS, Lee J, Malaviya A, Seung DY, Cho JH, Lee SY. Butanol production from renewable biomass: rediscovery of metabolic pathways and metabolic engineering. *Biotechnol J* 2012;**7**:186–98.

46. Zheng YN, Li LZ, Xian M, Ma YJ, Yang JM, Xu X, et al. Problems with the microbial production of butanol. *J Ind Microbiol Biotechnol* 2009;**36**:1127–38.

47. Inui M, Suda M, Kimura S, Yasuda K, Suzuki H, Toda H, et al. Expression of *Clostridium acetobutylicum* butanol synthetic genes in *Escherichia coli. Appl Microbiol Biotechnol* 2008;**77**:1305–16.

48. Lee JY, Jang YS, Lee J, Papoutsakis ET, Lee SY. Metabolic engineering of *Clostridium acetobutylicum* M5 for highly selective butanol production. *Biotechnol J* 2009;**4**:1432–40.

49. Harris LM, Desai RP, Welker NE, Papoutsakis ET. Characterization of recombinant strains of the *Clostridium acetobutylicum* butyrate kinase inactivation mutant: need for new phenomenological models for solventogenesis and butanol inhibition? *Biotechnol Bioeng* 2000;**67**:1–11.

50. Nair RV, Green EM, Watson DE, Bennett GN, Papoutsakis ET. Regulation of the sol locus genes for butanol and acetone formation in *Clostridium acetobutylicum* ATCC 824 by a putative transcriptional repressor. *J Bacteriol* 1999;**181**:319–30.

51. Chen CK, Blaschek HP. Acetate enhances solvent production and prevents degeneration in Clostridium beijerinckii BA101. *Appl Microbiol Biotechnol* 1999;**52**:170–3.

52. Tashiro Y, Takeda K, Kobayashi G, Sonomoto K, Ishizaki A, Yoshino S. High butanol production by *Clostridium saccharoperbutylacetonicum* N1-4 in fed-batch culture with pH-stat continuous butyric acid and glucose feeding method. *J Biosci Bioeng* 2004;**98**:263–8.

53. Jiang Y, Xu C, Dong F, Yang Y, Jiang W, Yang S. Disruption of the acetoacetate decarboxylase gene in solvent-producing *Clostridium acetobutylicum* increases the butanol ratio. *Metab Eng* 2009;**11**:284–91.

54. Atsumi S, Cann AF, Connor MR, Shen CR, Smith KM, Brynildsen MP, et al. Metabolic engineering of *Escherichia coli* for 1-butanol production. *Metab Eng* 2008;**10**:305–11.

55. Bond-Watts BB, Bellerose RJ, Chang MCY. Enzyme mechanism as a kinetic control element for designing synthetic biofuel pathways. *Nat Chem Biol* 2011;**7**:222–7.

56. Shen CR, Liao JC. Metabolic engineering of Escherichia coli for 1-butanol and 1-propanol production via the keto-acid pathways. *Metab Eng* 2008;**10**:312–20.

57. Steen EJ, Chan R, Prasad N, Myers S, Petzold CJ, Redding A, et al. Metabolic engineering of *Saccharomyces cerevisiae* for the production of n-butanol. *Microb Cell Fact* 2008;**7**:36.

58. Nielsen DR, Leonard E, Yoon SH, Tseng HC, Yuan C, Prather KLJ. Engineering alternative butanol production platforms in heterologous bacteria. *Metab Eng* 2009;**11**:262–73.

59. Berezina OV, Zakharova NV, Brandt A, Yarotsky SV, Schwarz WH, Zverlov VV. Reconstructing the clostridial n-butanol metabolic pathway in *Lactobacillus brevis. Appl Microbiol Biotechnol* 2010;**87**:635–46.

60. Lan EI, Liao JC. Metabolic engineering of cyanobacteria for 1-butanol production from carbon dioxide. *Metab Eng* 2011;**13**:353–63.

61. Lan EI, Liao JC. ATP drives direct photosynthetic production of 1-butanol in cyanobacteria. *Proc Natl Acad Sci U S A* 2012;**109**:6018–23.

62. Baez A, Cho KM, Liao JC. High-flux isobutanol production using engineered *Escherichia coli*: a bioreactor study with in situ product removal. *Appl Microbiol Biotechnol* 2011;**90**:1681–90.

63. Smith KM, Liao JC. An evolutionary strategy for isobutanol production strain development in *Escherichia coli*. *Metab Eng* 2011;**13**:674–81.

64. Smith KM, Cho KM, Liao JC. Engineering *Corynebacterium glutamicum* for isobutanol production. *Appl Microbiol Biotechnol* 2010;**87**:1045–55.

65. Li SS, Huang D, Li Y, Wen JP, Jia XQ. Rational improvement of the engineered isobutanol-producing *Bacillus subtilis* by elementary mode analysis. *Microb Cell Fact* 2012;**11**:101.

66. Li SS, Wen JP, Jia XQ. Engineering *Bacillus subtilis* for isobutanol production by heterologous Ehrlich pathway construction and the biosynthetic 2-ketoisovalerate precursor pathway overexpression. *Appl Microbiol Biotechnol* 2011;**91**:577–89.

67. Chen X, Nielsen KF, Borodina I, Kielland-Brandt MC, Karhumaa K. Increased isobutanol production in *Saccharomyces cerevisiae* by overexpression of genes in valine metabolism. *Biotechnol Biofuels* 2011;**4**:21.

68. Kondo T, Tezuka H, Ishii J, Matsuda F, Ogino C, Kondo A. Genetic engineering to enhance the Ehrlich pathway and alter carbon flux for increased isobutanol production from glucose by *Saccharomyces cerevisiae*. *J Biotechnol* 2012;**159**:32–7.

69. Lee WH, Seo SO, Bae YH, Nan H, Jin YS, Seo JH. Isobutanol production in engineered *Saccharomyces cerevisiae* by overexpression of 2-ketoisovalerate decarboxylase and valine biosynthetic enzymes. *Bioproc. Biosyst Eng* 2012;**35**:1467–75.

70. Matsuda F, Kondo T, Ida K, Tezuka H, Ishii J, Kondo A. Construction of an artificial pathway for isobutanol biosynthesis in the cytosol of *Saccharomyces cerevisiae*. *Biosci Biotechnol Biochem* 2012;**76**:2139–41.

71. Atsumi S, Higashide W, Liao JC. Direct photosynthetic recycling of carbon dioxide to isobutyraldehyde. *Nat Biotechnol* 2009;**27**:1177–U1142.

72. Li H, Opgenorth PH, Wernick DG, Rogers S, Wu TY, Higashide W, et al. Integrated electromicrobial conversion of CO_2 to higher alcohols. *Science* 2012;**335**:1596.

73. Huo YX, Cho KM, Rivera JGL, Monte E, Shen CR, Yan YJ, et al. Conversion of proteins into biofuels by engineering nitrogen flux. *Nat Biotechnol* 2011;**29**:346– U160.

74. Bastian S, Liu X, Meyerowitz JT, Snow CD, Chen MMY, Arnold FH. Engineered ketol-acid reductoisomerase and alcohol dehydrogenase enable anaerobic 2-methylpropan-1-ol production at theoretical yield in *Escherichia coli*. *Metab Eng* 2011;**13**:345–52.

75. Heider J, Mai XH, Adams MWW. Characterization of 2-ketoisovalerate ferredoxin oxidoreductase, a new and reversible coenzyme A-dependent enzyme involved in peptide fermentation by hyperthermophilic archaea. *J Bacteriol* 1996;**178**:780–7.

76. Tersteegen A, Linder D, Thauer RK, Hedderich R. Structures and functions of four anabolic 2-oxoacid oxidoreductases in *Methanobacterium thermoautotrophicum*. *Eur J Biochem* 1997;**244**:862–8.

77. Ozawa Y, Nakamura T, Kamata N, Yasujima D, Urushiyama A, Yamakura F, et al. Thermococcus profundus 2-ketoisovalerate ferredoxin oxidoreductase, a key enzyme in the archaeal energy-producing amino acid metabolic pathway. *J Biochem* 2005;**137**:101–7.

78. Lynd LR, van Zyl WH, McBride JE, Laser M. Consolidated bioprocessing of cellulosic biomass: an update. *Curr Opin Biotechnol* 2005;**16**:577–83.

79. Sonnleitner B, Fiechter A. Advantages of using thermophiles in biotechnological processes: expectations and reality. *Trends Biotechnol* 1983;**1**:74–80.

80. Pages S, Belaich A, Belaich JP, Morag E, Lamed R, Shoham Y, et al. Species-specificity of the cohesin-dockerin interaction between *Clostridium thermocellum* and *Clostridium cellulolyticum*: prediction of specificity determinants of the dockerin domain. *Proteins* 1997;**29**:517–27.

81. Perret S, Casalot L, Fierobe HP, Tardif C, Sabathe F, Belaich JP, et al. Production of heterologous and chimeric scaffoldins by *Clostridium acetobutylicum* ATCC 824. *J Bacteriol* 2004;**186**:253–7.

82. Mingardon F, Perret S, Belaich A, Tardif C, Belaich JP, Fierobe HP. Heterologous production, assembly, and secretion of a minicellulosome by *Clostridium acetobutylicum* ATCC 824. *Appl Environ Microbiol* 2005;**71**:1215–22.

CHAPTER 15

Influence of Particle Size on Direct Microbial Conversion of Hot Water-Pretreated Poplar by Clostridium thermocellum

John M. Yarbrough[1], Ashutosh Mittal[1], Yannick J. Bomble[1], Jessica Olstad[2], Edward J. Wolfrum[2], Sarah E. Hobdey[1], Michael E. Himmel[1], Todd B. Vinzant[1]

[1]*Biosciences Center, National Renewable Energy Laboratory (NREL), Golden, CO, USA;*
[2]*National Bioenergy Center, National Renewable Energy Laboratory (NREL), Golden, CO, USA*

Introduction

Particle size has been shown to play a significant role in attaining efficient enzymatic hydrolysis of biomass with free cellulase systems from commonly encountered fungal and bacterial enzymes.[1–4] An increase in enzymatic hydrolysis in response to biomass particle size has been interpreted as being directly related to the increase in the surface area-to-volume ratio, thereby improving enzyme accessibility. This concept has led to several pretreatment processes that incorporate more extensive particle size reduction.[1–3] However, the effective reduction and optimization of particle size is known to represent a significant process cost,[5] which suggests that secondary milling after pretreatment should be considered more broadly in biomass conversion schemes. It has also been suggested that particle size reduction of biomass has the same influence on microbial conversion as it does on enzymatic conversion.[6–8] For example, in the work reported by Shao et al., AFEX pretreated corn stover (PCS) that was knife milled and passed through a 500-μM sieve demonstrated an increase in cellulose and xylan conversion by *Clostridium thermocellum* compared with larger particles.[8] Unfortunately, this increase was attributed to particle size reduction and removal of hemicellulose by enzymatic hydrolysis using Multifect Xylanase and Multifect Pectinase (Novozymes) before the fermentation by *C. thermocellum*. Jin and coworkers performed similar studies on AFEX PCS for ethanol production using *Clostridium phytofermantans* and showed a correlation between particle size reduction and an increase in glucan and xylan hydrolysis.[6] The main difference between these two microorganisms is that *C. phytofermantans* has the native enzymatic activities to degrade and utilize cellulose and hemicellulose whereas *C. thermocellum* can degrade hemicellulose but lacks the metabolic machinery to utilize it. However, both studies conclude that particle size reduction positively affects direct microbial conversion of biomass.

Direct Microbial Conversion of Biomass to Advanced Biofuels. http://dx.doi.org/10.1016/B978-0-444-59592-8.00015-4

Our work revisits the effects of particle size on the microbial conversion of corn stover utilizing more cost-effective pretreatment methods that reduce the hemicellulose content of feedstocks. The most commonly used pretreatment techniques involve the use of mineral acids such as dilute sulfuric acid that is used in a range of concentrations (0.20–2% w/w) at temperatures from 100 to 200 °C.[9–13] Today, steam explosion is often combined with sulfuric acid[14] or SO_2[15] to enhance the subsequent enzymatic hydrolysis of the solid residue. We note that the application of chemicals for pretreatment causes corrosion problems in process hardware and requires expensive chemical recovery steps; thus, such processes are expensive and not environmentally friendly.[16]

Thus, more benign pretreatment technologies are sought for the fractionation and utilization of the individual components of biomass while maintaining high yields with minimal losses. In this context, hot water extraction is considered to be economically attractive because it avoids the use of expensive mineral acids to catalyze the hydrolysis reaction and is environmentally friendly.[16–19] In a recent study, Chen and coworkers have shown yields greater than 84% and 77% for glucose and xylose, respectively, during enzymatic hydrolysis of PCS using a novel pretreatment process utilizing a low-temperature, dilute alkaline deacetylation followed by disc refining under modest levels of energy consumption.[20] One of the important features of this pretreatment process is that it resulted in a significant reduction in biomass particle size and greatly enhanced particle surface area, a factor believed to be the most critical affecting the enzymatic digestibility of biomass.

Materials and Methods
Microorganism and Fermentations

Clostridium thermocellum was obtained from ATCC 27,405 and grown on ATCC 1193 growth medium consisting of 1.5 g/L KH_2PO_4, 4.2 g/L Na_2HPO_4, 0.5 g/L NH_4CL, 0.18 g/L $MgCl_2·6H_2O$, 0.5 mL/L of vitamin solution, 1.0 mL/L Resazurin (0.1%), 5.0 mL/L Wolf's modified mineral elixir, and 40.0 mL/L reducing solution (200 mL of 0.2 N NaOH, 2.5 g $Na_2S·9H_2O$ and 2.5 g L-cysteine·HCL). This media was titrated to pH 7.00 with a 1% loading of carbohydrate source ranging among cellobiose, Avicel PH101, and hot water-pretreated poplar. Bacterial cells were grown under anaerobic conditions in 100-mL serum fermentation bottles at a temperature of 60 °C with an orbital rotation of 125 rpm.

Substrate

Avicel PH101 (particle size ~50 μm) and poplar provided by the BioEnergy Science Center were used in this work. The BESC poplar was dry milled down to 6-mm particle size before being hot water pretreated at 180 °C for 40 min in a ZipperClave with particle distribution between 3 and 6 mm (Figure 1).

After pretreatment, the biomass was wet sieved into four particle fractions consisting of particle sizes ranging between 63 and 250 μm, 250 and 500 μm, and 500 and 1000 μm as well

Figure 1
Images showing the initial size of the starting material in comparison to the milled material before hot water pretreatment.

Table 1: Particle size distribution for each fraction with its corresponding dry weight and percentage of the fraction compared with the total weight

Mesh Range	Opening Range (μm)	Weight (g)	Estimated Dry Weight (g)	Yield (%)
+18 from large chunks	+1000	294.7	44.2	28.7
+18 from 6 mm milled	+1000	482.1	87.7	56.9
18–35	500–1000	38.5	11.5	7.5
35–60	250–500	25.6	7.7	5.0
60–230	63–250	10.2	3.1	2.0

as larger than 1 mm (restricted to 6 mm by the inner diameter of the serum bottles) (Figure 1(b)). Moisture analysis was performed for each of the particle size distribution fractions, and the total weight of biomass is shown in Table 1.

Compositional Analysis

The compositions of the native and pretreated feedstocks, as well as the particle size fractions, measured according to National Renewable Energy Laboratory Laps,[21] are shown in Table 2. All values are given as a percentage of the dry wood weight.

Digestion Assay and Analysis

Digestions were performed on never-dried poplar wood in duplicate in 100-mL serum fermentation bottles at a temperature of 60 °C with orbital rotation of 125 rpm for 96 h. The entire serum bottle was sacrificed for each time point. *Clostridium thermocellum* was initially grown on

Table 2: Carbon and nitrogen content (percent) for the initial poplar biomass and for
Clostridium thermocellum with the C:N ratio

Sample	Carbon	Nitrogen	C:N
Poplar	49.04	0.20	245:1
C. thermocellum	44.60	8.55	5.22:1

cellobiose and transferred to the serum bottles during log phase. To investigate the effects of *C. thermocellum* growth on the maximum conversion of glucan, gravimetric and compositional were utilized. For gravimetric measurements, the wet biomass was washed and then transferred to preweighted 50-mL Falcon tubes and then freeze dried. After this process, the Falcon tubes were weighed to obtain the mass of the substrate and microbial mass combined. For sugar analysis, samples from the compositional analyses were filtered through a 0.2-μm filter and then refrigerated until subjected to glucose analysis. Glucose yield was measured by the Carblow method, which uses high-performance liquid chromatography with an Aminex 87H column maintained at 65 °C. The mobile phase used was 0.2 μm filtered 0.01 N sulfuric acid solutions at a flow rate of 0.6 mL/min. The sample injection volume was 10 μL and the run time was 26 min. Finally, compositional analysis was performed as previously stated.

X-Ray Diffraction Measurements

X-ray diffraction (XRD) was performed to evaluate the crystalline structure of untreated and hot water-pretreated samples by using a Rigaku (Tokyo, Japan) Ultima IV Diffractometer with CuKα radiation having a wavelength $\lambda(K\alpha 1) = 0.15406$ nm generated at 40 kV and 44 mA. The diffraction intensities of air-dried samples placed on a quartz substrate were measured in the range of 8–42° 2θ using a step size of 0.02° at a rate of 2°/min. The crystallinity indexes (*CrI*) of the cellulose samples were calculated according to the method described by Segal et al.[22] using the formula

$$CrI = \left(\frac{I_{002} - I_{am}}{I_{002}} \right) \times 100$$

(1)

where I_{002} and I_{am} are the maximum and minimum intensity of diffraction at approximately $2\theta = 22.4$–$22.5°$ and $2\theta = 18.0$–$19.0°$, respectively. It should be noted that the values obtained here are relative, not absolute (i.e., other methods have been shown to give slightly different CrI values when compared with XRD).

Growth Studies

Clostridium thermocellum was grown on three different carbon sources (cellobiose, Avicel, and hot water-pretreated poplar) and the microbial growth was measured with a standard curve of OD_{600} versus gravimetric weight.

Nitrogen Analysis for Carbon:Nitrogen Ratio

The biomass substrates, including microbial mass, were washed to remove the media components followed by freeze-drying, and then they were weighed to yield the initial total mass. An aliquot of the dried mixture of substrate and microbial mass, m_{tot}, was used for carbon:nitrogen analysis. After this determination, the carbon and nitrogen mass fractions of the mixture were measured and were given the following values: x_C for the carbon mass fractions and x_N for the nitrogen mass fractions. From these data, the total mass of the carbon and nitrogen from the sample are calculated using Eqns (2) and (3):

$$C_{tot} = x_C \times m_{tot} \tag{2}$$

$$N_{tot} = x_N \times m_{tot} \tag{3}$$

The nitrogen in the mixture is assigned to the microorganism, $N_{bug} = N_{tot}$, and the amount of carbon in the organism is calculated from the amount of nitrogen found in the microorganism and the C:N ratio of the organism defined by Eqn (4):

$$C_{bug} = N_{tot} \times C:N_{bug} \tag{4}$$

Finally, the balance of the carbon is assigned to the substrate by Eqn (5):

$$C_{sub} = C_{tot} - C_{bug} \tag{5}$$

Thus, this allows for the loss of substrate to be determined by comparing the current substrate carbon content to the initial substrate carbon content. Table 2 shows the amount of carbon and nitrogen for the initial poplar sample and *C. thermocellum* along with the initial C:N ratios for both samples.

Results

Optimizing Growth Media for C. thermocellum *Growth on Cellobiose, Avicel, and Poplar*

To consistently grow *C. thermocellum* from a starter culture, growth studies were performed to optimize the media with the intention of transferring a seed culture to fresh media every 24 h. From the initial growth studies conducted with a cellobiose concentration of 0.5%, it was determined that *C. thermocellum* was entering the lag phase within 14 h after inoculation (Figure 2(a)). This situation resulted in a long lag phase before the cells entered into log phase, thus impeding our ability to maintain a healthy cell culture and to transfer cells in log phase.

Cellobiose concentration and pH were adjusted to determine their effects on extending the log phase past 24 h. Figure 2(a) shows the cells reaching an OD_{600} of 1.7 within 15 h and entering stationary phase. The short log phase is troublesome when trying to maintain an active culture and performing transfers every 24 h. From this data set, it was decided to monitor the effects of pH and concentration for different carbon sources.

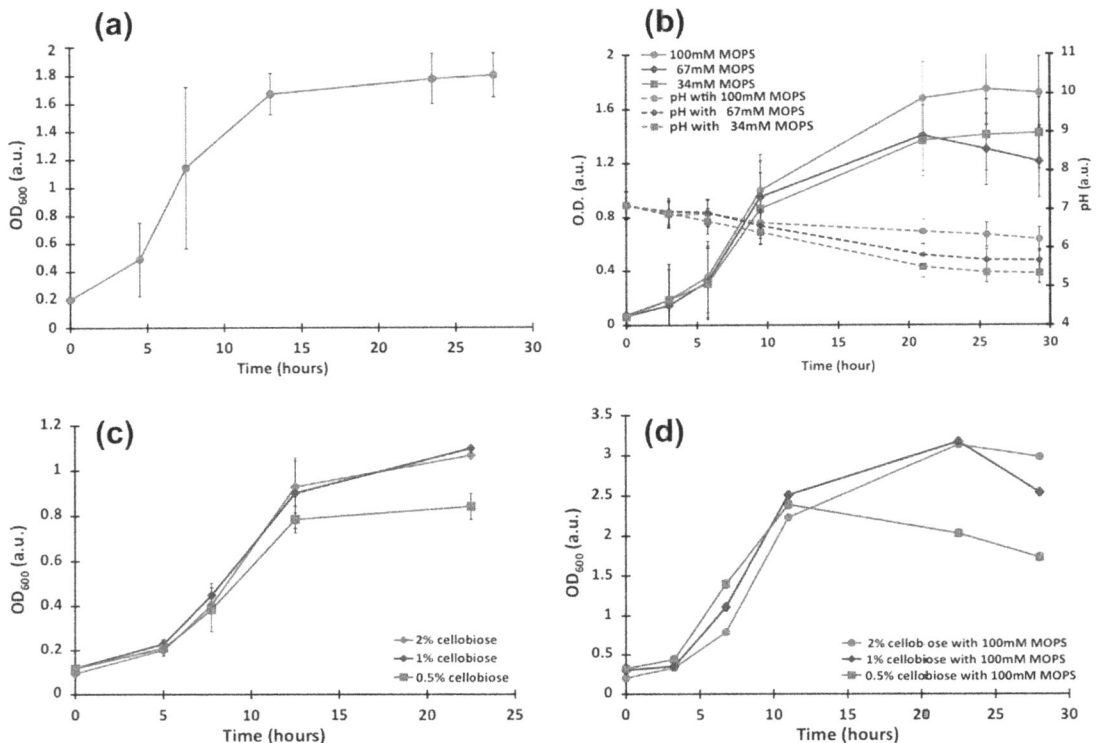

Figure 2

(a) Optimization study with the original *Clostridium thermocellum* growth reaching stationary phase within 15 h, (b) the effects of increasing buffering capacity on the growth, (c) the effects of different carbon source loadings, and (d) optimization of substrate loading with the optimal buffering capacity to obtain the ability to transfer at 24 h without entering stationary phase.

It was found that the buffering capacity of ATCC 1193 growth media was limited; therefore, MOPS buffer was chosen as the secondary buffering system for maintaining the pH close to neutrality, which is necessary for the ideal *C. thermocellum* growth. Three different concentrations of MOPS were used—100, 67, and 34 mM—with a cellobiose concentration of 0.5%; the respective growth curves are shown in Figure 2(b). It was found that 100 mM MOPS had the best buffering capacity because it showed just a 13% decrease in the pH during culturing as opposed to a 27% drop in pH without MOPS, as seen in Figure 2(b). More importantly, *C thermocellum* growth on MOPS increased by almost 50%, thereby doubling the amount of microbial mass produced in comparison to the initial growth in Figure 2(a).

A substrate concentration dependence study was performed to identify the optimal concentration of the carbon source (cellobiose) needed to achieve maximum growth of *C. thermocellum*. Three different concentrations were chosen: 0.5%, 1.0%, and 2.0% (Figure 2(c)). As this figure shows, an increase in the cellobiose concentration from 0.5% to 1.0% resulted in a 27%

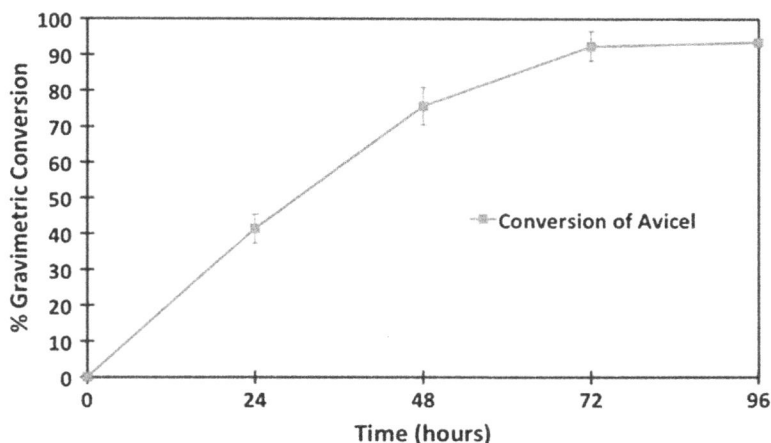

Figure 3
Reproducibility of *Clostridium thermocellum* growth on Avicel.

increase in the microbial OD_{600} whereas no increase in the microbial OD_{600} was observed with increasing the cellobiose concentration above 1.0%.

Once the growth conditions were optimized for both substrate loading and buffering conditions (Figure 2(d)), *C. thermocellum* was grown in triplicate on Avicel to verify that the media had been optimized for soluble and insoluble substrates. The reproducibility of *C. thermocellum* growth on Avicel is shown in Figure 3, which shows that over 90% of total gravimetric conversion of Avicel was achieved within 72 h with tight error bars. The question arises: How much of an influence does microbial mass play in the total mass of the remaining substrate seeing that there was only 90% conversion of Avicel within 72 h and not 100%?

To address this issue, the carbon and nitrogen content was measured from the Avicel digestion and the C:N ratios were calculated and subtracted from the initial dry weights of the Avicel or microbial mass. Figure 4 shows the mass of Avicel versus time (blue dashed line), the Avicel mass with the microbial mass subtracted (blue solid line), and the actual microbial mass (solid black line). Here, it can be seen that by 48 h, the microbial mass (as measured by C:N ratio) accounts for 23% of the mass of Avicel as measured by gravimetric analysis.

Particle Size Comparison (Poplar vs Avicel)

As previously stated, *C. thermocellum* does not have the ability to utilize the xylose fractions produced during the digestion of plant cell walls, which could result in potential xylose inhibition. Therefore, for the efficient digestion of the biomass utilizing *C. thermocellum* alone, pretreatment of the biomass is needed to solubilize and remove the majority of the hemicellulose fraction. The composition data shown in Table 3 clearly show that hot water pretreatment of poplar was able to reduce the amount of hemicellulose by almost 75%.

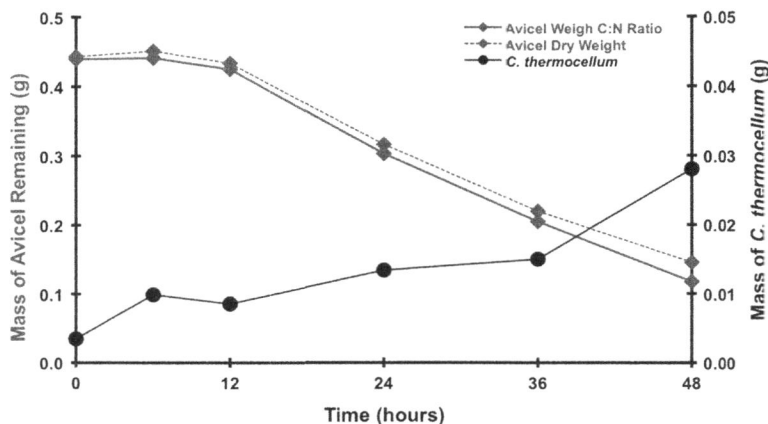

Figure 4
Residual Avicel mass plotted with (dotted blue line (dotted gray line in print versions)) and without (solid blue line (gray line in print versions)) correction for the microbial mass calculated via C:N ratio and microbial mass (solid black line).

Table 3: Compositional analysis of the native and pretreated poplar in addition to the compositional analysis of the four fractions that have statistically the same composition

Substrate	Lignin	Glucan	Xylan	Galactan	Arabinan
Native poplar	26.3	44.5	14.0	1.09	0.14
Pretreated poplar	30.8	60.4	5.9	1.25	0.28
Fraction 1 (63–250 μm)	30.4	57.5	7.9	0.73	0.36
Fraction 2 (250–500 μm)	29.2	60.0	7.2	0.68	0.19
Fraction 3 (500–1000 μm)	28.7	60.9	6.5	0.68	0.04
Fraction 4 (1–6 mm)	26.7	61.0	6.9	0.70	0.12
Standard deviation between fractions	1.53	1.63	0.59	0.02	0.14

To determine the influence of particle size on microbial conversion of biomass, *C. thermocellum* was grown on four different particle size fractions, ranging between 62 and 250 μm (Fraction 1), 250 and 500 μm (Fraction 2), 500 and 1000 μm (Fraction 3), and 1–6 mm (Fraction 4). As shown in Table 3, the compositions of all four fractions were statistically the same. During microbial conversion, the samples were taken at different time points by sacrificing the entire serum bottle. The dry weight of the remaining biomass was then measured by gravimetrical analysis. This practice eliminated any potential sampling errors related to using

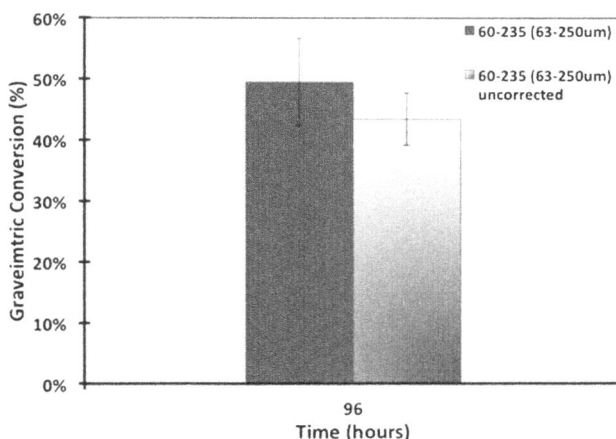

Figure 5

Comparison between the gravimetric weight of Fraction 1 with microbial mass (solid blue bar (gray bar in print versions)) and without microbial mass (light blue bar (light-gray bar in print versions)).

a syringe for collecting individual time points. Compositional analyses and C:N ratios were determined for the washed, dried biomass samples.

C:N ratios were measured and the microbial mass was calculated for each of the time points. The microbial mass accounted for anywhere between 3% and 11% of the total mass. Figure 5 shows the difference between the gravimetric conversion with and without correction for the microbial mass for Fraction 1. We found that microbial mass does, in fact, play a small role in the gravimetric measurements for actual biomass, but it can be accounted for by measuring the C:N ratio. This technique is also less cumbersome and time-consuming compared with classical compositional analysis.

The gravimetric conversion at 24, 72, and 96 h are significantly different between Fraction 1 and the other three fractions (specifically, a break point between Fraction 1 and Fraction 2 (Figure 6)). Fraction 1, with particle sizes less than 250 µm, showed the highest microbial conversion tested, with smaller and slower microbial conversions obtained using Fractions 2, 3, and 4. Results from the gravimetrical analysis (Figure 6) show that after 96 h, Fraction 1 had a total conversion of 50%. However, Fractions 2, 3, and 4 showed total conversions of 34%, 27%, and 27%, respectively. Compositional analysis performed (data not shown) on the biomass residue also showed slower glucan conversion with increasing particle size.

These findings demonstrate that the effect of substrate particle size on overall microbial conversion by *C. thermocellum* is substantial—an observation that is strengthened by the fact that all four fractions were statistically identical in composition. These findings are in agreement with the effects that particle size reduction has on free enzyme substrate conversion.[1–3] However, there are two caveats that are related to the most commonly used processes that

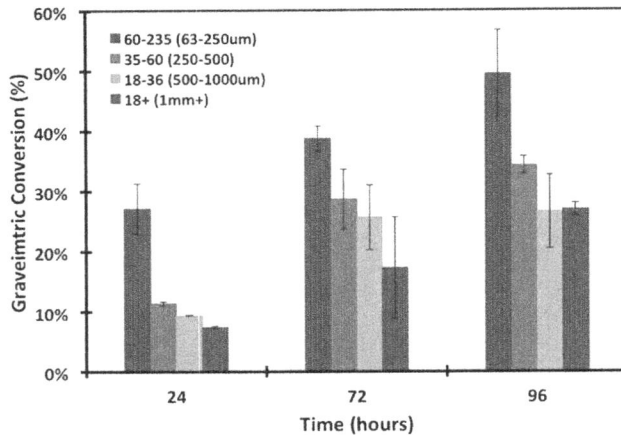

Figure 6

Relationships between substrate particle size and microbial conversion using gravimetric measurements. Fraction 1 had a total conversion of 50%. Fractions 2, 3, and 4 showed total conversions of 34%, 27%, and 27%, respectively.

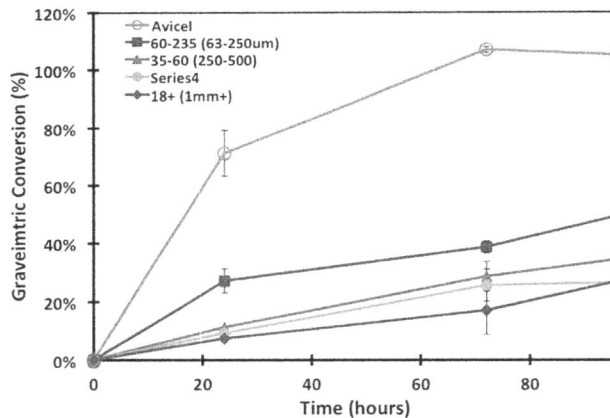

Figure 7

Microbial conversion of Avicel and the four biomass fractions.

have to be taken into consideration. First, only a small amount of Fraction 1 material could be generated after the pretreatment and sieving. Fraction 1 accounted for only 2% of the total weight of the material, as shown in Table 1. Fractions 2 and 3 had similar low yields (5% and 7.5%, respectively) when compared with Fraction 4, which accounts for 57% of the total weight. This finding shows that less than 10% of the total mass, when fractionated, allows for the fastest microbial conversion. Second, at the highest microbial conversion reached 50% (Fraction 1), *C. thermocellum* lacks the ability to completely utilize biomass as an efficient carbon source. In contrast, Avicel, which is a product of pure cellulose derived from wood

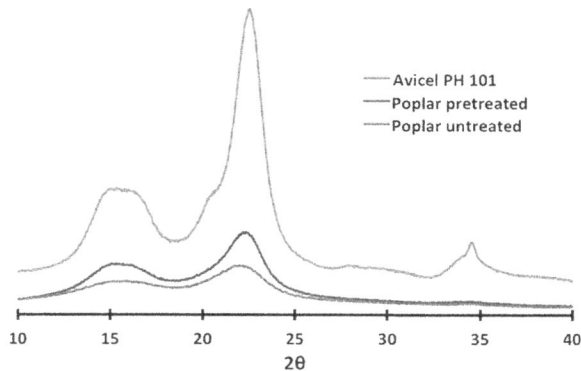

Figure 8
XRD diffractograms of Avicel (76.4% crystalline), hot water-pretreated poplar (59.7% crystalline),
and native poplar (46.1% crystalline).

pulp, was 100% digested by *C. thermocellum* within 72 h (Figure 7). Note that Avicel has a particle size of 50 μm, which is smaller than the average particle size of Fraction 1 (156 μm), and it has negligible amounts of lignin and hemicellulose but has a higher crystallinity than native and pretreated poplar as demonstrated in Figure 8.

One would expect higher conversion with lower crystallinity, but the significant differences between the microbial conversion on Avicel vs. pretreated poplar is due to the heterogeneous nature of pretreated poplar which includes mixed glucans, hemicellulose and lignin. Therefore, the overall composition and structure of the biomass may have more of an influence on microbial conversion for growing *C. thermocellum* cells than does the size of the biomass particles.

Conclusion

This work shows that the substrate particle size influences the overall biomass conversion by *C. thermocellum*, with particle sizes between 63 and 250 μm allowing the greatest conversion (50%) followed by the larger particle sizes. Unfortunately, the complex nature of biomass (i.e., composition, structure, porosity, etc.) probably plays a more dominant role in overall microbial conversion. This conclusion is supported by our observations that all four fractions were statistically the same in composition and that Avicel, which has a significantly higher crystallinity compared with cellulose in pretreated biomass, was digested completely within 48 h by *C. thermocellum*. It has been suggested that *C. thermocellum* is deficient in its ability to utilize cell wall hemicellulose, a result that is consistent with its known natural inability to metabolize C5 sugars. Thus, it is important to pretreat biomass in a way that sufficiently removes hemicellulose and yet promotes microbial growth. We conclude that it would be advantageous to genetically modify *C. thermocellum* to permit it to utilize all biomass sugars, leading to higher conversion yields. Further work is also needed to understand its hydrolytic enzyme complexes to enhance microbial conversion of biomass with *C. thermocellum*.

Acknowledgments

The BioEnergy Science Center is a U.S. Department of Energy (DOE) Bioenergy Research Center supported by the Office of Biological and Environmental Research in the DOE Office of Science. The authors also thank Rob Sykes for supplying the poplar material used in this experiment.

References

1. Chundawat SP, Venkatesh B, Dale BE. Effect of particle size based separation of milled corn stover on AFEX pretreatment and enzymatic digestibility. *Biotechnol Bioeng* 2007;**96**:219–31.
2. Mansfield SD, Mooney C, Saddler JN. Substrate and enzyme characteristics that limit cellulose hydrolysis. *Biotechnol Progr* 1999;**15**:804–16.
3. Pedersen M, Meyer AS. Influence of substrate particle size and wet oxidation on physical surface structures and enzymatic hydrolysis of wheat straw. *Biotechnol Progr* 2009;**25**:399–408.
4. Yeh A-I, Huang Y-C, Chen SH. Effect of particle size on the rate of enzymatic hydrolysis of cellulose. *Carbohydr Polym* 2010;**79**:192–9.
5. Himmel ME, Tucker MP, Baker J, Rivard C, Oh KK, Grohmann K. Comminution of biomass: hammer and knife mills. *Biotechnol Bioeng Symp* 1985;**15**:39–58.
6. Jin M, Balan V, Gunawan C, Dale BE. Consolidated bioprocessing (CBP) performance of Clostridium phytofermentans on AFEX treated corn stover for ethanol production. *Biotechnol Bioeng* 2011;**108**:1290–7.
7. Lynd LR, Weimer PJ, Van Zyl WH, Pretorius IS. Microbial cellulose utilization: fundamentals and biotechnology. *Microbiol Mol Biol Rev* 2002;**66**:506–77.
8. Shao X, Jin M, Guseva A, Liu C, Balan V, Hogsett D, et al. Conversion for Avicel and AFEX pretreated corn stover by *Clostridium thermocellum* and simultaneous saccharification and fermentation: Insights into microbial conversion of pretreated cellulosic biomass. *Bioresour Technol* 2011;**102**:8040–5.
9. Esteghlalian A, Hashimoto AG, Fenske JJ, Penner MH. Modeling and optimization of the dilute sulfuric acid pretreatment of corn stover, poplar and switchgrass. *Bioresour Technol* 1997;**59**:129–36.
10. Lloyd TA, Wyman CE. Combined sugar yields for dilute sulfuric acid pretreatment of corn stover followed by enzymatic hydrolysis of the remaining solids. *Bioresour Technol* 2005;**96**:1967–77.
11. Neureiter M, Danner H, Fruhauf S, Kromus S, Thomasser C, Braun R, et al. Dilute acid hydrolysis of presscakes from silage and grass to recover hemicellulose-derived sugars. *Bioresour Technol* 2004;**92**:21–9.
12. Aguilar R, Ramirez JA, Garrote G, Vazquez M. Kinetic study of the acid hydrolysis of sugar cane bagasse. *J Food Eng* 2002;**55**:309–18.
13. Lee YY, Iyer P, Torget RW. Dilute-acid hydrolysis of lignocellulosic biomass. *Adv Biochem Eng/Biotechnol* 1999;**65**:93–115.
14. Emmel A, Mathias AL, Wypych F, Ramos LP. Fractionation of *Eucalyptus grandis* chips by dilute acid-catalysed steam explosion. *Bioresour Technol* 2003;**86**:105–15.
15. Shevchenko SM, Chang K, Robinson J, Saddler JN. Optimization of monosaccharide recovery by post-hydrolysis of the water-soluble hemicellulose component after steam explosion of softwood chips. *Bioresour Technol* 2000;**72**:207–11.
16. Garrote G, Dominguez H, Parajó JC. Mild autohydrolysis: an environmentally friendly technology for xylooligosaccharide production from wood. *J Chem Technol Biotechnol* 1999;**74**:1101–9.
17. Garrote G, Dominguez H, Parajo J. Hydrothermal processing of lignocellulosic materials. *Eur J Wood Wood Prod* 1999;**57**:191–202.
18. Mittal A, Chatterjee SG, Scott GM, Amidon TE. Modeling xylan solubilization during autohydrolysis of sugar maple and aspen wood chips: reaction kinetics and mass transfer. *Chem Eng Sci* 2009;**64**:3031–41.

19. Mittal A, Scott GM, Amidon TE, Kiemle DJ, Stipanovic AJ. Quantitative analysis of sugars in wood hydrolyzates with 1H NMR during the autohydrolysis of hardwoods. *Bioresour Technol* 2009;**100**:6398–406.
20. Chen X, Shekiro J, Pschorn T, Sabourin M, Tao L, Elander R, et al. A highly efficient dilute alkali deacetylation and mechanical (disc) refining process for the conversion of renewable biomass to lower cost sugars. *Biotechnol Biofuels* 2014;**7**:98.
21. Sluiter A, Hames B, Ruiz R, Scarlata C, Sluiter J, Templeton D. *Laboratory Analytical Procedure National Renewable Energy Laboratory*. Golden, CO; 2006.
22. Segal L, Creely JJ, Martin AE, Conrad CM. An empirical method for estimating the degree of crystallinity of native cellulose using the X-ray diffractometer. *Text Res J* 1959;**29**:786–94.

CHAPTER 16

Clostridium thermocellum: *Engineered for the Production of Bioethanol*

Steven D. Brown[1,2,3], Kyle B. Sander[1,2,3], Chia-Wei Wu[1], Adam M. Guss[1,2,3]

[1]*Biosciences Division, Oak Ridge National Laboratory, Oak Ridge, TN, USA;* [2]*BioEnergy Science Center, Oak Ridge National Laboratory, Oak Ridge, TN, USA;* [3]*Bredesen Center for Interdisciplinary Research and Graduate Education, University of Tennessee, Knoxville, TN, USA*

Biotechnological Interest *in* Clostridium thermocellum

Plant biomass is a potentially scalable source of feedstocks to produce sustainable fuels and chemicals and to displace petroleum products.[1,2] Biological fermentation of plant-derived biomass is a promising and leading technology route for liquid fuel production, but there are many challenges to overcome. Bioconversion of lignocellulose requires difficult deconstruction of the plant biomass and transformation of the resulting sugars into the preferred fermentation products, presenting a significant challenge to the developing biofuel industry. Thermal and/or chemical pretreatment of biomass and subsequent addition of enzymes to hydrolyze cellulosic biomass polymers to simple fermentable sugars add substantial process economic costs.[3] One of the potentially most transformative processing options, termed *consolidated bioprocessing* (CBP), combines all biologically catalyzed steps into one unit operation so that enzyme production, biomass hydrolysis, and sugar fermentation occur in a single reactor (see reviews[3-10]). CBP would lower capital and process costs for cellulosic biofuel production. *Clostridium thermocellum* was recognized early on for its efficient degradation and utilization of cellulose, and it is a candidate CBP biocatalyst for the production of cellulosic ethanol.

C. thermocellum *Characteristics*

C. thermocellum is a rod-shaped Gram-positive bacterium and a member of the Firmicutes phylum that has been studied extensively for its ability to hydrolyze lignocellulosic biomass.[3] It hydrolyzes lignocellulosic material faster than many other microorganisms through the synergistic action of hydrolytic enzymes bound to a backbone scaffold protein, collectively called the *cellulosome*, which are tethered to the outer cell wall of the bacteria. The cellulosome concept was first conceived from the recognition that *C. thermocellum* cellulases and

Direct Microbial Conversion of Biomass to Advanced Biofuels. http://dx.doi.org/10.1016/B978-0-444-59592-8.00016-6

associated polysaccharide-degrading enzymes are packaged in organized, high-molecular-weight, cellulolytic enzyme complexes.[11,12] Cellulases and cellulosomes have been reviewed extensively.[13–18]

C. thermocellum naturally produces acetic acid, lactic acid, formic acid, ethanol, CO_2, and H_2 as fermentation products. It is an obligate anaerobe and thermophile that grows optimally at temperatures near 55–60 °C.[19,20] It is of biotechnologic interest because of its native ability to hydrolyze cellulose and convert cellodextrins into ethanol. However, *C. thermocellum* and microorganisms capable of direct microbial conversion of biomass often display low ethanol yields, and these organisms have yet to be shown to be capable of effective substrate conversion at the high substrate loadings required by industry. Although *C. thermocellum* is able to solubilize xylan, it lacks an identifiable xylose isomerase and xylulokinase, explaining its inability to catabolize xylose or xylooligomers.[6] The development of new genetic tools and new insights into physiology and regulation through application of classical and high-throughput "omics" techniques and models offers new opportunities for industrial strain development.

Ecology and Isolates

The main ecological role of *C. thermocellum* is the degradation of cellulose (for detailed consideration of ecological aspects of cellulose-degrading communities, see Ref. 3). In nature, this important step in the carbon cycle (hydrolysis of cellulosic carbon) is largely performed by cellulolytic bacteria and/or fungi, both of which produce their own cellulolytic enzymes. Cellulosomal attachment to the cell and substrate ensures that cells are deliberately located close to cellulose material relative to other ecological community members such that they may provide hydrolyzed cellulose products to themselves and residual material to the rest of the community. This role is made evident by its synthesis and use of xylanase enzymes, although the organism itself cannot metabolize five-carbon sugars. Because of this, it is thought that *C. thermocellum* ubiquitously inhabits environments where decaying plant biomass is present, such as compost heaps, soils, sediments, and the digestive system of animals that feed on cellulose-laden plant material (e.g., grass, hay).[20–22] *C. thermocellum* strain YS was isolated from a sample derived from hot springs at Yellowstone National Park in the United States and was integral in describing the original cellulosome concept.[11] Strains similar to ATCC 27405 have been found in intestinal microflora, soil, and sediments,[20] and recently strain BC1 was isolated from a thermophilic biowaste compost treatment in Germany.[23] Strain BC1 has been reported to have higher cellulose-degrading efficacy compared with type strain ATCC 27405, growth up to 67 °C, and an expanded substrate range being able grow glucose and sorbitol, which shows the value of isolating and comparing different strains.

In natural environments, *C. thermocellum* and other anaerobic bacteria belong to communities that form trophic relationships, and understanding these interactions informs strain development strategies for biotechnology. These anaerobic communities are thought to account for only a small percentage of cellulose degradation and carbon cycling.[3] Of that small percentage, *C. thermocellum* likely plays only a relatively minor role because it is a thermophile and requires such a specialized growth environment. Cellulolytic bacteria degrade cellulose to fermentable sugars and make them available for themselves, and residual material is released to the community. Other fermentative bacteria metabolize hydrolyzed carbohydrates released by cellulolytic bacteria. Methanogens and sulfur-reducing bacteria consume H_2 produced through the redox balancing and hydrogenase activity of fermentative bacteria. Indeed, early defined *C. thermocellum/Methanobacterium thermoautotrophicum* co-culture studies showed that there was a shorter lag phase for co-cultures grown on cellulose compared with the monoculture, product profiles could be shifted by the presence of a partner microorganism,[24] and hydrogenase has been suggested as a potential gene deletion target.[3] Still other bacteria metabolize fermentation products produced by the community that would otherwise accumulate to inhibitory levels. Because there is little to no oxygen present in these environments, another defining characteristic of these communities is the diverse array of terminal electron acceptors used by the different community members. Through the concerted action of the community, carbon from lignocellulosic material is completely degraded to CO_2 and methane, completing this step of the global carbon cycle.

Physiology, Metabolism, and Ethanol Tolerance

C. thermocellum prefers to assimilate cellodextrins in the range of approximately two to six glycosyl units as a strategy to conserve energy for sugar uptake during growth on cellulose.[25] Cellodextrins enter the cell via ATP-dependent ABC transport systems rather than the phosphorotransferase system transporters used by many organisms, and once inside, a phosphate anion acts as a nucleophile for phosphorolytic cleavage via cellodextrin phosphorylase,[26–28] resulting in a cellodextrin of length $(n-1)$ and a molecule of glucose-1-phosphate (G1P). On the basis of genomic sequence,[29] proteomics,[30] and enzymology,[31,32] *C. thermocellum* likely catabolizes the resulting glucose and G1P by a modified Embden-Meyerhof-Parnas pathway; however, multiple steps vary from the canonical pathway. Glucose is converted to glucose-6-phosphate (G6P) via a GTP-dependent glucose kinase rather than an ATP-dependent one,[32] and G1P is isomerized to a second molecule of G6P by phosphoglucomutase. Often, fructose-6-phosphate is phosphorylated using an ATP-dependent phosphofructokinase (PFK); *C. thermocellum* encodes two ATP-dependent and one pyrophosphate (PPi)-dependent PFK, although the relative importance of each is currently unknown. Proteomic analysis suggests that all three enzymes are highly abundant,[30] but only PPi-dependent activity was detected in cell extracts.[32] Another atypical difference includes a phosphoglycerate kinase that can use either GDP or ADP. The effect of these differences on metabolic flux is currently unknown, as is the effect on metabolic engineering strategies.

Perhaps most interestingly, *C. thermocellum* lacks an identifiable pyruvate kinase, and the metabolic route from phosphoenolpyruvate (PEP) to pyruvate is unclear. *C. thermocellum* encodes a PEP synthase and a pyruvate-phosphate dikinase, both of which typically catalyze the reverse reaction of PEP synthesis from pyruvate. An alternative route was proposed in which PEP is converted to oxaloacetate, which is then reduced to malate by malate dehydrogenase using NADH as the electron donor and is then oxidatively decarboxylated to pyruvate by malic enzyme using $NADP^+$ as the electron acceptor.[33] Although this provides a route for carbon from PEP to pyruvate, it also has the effect of transhydrogenation, transferring electrons from NADH to NADPH. Recent proteomic,[30] genetic,[34] and biochemical evidence lends support to the hypothesis that this "malate shunt" pathway is active in *C. thermocellum*, although the amount of flux through this pathway is still unknown.

Pyruvate is the first major branch point in fermentation. Pyruvate can be reduced to lactate via lactate dehydrogenase, which is allosterically activated by fructose-1,6-bisphosphate.[35] Thus, *C. thermocellum* produces lactate when sugar uptake outpaces flux through glycolysis. However, under carbon-limiting conditions, most pyruvate is converted to acetyl-CoA either via pyruvate-ferredoxin oxidoreductase, which produces CO_2 and reduces ferredoxin, or via pyruvate-formate lyase (PFL), which produces formate. The acetyl-CoA can then be converted to acetate with concomitant production of ATP, or it can be reduced to acetaldehyde and then to ethanol, reoxidizing the nicotinamide cofactors that were reduced during glycolysis.

When *C. thermocellum* produces acetate, an electron sink is needed to balance redox reactions. One route is production of formate via PFL. Another is reduction of $2H^+$ via hydrogenase. *C. thermocellum* encodes four clusters of hydrogenases. Two putative electron bifurcating hydrogenases[36] may reduce ferredoxin and NADH together to reduce $4H^+$ to $2H_2$. *C. thermocellum* also encodes an energy-conserving, membrane-bound ferredoxin-dependent hydrogenase, Ech. Although H_2 is a major fermentation product, the relative importance of each hydrogenase is unknown. In addition to traditional mixed acid fermentation products, *C. thermocellum* also produces various other products, including free amino acids, pyruvate, malate, and uracil.[37] The reason these products are made is currently unclear, but it has been hypothesized that it is the result of overflow metabolism, possibly due to redox imbalance.[38] *C. thermocellum* lacks the oxidative pentose phosphate pathway and the Entner-Doudoroff pathway, raising the question of how NADPH is generated for biosynthetic reactions. Although the malate shunt described above is one possibility, another possible source is the putative NADH-ferredoxin:$NADP^+$ oxidoreductase.[39]

High product titer is an essential industrial consideration.[40] As mentioned above, microorganisms capable of directly utilizing plant biomass often have low ethanol yields, and in the case of *C. thermocellum*, the highest titer of ethanol produced by wild-type *C. thermocellum* is reported as less than 30 g/L.[41] Mutants have been selected to be tolerant to as much as 80 g/L of ethanol,[41–43] and *C. thermocellum* ethanol tolerance has recently been reviewed.[4] The *C. thermocellum* wild-type ATCC 27405 strain and a derived ethanol-adapted (EA) culture were resequenced to identify important mutations related to the mutant phenotypes.[44] EA was found to have a mutated *adhE* gene that encodes a bifunctional acetaldehyde-CoA/alcohol dehydrogenase. Genetic analysis demonstrated that this *adhE* mutation alone conferred an ethanol-tolerant phenotype by transferring the mutation to *C. thermocellum* strain DSM 1313. Biochemical and structural studies indicated and the resulting strain showed a loss of NADH-dependent activity with concomitant acquisition of NADPH-dependent activity, likely altering electron flow in the mutant. Although more tolerant of ethanol, this strain did not have productivity advantages, which suggests that other metabolic bottlenecks need to be overcome. However, by transferring the *adhE* mutation to the more genetically tractable *C. thermocellum* DSM 1313, an ethanol-tolerant platform strain was created that could be used for further metabolic engineering.

Most *C. thermocellum* studies to date have been conducted using model substrates, but it will become increasingly important that plant biomass substrates be used and studied directly if *C. thermocellum* is to become an industrial CBP organism. In one recent example, downregulation of the lignin pathway gene for caffeic acid 3-*O*-methyltransferase yielded switchgrass (*Panicum virgatum*) that gave better bioconversion results after dilute acid pretreatment. However, *C. thermocellum* fermentation of the modified switchgrass was inhibited, unlike a *Saccharomyces cerevisiae* simultaneous saccharification and fermentation, until a hot-water extraction process was incorporated. A better understanding of the organism's physiology during growth on industrial substrates will facilitate development of more robust strains. When *C. thermocellum* inhibition by a *Populus* hydrolysate was modeled and individual inhibitors tested, 4-hydroxybenzoic acid was found to be the most inhibitory compound, followed by galacturonic acid, which suggests future avenues for strain improvements.[45]

Few studies have examined the mechanisms by which *C. thermocellum* coordinates and regulates physiology under different conditions. The LacI family transcriptional regulator *glyR3* is a negative regulator of the *celC* operon originally identified as containing the *celC*, *glyR3*, and *licA* genes and inducible by laminaribiose.[46] Another important class of regulators that have been identified and begun to be characterized are *C. thermocellum* sigma and anti-sigma factors.[47] It is interesting to note that strain ATCC 27405 can enter a dormant L-form state that is different from the spore form.[48] Further investigations into the genetic control of the transition into L-form and the resulting metabolic effects will shed light on mechanisms of physiological control in *C. thermocellum*. More broadly, the combination of genetic, biochemical, and systems biology studies offers the prospect of more rapid insights into the regulation of *C. thermocellum* physiology.

Genome Sequences

A genome sequence underpins systems biology studies; it is now required for metabolic engineering and is able to be rapidly attained with recent advances in next-generation DNA sequencing technologies. The ATCC 27405 type strain was the first to have its genome sequence determined for this species, and the classical Sanger method of DNA sequencing was used by the U.S. Joint Genome Institute (GenBank accession number CP000568). Professor J.H. David Wu (University of Rochester) and Dr. Michael E. Himmel (National Renewable Energy Laboratory) submitted the proposal to generate the ATCC 27405 genome sequence. Professor Wu's laboratory supplied DNA for the ATCC 27405 genome project and the first draft sequence was available to the public in November 2003; however, repetitive sequences made closing this genome difficult and the genome sequence was not finished until February 2007. The Glimmer[49] and Critica[50] gene prediction algorithms were originally used and combined to predict gene models, which was followed by a round of manual curation. More recently, an improved gene prediction algorithm was applied to the ATCC 27405 genome and its annotation was updated (GenBank accession number CP000568.1).[51] A comprehensive comparison of different annotation versions can be found at http://genome. ornl.gov/microbial/cthe/. As algorithms continue to improve and novel features such as small regulatory RNAs are discovered and identified, it is likely there will be refinements to genomes.

Since the first *C. thermocellum* genome sequence was generated, there has been a revolution in DNA sequencing technologies.[52] Twenty genome sequences for Clostridia species across multiple genera were recently determined,[53] two of which were for *C. thermocellum* strains JW20 (4150) and LQRI (DSM 2360). Finished and draft genomes have been described for *C. thermocellum* strains DSM 1313,[54] YS and derivative strain AD2,[55] and strain BC1.[23] The genome sequence for strain ATCC 27405 has been used to design oligonucleotides for strain DSM 1313, indicating that they are closely related,[56] which was confirmed by subsequent *in silico* genome comparisons.[4] *C. thermocellum* DSM 1313 is the background strain for one recently developed genetic system (see below). A summary of several key genome features is provided for wild-type strains for which the genome sequences are available (Table 1). Although there are strain-level differences in gene content for encoding transposes and restriction systems,[4] many of the differences in genome sizes and the number of predicted

Table 1: Summary statistics for wild-type *C. thermocellum* genome sequences

Strain	Status	Genome Size (bp)	% G+C	Total Genes	Protein Coding Genes	rRNA Operons	Ref.[a]
ATCC 27405	Finished	3,84,3301	39	3,335	3,236	4	51
DSM 1313	Finished	3,56,1619	39	3,102	3,031	4	29
YS	Draft	3,84,3301	39	3,081	3,026	1	55
JW20 (DSM 4150)	Draft	3,32,1980	39	3,027	2,979	3[b]	53
LQRI (DSM 2360)	Draft	3,45,4608	39	3,147	3,091	1	53
BC1	Draft	3,45,4918	39	3,159	3,095	4	23

[a]With the exception of strain BC1, data were obtained from the Integrated Microbial Genomes database on September 7, 2013.
[b]Three 16S rDNA genes were identified for strain JW20, but only single copies of the 5S and 23S genes were noted.

genes likely reflect differences in sequencing technologies, assembly methods, and gene prediction algorithms. Longer read technologies continue to develop, and we expect that such approaches will be useful to improve genome assemblies,[57] which will facilitate comparative genomic studies. Future comparative genomic studies may permit more refined bioinformatics predictions for genes, operons, and *cis*-regulatory motifs and insights into phenotypic differences reported for strain BC1 or others such as hypercellulase production.[58]

These *C. thermocellum* genome sequences have been leveraged to produce a genome-scale metabolic model.[59] This model consisted of 577 reactions, 525 intracellular metabolites, and 432 genes. In addition to providing a tool to predict modifications that could improve fuel production, it also highlighted gaps in metabolic pathways. Because these missing reactions are part of essential metabolic pathways, they either represent incorrectly annotated genes or situations in which *C. thermocellum* uses an unusual pathway. Future studies will be needed to resolve this question. The Roberts model was further updated using RNAseq data, further improving this tool.[60]

Transcriptomics and Proteomics

To better understand *C. thermocellum* physiology and to inform metabolic engineering strategies, transcriptomic and proteomic studies have been undertaken. After the ATCC 27405 genome sequence became available, a whole genome DNA microarray was developed using oligonucleotide probes that represented <95% of the putative protein-coding genes,[61] which was subsequently applied to generate transcriptional profiles for strain ATCC 27405 during early-exponential to late-stationary growth phase in cellulose fermentations.[62] These analyses showed that when batch cultures entered the stationary phase, cell growth slowed in concert with decreased expressions of genes involved in energy production, translation, glycolysis, and metabolisms of amino acid, nucleotide, and coenzyme. Cells instead increased expressions of genes involved in chemotaxis, flagella biosynthesis, signal transduction, and enzymes responsible for digesting plant polysaccharides. *C. thermocellum* ATCC 27405 global gene expression patterns were also profiled in cellulose or cellobiose-limited chemostats with various controlled growth rates.[63] A set of 348 genes were consistently regulated among different growth rates and carbon sources. Statistical analysis of differential expression patterns suggested that growth rate is a dominant factor controlling global gene expression.[63]

C. thermocellum gene and protein expression differences to ethanol stress were examined in a time course study that compared treated to untreated control cells for insights into the response of this fermentation end product.[64] The most upregulated gene and proteins, when compared with untreated wild-type strain, were genes involved in nitrogen metabolism, including urea ABC transporter component genes Cthe_1819 to Cthe_1823. Expression of *C. thermocellum ureABCDEFG* genes into *Thermoanaerobacterium saccharolyticum* increased the titer of ethanol production,[65] suggesting an important link between ethanol production and nitrogen metabolism. This also demonstrates the importance of systems biology studies to inform strain improvement.

In an industrial setting, *C. thermocellum* will need to tolerate inhibitors present in pretreated plant biomass and perturbations in process conditions. Therefore, general and specific physiological and regulatory stress responses were investigated when *C. thermocellum* was grown to midexponential phase and shocked with either furfural to a final concentration of 3 g/L or with a 10 °C heat shock to 68 °C.[66] After 10, 30, 60, and 120 min of postshock, samples were obtained from treated and untreated fermentations for transcriptomic and fermentation product analyses. Urea uptake genes again had higher expression levels after furfural stress, but not to the same degree as after ethanol stress, and these genes were largely unaffected by heat shock. The greatest transcriptomic response to furfural stress was in sulfate transport and for sulfate assimilatory pathway enzymes; however, these transcripts were also changed late after heat and ethanol stress. This study showed that *C. thermocellum* has a complex and dynamic transcriptional response to different stressors that involved genes involved in sulfur and nitrogen assimilation, protection against oxidative stress, electron transfer, detoxification, and DNA repair and the use of different regulatory networks to control and coordinate adaptation.

Early proteomic analyses identified cellulosome compositions under various conditions,[67–71] and these studies have been reviewed.[4] Most predicted cellulosome proteins have now been detected by mass spectrometry,[64] and proteomic[30] and enzyme analyses[31] of core metabolism have informed key aspects of the organism's physiology, as described above. The application of mass-spectrometry-based proteomics techniques to investigate mutant strains and cells grown on industrially relevant substrates will be useful to explore metabolism and regulation in greater detail.

C. thermocellum *Genetic Tools and Metabolic Engineering*

Metabolic engineering requires the ability to genetically modify the organism of interest. The first example of transformation was achieved using a custom electroporator and cuvettes,[72,73] and until recently progress on genetic tool development to facilitate targeted mutagenesis was slow. An *Mbo*I-like restriction system, which recognizes and cleaves unmethylated GATC sequences, was discovered in type strain ATCC 27405.[74] *Escherichia coli* naturally methylates this sequence with Dam methylase, minimizing this restriction system as a barrier to transformation. Strain DSM 1313 is also predicted to encode a *Mbo*I-like restriction system, and Dam methylated DNA transforms this strain more efficiently than unmethylated DNA,[75] supporting this assertion. Furthermore, transformation efficiency has been improved by isolation of DNA from *E. coli* strains lacking Dcm methylase, leading to the suggestion that a putative type IV restriction system found in *C. thermocellum* DSM 1313 may hinder transformation of *C. thermocellum* by cleavage of DNA methylated at Dcm sites.[75]

Significant progress has been made recently on developing methodologies for manipulation of the *C. thermocellum* chromosome. Detailed protocols for transforming *C. thermocellum* and creating deletion strains have been described, and deletion experiments are routinely performed in strain DSM 1313 because of increased transformation efficiencies.[76] The chloramphenicol acetyltransferase gene (*cat*) was demonstrated as a positive selectable marker conferring resistance to thiamphenicol using derivatives of plasmid pNW33N,[56] as has the *kan* gene, which confers resistance to neomycin.[71] Counterselectable markers have also been developed, including *pyrF* for sensitivity to 5-fluoroortic acid,[56] *hpt* for sensitivity to 8-azahypoxanthine, and *tdk* for sensitivity to fluorodeoxyuracil.[77] The use of these markers has allowed for the creation of gene deletions,[71,77,78] which is significant and necessary to advance our fundamental understanding of *C. thermocellum* physiology/regulation and to advance strain developments.

Utilization of new genetic tools has allowed for manipulation of central metabolism, increasing flux of cellulosic substrates to ethanol. Tripathi and co-workers developed a first-generation gene deletion tool and succeeded in deleting the phosphotransacetylase (*pta*), blocking acetic acid production and increasing ethanol yield by approximately 10% during growth on cellobiose.[56] Further tool development and gene deletion allowed for deletion of lactate dehydrogenase (*ldh*) and *pta* in the same strain.[77] Although the single *ldh* and *pta* mutants each produced only approximately 20% more ethanol than the wild-type strain, combination of these mutations into a single strain resulted in an approximately 50% increase in ethanol production over the wild-type strain. Not surprisingly, this strain grew poorly, and it was evolved by serial transfer for 2000 h to select for a faster growing strain. This evolved strain produced 4-fold more ethanol than the wild-type strain. Argyros et al. (2011) described the highest ethanol titer from a thermophilic cellulose fermentation to date using a co-culture of *C. thermocellum* and *T. saccharolyticum*, both of which had been engineered to be organic acid-deficient strains by interrupting pathways for lactate and acetate production.[77] The co-culture produced in 38 g/L ethanol in 146 h from 92 g/L of Avicel, and acetic and lactic acids were not detected. Although it is unlikely that an industrial process will utilize a coculture, thermophilic, cellulolytic bacteria are amenable to metabolic engineering and substantial progress can be made. Although not yet sufficient for industrial purposes, these advances in *C. thermocellum* metabolic engineering give confidence that additional modifications will allow this organism to be engineered for greater productivity.

A targeted mutagenesis system built on the mobile group II intron has been a very effective tool in various mesophilic bacteria, including several clostridia species.[79] A mobile group II intron from the thermophilic cyanobacterium *Thermosynechococcus elongatus* has been used to construct a thermophilic "targetron" system, which has been applied to interrupt six *C. thermocellum* genes (*cipA* [Clo1313_0627], cellulosome scaffoldin protein; *hfat* [also called *pfl*; Clo1313_2343], hypothetical formate acetyltransferase; *hyd* [Clo1313_0554], hydrogenase; *ldh* [Clo1313_1160], lactate dehydrogenase; *pta* [Clo1313_1185], phosphotransacetylase; *pyrF* [Clo1313_1266], orotidine 5'-phosphate decarboxylase).[80] The advent of newly developed tools for *C. thermocellum* promises a new era of rapid metabolic engineering in this organism.

Outlook

Being able to produce liquid biofuels and chemicals at large industrial scales using plant-derived biomass has yet to be realized. Future directions for *C. thermocellum* research and development that will likely be productive include systems biology studies of mutant strains for a deeper understanding of physiology and regulation, detailed metabolic flux studies, introduction of genes and pathways to transport and utilize five-carbon sugars present in biomass, and examination of strains grown at high loadings for substrates relevant to industry and at scale. Substantial progress has been made recently in tools and understanding for metabolically engineering promising CBP candidates including *C. thermocellum*, making now an exciting time.

Acknowledgment

The BioEnergy Science Center is a U.S. Department of Energy (DOE) Bioenergy Research Center supported by the Office of Biological and Environmental Research in the DOE Office of Science.

References

1. Himmel ME, Ding S-Y, Johnson DK, Adney WS, Nimlos MR, Brady JW, et al. Biomass recalcitrance: engineering plants and enzymes for biofuels production. *Science* 2007;**315**:804–7.
2. Himmel ME, Picataggio SK. *Biomass recalcitrance: deconstructing the plant cell wall for bioenergy.* Blackwell Publishing Ltd; 2009. p. 1–6.
3. Lynd LR, Weimer PJ, van Zyl WH, Pretorius IS. Microbial cellulose utilization: fundamentals and biotechnology. *Microbiol Mol Biol Rev* 2002;**66**:506–77.
4. Blumer–Schuette SE, Brown SD, Sander KB, Bayer EA, Kataeva I, Zurawski JV, et al. Thermophilic lignocellulose deconstruction. *FEMS Microbiology Reviews* 2014;**38**:393–48.
5. Carere CR, Sparling R, Cicek N, Levin DB. Third generation biofuels via direct cellulose fermentation. *Int J Mol Sci* 2008:1342–60.
6. Demain AL, Newcomb M, Wu JHD. Cellulase, clostridia, and ethanol. *Microbiol Mol Biol Rev* 2005;**69**:124–54.
7. Lynd LR, van Zyl WH, McBride JE, Laser M. Consolidated bioprocessing of cellulosic biomass: an update. *Curr Opin Biotechnol* 2005;**16**:577–83.
8. Maki M, Leung KT, Qin W. The prospects of cellulase-producing bacteria for the bioconversion of lignocellulosic biomass. *Int J Biol Sci* 2009;**5**:500–16.
9. Olson DG, McBride JE, Shaw AJ, Lynd LR. Recent progress in consolidated bioprocessing. *Curr Opin Biotechnol* 2012;**23**:396–405.
10. la Grange DC, den Haan R, van Zyl WH. Engineering cellulolytic ability into bioprocessing organisms. *Appl Microbiol Biotechnol* 2010;**87**:1195–208.
11. Bayer EA, Kenig R, Lamed R. Adherence of *Clostridium thermocellum* to cellulose. *J Bacteriol* 1983;**156**:818–27.

12. Lamed R, Setter E, Bayer EA. Characterization of a cellulose-binding, cellulase-containing complex in *Clostridium thermocellum*. *J Bacteriol* 1983;**156**:828–36.

13. Bayer EA, Belaich JP, Shoham Y, Lamed R. The cellulosomes: multienzyme machines for degradation of plant cell wall polysaccharides. *Ann Rev Microbiol* 2004;**58**:521–54.

14. Fontes C, Gilbert HJ. In: Kornberg RD, Raetz CRH, Rothman JE, Thorner JW, editors. *Ann Rev Biochem*, vol. 79; 2010, p. 655–81.

15. Gilbert HJ. Cellulosomes: microbial nanomachines that display plasticity in quaternary structure. *Mol Microbiol* 2007;**63**:1568–76.

16. Garcia-Martinez DV, Shinmyo A, Madia A, Demain AL. Studies on cellulase production by *Clostridium thermocellum*. *Appl Microbiol Biotechnol* 1980;**9**:189–97.

17. Schwarz WH. The cellulosome and cellulose degradation by anaerobic bacteria. *Appl Microbiol Biotechnol* 2001;**56**:634–49.

18. Sheehan J, Himmel M. Enzymes, energy, and the environment: a strategic perspective on the U.S. Department of Energy's research and development activities for bioethanol. *Biotechnol Prog* 1999;**15**:817–27.

19. Lamed R, Zeikus JG. Ethanol production by thermophilic bacteria: relationship between fermentation product yields of and catabolic enzyme activities in *Clostridium thermocellum* and *Thermoanaerobium brockii*. *J Bacteriol* 1980;**144**:569–78.

20. McBee RH. The characteristics of *Clostridium thermocellum*. *J Bacteriol* 1954;**67**:505–6.

21. Izquierdo JA, Sizova MV, Lynd LR. Diversity of bacteria and glycosyl hydrolase family 48 genes in cellulolytic consortia enriched from thermophilic biocompost. *Appl Environ Microbiol* 2010;**76**:3545–53.

22. Zverlov VV, Schwarz WH. Bacterial cellulose hydrolysis in anaerobic environmental subsystems—*Clostridium thermocellum* and *Clostridium stercorarium*, thermophilic plant-fiber degraders. *Ann NY Acad Sci* 2008;**1125**:298–307.

23. Koeck DE, Wibberg D, Koellmeier T, Blom J, Jaenicke S, Winkler A. Draft genome sequence of the cellulolytic *Clostridium thermocellum* wild-type strain BC1 playing a role in cellulosic biomass degradation. *J Biotechnol* 2013;**168**:62–63. doi:10.1016/j.jbiotec.2013.08.011.

24. Weimer PJ, Zeikus JG. Fermentation of cellulose and cellobiose by *Clostridium thermocellum* in the absence of Methanobacterium thermoautotrophicum. *Appl Environ Microbiol* 1977;**33**:289–97.

25. Zhang Y-HP, Lynd LR. Cellulose utilization by *Clostridium thermocellum*: bioenergetics and hydrolysis product assimilation. *Proc Natl Acad Sci USA* 2005;**102**:7321–5.

26. Strobel HJ, Caldwell FC, Dawson KA. Carbohydrate transport by the anaerobic thermophile *Clostridium thermocellum* LQRI. *Appl Environ Microbiol* 1995;**61**:4012–5.

27. Alexander JK. Purification and specificity of cellobiose phosphorylase from *Clostridium thermocellum*. *J Biol Chem* 1968;**243**:2899–904.

28. Nochur SV, Jacobson GR, Roberts MF, Demain AL. Mode of sugar phosphorylation in *Clostridium thermocellum*. *Appl Biochem Biotechnol* 1992;**33**:33–41.

29. Feinberg L, Foden J, Barrett T, Davenport KW, Bruce D, Detter C, et al. Complete genome sequence of the *cellulolytic thermophile Clostridium thermocellum* DSM1313. *J Bacteriol* 2011;**193**:2906–7.

30. Rydzak T, McQueen PD, Krokhin OV, Spicer V, Ezzati P, Dwivedi RC, et al. Proteomic analysis of *Clostridium thermocellum* core metabolism: relative protein expression profiles and growth phase-dependent changes in protein expression. *BMC Microbiol* 2012;**12**:214.

31. Rydzak T, Levin DB, Cicek N, Sparling R. Growth phase-dependent enzyme profile of pyruvate catabolism and end-product formation in *Clostridium thermocellum* ATCC 27405. *J Biotechnol* 2009;**140**:169–75.

32. Zhou J, Olson DG, Argyros DA, Deng Y, van Gulik WM, van Dijken JP, et al. Atypical glycolysis in *Clostridium thermocellum*. *Appl Environ Microbiol* 2013;**79**:3000–8.

33. Lamed R, Zeikus JG. Thermostable, ammonium-activated malic enzyme of *Clostridium thermocellum*. *Biochim Biophys Acta* 1981;**660**:251–5.

34. Deng Y, Olson DG, Zhou J, Herring CD, Joe SA, Lynd LR. Redirecting carbon flux through exogenous pyruvate kinase to achieve high ethanol yields in *Clostridium thermocellum*. *Metab Eng* 2013;**15**:151–8.

35. Özkan M, Yllmaz EI, Lynd LR, Özcengiz G. Cloning and expression of the *Clostridium thermocellum* L-lactate dehydrogenase gene in *Escherichia coli* and enzyme characterization. *Can J Microbiol* 2004;**50**:845–51.

36. Schut GJ, Adams MWW. The iron-hydrogenase of *Thermotoga maritima* utilizes ferredoxin and NADH synergistically: a new perspective on anaerobic hydrogen production. *J Bacteriol* 2009;**191**:4451–7.

37. Ellis LD, Holwerda EK, Hogsett D, Rogers S, Shao X, Tschaplinski T, et al. Closing the carbon balance for fermentation by *Clostridium thermocellum* (ATCC 27405). *Bioresour Technol* 2012;**103**:293–9.

38. van der Veen D, Lo J, Brown SD, Johnson CM, Tschaplinski TJ, Martin M, et al. Characterization of *Clostridium thermocellum* strains with disrupted fermentation end-product pathways. *J Ind Microbiol Biotechnol* 2013;**40**:725–34.

39. Wang S, Huang H, Moll J, Thauer RK. NADP+ reduction with reduced ferredoxin and NADP+ reduction with NADH are coupled via an electron-bifurcating enzyme complex in *Clostridium kluyveri*. *J Bacteriol* 2010;**192**:5115–23.

40. Stephanopoulos G. Challenges in engineering microbes for biofuels production. *Science* 2007;**315**:801–4.

41. Rani KS, Swamy MV, Sunitha D, Haritha D, Seenayya G. Improved ethanol tolerance and production in strains of *Clostridium thermocellum*. *World J Microbiol Biotechnol* 1996;**12**:57–60.

42. Lynd LR, Weimer PJ, van Zyl WH, Pretorius IS. Microbial cellulose utilization: fundamentals and biotechnology. *Microbiol Mol Biol Rev* 2002;**66**:506.

43. Williams TI, Combs JC, Lynn BC, Strobel HJ. Proteomic profile changes in membranes of ethanol-tolerant *Clostridium thermocellum*. *Appl Microbiol Biotechnol* 2007;**74**:422–32.

44. Brown SD, Guss AM, Karpinets TV, Parks JM, Smolin N, Yang S, et al. Mutant alcohol dehydrogenase leads to improved ethanol tolerance in *Clostridium thermocellum*. *Proc Natl Acad Sci USA* 2011;**108**:13752–7.

45. Linville JL, Rodriguez Jr M, Mielenz JR, Cox CD. Kinetic modeling of batch fermentation for *Populus* hydrolysate tolerant mutant and wild type strains of *Clostridium thermocellum*. *Bioresour Technol* 2013;**147**:605–13.

46. Newcomb M, Chen CY, Wu JHD. Induction of the *celC* operon of *Clostridium thermocellum* by laminaribiose. *Proc Natl Acad Sci USA* 2007;**104**:3747–52.

47. Kahel-Raifer H, Jindou S, Bahari L, Nataf Y, Shoham Y, Bayer EA, et al. The unique set of putative membrane-associated anti-sigma factors in *Clostridium thermocellum* suggests a novel extracellular carbohydrate-sensing mechanism involved in gene regulation. *FEMS Microbiol Lett* 2010;**308**:84–93.

48. Mearls EB, Izquierdo JA, Lynd LR. Formation and characterization of non-growth states in *Clostridium thermocellum*: spores and L-forms. *BMC Microbiol* 2012;**12**:180.

49. Delcher A, Bratke K, Powers E, Salzberg S. Identifying bacterial genes and endosymbiont DNA with glimmer. *Bioinformatics* 2007;**23**:673–9.

50. Badger J, Olsen G. CRITICA: coding region identification tool invoking comparative analysis. *Mol Biol Evol* 1999;**16**:512–24.

51. Wilson CM, Rodriguez Jr M, Johnson CM, Martin SL, Chu TM, Wolfinger RD, et al. Global transcriptome analysis of *Clostridium thermocellum* ATCC 27405 during growth on dilute acid pretreated *Populus* and switchgrass. *Biotechnol. Biofuels* 2013;**6**:179.

52. Mardis ER. Next-generation DNA sequencing methods. *Ann Rev Gen Hum Genet* 2008;**9**:387–402.

53. Hemme CL, Mouttaki H, Lee Y-J, Goodwin L, Lucas S, Copeland A, et al. Genome announcement: sequencing of multiple Clostridia genomes related to biomass conversion and biofuels production. *J Bacteriol* 2010;**192**:6494–6.

54. Feinberg L, Foden J, Barrett TF, Davenport KW, Bruce D, Detter C, et al. Complete genome sequence of the cellulolytic thermophile *Clostridium thermocellum* DSM1313. *J Bacteriol* 2011. JB.00322–00311.

55. Brown SD, Lamed R, Morag E, Borovok I, Shoham Y, Klingeman DM, et al. Draft genome sequences for *Clostridium thermocellum* wild-type strain YS and derived cellulose adhesion-defective mutant strain AD2. *J Bacteriol* 2012;**194**:3290–1.

56. Tripathi SA, Olson DG, Argyros DA, Miller BB, Barrett TF, Murphy DM, et al. Development of *pyrF*-based genetic system for targeted gene deletion in *Clostridium thermocellum* and creation of a *pta* mutant. *Appl Environ Microbiol* 2010;**76**:6591–9.

57. Brown SD, Nagaraju S, Utturkar SM, De Tissera S, Segovia S, Mitchell W, et al. Comparison of single-molecule sequencing and hybrid approaches for finishing the genome of *Clostridium autoethanogenum* and analysis of CRISPR systems in industrial relevant Clostridia. *Biotechnol. Biofuels* 2014;**7**:40.

58. Mori Y. Isolation of mutants of *Clostridium thermocellum* with enhanced cellulase production. *Agric Biol Chem* 1990;**54**:825–6.

59. Roberts SB, Gowen CM, Brooks JP, Fong SS. Genome-scale metabolic analysis of *Clostridium thermocellum* for bioethanol production. *BMC Syst Biol* 2010;**4**:31.

60. Gowen CM, Fong SS. Genome-scale metabolic model integrated with RNAseq data to identify metabolic states of *Clostridium thermocellum*. *Biotechnol J* 2010;**5**:994.

61. Brown S, Raman B, McKeown C, Kale S, He Z, Mielenz J. Construction and evaluation of a *Clostridium thermocellum* ATCC 27405 whole-genome oligonucleotide microarray. *Appl Biochem Biotechnol* 2007;**137–140**:663–74.

62. Raman B, McKeown CK, Rodriguez Jr M, Brown SD, Mielenz JR. Transcriptomic analysis of *Clostridium thermocellum* ATCC 27405 cellulose fermentation. *BMC Microbiol* 2011;**11**:134.

63. Riederer A, Takasuka TE, Makino S, Stevenson DM, Bukhman YV, Elsen NL, et al. Global gene expression patterns in *Clostridium thermocellum* as determined by microarray analysis of chemostat cultures on cellulose or cellobiose. *Appl Environ Microbiol* 2011;**77**:1243–53.

64. Yang S, Giannone RJ, Dice L, Yang ZK, Engle NL, Tschaplinski TJ, et al. *Clostridium thermocellum* ATCC27405 transcriptomic, metabolomic and proteomic profiles after ethanol stress. *BMC Genomics* 2012;**13**:336.

65. Shaw AJ, Covalla SF, Miller BB, Firliet BT, Hogsett DA, Herring CD. Urease expression in a *Thermoanaerobacterium saccharolyticum* ethanologen allows high titer ethanol production. *Metab Eng* 2012;**14**:528–32.

66. Wilson CM, Yang S, Rodriguez Jr M, Ma Q, Johnson CM, Dice L, et al. Clostridium thermocellum transcriptomic profiles after exposure to furfural or heat stress. *Biotechnol Biofuels* 2013;**6**:131.

67. Morag E, Lapidot A, Govorko D, Lamed R, Wilchek M, Bayer EA, et al. Expression, purification, and characterization of the cellulose-binding domain of the scaffoldin subunit from the cellulosome of *Clostridium thermocellum*. *Appl Environ Microbiol* 1995;**61**:1980–6.

68. Zverlov VV, Kellermann J, Schwarz WH. Functional subgenomics of *Clostridium thermocellum* cellulosomal genes: identification of the major catalytic components in the extracellular complex and detection of three new enzymes. *Proteomics* 2005;**5**:3646–53.

69. Gold ND, Martin VJ. Global view of the *Clostridium thermocellum* cellulosome revealed by quantitative proteomic analysis. *J Bacteriol* 2007;**189**:6787–95.

70. Raman B, Pan C, Hurst G, Rodriguez M, McKeown C, Lankford P, et al. Impact of pretreated switchgrass and biomass carbohydrates on *Clostridium thermocellum* ATCC 27405 cellulosome composition: a quantitative proteomic analysis. *PLoS One* 2009;**4**:1–13.

71. Olson DG, Tripathi SA, Giannone RJ, Lo J, Caiazza NC, Hogsett DA, et al. Deletion of the Cel48S cellulase from *Clostridium thermocellum*. *Proc Natl Acad Sci USA* 2010;**107**:17727–32.

72. Tyurin MV, Desai SlG, Lynd LR. Electrotransformation of *Clostridium thermocellum*. *Appl Environ Microbiol* 2004;**70**:883–90.

73. Tyurin MV, Sullivan CR, Lynd LR. Role of spontaneous current oscillations during high-efficiency electrotransformation of thermophilic anaerobes. *Appl Environ Microbiol* 2005;**71**:8069–76.

74. Klapatch TR, Demain AL, Lynd LR. Restriction endonuclease activity in *Clostridium thermocellum* and *Clostridium thermosaccharolyticum*. *Appl Microbiol Biotechnol* 1996;**45**:127–31.

75. Guss AM, Olson DG, Caiazza NC, Lynd LR. Dcm methylation is detrimental to plasmid transformation in *Clostridium thermocellum*. *Biotechnol Biofuels* 2012;**5**:30.

76. Olson DG, Lynd LR. In: Gilbert, HJ. editor. *Methods Enzymol*, vol. 510. Academic Press; 2012, p. 317–30.

77. Argyros DA, Tripathi SA, Barrett TF, Rogers SR, Feinberg LF, Olson DG, et al. High ethanol titers from cellulose using metabolically engineered thermophilic, anaerobic microbes. *Appl Environ Microbiol* 2011;**77**:8288–94.

78. Waller BH, Olson DG, Currie DH, Guss AM, Lynd LR. Exchange of type II Dockerin-containing subunits of the *C. thermocellum* cellulosome as revealed by SNAP-tags. *FEMS Microbiol Lett* 2012;**333**:46–53.

79. Heap JT, Pennington OJ, Cartman ST, Carter GP, Minton NP. The ClosTron: a universal gene knock-out system for the genus *Clostridium*. *J Microbiol Methods* 2007;**70**:452–64.

80. Mohr G, Hong W, Zhang J, Cui G-z, Yang Y, Cui Q, et al. A targetron system for gene targeting in thermophiles and its application in *Clostridium thermocellum*. *PLoS One* 2013;**8**:e69032.

CHAPTER 17

Omics Approaches for Designing Biofuel Producing Cocultures for Enhanced Microbial Conversion of Lignocellulosic Substrates

David B. Levin[1], Tobin J. Verbeke[2], Riffat Munir[1], Rumana Islam[1], Umesh Ramachandran[1], Sadhana Lal[1], John Schellenberg[2], Richard Sparling[2]

[1]*Department of Biosystems Engineering, University of Manitoba, Winnipeg MB, Canada;*
[2]*Department of Microbiology, University of Manitoba, Winnipeg MB, Canada*

Introduction

Consolidated bioprocessing (CBP) is a system in which enzyme production, substrate hydrolysis, and fermentation are accomplished in a single process step by lignocellulolytic microorganisms. CBP offers the potential for lower biofuel production costs due to simpler feedstock processing, lower energy inputs, and higher conversion efficiencies than separate hydrolysis and fermentation processes, and it is an economically attractive near-term goal for "next-generation" cellulosic biofuel production.[1–4] Cellulolytic species of clostridia have received significant attention because of their potential to directly utilize cellulose as a carbon source and convert lignocellulosic biomass into industrially valuable chemicals.

In nature, anaerobic degradation of cellulosic materials (e.g., in composts) typically calls into play complex communities of microorganisms excreting a broad arsenal of hydrolytic enzymes to tackle the diverse polysaccharides that compose lignocellulose under thermophilic[5] and mesophilic[6] conditions. From such communities, mesophilic and thermophilic species have been isolated (e.g., Sizova et al.[7]) and are characterized mainly as strictly anaerobic spore-formers with many expressing multienzyme extracellular complexes, known as a cellulosome, which is able to rapidly and efficiently hydrolyze cellulose.[2,3] However, it is important to remain mindful that the physiological conditions under which these species have been characterized do not necessarily reflect the environments in which these bacteria have evolved. Rather, the metabolism of these bacteria has evolved to simultaneously exploit a narrow ecological niche and to function as part of a diverse microbial ecosystem. Guedon et al.[8] contend that cellulolytic

Direct Microbial Conversion of Biomass to Advanced Biofuels. http://dx.doi.org/10.1016/B978-0-444-59592-8.00017-8

species, such as the mesophile *Clostridium cellulolyticum*, are not adapted to utilize carbon sources and other nutrients in excess because many natural ecosystems rarely contain all of the nutrients required in saturating quantities. *Clostridium thermocellum* is incapable of utilizing the products of hemicellulose or lignin hydrolysis and instead is restricted to catabolizing the oligosaccharides released during cellulose decomposition.

Although some monocultures of cellulolytic bacteria have been studied extensively, they have their limitations, and cocultures of *C. thermocellum* with saccharolytic bacteria such as *Clostridium thermosaccharolyticum*,[9] *Clostridium thermohydrosulfuricum*,[10,11] *Caldicellulosiruptor bescii*,[12] and *Thermoanaerobacter pseudethanolicus*[13] (formerly *C. thermohydrosulfuricum* 39E[14]) have reported increased rates of cellulose hydrolysis and higher ethanol titers than observed in monocultures. Indeed, the isolation of *C. thermocellum* was fraught with challenges resultant from a proclivity to exist in stable, synergistic relationships with other species.[11] Naturally occurring cocultures of cellulose fermenting mesophiles have also been described.[15]

Most of the research on cocultures until now has been empirical. Development of "designer consortia" for novel biorefining processes may be a highly productive area of research for H_2 and/or ethanol production coupled to synthesis of value-added co-products. A comprehensive understanding of lignocellulose fermentation may require, in future studies, the utilization of holistic approaches that involve meta-genomic, -proteomic, and -transcriptomic methodologies. Nevertheless, understanding the role and behavior of the individual participants in lignocellulosic hydrolysis/fermentation remains a necessary prerequisite. In this chapter, we discuss the application of "omics" (bioinformatics, genomics, proteomics, and transcriptomics) to identify potential synergistic partners for the design of cocultures of bacteria that display higher rates of substrate conversion and enhanced yields of desired end products.

Synergistic Cocultures for Fermentation of Lignocellulosic Substrates

Growing evidence indicates that cocultures of surface-attached microorganisms are more effective in substrate conversion and can synthesize end products such as ethanol or H_2 in greater amounts than monocultures. Cocultures of the thermophilic, cellulolytic bacterium, *C. thermocellum* strain YM4 and the xylanolytic bacterium, *C. thermohydrosulfuricum* strain YM3, degraded crystalline cellulose (Avicel) more rapidly than monocultures of *C. thermocellum*.[11] Cocultures of *C. thermocellum* LQRI and either *T. pseudethanolicus* 39E or *Thermoanaerobacter* sp. X514 were more effective in utilizing cellulose and produced more ethanol in cocultures than *C. thermocellum* monocultures.[13] These studies suggest that defined cocultures can synergistically increase rates of substrate conversion and H_2 and/or ethanol production and yields.

More recently, the volumetric H_2 production and the H_2 yield more than doubled in cocultures of *C. thermocellum* strain JN4 and *T. thermosaccharolyticum* GD17.[16] *C. thermocellum* JN4 can decompose cellulose and xylans, but it cannot utilize glucose or xylose produced by the degradation, respectively. In contrast, *T. thermosaccharolyticum* GD17 can utilize xylose for growth. The experiment was performed at a pH of 4.4 and a temperature of 60 °C in a batch reactor. The hydrogen yield from monocultures of *C. thermocellum* was approximately 0.8 mol H_2/mol glucose, with lactate as the main product. H_2 production increased approximately 2-fold and H_2 yield increased to 1.8 mol H_2/mol glucose when *C. thermocellum* JN4 was cocultured with *T. thermosaccharolyticum* GD17. Butyrate was the most abundant byproduct, and lactate was not detected at the end of the fermentation reaction in the coculture.

Coculture experiments with *C. thermocellum* DSM 1237 and the noncellulolytic H_2-producing bacterium, *Clostridium thermopalmarium* DSM 5974 were reported by Geng et al.[17] The coculture produced nearly double the amount of H_2 produced by *C. thermocellum* monocultures, with ethanol and acetate as the main soluble fermentation end products. However, the primary soluble fermentation end product in the cocultures was butyrate (produced by *C. thermopalmarium*). Nevertheless, these results support the synergies between cellulolytic anaerobes and high-yield biohydrogen producers.

Another example of technical and economical efficiencies attained by the synergistic cocultures was provided by Li and Liu,[18] who investigated fermentative H_2 production by a coculture of *C. thermocellum* and *C. thermosaccharolyticum* utilizing corn stalk waste. *C. thermosaccharolyticum* is a noncellulolytic, high H_2-producing strain. The experiments were conducted in batch and continuous-flow modes at a temperature of 55 °C and a pH of 7.2. Residual cellobiose, glucose, and xylose were detected in *C. thermocellum* monocultures after fermentation of the corn silk waste substrate. However, neither cellobiose nor xylose were detected at either the end of the *C. thermocellum* + *C. thermosaccharolyticum* coculture fermentations because *C. thermosaccharolyticum* had utilized these substrates for growth, producing the H_2 and acetate as fermentation end products. The hydrogen yield in the coculture batch fermentation reached 68.2 mL/g cornstalk, which was 94.1% higher than that in the monoculture, and the rate of hydrogen production reached 14.1 mL H_2/L/h. A higher H_2 yield of 74.9 mL/g cornstalk, as well as a higher production rate of 18.5 mL H_2/L/h, were achieved using the optimized coculture method in a scaled-up reactor compared with that produced from the anaerobic bottles. A coculture of *C. thermocellum* and *C. bescii* was able to increase yields of end products from the fermentation of raw switchgrass over either monoculture,[12] indicating that the coculture may have enhanced the accessibility of *C. thermocellum* to the cellulose through the more effective removal of hemicellulose by *C. bescii*.

Predicting Synergistic Cocultures
Taking Advantage of "Omics" to Understand Microbial Complementarity

There has been a concerted effort to discover, characterize, and sequence organisms capable of lignocellulose fermentation in recent years. Because of the large and increasing number of genome sequences of relevant organisms in the databases, genomic approaches investigating biofuel-relevant organisms have allowed significant comparative insight in how these organisms produce end products of interest,[19] and they have been used for flux modelling[20] with organisms such as *C. thermocellum*. Genomic information can also allow for significant insight when designing cocultures. Nevertheless, genome sequences indicate the putative potential of a cell's capabilities. The subset of genes that are actually expressed under specific growth conditions through the use of other "omics" tools such as proteomics and transcriptomics provides further insights into what the cells are capable of doing under specific growth conditions. Complementarity among microorganisms in communities can occur at multiple levels, including complementation of complex substrate hydrolysis, complementation of carbon metabolism and soluble substrate utilization, and nutrient complementation. Omics approaches can also be a powerful tool in understanding global and specific regulatory mechanisms related to sugar usage, and the resulting data can be used to make informed decisions when designing cocultures.

Complementarity in Glycoside Hydrolases and Hydrolysis of Complex Substrates

The recalcitrant nature of lignocellulose and its resistance to enzymatic hydrolysis are often viewed as one of the major limitations to overcome to improve lignocellulosic biofuel production in CBP systems.[21,22] Although many studies investigate the native hydrolytic capabilities of biofuel-producing strains in monoculture,[23-26] co-culturing strains with their own unique capacities has the potential for synergistic relationships to develop. Analysis of genome content can be used to identify the "maximal potential" within a given strain to contribute to lignocellulose hydrolysis in a CBP system. This process can be facilitated by first identifying all of the carbohydrate active enzymes (CAZymes) in an organism, for which web-based tools[27] as well as a CAZyme database[28] exist.

Although many of the identified CAZyme sequences in biofuel-relevant strains (Table 1) have not been functionally characterized, inferences into enzyme function can be predicted based on functionalities assigned to each CAZyme class, which allows for the identification of potential hydrolytic limitations. For example, the study by Verbeke et al.[29] correlated the presence or absence of glycoside hydrolase (GH) 10 enzymes (endo-xylanases) in *Thermoanaerobacter* genomes (Table 1) with the ability or inability, respectively, of the same strains to grow on xylan. Similar observations involving *Caldicellulosiruptor* spp. were made by Blumer-Schuette et al.[25] In that study, the absence of GH48 domain-containing proteins in *Caldicellulosiruptor hydrothermalis*, *Caldicellulosiruptor kristjanssonii*, and *Caldicellulosiruptor owensensis* (Table 1) correlated with their reduced ability to grow on microcrystalline cellulose despite containing other cellulose-degrading enzymes, which permitted growth on carboxymethylcellulose. Thus, the absence of specific lignocellulose-relevant CAZymes in the genomes of specific strains can be viewed as a requirement for those CAZymes, or functional equivalents, in the genomes of a potential coculture partner for the purposes of maximizing hydrolysis of the lignocellulosic biomass.

Although genome analysis can be used to identify potential synergies or limitations in the hydrolytic capabilities of strains, these analyses can be further supplemented and focused through the use of cellular localization software such as PSortB[30] or any of multiple signal peptide prediction programs (for a comparison of programs, see Choo et al.[31]). Proteomic analysis of *Clostridium phytofermentans* grown on hemicellulose (Birch wood xylan) or cellulose found good correlation between proteins observed in the extracellular "secretome" (Table 2) and those predicted to be secreted in silico. The presence of dockerin domains in cellulosome-possessing strains, or S-layer homology domains (pfam classification PF00395), are also signatures of extracellular localization. Although the relationship between extracellular (cell-bound or secreted) CAZymes and hydrolysis of insoluble lignocellulose polymers seems obvious, distinguishing between intracellular and extracellular CAZymes is often overlooked when reporting an organism's CAZyme. In such cases, the hydrolytic potential of an organism may be ultimately be overestimated as a result. The presence of GHs in the cytoplasm should not be surprising because it has been shown that several organisms associated with lignocellulose fermentation take up oligosaccharides into the cell,[32,33] making the presence of such enzymes essential in the cytoplasm (e.g., β-glucocidases in *C. thermocellum* or β-xylosidases *T. pseudethanolycus* JW200).

Gene expression data can provide further important insights into realized, as well as unfilled, hydrolytic potential that can influence strain selection in designer cocultures (Table 2). For example, quantitative proteomic analysis of *C. thermocellum* by Raman et al. (2009) identified that xylanase expression decreased when grown on pretreated switchgrass, which contains xylan, in comparison to growth on cellulose (Table 2). A similar result was observed in *C. cellulolyticum*,[24] another cellulosome-producing *Clostridium* species. Comparative proteomic analysis of cellulosomes derived from *C. cellulolyticum* grown on cellulose, oat spelt xylan, or wheat straw identified that xylanases from the oat spelt xylan cellulosomes had fewer xylanase peptides identified than cellulosomes from either the cellulose or wheat straw conditions. The apparent downregulation of xylanases is further supported by the observation that the cellulosomes showing the least hydrolytic activity toward xylan were derived from the oat spelt xylan grown cells. Thus, it appears that in *C. thermocellum* and *C. cellulolyticum*, the production of xylanases is not tied to the presence of xylan; therefore, xylan hydrolysis in lignocellulosic biomass may be limited when using these strains in monoculture.

In contrast, cellulose hydrolysis activity in *C. cellulolyticum* cellulosomes was highest on cellulose-grown cells whereas wheat straw hydrolysis was highest for cellulosomes derived from wheat straw-grown cells.[24] Thus, under these conditions, the nature of the polymer hydrolyzed correlates intimately with the resultant hydrolysis proteins produced. Similar findings were observed for *C. thermocellum*, in which decreased exo- and endoglucanase expression levels were observed for cells grown on the disaccharide cellobiose in comparison to polymeric cellulose.[23]

Table 1: Analysis of extracellular[a] CAZy designated glycoside hydrolase classes related to lignocellulose hydrolysis in the genomes of select Firmicutes

CAZyme Module	Caldicellulosiruptor bescii DSM 6725	Caldicellulosiruptor kristjanssonii 177R1B	Caldicellulosiruptor obsidiansis OB47	Caldicellulosiruptor saccharolyticus DSM 8903	Clostridium cellulolyticum H10	Clostridium cellulovorans 743B	Clostridium phytofermentans ISDg	Clostridium thermocellum ATCC 27405	Clostridium stercorarium DSM 8532	Thermoanaerobacter italicus Ab9	Thermoanaerobacter mathranii subsp. mathranii A3	Thermoanaerobacter thermohydrosulfuricus WC1	Thermoanaerobacterium saccharolyticum JW/SL-YS485	Thermoanaerobacterium thermosaccharolyticum DSM 571	Thermoanaerobacterium xylanolyticum LX-11
GH2					X										
GH3									X						
GH5	X	X	X	X	X	X		X					X	X	
GH8						X		X							
GH9	X	X	X	X	X	X	X	X	X						
GH10	X	X	X	X	X	X	X	X	X	X	X	X	X	X	X
GH11	X				X	X	X	X	X				X		
GH16			X												X
GH26						X	X	X					X	X	
GH27						X	X	X	X						
GH31					X			X							
GH42								X							
GH43									X		X				
GH44		X				X	X	X							
GH48	X		X	X					X						
GH51					X	X	X	X							
GH52						X				X	X	X	X	X	X
GH53									X				X		
GH54						X		X							
GH62					X										
GH74		X	X	X	X	X	X								
GH81															X
GH98					X										

[a]Extracellular localization for glycoside hydrolases within each CAZyme class predicted using PSortB 3.0[30] and using the final subcellular localization prediction reported for each CAZyme. Individual sequences for each strain were identified by accessing the CAZy database.[28]

Table 2: Observed[a] glycoside hydrolase encoding genes involved in lignocellulose hydrolysis in select Firmicutes with available gene expression data

CAZyme Module	*Caldicellulosiruptor bescii* DSM 6725	*Caldicellulosiruptor obsidiansis* OB47	*Clostridium cellulolyticum* H10	*Clostridium phytofermentans* ISDg	*Clostridium thermocellum* ATCC 27405
GH2	Athe_1859[A]	COB47_1671[A]	Ccel_1239[S]		Cthe_0405[C,CX,CP,CPX]
GH5			Ccel_2337[C,S,X]		Cthe_0536[C,CX,CP,CPX,CB,Z]
			Ccel_1099[C,S,X]		Cthe_0797[CX,CP,CPX,SWG]
			Ccel_0840[C,S]		Cthe_0821[C,CX,CP,CPX,SWG,Z]
					Cthe_1472[C,CX,CPX,SWG,CB,Z]
					Cthe_2147[C,CX,CP,CPX,SWG,CB]
					Cthe_2193[C,CX,CP,Z]
					Cthe_2872[C,CP,CPX,SWG,Z]
GH8					Cthe_0269[C,CX]
GH9	Athe_1867[A]	COB47_1669[A]	Ccel_0231[C,X,S]	Cphy_3367[X,C]	Cthe_0412[C,CX,CP,CPX,SWG]
	Athe_1865[A]	COB47_1673[A]	Ccel_0731[C,X,S]		Cthe_0413[C,CX,CPX,CB,SWG]
			Ccel_0732[C,X,S]		Cthe_0433[C,CX,CP,CPX,SWG,CB,Z]
			Ccel_0734[C,S]		Cthe_0578[C,CX,CP,SWG]
			Ccel_0735[C,S,X]		Cthe_0624[C,CX,CP,CPX,SWG,Z]
			Ccel_0737[C,S,X]		Cthe_0745[C,CX,CP,CPX,SWG]
			Ccel_0753[C,S,X]		Cthe_0825[C,CX,CP,CPX,CB,SWG]
			Ccel_1249[C,S,X]		Cthe_2760[C,CX,CP,CPX,SWG,Z]
			Ccel_1648[C,X,S]		Cthe_2761[C,CX,CP,CPX,SWG,Z]
			Ccel_2392[C,X,S]		
			Ccel_2621[C,X,S]		
GH10	Athe_1857[A]	COB47_1671[A]	Ccel_0931[C,S,X]	Cphy_0624[X,C]	Cthe_0912[C,CX,CP,CPX,CB]
			Ccel_1230[S]	Cphy_1510[X,C]	Cthe_1838[C, CX,CP,CPX,SWG,CB,Z]
			Ccel_2319	Cphy_2108[X,C]	Cthe_1963[C,CX,CP,CPX,CB,Z]
			Ccel_2320	Cphy_3010[X]	Cthe_2590[C,CX,SWG,CB,z]
				Cphy_3862[X,C]	

Table 2: Observed[a] glycoside hydrolase encoding genes involved in lignocellulose hydrolysis in select Firmicutes with available gene expression data—cont'd

CAZyme Module	Caldicellulosiruptor bescii DSM 6725	Caldicellulosiruptor obsidiansis OB47	Clostridium cellulolyticum H10	Clostridium phytofermentans ISDg	Clostridium thermocellum ATCC 27405
GH11			Ccel_0750^{C,X,S}	Cphy_2105^{X,C}	Cthe_2972^{C,CX,CP,CPX,Z}
GH27				Cphy_1071^{X,C} Cphy_2128^{X,C}	Cthe_0032^{C,CX,CP,CPX,SWG,CB,Z} Cthe_1472^{C,CX,CPX,SWG,CB,Z} Cthe_2811^{C,CX,CP,CPX,SWG}
GH31			Ccel_0649^{C,X,Z}		Cthe_2139^{CX} Cthe_3012^{C,CP,CPX,CB}
GH42				Ccel_1229^{S}	Cthe_0661^{C}
GH44				Ccel_1231^{S} Ccel_1235^{S}	Cthe_1271^{C,CP,CPX,CB,Z}
GH48	Athe_1867^{A}	COB47_1664^{A} COB47_1673^{A}			
GH51			Ccel_0729^{C,X,S}	Cphy_3368^{X,C}	Cthe_2089^{C,CX,CP,CPX,SWG}
GH54					Cthe_1400^{C,CX,CP,CPX,CB,Z}
Reference	34	34	24	35	23

Note: Identified locus tags correspond only to observed gene sequences. Additional sequences within specific CAZy classes exist, although they are not identified.
[a]Observed through either transcriptomic or proteomic analysis during growth on a specific substrate. Substrates identified as superscripts as follows: A = Avicel; C = cellulose; X = xylan; S = wheat straw; P = pectin; SWG = switchgrass; CB = cellobiose; Z = Z-trim (60% cellulose +16% hemicellulose).

Analysis of the *Clostridium phytofermentans* secretome identified that cellulases and hemicel-lulases were upregulated on cells grown on cellulose or hemicellulose in contrast to glucose-grown cells, but under all conditions, the hydrolytic activities of the secretome were higher for hemicellulose hydrolysis rather than cellulose hydrolysis.[35] Thus, although *C. phytofermentans* is able to utilize cellulose, under the conditions tested, the enzymatic machinery produced was more suited toward hemicellulose degradation.

As is shown in Table 2, secretome analysis identifies that, in all cellulolytic *Clostridium* strains for which expression data are available, independent of growth substrate, the expression of cellulose-acting and hemicellulose-acting CAZymes occurs simultaneously. Similar findings were observed in *C. bescii* and *Caldicellulosiruptor obsidiansis*, in which, despite having a significant number of CAZymes annotated (Table 1), only a few CAZyme-related proteins were observed in the secretome (Table 2) of cellulose-grown cells.[34] Thus, it is possible to infer that the expression of diversely acting enzymes may have evolved as a means of coordinating biomass hydrolysis activity between enzyme complexes. This inference is supported by the study of Blumer-Schuette et al.,[25] which found that found that the Avicel-induced secretomes closely matched the xylose-induced secretomes of *Caldicellulosiruptor* spp. that were also capable of degrading microcrystalline cellulose. In contrast, the strains with reduced growth on microcrystalline cellulose—*C. hydrothermalis*, *C. kristjanssonii*, and *C. owensensis*—had more pronounced differences in their Avicel- vs. xylose-induced secretomes. It is also interesting to note that the potential coordination of divergently functioning CAZymes observed in *Caldicellulosiruptor* spp. mimics the expression profiles of *Clostridium* spp., such as *C. thermocellum* and *C. cellulolyticum*, despite apparently secreting far fewer CAZymes (Table 2).

The production of a cocktail of divergently functioning enzymes in cellulolytic Firmicutes, rather than enzymes tailored to a specific function, may serve as a model for designing synergistic cocultures. This is the logic under which the *C. thermocellum-C. bescii* was built and tested.[12] It may be advantageous to couple a strain showing either low expression levels and/or low secretome activity levels for specific lignocellulose polymers to strains proficient in that process. However, using the hydrolysis profiles to develop cocultures is currently limited because only a few organisms have expression data profiles (Table 2) and even fewer have coupled secretome activity analysis to those profiles.

Underlying the differences in expression profiles and hydrolytic potential observed in biofuel-relevant Firmicutes are complex regulatory networks, the nature of which has only begun to be elucidated. Understanding and manipulating these networks may have significant implications for the industrial implementation of any of these strains in mono- or co-culture. However, identifying these mechanisms is further complicated by the realization that significant differences exist amongst the Firmicutes, which have implications in terms of extracellular CAZyme expression. For example, glucose and cellobiose downregulate cellulase expression for *C. phytofermentans*[35] and *C. thermocellum*,[23] respectively, as suggested by the proteomic profile for each strain. In contrast, secretome exo- and endoglucanase expression levels were highest in *C. obsidiansis* on cellobiose-grown cells in comparison to cellulose-grown cells.[26] Thus, at least in the case of *C. obsidiansis*, cellulose hydrolysis products seem to induce, or alternatively prevent repression, of cellulose-encoding sequences.

In *C. thermocellum*, the *celC* cellulase operon is reported to be under the regulatory control of the dissacharide laminaribiose,[36] whereas the study by Raman et al.[23] has suggested that pectin may play a role in *C. thermocellum* xylanase expression. Thus, in these cases, regulation does not seem to be connected to the carbohydrate upon which the enzymes produced act. Further, the *cip-cel* gene cluster in *C. cellulolyticum*, which are major components of the *C. cellulolyticum* cellulosome, are additionally controlled through carbon catabolite repression (CCR) mechanisms.[37] Thus, given the drastic regulatory differences between strains, using expression profiling data to begin to unravel lignocellulose hydrolysis-related regulons is an important component in further development of these organisms.

Omics tools can help to shed significant insight into the potential, both realized and unfulfilled, that an organism can hydrolyze lignocellulosic biomass. Although data sets are currently available for only a few organisms (Table 2), further investigation into those same strains, as well as diverse ones, will help to identify these limitations that can be addressed through the use of an appropriate coculture partner. Further, the use of omic data provides the possibility to evaluate expression profiles of strains in coculture and identify differential expression patterns related to hydrolysis that could be attributed to coculturing itself. Although this approach has not yet been discussed in the literature, the potential it holds may rapidly advance the development of industry-ready microorganisms.

Carbohydrate Utilization in Firmicutes

The hydrolysis of the cellulose and hemicellulose fractions of lignocellulosic biomass will generate a mixed pool of saccharides available for fermentation. Ideally, simultaneous conversion of the resultant hydrolysis products into biofuels will occur with no distinct preference for one substrate over another. However, CCR, in which the presence of one carbon source exerts a regulatory effect on the expression of genes and gene products associated with the utilization of alternative carbon sources,[38] may permit only sequential, and not simultaneous, utilization.

CCR has been reported in many Firmicutes, including biofuel-relevant strains such as *C. cellulolyticum*[37] and *Thermoanaerobacterium saccharolyticum* M2476.[39] In contrast, other Firmicutes, such as *Caldicellulosiruptor saccharolyticus* DSM 8903[40,41] and *Thermoanaerobacter* sp. X514,[42] have been shown to simultaneously co-utilize lignocellulose-relevant saccharides, suggesting a lack of CCR mechanisms in these organisms. In other organisms, such as *T. pseudethanolicus* 39E, glucose has been shown to have a repressive effect on maltodextrin utilization,[43] but not inhibit xylose utilization or synthesis of xylose-related gene products.[44,45] Thus, in these organisms, understanding the specific influences of the CCR regulon is important in understanding carbon flux pathways and potential limitations in substrate usage. In addition, the global nature of CCR requires that omics approaches be implemented to fully understand the metabolic potential of cells in the presence of mixed carbohydrates.

Within the Firmicutes associated with lignocellulose fermentation, there is little biochemical confirmation of the elements of CCR, and we need to anchor our current analysis in other members of the phylum. In Firmicutes, mechanisms of CCR include (1) inducer exclusion, (2) global regulatory control, and (3) specific transcriptional control (for more in-depth reviews of these mechanisms see Ref. 38,46–48). Central to all of these processes is a single protein, HPr (histidine-containing protein), belonging to protein family (pfam) classification 00381.[49] The HPr protein is multifunctional in Firmicutes, and its function in vivo is dependent on its phosphorylation state. When phosphorylated at the His[15] residue by Enzyme I, P-His-HPr acts as a phosphor-carrier protein involved with phosphotransferase system (PTS)-mediated transport. The P-His-HPr transfers its phosphate to a PTS-EIIA protein, which subsequently donates the phosphate to the sugar being transported. Alternatively, when phosphorylated at a conserved Ser[46] residue, the HPr protein plays multiple roles in CCR (discussed below).

A functionally similar HPr-like protein termed *crh* (catabolite repression HPr) is known to exist in some strains and plays a role in CCR similar to P-Ser-HPr.[50,51] Similar to HPr, crh also contains a conserved Ser[46] residue that, when phosphorylated, transforms it into an effector molecule for CCR regulatory proteins. However, unlike HPr, crh contains no His[15] residue and it is not involved in PTS-mediated transport.

Phosphorylation at the Ser[46] residue is catalyzed by a bidirectional HPr kinase/phosphatase (HPrK/P) (pfam07475),[49] the activity of which is modulated by the allosteric activator fructose-1,6-bisphosphate. When activated, the HPr kinase phosphorylates the Ser[46] residue of HPr or crh through the consumption of an ATP. Thus, when intracellular fructose-1,6-bisphosphate concentrations, as well as ATP levels, are high (indicative of the rapid metabolism of a preferred carbon source), P-Ser-HPr and/or P-Ser-crh are readily formed and a CCR effect is observed. Upon depletion of the carbon source, intracellular fructose-1,6-bisphosphate and ATP levels decline and the HPrK/P dephosphorylates P-Ser-HPr, alleviating the CCR effect.

P-Ser-HPr is involved with inducer exclusion as a CCR mechanism. The molecule interacts with transport permeases, preventing transport of alternative carbon sources, which may serve as inducing molecules for their own catabolism.[52] This mechanism has been demonstrated in vivo in *Lactobacillus casei*,[53] in which the addition of glucose immediately arrests maltose uptake in maltose-growing cells. A similar effect was observed in an *L. casei* catabolite control protein A (*ccpA*—see below) mutant, suggesting that the observed CCR was not dependent on *ccpA*. However, the CCR effect was not observed when glucose was added to maltose-growing cells containing an HPr mutation (Ser-46-Ala), showing that P-Ser-HPr plays a direct role in maltose transport. In *Lactobacillus brevis*, it was additionally noted that the P-Ser-HPr bound to inside-out membrane vesicles containing lactose permease protein.[52] Thus, further evidence was provided that the P-Ser-HPr-dependent allosteric regulation of specific transport permeases is involved with inducer exclusion.

In Firmicutes, P-Ser-HPr or P-Ser-crh are also effector molecules for the global transcriptional regulator ccpA, a *lacI/gal*R-family transcriptional repressor protein.[54] The binding of either P-Ser-HPr[55] or P-Ser-crH[51,56] induces a conformational change to ccpA, allowing it to bind consensus catabolite responsive element (*cre*) operator sequences and repress transcriptional activity.

Despite this global response mechanism being reported in numerous Firmicutes, using genomics and bioinformatics to identify homologous gene sequences and gene products is a challenge. First, multiple *lacI/gal*R-family transcriptional repressor paralogs are common within any given genome. Thus, differentiating between sequences involved with specific regulation and the global-ccpA regulator sequence is difficult to do through sequence homology alone. Second, reported *cre* sequences are not only divergent between strains, but the consensus sequences are often degenerate in nature.[57–59] Thus, searching for homologous regions in specific genomes requires the development of custom scripts, which are not publicly accessible. Given the degenerate nature of the *cre* consensus sequences, solely homology-based findings should be also viewed cautiously in the absence of experimental evidence.

The specific transcriptional control result in CCR is mediated through antiterminator proteins containing PTS-regulatory domains (PRDs) (pfam00874) such as the LicT protein in *Bacillus subtilis*.[60] In the absence of PTS-transported sugars, Domain 1 of the PRD protein is phosphorylated by the EIIB protein of the PTS complex, which prevents dimerization and activation of licT. In the presence of PTS-transported sugars, Domain 1 of the PRD protein dephosphorylates by donating the phosphate back to the EIIB protein. At the same time, P-His-HPr phosphorylates Domain 2 of the PRD protein, allowing licT to dimerize and act as an antiterminator of the cognate genes under its regulatory control. Although this is a more specific mode of regulation than *ccpA*-dependent regulation, it also allows for the preferential use of PTS sugars.[48]

Understanding CCR in lignocellulosic biofuel-producing microorganisms is an important component in identifying and developing strategies toward maximizing carbohydrate conversion efficiencies in these organisms. The complexities associated with understanding sugar usage regulons in a single organism are further magnified by potential interstrain interactions that develop in cocultures and may prove to be a significant challenge in developing synergistic cocultures. It is possible to imagine how the hydrolysis products of a single strain may influence the carbon metabolism profile of the coculture partner differently than the innate hydrolysis capabilities found in the coculture partner itself. For example, when grown in monoculture, *C. thermocellum* is known to hydrolyze xylan, but its inability to utilize the resulting products allows for the accumulation of xylodextrins in the fermentation medium.[10] The free xylose or xylodextrins are available for use by an appropriate coculture partner, although their usage may be dependent on the simultaneous availability of cellulose hydrolysis products (also generated through *C. thermocellum*-mediated hydrolysis), which may exert a CCR effect on the coculture partner. Thus, understanding sugar usage preferences, and coupling this knowledge with an understanding of the lignocellulose hydrolysis patterns of strains in coculture, may provide valuable insights into strain selection for coculture design.

Although simple growth studies can be used to identify sugar utilization preferences in microorganisms, genomic and expression profiling facilitate the identification of potential CCR regulons. Although purely in silico analysis, without supporting experimental data, cannot identify which genes are under CCR regulatory control, the identification of *hpr* and *crh* homologs (and whether or not they are expressed under specific growth conditions) can provide insights into molecular engineering targets for purposes of alleviating CCR effects. However, it is important to note that although gene deletion of crh is a suitable engineering strategy, the multiple roles of HPr in vivo do not allow for a similar strategy for mitigating the CCR effects of HPr. HPr knockout strains may alleviate CCR, but they may also simultaneously lose the ability to transport sugar through PTS-mediated mechanisms. However, gene mutation strategies in which the Ser^{46} residue is mutated to prevent phosphorylation of HPr have proven to be successful approaches to remove CCR effects.[39,53]

Omics approaches can be a powerful tool in understanding global and specific regulatory mechanisms related to sugar usage,[42] and the resulting data can be used to make informed decisions when designing cocultures. However, phenotypes observed in monoculture may not be maintained in coculturing approaches. Thus, it is important that similar omics approaches be implemented with cocultures as one approach of understanding interstrain dynamics and further improvement of lignocellulosic biofuel production.

Nutrient Complementation in Cellulolytic Cocultures

No naturally occurring organism identified to date is capable of effective depolymerization of lignocelluloses and complete utilization of derived soluble oligomers or monomers, which is the central theme of CBP. However, robust co-existence of microbial consortia are abundant in nature, which accomplish more challenging tasks through mutually beneficial interactions between different species such as exchange of growth factor or metabolites.[61] Therefore, co-culturing strategies are being adopted in which complementary or synergistic phenotypes in distinct organisms can improve the conversion of lignocellulose into biofuels and/or value-added products.[11,16,62–65]

In silico analyses of amino acid synthesis pathways in known cellulose and biofuel-producing Firmicutes using the Integrated Microbial Genomics (IMG) database[66,67] have been done for 27 mesophiles, 19 thermophiles, and 4 hyperthermophiles (Table 3) to identify the potential for metabolite complementarity for strains in coculture. Among these, glutamine and glutamate are the immediate products of ammonia assimilation and essential nitrogen donors for the synthesis of other intermediates. Amino acids are not only protein precursors, but also precursors for numerous other crucial compounds, such as polyamines, S-adenosylmethionine, pantothenic acid, and nucleotides.[68] Very little information is available in the literature for the pathways of amino acid metabolism and their regulation. The availability of genome sequences has the potential to increase our knowledge of amino acid synthesis in bacteria and facilitate development of cost-reduced minimal media.

Most of the selected Firmicutes associated with lignocellulose fermentation are auxotroph for L-lysine except *C. cellulolyticum* H10, *C. thermocellum* ATCC 27405, *C. thermocellum* DSM 2360, and *Caldicellulosiruptor saccharolyticus* DSM 8903. All listed mesophilic and thermophilic Firmicutes are prototrophic for L-glutamate and L-glutamine and auxotrophic for glycine. In the case of L-histidine, only *Bacillus pumilus* SAFR-032 is prototrophic. Analysis showed the presence of L-proline metabolism in very few organisms, such as *C. phytofermentans* ISDg, *C. cellulolyticum* H10, *Cohnella panacarvi* Gsoil 349, *B. pumilus* SAFR-032, *C. thermocellum* ATCC 27405, *T. pseudethanolicus* 39E, and *Thermoanaerobacter* sp. X514, are prototrophic whereas *Clostridiales* sp. SS3/4 and *Ruminococcus* sp. 18P13 are auxotrophic. Most mesophilic, thermophilic, and hyperthermophilic bacteria are auxotrophic for L-alanine, L-aspartate, L-phenylalanine, L-tyrosine, L-tryptophan, L-arginine, L-asparagine, L-cystein, L-isoleucine, L-leucine, L-serine, L-threonine, and L-valine (Table 4). However, caution has to be taken when looking at such analyses because they can occasionally be misleading. For example, *C. thermocellum* has been shown to grow in amino-acid-free minimal media[12,69] despite genomic analyses to the contrary. Metabolic interactions and cross-feeding of growth nutrients between some biofuel producing cocultures that utilized lignocellulose-derived substrates are briefly discussed below.

By culturing on defined minimal media, it was confirmed that *C. thermocellum*, the fastest known cellulose degrader, is impaired of biosynthesizing four vitamins: biotin, pyridoxamine, vitamin B_{12}, and *p*-aminobenzoic acid.[69] Cocultures of *C. thermocellum* with noncellulolytic *Thermoanaerobacter* strains (X514 or 39E) resulted in 194–440% improvement in ethanol production in comparison to *C. thermocellum* monocultures. The presence of a complete vitamin B12 biosynthesis pathway in strain X514, in contrast to *T. pseudethanolicus* 39E, allowed the *C. thermocellum* X514 coculture to produce 62% more ethanol compared with the *C. thermocellum* 39E coculture.[62] The significance of de novo B12 synthesis was further supported by the realization that the exogenous addition of B12 to culture medium of *C. thermocellum* 39E cocultures showed improved ethanol production comparable to that of

Table 3: Cellulolytic and noncellulolytic mesophilic, thermophilic, and hyperthermophilic Firmicutes

Genome Status	Genome Name	Temperature Optimum	Metabolism
Permanent draft	*Clostridium alkalicellulosi* Z-7026, DSM 17461	35 °C	Cellulose degrader
Draft	*Clostridium termitidis* CT1112, DSM 5398	37 °C	Cellulose degrader
Finished	*Clostridium phytofermentans* ISDg	37 °C	Cellulose degrader, ethanol production, acetate producer
Finished	*Clostridium cellulolyticum* H10	35 °C	Cellulose degrader
Draft	*Clostridium cellulovorans* 743B, ATCC 35296	37 °C	Cellulose degrader
Draft	*Clostridium papyrosolvens* DSM 2782	25 °C	Cellulose degrader, xylan degrader
Finished	*Clostridium saccharolyticum* WM1, DSM 2544	37 °C	Cellulose degrader, ethanol production
Draft	*Clostridium carboxidivorans* P7, DSM 15243	37–40 °C	Solvent producer, acetogenic
Finished	*Clostridium ljungdahlii* PETC, DSM 13528	37 °C	Ethanol production, acetogen
Finished	*Clostridiales* sp. SM4/1	40 °C	Cellulose degrader
Finished	*Clostridiales* sp. SSC/2	40 °C	Cellulose degrader
Finished	*Clostridiales* sp. SS3/4	40 °C	Cellulose degrader
Draft	*Clostridium* sp. URNW	37 °C	Cellobiose-degrading, hydrogen production, acetate producer
Finished	*Butyrivibrio fibrisolvens* 16/4	37 °C	Cellulose degrader
Finished	*Ruminococcus* sp. 18P13	40 °C	Cellulose degrader
Finished	*Ruminococcus* sp. SR1/5	40 °C	Cellulose degrader
Finished	*Ruminococcus albus* 7	40 °C	Cellulose degrader
Draft	*Ruminococcus albus* 8	40 °C	Cellulose degrader
Finished	*Ruminococcus torques* L2-14	40 °C	Cellulose degrader
Finished	*Ruminococcus obeum* A2-162	40 °C	Cellulose degrader
Finished	*Ruminococcus bromii* L2-63	40 °C	Cellulose degrader, ethanol production
Draft	*Ruminococcus flavefaciens* FD-1	37 °C	Cellulose degrader
Finished	*Eubacterium siraeum* V10Sc8a	40 °C	Cellulose degrader
Permanent draft	*Eubacterium cellulosolvens* 6	40 °C	Cellulose degrader
Draft	*Marvinbryantia formatexigens* I-52, DSM 14469	37 °C	Cellulose degrader
Permanent Draft	*Cohnella panacarvi* Gsoil 349, DSM 18696	30 °C	Xylan degrader
Finished	*Bacillus pumilus* SAFR-032	37 °C	Biomass degrader
Draft	*Clostridium stercorarium* BW, DSM 8532	65 °C	Cellulose and xylan degrader
Finished	*Clostridium thermocellum* ATCC 27405	60 °C	Cellulose degrader, ethanogenic

Continued

Table 3: Cellulolytic and noncellulolytic mesophilic, thermophilic, and hyperthermophilic Firmicutes—cont'd

Genome Status	Genome Name	Temperature Optimum	Metabolism
Draft	*C. thermocellum* DSM 2360	60 °C	Cellulose degrader, ethanol production, ethanogenic
Finished	*C. thermocellum* LQ8, DSM 1313	60 °C	Cellulose degrader, ethanogenic, ethanol production
Finished	*Clostridium clariflavum* EBR 45, DSM 19732	55 °C	Cellulose degrader
Draft	*Thermoanaerobacter ethanolicus* JW 200	60 °C	Xylose consumer, ethanol production
Finished	*Thermoanaerobacter pseudethanolicus* 39E, ATCC 33223	65 °C	Sugars fermentor, iron reducer, ethanol production
Draft	*Thermoanaerobacter thermohydrosulfuricus* WC1	60 °C	Xylan degrader
Finished	*Thermoanaerobacter brockii finnii* Ako-1, DSM 3389	65 °C	Saccharolytic
Finished	*Thermoanaerobacter italicus* Ab9, DSM 9252	70 °C	Saccharolytic
Finished	*Thermoanaerobacterium xylanolyticum* LX-11	60 °C	Saccharolytic
Finished	*Thermoanaerobacter* sp. X514	60 °C	Solvent producer
Finished	*Thermoanaerobacterium thermosaccharolyticum* DSM 571	60 °C	Cellulose degrader
Finished	*Thermoanaerobacterium saccharolyticum* JW/SL-YS485, DSM 8691	30–66 °C	Xylan degrader
Finished	*Caldicellulosiruptor lactoaceticus* 6A, DSM 9545	68–75 °C	Cellulose degrader, biomass degrader, nitrogen producer
Finished	*Caldicellulosiruptor bescii* Z-1320, DSM 6725	75 °C	Cellulose degrader
Draft	*Caldicellulosiruptor lactoaceticus* 6A, DSM 9545	68–75 °C	Cellulose degrader, biomass degrader, nitrogen producer
Finished	*Caldicellulosiruptor kronotskyensis* 2002	70 °C	Cellulose degrader, biomass degrader, nitrogen producer
Finished	*Caldicellulosiruptor saccharolyticus* DSM 8903	65 °C	Cellulose degrader, biomass degrader, nitrogen producer
Finished	*Caldicellulosiruptor kristjanssonii* 177R1B, DSM 12137	78 °C	Cellulose degrader, nitrogen producer, biomass degrader
Finished	*Caldicellulosiruptor hydrothermalis* 108	79 °C	Cellulose degrader, biomass degrader, nitrogen producer
Finished	*Caldicellulosiruptor obsidiansis* OB47	79 °C	Cellulose degrader, biomass degrader, nitrogen producer
Finished	*Caldicellulosiruptor owensensis* OL	79 °C	Cellulose degrader, biomass degrader, nitrogen producer

Table 4: In silico analyses of amino acid synthesis of mesophilic, thermophilic, and hyperthermophilic Firmicutes from IMG/ER

Genome Name	Lys	Glu	Ala	Asp	Phe	Tyr	Trp	His	Gly	Arg	Asn	Cys	Gln	Ile	Leu	Pro	Ser	Thr	Val
Mesophiles																			
C. alkalicellulosi Z-7026, DSM 17461	–	P	–	–	A	A	A	A	–	A	P	–	P	A	A	–	A	A	A
C. termitidis CT1112, DSM 5398	–	P	P	A	A	A	A	A	A	A	P	–	P	A	A	–	A	A	A
C. phytofermentans ISDg	–	P	P	P	A	A	A	A	A	P	P	P	P	P	P	P	A	–	P
C. cellulolyticum H10	P	P	P	P	A	A	P	A	A	P	P	–	P	A	A	P	A	A	P
C. cellulovorans 743B, ATCC 35296	A	P	P	P	A	A	A	A	A	–	P	–	–	A	A	–	–	–	A
C. papyrosolvens DSM 2782	–	P	P	A	A	A	A	A	A	–	P	–	–	A	A	–	A	A	P
C. saccharolyticum WM1, DSM 2544	–	P	–	P	A	A	A	A	A	–	P	P	–	–	A	–	A	A	–
Clostridiales sp. SSC/2	–	P	P	A	A	A	A	A	A	A	P	–	–	A	A	–	A	A	A
Clostridiales sp. SM4/1	A	P	A	A	A	A	A	A	A	A	P	A	–	A	A	–	A	A	A
Clostridiales sp. SS3/4	–	P	–	A	A	A	A	A	A	A	P	–	–	A	A	A	–	A	A
Clostridium sp. URNW	A	P	–	–	A	A	A	A	A	A	P	–	–	A	A	–	–	–	A
Butyrivibrio fibrisolvens 16/4	–	–	A	P	A	A	A	A	A	A	–	–	P	–	A	–	–	A	–
Ruminococcus sp. 18P13	–	P	A	A	A	A	A	A	A	–	–	A	P	A	A	A	A	A	A
Ruminococcus sp. SR1/5	A	P	–	A	A	A	A	A	A	A	P	–	–	A	A	–	–	A	A
R. albus 7	–	P	A	A	A	A	A	A	A	–	P	P	P	A	A	–	–	–	A
R. albus 8	–	P	A	A	A	A	A	A	A	–	P	P	P	A	A	A	–	–	A
R. torques L2-14	–	P	–	A	A	A	A	A	A	A	P	–	–	A	A	–	–	A	A
R. obeum A2-162	–	P	–	–	A	A	A	A	A	A	P	–	–	A	A	–	A	A	A
R. bromii L2-63	A	P	–	A	A	A	A	A	A	A	A	–	P	A	A	–	A	A	A
R. flavefaciens FD-1	A	P	A	A	A	A	A	A	A	–	P	P	P	P	A	–	–	A	P
E. siraeum V10Sc8a	A	–	A	–	A	A	A	A	A	A	–	–	–	A	A	A	A	A	A
E. cellulosolvens 6	–	P	P	A	A	A	A	A	A	A	P	–	–	A	A	–	–	A	A
M. formatexigens I-52, DSM 14469	–	P	–	P	A	A	A	A	A	–	P	–	–	–	A	–	A	A	–

Continued

Table 4: In silico analyses of amino acid synthesis of mesophilic, thermophilic, and hyperthermophilic Firmicutes from IMG/ER—cont'd

Genome Name	Lys	Glu	Ala	Asp	Phe	Tyr	Trp	His	Gly	Arg	Asn	Cys	Gln	Ile	Leu	Pro	Ser	Thr	Val
Mesophiles																			
C. carboxidivorans P7, DSM 15243	A	P	P	–	A	A	A	A	A	A	P	–	–	A	A	–	A	–	A
C. ljungdahlii PETC, DSM 13528	–	P	–	–	A	A	A	A	A	A	P	–	–	A	A	–	A	–	A
C. panacarvi Gsoil 349, DSM 18696	–	P	–	P	A	A	A	A	A	–	–	–	–	A	P	P	A	–	A
B. pumilus SAFR-032	A	P	P	P	P	P	P	P	P	P	P	P	P	P	P	P	A	P	P
Thermophiles and hyperthermophiles																			
T. ethanolicus JW 200	A	P	–	–	A	A	A	A	A	A	P	–	–	A	A	–	A	–	A
C. thermocellum ATCC 27405	P	P	P	P	A	A	P	A	A	P	P	P	P	–	P	P	A	A	P
C. thermocellum DSM 2360	P	P	P	A	A	A	A	A	A	P	P	P	P	A	A	–	A	A	P
C. thermocellum LQ8, DSM 1313	–	P	P	A	A	A	A	A	A	P	P	P	P	A	A	–	A	A	A
C. clariflavum EBR 45, DSM 19732	–	P	–	A	A	A	A	A	–	–	P	–	–	A	A	–	A	–	A
C. stercorarium BW, DSM 8532	–	P	P	P	A	A	A	A	–	–	P	–	P	A	A	–	A	A	A
T. pseudethanolicus 39E, ATCC 33223	A	P	P	P	P	P	P	P	A	P	P	P	P	–	P	P	P	P	P
T. thermohydrosulfuricus WC1	A	P	–	–	A	A	A	A	A	A	P	–	–	A	A	–	A	–	A
T. saccharolyticum JW/ SL-YS485, DSM 8691	A	P	P	–	A	A	A	A	A	A	P	–	–	A	A	–	A	–	A

Organism																
T. brockii finnii Ako-1, DSM 3389	A	P	–	A	A	A	A	A	A	P	–	–	A	A	–	A
T. italicus Ab9, DSM 9252	A	P	–	A	A	A	A	A	–	P	–	–	A	A	–	A
T. xylanolyticum LX-11	A	P	–	A	A	A	A	A	A	P	–	–	A	A	–	A
Thermoanaerobacter sp. X514	A	P	P	P	P	P	A	P	–	P	P	–	P	A	P	P
T. thermosaccharolyticum DSM 571	A	P	–	A	A	A	A	A	A	P	–	–	A	A	–	A
C. lactoaceticus 6A, DSM 9545	A	P	–	A	A	A	A	–	A	P	–	P	A	A	–	A
C. bescii Z-1320, DSM 6725	–	P	–	A	A	A	A	A	–	P	–	P	P	–	–	P
C. lactoaceticus 6A, DSM 9545	A	P	–	A	A	A	A	A	A	P	–	P	A	A	–	A
C. kronotskyensis 2002	–	P	–	A	A	A	A	–	A	P	P	P	A	A	–	A
C. saccharolyticus, DSM 8903	P	P	P	A	A	A	A	A	A	P	P	P	P	–	P	P
C. kristjanssonii 177R1B, DSM 12137	–	P	–	A	A	A	A	–	A	P	–	P	A	A	–	A
C. hydrothermalis 108	–	P	–	A	A	A	A	–	A	P	–	P	A	A	–	A
C. obsidiansis OB47	–	P	–	A	A	A	A	–	A	P	–	P	A	A	–	–
C. owensensis OL	–	P	–	A	A	A	A	–	A	P	–	P	A	A	–	A

A = auxotrophic, P = prototrophic, L-Lysine = Lys, L-glutamate = Glu, L-alanine = Ala, L-aspartate = Asp, L-phenylalanine = Phe, L-tyrosine = Tyr, L-tryptophan = Trp, L-histidine = His, Glycine = Gly, L-arginine = Arg, L-asparagine = Asn, L-cysteine = Cys, L-glutamine = Gln, L-isoleucine = Ile, L-leucine = Leu, L-proline = Pro, L-serine = Ser, L-threonine = Thr, L-valine = Val.

C. thermocellum X514 cocultures.[70] Exchange of substrate and growth factors was also observed between *Clostridium* strain C7, a mesophilic cellulose degrader, and *Klebsiella* strain W1, a noncellulolytic bacterium. On defined medium, *Klebsiella* utilized soluble sugars released by *Clostridium* and excreted biotin and *p*-aminobenzoic acid that were required for the growth of *Clostridium*.[71]

Two extreme thermophiles, *C. saccharolyticus* DSM 8903 and *C. kristjanssonii* DSM 12137, exhibited a stable coculture on glucose and xylose for 70 days in a chemostat at different dilutions. The H_2 yield of 3.7 mol/mol glucose obtained from the combined culture was higher than those from monocultures by either organism. When *C. kristjanssonii* was grown on glucose with and without the addition of cell-free culture broth of *C. saccharolyticus*, the lag phase of *C. kristjanssonii* was shortened with 18% higher biomass yield. On the basis of this observation, it was concluded that a growth enhancement compound for *C. kristjanssonii* was supplied by *C. saccharolyticus* growth supernatant.[65] This compound is possibly related to the biosynthesis of one or more of the four amino acids L-Aspartate, L-leucine, L-valine, and L-threonine because *C. saccharolyticus* is a prototroph and *C. kristjanssonii* is an auxotroph as revealed by the amino acid metabolism data presented in Table 4. However, detailed genomic- and proteomic-level investigation involving the amino acid biosynthesis pathways of these organisms is required to confirm such assumptions.

Mutually beneficial interactions have been reported between aerobe-anaerobe and chemo-photoheterotroph organisms.[63,72] During an investigation on a stable consortium of five bacterial strains, synergistic relationships were detected among an anaerobic cellulolytic bacterium (*C. straminisolvens* CSK1) and two strains of aerobic bacteria (*Pseudoxanthomonas* sp. strain M1-3 and *Brevibacillus* sp. strain M1-5). The aerobes introduced an anaerobic condition whereas the anaerobe supplied metabolites (acetate and glucose), and cellulose degradation was more efficient in the presence of these aerobes, resembling perfect conditions for symbiosis.[72] The cellulosic hydrogen production rate by *C. cellulolyticum* H10 doubled with 1.6-times more total accumulation when cocultured with *Rhodopseudomonas palustris* CGA676, a photoheterotrophic facultative aerobe, as a result of the higher (2-fold) growth rate and cell density (2.6 fold) of *C. cellulolyticum* compared with its monoculture. Removal of acetate and pyruvate, two major metabolites of *C. cellulolyticum*, by *R. palustris*, was identified as the beneficial effect of the co-culture that reduced end product inhibition and pH drops.[63]

A sequential culture of *Zymomonas anaerobia* and *C. thermocellum* was attempted by growing each culture separately for 3 days and then inoculating *C. thermocellum* cultures with *Z. anaerobia* followed by incubation of the co-culture at 37 °C. Ethanol yield with the co-culture on 1% cellulose was almost 9 times greater (2.7 g/L) than the value obtained from *C. thermocellum* alone.[73] Ethanol produced by the co-culture from the steam-exploded wood fraction was similar to that from equivalent amounts of solka floc.[74] Under a high pH environment (pH = 9), synergistic effects of a coculture consisting of *C. themocellum* and *Clostridium thermolacticum* were studied through fermentation of lignocellulose derivatives (xylose, cellobiose, and cellulose) into ethanol. The lag period of fermentation was always shorter in this co-culture compared with monocultures and consistently yielded several fold more ethanol than monocultures.[75] Enhanced production was witnessed by these studies, and only cross-feeding of growth substrate could not sufficiently explain the underlying cause. Because none of these studies explored the possible sharing of metabolites released by co-culturing species, in-depth analyses applying omics tools such as metabolomics is warranted. Indeed, a recent analysis of secreted metabolites in *C. thermocellum* ATCC 27405 has demonstrated significant secretion of a broad range of amino acids into medium when cells are grown on Avicel.[76] This excretion may facilitate the growth of other organisms in their environment, permitting growth of both in a minimal medium.

Interdependence of participants in cellulose-degrading cocultures, as discussed above, mimics the syntrophy of anaerobic microflora in a natural environment where the conversion is slow but efficient. In addition to enhanced rates of cellulose degradation and biofuels production, cross-feeding between co-culturing organisms may allow (1) elimination of vitamins, reducing agents and/or pH control chemicals from growth medium and (2) utilization of undesired metabolites and leftover substrates. In turn, these would result in medium cost-savings and waste minimization, which translates to a more economically competitive bioprocess.

Regulation of Microbial Interactions: Quorum Sensing

Quorum sensing (QS) is a process of cell-to-cell communication during which individual cells respond to the extracellular signaling molecules called autoinducers or small molecules (SMs) by regulating their phenotype.[77,78] Microbial cells secrete these SMs into the environment, and the SMs bind to sensory proteins, causing an effect on the transcription and translation of certain proteins either directly or indirectly. To regulate various physiological functions, this type of control of gene expression in response to cell density is utilized by gram-positive and gram-negative bacteria. Although common signal themes exist, there are differences in the design of extracellular signals, signal transduction apparatuses, and the biochemical pathways related to signal detection that have permitted QS systems to evolve accordingly depending on their specific uses. Current studies have shown modulation of inter- and intraspecies cell-to-cell communication via QS and further facilitating bacteria to architect complex community structures.[77]

Table 5 summarizes the known QS systems found in Clostridia along with their functional regulation. Cell-to-cell signaling among bacterial populations or individuals enables their ability to exhibit complex regulatory function. The exchange of dedicated signal molecules between or within populations facilitates communication in natural consortia.[61] Bacterial cells within populations communicate and coordinate growth, whether during infections or

Table 5: Summary of all of the known different types of QS systems found in Clostridia with its specific genes associated with corresponding functional regulation

Microorganism	QS System	Genes	Functions	References
Clostridium acetobutylicum	Agr	agrBCDA spo0A	Solventogenesis, sporulation	81–83
Clostridium acetobutylicum	Agr	agrBCDA	Energy storage molecule granule formation	81,82
Clostridium perfringens	LuxS	Luxs/A1-2; pfo	Toxin production: a, k, and f toxins	84
Clostridium perfringens	Agr	agrBCDA/spo0A	Toxin production: Perfringolysin O and a-toxin; enterotoxin CPE and b2 toxin; sporulation	80,85
Clostridium botulinum	Agr	agr-1 agr-2	Sporulation; neurotoxin production	86
Clostridium saccharoperbutylacetonicum	Agr-like	sol	Metabolic switch: acidogenesis to solventogenesis	87
Thermoanaerobacter tencongensis	Agr	agrBCDA	Biofilm formation	80,88
Desulfitobacterium hafniensis Y51	Agr	agrBCDA	Biofilm formation in cocultures	78,89
Clostridium difficle	LuxS	Luxs/A1-2; pfo	Toxin production and synthesis of other virulence factors	77

in biofilms, by exchanging acyl-HSL signal molecules (gram-negative bacteria) or small peptides (gram-positive bacteria).[79] The most widespread QS system observed in the Firmicutes is the *agr* signal transduction system. The *agr* cell-to-cell signaling system is evolutionarily conserved across Firmicutes, with its general functional regulations being the same in different species, but the individual regulated genes are slightly different.[80] Recent bioinformatics analysis indicates the occurrence of an *agr*-type QS system in most of the sequenced species of Clostridia, although experimental reports have confirmed only in a few species until now.[80] Using comparative genomics, an *agr*-type QS system was predicted to be functional in the three strains of *Clostridium perfringens* ATCC 13124, strain 13, and *C. perfringens* SM101 that also contained an AgrB homologue that is not associated with a two-component QS system. Hence, the researchers suggested that a complete two-component sensing system might be encoded elsewhere in the genome rather than being associated with *agrB*.[80] All of the three strains of *C. perfringens* consisted of a single copy of AgrD and two copies of AgrB domains. Further investigations suggested the co-occurrence of one of the two paralogues of AgrB with a histidine sensor kinase in *C. perfringens* ATCC 13124 and strain 13. An uncharacterized transmembrane protein with conserved sequences was found closer to the *agr* locus in the genome of the three strains. Short open reading frames with functional similarity and not detectable sequence similarity to *agrD* were identified in all of the three strains of *C. perfringens* as well as *Bacillus halodurans*.[80]

Whole genome analysis of different strains of Group I *Clostridium botulinum* demonstrated the presence of two specific *agr* loci (*agr-1* and *agr-2*), each encoding putative proteins similar to AgrB and AgrD of the *Staphylococcus aureus agr* QS system.[86] The genome of *Clostridium acetobutylicum* encodes a peptide-based QS system that involves a putative *agr* locus, *agrBDCA*, similar to that observed in Staphylococcci.[81,90] In *C. acetobutylicum*, the transitional-state regulator *abrB* has three homologs: *abrB0310*, *abrB1941*, and *abrB3647*.[83] Recent studies involving full genome microarray analysis have suggested *abrB0310* could be the true transitional-state regulator, and it showed the highest expression level just after the onset of sporulation.[91] In comparison to *B. subtilis*, maximum expression of *abrB* is achieved just 2 h before the onset of sporulation.[92] The researchers speculate that the microarray probe was picking up signals for all of the three genes.[91] Hence, in-depth proteomic analysis is required to confirm the expression profiles of the *abrB* genes.

Comparative genomic study in *Caldanaerobacter subterraneus* subsp. *tengcongensis*, *Desulfitobacterium hafniense* Y51, and *Moorella thermoacetica* ATCC 39070 showed the presence of *agrB* next to a gene encoding the response regulator and a gene that encodes histidine kinase, suggesting a functional peptide QS circuit.[78,80,88] Desulfitobacteria is widely used in anaerobic bioremediation processes because it is known to utilize different substances such as metals, sulfite, nitrate, and halogenated compounds as electron acceptors. Experiments have shown the occurrence of *D. hafniense* in a biofilm with sulfate-reducing bacteria in anaerobic fixed-film reactors.[89] If cell-to-cell signaling occurs through the expression of the *agr* locus, then modulation of biofilm formation occurs in cocultures of *D. hafniense*.[80]

Bacteria in consortia or cocultures can perform cell-to-cell signaling through sharing metabolites. By exchanging metabolic intermediates, the microbes in a consortium could exert positive and negative control over each other's activities. To design a suitable coculture, understanding, and potentially engineering, cell-to-cell signaling is an essential process to address. Hence, in-depth knowledge of the different components of the QS system and their regulation is very essential.

Engineering a signal response system with specific signal molecules could also be used for studying bacterial behaviors in a given population or to mimic their interactions under controlled physiological conditions. Apart from exchanging specific signal molecules between cells, cell-to-cell signaling could also involve the exchange of metabolites involved in growth and metabolism.[93] Bacterial signaling and metabolism are closely interconnected because bacteria are actively striving to enhance their metabolic state with a corresponding increase in cell density. Hence, it is very complex to distinguish the regulation of gene expression on the basis of signaling function from that of the differential expression of genes stimulated by the metabolic activity of the cell.

Conclusions

The number of isolated organisms with potential for use in lignocellulose fermentation is large and still growing. There are many instances in which genomic databases, such as IMG,[66,67] provide sequence information for organisms that have not been characterized beyond an initial isolation paper. With the number of poorly characterized organisms available for further study, a triage using the sequence information may be useful to focus our attention on organisms for which the potential interest may not be as apparent from a quick initial characterization. Genomics as a tool to gain insight into the physiology of these organisms is still a rather blunt instrument because many of the genes are still somewhat poorly characterized, leading to ambiguous annotations, which means that genomic interpretation should still be treated with caution. Despite its current shortcomings, genomic analysis can provide some insight into the physiological potential of these Firmicutes. Furthermore, whole organism analysis tools such as transcriptomics and proteomics do provide insights into the specific complements of gene products available to cells under specific growth conditions. Although this knowledge should allow us to better exploit specific organisms, the use of monocultures for CBP of specific raw lignocellulosic substrates does limit us to the capabilities of that single organism.

The underlying concept in the development of designer microbial consortia for CBP is to mimic synergistic physiologies that exist in natural communities while simultaneously limiting the complexities so as maintain an operational understanding of the system to ensure process consistency. Achieving this is a seemingly daunting task, particularly given that the physiology of most lignocellulosic biofuel-producing organisms is poorly understood in monoculture, let alone in combination with other microorganisms. However, the increasing availability of "omic" data for these strains allows, for the first time, detailed insights into global cellular processes that have previously not been possible. Although informative in monoculture studies, these data sets can be further mined for the identification of potential limitations in biofuel-relevant physiological processes in monoculture, which can be addressed through the careful selection of a co-culture partner with complementary physiology (e.g., complementary lignocellulose hydrolysis or nutritional requirements).

Natural microbial communities have allowed for individual strains to adopt niche diversification and specialization strategies in which survival is dependent on community-level dynamics. A current challenge in moving past purely empirical observations of co-culture phenotypes is the lack of understanding of interstrain interactions. Thus, the ability to couple omics data sets from experiments from divergent organisms (e.g., GH expression/activity with carbon utilization profiles) may explain observed co-culture phenotypes and eventually permit predictive insights into the expected phenotype of proposed cocultures. This can be even further advanced by improving our understanding of interspecies "cross-talk" mechanisms (e.g., QS or signaling molecule production). Understanding the signaling components that induce physiological changes, and being able to manipulate them to better mimic observed naturally occurring synergies, may be a significant advancement in improving lignocellulosic biofuel production.

In addition, although catabolic processes (e.g., biomass hydrolysis, carbon utilization, end product catabolism, etc.) are often the focal point in many lignocellulosic biofuel production studies, omics data also allow for a deeper investigation into pertinent anabolic processes. In laboratory settings, the functional "role" of microbial communities is replaced by providing all components needed for growth in the culture medium itself. However, these have inherent disadvantages in terms of cost and process simplification at industrial levels. Coupling auxotrophic strains with prototrophic strains for essential anabolic processes may help to reduce or even avoid the addition of costly medium components. Identification of these anabolic processes can be greatly enhanced through genomics as well as targeted or "deep"-expression profiling strategies.

The use of omics approaches allows for organisms to be viewed as whole entities and not as products of individual physiological processes. In doing so, informed decisions in regards to designer co-culture construction may be facilitated to an extent not possible through pure empirical experimentation. Further, the implementation of these strategies in co-culture experiments, and not just monoculture experiments, can help to unravel some of the complexities associated with interstrain interactions at a depth not previously possible.

References

1. Lynd LR, Elander RT, Wyman CE. Likely features and costs of mature biomass ethanol technology. *Appl Biochem Biotechnol* 1996;**57–58**:741–61.
2. Lynd LR, Weimer PJ, van Zyl WH, Pretorius IS. Microbial cellulose utilization: fundamentals and biotechnology. *Microbiol Mol Biol Rev* 2002;**66**:506–77.
3. Lynd LR, van Zyl WH, McBride LE, Laser M. Consolidated bioprocessing of cellulosic biomass: an update. *Curr Opin Biotechnol* 2005;**16**(5):577–83.
4. Lynd LR, Laser MS, Bransby D, Dale BE, Davison B, Hamilton R, et al. How biotech can transform biofuels. *Nat Biotechnol* 2008;**26**:169–72.
5. Izquierdo JA, Sizova MV, Lynd LR. Diversity of bacteria and glycosyl hydrolase family 48 genes in cellulolytic consortia enriched from thermophilic biocompost. *Appl Envir Microbiol* 2010;**76**:3545–53.
6. van der Lelie D, Taghavi S, McCorkle SM, Li L-L, Malfatti SA, Monteleone D, et al. The metagenome of an anaerobic microbial community decomposing poplar wood chips. *PLoS One* 2012;**7**:e36740.
7. Sizova MV, Izquierdo JA, Panikov NS, Lynd LR. Cellulose- and xylan-degrading thermophilic anaerobic bacteria from biocompost. *Appl Envir Microbiol* 2011;**77**:2282–91.
8. Guedon E, Payot S, Desvaux M, Petitdemange H. Carbon and electron flow in *Clostridium cellulolyticum* grown in chemostat culture on synthetic medium. *J Bacteriol* 1999;**181**:3262–9.
9. Wang DI, Biosic I, Fang Y, Wang JD. Direct microbiological conversion of cellulosic biomass to ethanol. In: *Proceedings of the 3rd annual biomass energy systems conference*. Springfield (VA): National Technical Information Service; 1979. p. 61–7.
10. Ng TK, Ben-Bassat A, Zeikus JG. Ethanol production by thermophilic bacteria: fermentation of cellulosic substrates by cocultures of *Clostridium thermocellum* and *Clostridium thermohydrosulfuricum*. *Appl Envir Microbiol* 1981;**41**(6):1337–43.

11. Mori Y. Characterization of a symbiotic coculture of *Clostridium thermohydrosulfuricum* YM3 and *Clostridium thermocellum* YM4. *Appl Envir Microbiol* 1990;**56**(1):37–42.

12. Kridelbaugh DM, Nelson J, Engle NL, Tschaplinski TJ, Graham DE. Nitrogen and sulfur requirements for *Clostridium thermocellum* and *Caldicellulosiruptor bescii* on cellulosic substrates in minimal nutrient media. *Bioresour Technol* 2013;**130**:125–35.

13. He Q, Hemme CL, Jiang H, He Z, Zhou J. Mechanisms of enhanced cellulosic bioethanol fermentation by co-cultivation of *Clostridium* and *Thermoanaerobacter* spp. *Bioresour Technol* 2011;**102**(20):9586–92.

14. Onyenwoke RU, Kevbrin VV, Lysenko AM, Wiegel J. *Thermoanaerobacter pseudethanolicus* sp. nov., a thermophilic heterotrophic anaerobe from yellowstone national park. *Int J Syst Evol Microbiol* 2007; **57**:2191–3.

15. Murray WD. Symbiotic relationship of *Bacteroides cellulosolvens* and *Clostridium saccharolyticum* in cellulose fermentation. *Appl Envir Microbiol* 1986;**51**(4):710–4.

16. Liu Y, Yu P, Song X, Qu Y. Hydrogen production from cellulose by coculture of *Clostridium thermocellum* JN4 and *Thermoanaerobacterium thermosaccharolyticum* GD17. *Int J Hydrog Energy* 2008;**33**:2927–33.

17. Geng A, He Y, Qian C, Yan X, Zhou Z. Effect of key factors on hydrogen production from cellulose in a co-culture of *Clostridium thermocellum* and *Clostridium thermopalmarium*. *Bioresour Technol* 2010; **101**:4029–33.

18. Li Q, Liu C. Co-culture of *Clostridium thermocellum* and *Clostridium thermosaccharolyticum* for enhancing hydrogen production via thermophilic fermentation of cornstalk waste. *Int J Hydrog Energy* 2012; **37**:10648–54.

19. Carere RC, Rydzak T, Verbeke TJ, Cicek N, Levin DB, Sparling R. Linking genome content to biofuel production yields: a meta-analysis of major catabolic pathways among select H$_2$ and ethanol-producing bacteria. *BMC Microbiol* 2012;**12**:295.

20. Roberts SB, Christopher MG, Brooks JP, Fong SS. Genome-scale metabolic analysis of *Clostridium thermocellum* for bioethanol production. *BMC Syst Biol* 2010;**4**:31.

21. Chundawat SPS, Beckham GT, Himmel ME, Dale BE. Deconstruction of lignocellulosic biomass to fuels and chemicals. *Annu Rev Chem Biomol Eng* 2011;**2**:121–45.

22. Das SP, Ravindran R, Ahmed S, Das D, Goyal D, Fontes CM, et al. Bioethanol production involving recombinant *C. thermocellum* hydrolytic hemicellulase and fermentative microbes. *Appl Biochem Biotechnol* 2012;**167**:1475–88.

23. Raman B, Pan C, Hurst GB, Rodriguez Jr M, McKeown CK, Lankford RK, et al. Impact of pretreated switchgrass and biomass carbohydrates on *Clostridium thermocellum* ATCC 27405 cellulosome composition: a quantitative proteomic analysis. *PLoS One* 2009;**4**:e5271.

24. Blouzard JC, Coutinho PM, Fierobe HP, Henrissat B, Lignon S, Tardif C, et al. Modulation of cellulosome composition in *Clostridium cellulolyticum*: adaptation to the polysaccharide environment revealed by proteomic and carbohydrate-active enzyme analyses. *Proteomics* 2010;**10**:541–54.

25. Blumer-Schuette SE, Lewis DL, Kelly RM. Phylogenetic, microbiological, and glycoside hydrolase diversities within the extremely thermophilic, plant biomass-degrading genus *Caldicellulosiruptor*. *Appl Envir Microbiol* 2010;**76**:8084–92.

26. Lochner A, Giannone RJ, Keller M, Antranikian G, Graham DE, Hettich RL. Label-free quantitative proteomics for the extremely thermophilic bacterium *Caldicellulosiruptor obsidiansis* reveal distinct abundance patterns upon growth on cellobiose, crystalling cellulose, and switchgrass. *J Proteome Res* 2011;**10**:5302–14.

27. Park BH, Karpinets TV, Syed MH, Leuze MR, Uberbacher EC. CAZymes analysis Toolkit (CAT): web service for searching and analyzing carbohydrate-active enzymes in a newly sequenced organism using CAZy database. *Glycobiology* 2010;**20**:1574–84.

28. Cantarel BL, Coutinho PM, Rancurel C, Bernard T, Lombard V, Henrissat B. The carbohydrate-active enZymes database (CAZy): an expert resource for Glyocgenomics. *Nucleic Acids Res* 2009;**37**:D233–8.

29. Verbeke TJ, Zhang X, Henrissat B, Spicer V, Rydzak T, Krokhin OV, et al. Genomic evaluation of *Thermoanaerobacter* spp. for the construction of designer co-cultures to improve lignocellulosic biofuel production. *PLoS One* 2013;**8**(3):e59362. http://dx.doi.org/10.1371/journal.pone.0059362. Accepted: PONE-D-12–33595.

30. Yu NY, Wagner JR, Laird MR, Melli G, Rey S, Lo R, et al. PSORTb 3.0: improved protein subcellular localization prediction with refined localization subcategories and predictive capabilities for all prokaryotes. *Bioinformatics* 2010;**26**:1608–15.

31. Choo KH, Tan TW, Ranganathan S. A comprehensive assessment of N-terminal signal peptides prediction methods. *BMC Bioinforma* 2009;**10**:S2.

32. Wiegel J, Mothershed CP, Puls J. Differences in xylan degradation by various noncellulolytic thermophilic anaerobces and *Clostridium thermocellum*. *Appl Envir Microbiol* 1985;**49**:656–9.

33. Strobel HJ, Caldwell FC, Dawson KA. Carbohydrate transport by the anaerobic thermophile *Clostridium thermocellum* LQRI. *Appl Envir Microbiol* 1995;**61**:4012–5.

34. Lochner A, Giannone RJ, Rodriguez Jr M, Shah MB, Mielenz JR, Keller M, et al. Use of label-free quantitative proteomics to distinguish the secreted cellulolytic systems of *Caldicellulorisuptor bescii* and *Caldicellulosiruptor obsidiansis*. *Appl Envir Microbiol* 2011;**77**:4042–54.

35. Tolonen AC, Haas W, Chilaka AC, Aach J, Gygi SP, Church GM. Proteome-wide systems analysis of a cellulosic biofuel-producing microbe. *Mol Syst Biol* 2011;**7**:461.

36. Newcomb M, Chen C-Y, Wu JHD. Induction of the *celC* operon of *Clostridium thermocellum* by laminaribiose. *Proc Natl Acad Sci USA* 2007;**104**:3747–52.

37. Abdou L, Boileau C, de Philip P, Pagès S, Fiérobe HP, Tardif C. Transcriptional regulation of the *Clostridium cellulolyticum cip-cel* operon: a complex mechanism involving a catabolite-responsive element. *J Bacteriol* 2008;**190**:1499–506.

38. Brückner R, Titgemeyer F. Carbon catabolite repression in bacteria: choice of the carbon source and autoregulatory limitation of sugar utilization. *FEMS Microbiol Lett* 2002;**209**:141–8.

39. Tsakraklides V, Shaw AJ, Miller BB, Hogsett DA, Herring CD. Carbon catabolite repression in *Thermoanaerobacterium saccharolyticum*. *Biotechnol Biofuels* 2012;**5**:85.

40. van de Werken HJG, Verhaart MRA, VanFossen AL, Willquist K, Lewis DL, Nichols JD, et al. Hydrogenomics of the extremely thermophilic bacterium *Caldicellulosiruptor saccharolyticus*. *Appl Envir Microbiol* 2008;**74**:6720–9.

41. VanFossen AL, Verhaart MRA, Kengen SMW, Kelly RM. Carbohydrate utilization patterns for the extremely thermophilic bacterium *Caldicellulosiruptor saccharolyticus* reveals broad growth substrate preferences. *Appl Envir Microbiol* 2009;**75**:7718–24.

42. Lin L, Song H, Tu Q, Qin Y, Zhou A, Liu W, et al. The *Thermoanaerobacter* glycobiome reveals mechanisms of pentose and hexose co-utilization in Bacteria. *PLoS Genet* 2011;**7**:e1002318.

43. Hyun HH, Zeikus JG. Regulation and genetic enhancement of glucoamylase and pullulanase production in *Clostridium thermohydrosulfuricum*. *J Bacteriol* 1985;**164**:1146–52.

44. Erbeznik M, Dawson KA, Strobel HJ. Cloning and characterization of transcription of the *xylAB* operon in *Thermoanaerobacter ethanolicus*. *J Bacteriol* 1998;**180**:1103–9.

45. Jones CR, Ray M, Strobel HJ. Transcriptional analysis of the xylose ABC transport operons in the thermophilic anaerobe *Thermoanaerobacter ethanolicus*. *Curr Microbiol* 2002;**45**:54–62.

46. Warner JB, Lolkema JS. CcpA-dependent carbon catabolite repression in Bacteria. *Microbiol Mol Biol Rev* 2003;**67**:475–90.

47. Deutscher J. The mechanisms of carbon catabolite repression in bacteria. *Curr Opin Microbiol* 2008;**11**:87–93.

48. Görke B, Stülke J. Carbon catabolite repression in bacteria: many ways to make the most out of nutrients. *Nat Rev Microbiol* 2008;**6**:613–24.

49. Comas I, González-Candelas F, Zúñiga M. Unraveling the evolutionary history of the phosphory-transfer chain of the phosphoenolpyruvate: phosphotransferase system through phylogenetic analyses and genome context. *BMC Evol Biol* 2008;**8**:147.

50. Galinier A, Haiech J, Kilhoffer MC, Jaquinod M, Stülkes J, Deutscher J, et al. *The Bacillis subtilis* crh gene encodes a HPr-like protein involved in carbon catabolite repression. *Proc Natl Acad Sci USA* 1997;**94**:8439–44.

51. Schumacher MA, Seidel G, Hillen W, Brennan RG. Phosphoprotein Crh-Ser[46]-P displays altered binding to CcpA to effect carbon catabolite regulation. *J Biol Chem* 2006;**10**:6793–800.

52. Ye JJ, Saier Jr MH. Cooperative binding of lactose and the phosphorylated phosphocarrier protein HPr(Ser-P) to the lactose/H+ symport permease of *Lactobacillus brevis. Proc Natl Acad Sci USA* 1995;**92**:417–21.

53. Viana R, Monodero V, Dossonnet V, Vadeboncoeur C, Pérez-Martínez G, Deutscher J. Enzyme I and HPr from *Lactobacillus casei*: their role in sugar transport, carbon catabolite repression and inducer exclusion. *Mol Microbiol* 2000;**36**:570–84.

54. Henkin TM. The role of the CcpA transcriptional regulator in carbon metabolism in *Bacillus subtilis. FEMS Microbiol Lett* 1996;**135**:9–15.

55. Schumacher MA, Allen GS, Diel M, Seidel G, Hillen W, Brennan RG. Structural basis for allosteric control of the transcription regulator CcpA by the phosphoprotein HPr-Ser46-P. *Cell* 2004;**118**:731–41.

56. Galinier A, Deutscher J, Martin-Verstraete I. Phosphorylation of either Crh or HPr mediates binding of CcpA to the *Bacillus subtilis xyn cre* and catabolite repression of the *xyn* operon. *J Mol Biol* 1999;**286**:307–14.

57. Monedero V, Gosalbes MJ, Pérez-Martínez G. Catabolite repression in *Lactobacillus casei* ATCC 393 is mediated by ccpA. *J Bacteriol* 1997;**179**:6657–64.

58. Miwa Y, Nakata A, Ogiwara A, Yamamoto M, Fujita Y. Evaluation and characterization of catabolite-responsive elements (*cre*) of *Bacillus subtilis. Nucleic Acids Res* 2000;**28**:1206–10.

59. Antunes A, Camiade E, Monot M, Courtois E, Barbut F, Sernova NV, et al. Global transcription control by glucose and carbon regulator CcpA in *Clostridium difficile. Nucleic Acids Res* 2012;**40**:10701–8.

60. Stülke J, Arnaud M, Rapoport G, Martin-Verstraete I. PRD – a protein domain involved in PTS-dependent induction and carbon catabolite repression of catabolic operons in bacteria. *Mol Microbiol* 1998;**28**: 865–74.

61. Brenner K, You L, Arnold FH. Engineering microbial consortia: a new frontier in synthetic biology. *Trends Biotechnol* 2008;**26**(9):483–9.

62. He Q, Hemme CL, Jiang H, He Z, Zhou J. Mechanisms of enhanced cellulosic bioethanol fermentation by co-cultivation of *Clostridium* and *Thermoanaerobacter* spp. *Bioresour Technol* 2011;**102**(20):9586–92.

63. Jiao Y, Navid A, Stewart BJ, McKinlay JB, Thelen MP, Pett-Ridge J. Syntrophic metabolism of a co-culture containing *Clostridium cellulolyticum* and *Rhodopseudomonas palustris* for hydrogen production. *Int J Hydrog Energy* 2012;**37**(16):11719–26.

64. Wang A, Ren N, Shi Y, Lee D. Bioaugmented hydrogen production from microcrystalline cellulose using co-culture—*Clostridium acetobutylicum* X9X9 and *Ethanoigenens harbinense* B49B49. *Int J Hydrog Energy* 2008;**33**(2):912–7.

65. Zeidan AA, Rådström P, Van Niel EWJ. Stable coexistence of two *Caldicellulosiruptor* species in a de novo constructed hydrogen-producing co-culture. *Microb Cell Fac* 2010;**9**(1):102.

66. Markowitz VM, Chen IMA, Palaniappan K, Chu K, Szeto E, Grechkin Y, et al. The integrated microbial genomes system: an expanding comparative analysis resource. *Nucleic Acids Res* 2010;**38**:D382–90.

67. Markowitz VM, Chen IMA, Palaniappan K, Chu K, Szeto E, Grechkin Y, et al. IMG: the integrated microbial genomes database and comparative analysis system. *Nucleic Acids Res* 2012;**40**(D1):D115–22.

68. Reitzer L. Amino acid synthesis. In: *Encyclopedia of microbiology.* 3rd ed. Academic press; 2009. p. 1–17.

69. Johnson EA, Madia A, Demain AL. Chemically defined minimal medium for growth of the anaerobic cellulolytic thermophile *Clostridium thermocellum. Appl Envir Microbiol* 1981;**41**(4):1060–2.

70. Hemme CL, Fields MW, He Q, Deng Y, Lin L, Tu Q, et al. Correlation of genomic and physiological traits of thermoanaerobacter species with biofuel yields. *Appl Envir Microbiol* 2011;**77**(22):7998–8008.

71. Cavedon K, Canale-Parola E. Physiological interactions between a mesophilic cellulolytic *Clostridium* and a non-cellulolytic bacterium. *FEMS Microbiol Lett* 1992;**86**(3):237–45.

72. Kato S, Haruta S, Cui ZJ, Ishii M. Stable coexistence of five bacterial strains as a cellulose-degrading community. *Appl Envir Microb* 2005;**71**(11):7099–106. http://dx.doi.org/10.1128/AEM.71.11.7099.

73. Saddler JN, Chan MKH, Louis-Seize G. One step process for the conversion of cellulose to ethanol using anaerobic microorganisms in mono- and coculture. *Biotechnol* 1981;**3**(6):321–6.

74. Saddler JN, Chan MKH. Optimization of *Clostridium thermocellum* growth on cellulose and pretreated wood substrates. *Eur J Appl Microbiol Biotechnol* 1982;**16**(2–3):99–104.

75. Xu L, Tschirner U. Improved ethanol production from various carbohydrates through anaerobic thermophilic co-culture. *Bioresour Technol* 2011;**102**(21):10065–71.

76. Ellis LD, Holwerda EK, Hogsett D, Rogers S, Shao X, Tschaplinski T, et al. Closing the carbon balance for fermentation by *Clostridium thermocellum* (ATCC 27405). *Bioresour Technol* 2012;**103**:293–9.

77. Bassler BL. How bacteria talk to each other: regulation of gene expression by quorum sensing. *Curr Opin Microbiol* 1999;**2**(6):582–7.

78. Wuster A, Babu MM. Chemical molecules that regulate transcription and facilitate cell-to-cell communication. In: *Wiley encyclopedia of chemical biology*. Hoboken (New Jersey): John Wiley and Sons, Inc; 2008.

79. Bassler BL. Small talk. Cell-to-cell communication in bacteria. *Cell* 2002;**109**(4):421–4.

80. Wuster A, Babu MM. Conservation and evolutionary dynamics of the agr cell-to-cell communication system across firmicutes. *J Bacteriol* 2008;**190**(2):743–6.

81. Steiner E, Scott J, Minton NP, Winzer K. An agr quorum sensing system that regulates granulose formation and sporulation in *Clostridium acetobutylicum*. *Appl Envir Microbiol* 2012;**78**(4):1113–22.

82. Alsaker KV, Papoutsakis ET. Transcriptional program of early sporulation and stationary-phase events in *Clostridium acetobutylicum*. *J Bacteriol* 2005;**187**(20):7103–18.

83. Nolling J, Breton G, Omelchenko MV, Makarova KS, Zeng Q, Gibson R, et al. Genome sequence and comparative analysis of the solvent-producing bacterium *Clostridium acetobutylicum*. *J Bacteriol* 2001;**183**(16):4823–38.

84. Ohtani K, Hayashi H, Shimizu T. The luxS gene is involved in cell-cell signalling for toxin production in *Clostridium perfringens*. *Mol Microbiol* 2002;**44**(1):171–9.

85. Li J, Chen J, Vidal JE, McClane BA. The Agr-like quorum-sensing system regulates sporulation and production of enterotoxin and beta2 toxin by *Clostridium perfringens* type A non-food-borne human gastrointestinal disease strain F5603. *Infect Immun* 2011;**79**(6):2451–9.

86. Cooksley CM, Davis IJ, Winzer K, Chan WC, Peck MW, Minton NP. Regulation of neurotoxin production and sporulation by a putative agrBD signaling system in proteolytic *Clostridium botulinum*. *Appl Envir Microbiol* 2010;**76**(13):4448–60.

87. Kosaka T, Nakayama S, Nakaya K, Yoshino S, Furukawa K. Characterization of the sol operon in butanol-hyperproducing *Clostridium saccharoperbutylacetonicum strain N1-4* and its degeneration mechanism. *Biosci Biotechnol Biochem* 2007;**71**(1):58–68.

88. Bassler BL, Losick R. Bacterially speaking. *Cell* 2006;**125**(2):237–46.

89. Villemur R, Lanthier M, Beaudet R, Lepine F. The *Desulfitobacterium* genus. *FEMS Microbiol Rev* 2006;**30**(5):706–33.

90. Novick RP, Geisinger E. Quorum sensing in staphylococci. *Annu Rev Genet* 2008;**42**:541–64.

91. Alsaker KV, Papoutsakis ET. Transcriptional program of early sporulation and stationary-phase events in *Clostridium acetobutylicum*. *J Bacteriol* 2005;**187**(20):7103–18.

92. Jiang M, Shao W, Perego M, Hoch JA. Multiple histidine kinases regulate entry into stationary phase and sporulation in *Bacillus subtilis*. *Mol Microbiol* 2000;**38**(3):535–42.

93. Bulter T, Lee SG, Wong WW, Fung E, Connor MR, Liao JC. Design of artificial cell-cell communication using gene and metabolic networks. *Proc Natl Acad Sci USA* 2004;**101**(8):2299–304.

CHAPTER 18

Engineering Synthetic Microbial Consortia for Consolidated Bioprocessing of Ligonocellulosic Biomass into Valuable Fuels and Chemicals

Jeremy J. Minty, Xiaoxia N. Lin

Department of Chemical Engineering, University of Michigan, Ann Arbor, MI, USA

Introduction

Lignocellulosic biomass is an abundant and renewable resource. More than half of the carbon in the biosphere is present in the form of cellulose, with approximately 1 trillion tons of cellulose synthesized and degraded each year globally.[1] Unlike sugarcane or starch-based feedstocks, it is possible to produce and use lignocellulosic biomass without affecting food supplies. For example, marginal lands unsuitable for food production can provide lignocellulosic biomass via harvesting wild vegetation[2] or through intentional cultivation of robust bioenergy crops such as switchgrass (*Miscanthus* sp.). In addition to bioenergy crops, many underused waste streams are rich in cellulose (such as crop residues and municipal solid waste) and could serve as potential feedstocks for biofuel production. It is estimated that the United States is capable of sustainably producing 1.4 billion tons of lignocellulosic biomass annually, enough to replace 30% or more of our current petroleum consumption.[3]

Lignocellulose, serving as the primary structural component of plant cell walls, is composed of cellulose, hemicellulose, and lignin. Cellulose is a polysaccharide of glucose monomers linked by β-1,4 glucosidic bonds, while hemicellulose is a polysaccharide of mixed composition and structure, containing a large proportion of pentose sugars linked by β-1,4 bonds. Lignin has a complex structure and tends to be hydrophobic with a high proportion of aromatic groups. Bioprocessing of lignocellulose into fuels and chemicals typically comprises four main steps: pretreatment, saccharification (usually via enzymatic hydrolysis), fermentation of soluble hexose (C6) and pentose (C5) saccharides, and downstream processing (e.g., product separation) (Figure 1). In the first step, lignocellulosic biomass is subjected to mechanical and/or thermochemical treatments to improve digestibility. Pretreatment alters the

Direct Microbial Conversion of Biomass to Advanced Biofuels. http://dx.doi.org/10.1016/B978-0-444-59592-8.00018-X

Microbial conversion of lignocellulose

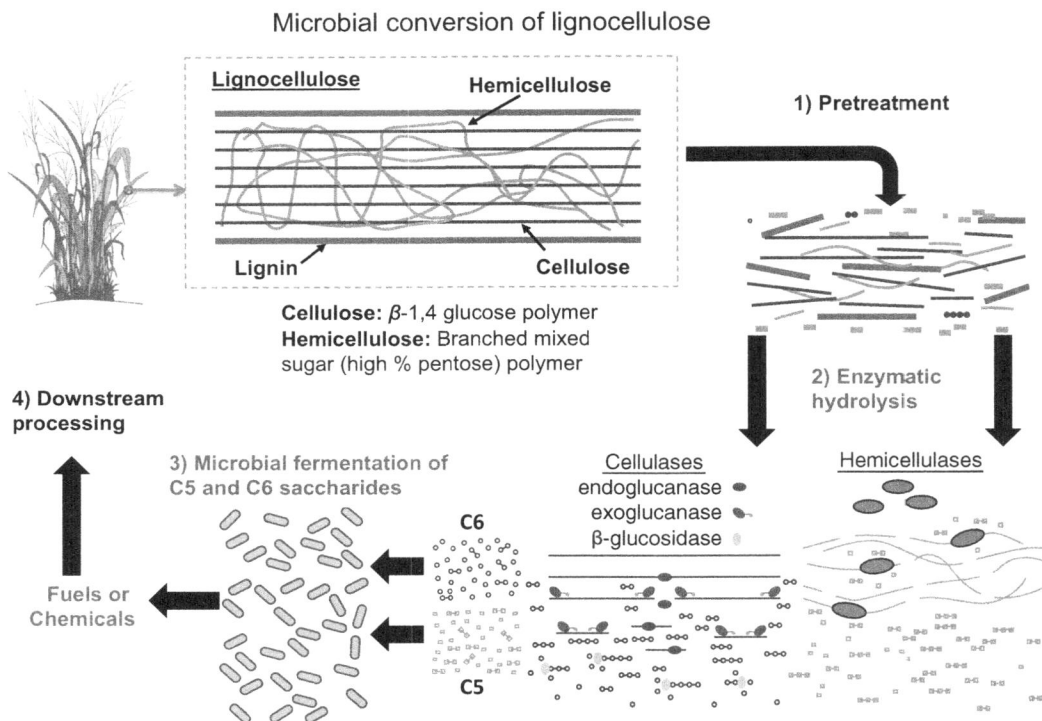

Figure 1

Bioprocessing of lignocellulosic feedstocks into biofuels. *Plant clipart from published materials by the Office of Biological and Environmental Research of the U.S. Department of Energy Office of Science.*

microstructure of lignocellulosic biomass, often via redistribution or removal of lignin and reduction of cellulose crystallinity, leading to improved hydrolysis rates.[4] Pretreatment contributes substantially to overall processing costs. Cellulolytic microbes produce sophisticated and synergistic enzyme systems called cellulases and hemicellulases that effectively hydrolyze cellulose and hemicellulose, respectively.[4] In the saccharification step, cellulases and hemicellulases are used to hydrolyze insoluble cellulose and hemicellulose into soluble C6 and C5 saccharides, respectively.[5] Cellulose is a structurally complex material that is highly recalcitrant to degradation, making enzymatic hydrolysis a limiting step in microbial biofuel production.[4] In the fermentation step, monosaccharide and oligosaccharide hydrolysis products are metabolized by microbes into biofuel molecules, such as ethanol.[5] Many microbes are unable to metabolize pentose sugars, and microbes that are capable of pentose metabolism generally consume these sugars diauxically, with hexoses consumed in preference to pentoses.[6] Additionally, the toxicity of biofuel products and inhibitory compounds present in the feedstock often limits fermentation productivity and titers. Achieving cost-effective pretreatment, efficient cellulose hydrolysis, co-utilization of hexose/pentose saccharides, and mitigating the toxic effects of biofuel products on microbes are all vital requirements for economically viable cellulosic biofuel production.[4,6]

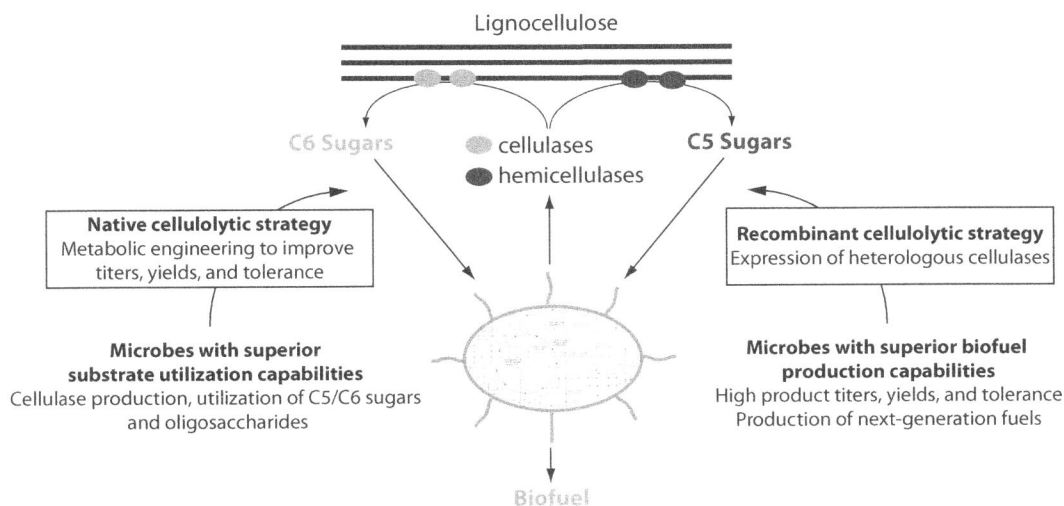

Figure 2
Two strategies for engineering a single microorganism for consolidated bioprocessing. (*Adapted from Ref. 5.*).

Process configurations for lignocellulosic biofuel production can be categorized based on the extent to which the biologically mediated steps (cellulase production, enzymatic saccharification, and fermentation of soluble saccharides) are consolidated.[4] Most cellulosic biofuel processes presently under commercial development use process configurations known as simultaneous saccharification and fermentation (SSF) or simultaneous saccharification and co-fermentation (SSCF).[5] SSF features a dedicated step for enzyme production. Cellulases and hemicellulases produced in the enzyme production step are then combined with pretreated biomass and microbes for SSF to biofuel. Hexose and pentose saccharides are co-fermented in a single step for SSCF, while in SSF, hexose and pentose conversions occur in separate bioreactors. Having a dedicated step for enzyme production contributes substantially to total processing costs. Consolidated bioprocessing (CBP) is a promising process configuration that integrates all biochemical transformations—cellulase/hemicellulose production, saccharification, hexose fermentation, and pentose fermentation—into a single step and may thus significantly improve process economics.[5] In addition to reducing processing costs via consolidation, CBP may provide other benefits, such as increased hydrolysis rates due to synergy between microbes and enzymes during fermentation.[5]

Engineering Single Microorganisms to Enable CBP

Since the concept of CBP was developed, the main paradigm in CBP research has been engineering microbes that incorporate all required biological functionalities into a single host. Two broad approaches are being pursued, the native cellulolytic strategy and the recombinant cellulolytic strategy (Figure 2). In the native cellulolytic strategy, anaerobic cellulolytic microbes (which naturally ferment cellulose to alcohols and mixed organic acids) are engineered to improve alcohol production.[5] Most natively cellulolytic species under consideration for CBP development produce ethanol. Cellulolytic *Clostridia* species, including *Clostridium phytofermentans*, *Clostridium cellulolyticum*, and *Clostridium thermocellum*, are among the best developed native hosts,[7] although other anaerobic cellulolytic bacteria and fungi are being investigated. The native cellulolytic strategy is hindered by limited molecular biology tools for candidate microbes, substantially restricting genetic engineering efforts.[7] Strain improvement approaches have therefore generally been limited to small-scale metabolic engineering efforts to improve ethanol yields (i.e., knocking out competing pathways to direct metabolism toward ethanol production) and developing strains with improved ethanol tolerance through adaptive evolution.[7] *Clostridium thermocellum* was recently engineered to produce ethanol with high titer (38 g/L) and yield (82% of theoretical) from microcrystalline cellulose (in co-culture with *Thermoanaerobacterium saccharolyticum*).[8] These results represent the current state-of-the-art and are close to projected performance targets for economically viable ethanol production (50 g/L titer and 90% theoretical yield).[7] However, this study did not use real lignocellulosic biomass or industrially relevant culture conditions. Additionally, the paucity of genetic tools for *C. thermocellum* will make it difficult to use this species as a platform for production of non-ethanol biofuels. While the native cellulolytic approach has mostly focused on ethanol production, *C. cellulolyticum* was recently engineered to produce isobutanol via a heterologous pathway.[9] This study represents an important proof-of-concept for producing next-generation biofuels using the native cellulolytic strategy; however, the reported isobutanol titer was low (660 mg/L), and *C. cellulolyticum* has limited potential as a CBP host due to poor cellulose utilization.[10]

In the recombinant cellulolytic strategy, microbes that are amenable to genetic manipulation and have good biofuel production properties (e.g., biofuel production strains of *Escherichia coli*, *Saccharomyces cerevisiae*, *Bacillus subtilis*, etc.) are engineered to produce cellulases/hemicellulases.[5] Since *E. coli* and *S. cerevisiae* have been metabolically engineered to produce a variety of biofuel molecules,[6] the recombinant cellulolytic strategy could be used to produce non-ethanol fuel molecules with desirable properties. However, engineering heterologous cellulase/hemicellulase production is extremely difficult, because typical cellulose systems comprise dozens of different synergistically acting enzymes, which need to be co-expressed and secreted from non-native hosts. Numerous proof-of-concept demonstrations of the recombinant cellulolytic approach have been reported; however, to date, cellulose conversion and product titers/yields remain low.[7] The most significant advances in expressing heterologous cellulases have been with *S. cerevisiae*. Several examples of *S. cerevisiae* strains engineered to secrete cellulases and produce ethanol from cellulose have been reported, although most of these studies use amorphous cellulose (easily digestible model substrate that is much less recalcitrant than real lignocellulose), with the exception of a recent development that achieved hydrolysis of microcrystalline cellulose through display of mini-cellulosomes on the cell surface of *S. cerevisiae*.[11] These works have only achieved ethanol titers in the range of 1 to 10 g/L, due in part to the difficulty in engineering recombinant cellulolytic capabilities.[7] Beyond the work done with *S. cerevisiae*, another notable advancement is the development of a *B. subtilis* strain engineered to secrete cellulase and produce lactate as a fermentation product, with titers up to 3.1 g/L and yields 64% of the theoretical maximum. In addition to model organisms, there is interest in engineering other biofuel producing microbes to express cellulases. For example, the highly ethanologenic bacterium *Zymomonas mobilis* has been engineered to secrete cellulases, with recent work demonstrating ethanol titers up to 4% (v/v) using NaOH pretreated sugarcane bagasse as a substrate.[12] There has also been progress in engineering butanologenic *C. acetobutylicum* to produce extracellular cellulases, although to date none of the resulting strains are able to grow on cellulose as the sole carbon source.[13]

Engineered Synthetic Microbial Consortia for CBP

Due to inherent challenges in both the recombinant and native cellulolytic strategies, the CBP approach of trying to incorporate all required biological functionalities into a single super-organism has hitherto proven to be difficult. Despite intensive research efforts spanning several decades, there have been few reports of achieving commercially viable product yields and titers.[5] An alternative approach that has gained rapidly increasing attention in the past several years aims to divide the required biochemical functions into different hosts, which when co-cultured together form a synthetic microbial consortium capable of single-step conversion of cellulose to desired products. This approach is inspired in part by natural microorganisms, which play many important roles, ranging from participating in global biogeochemical cycles to assisting animals with food digestion,[14] and live in synergistic multispecies communities where individual species with specialized roles cooperate to survive and thrive together.[15] In this section, we review the current state-of-art in engineering of synthetic microbial consortia for CBP of lignocellulosic biomass to biofuels and biochemicals.

Synthetic Consortia of Saccharification and Fermentation Specialists

The two major functionalities required for CBP are saccharification of lignocellulose and fermentation of monosaccharide/oligosaccharides into desired molecules. Accordingly, a general scheme for enabling CBP using synthetic microbial consortia has emerged that entails dividing these two biological functions between two microbial specialists: a saccharolytic specialist, which produces saccharifying enzymes to hydrolyze biomass carbohydrates into soluble saccharides; and a fermentation specialist, which metabolizes soluble saccharides into desired products (Figure 3). From an engineering standpoint, pursuing a microbial consortium approach of compartmentalizing each of these functionalities into separate organisms appears

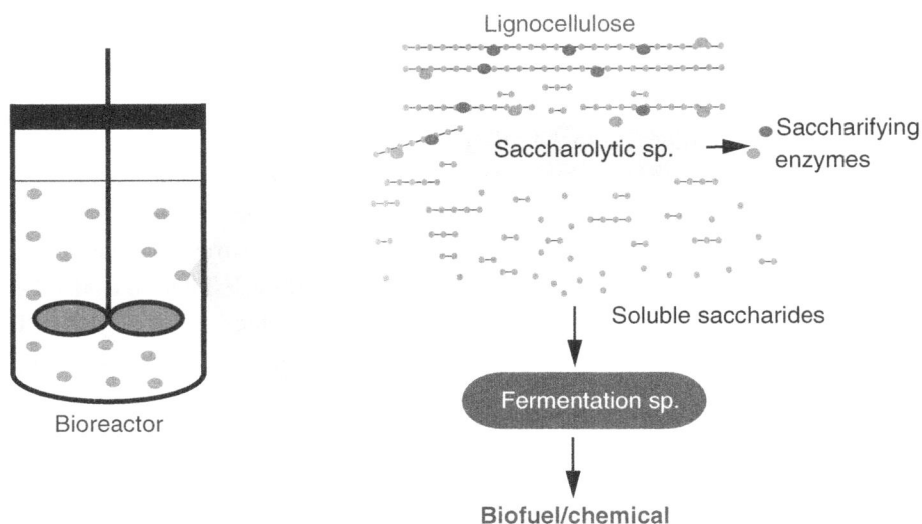

Figure 3

General scheme of consolidated bioprocessing (CBP) consortia consisting of saccharolytic/fermentation (S/F) bicultures for CBP of carbohydrates into fuels or other valuable chemicals. The saccharolytic specialist produces saccharifying enzymes that hydrolyze insoluble biomass polysaccharides into soluble monosaccharides and oligosaccharides, which are then metabolized by the fermentation specialist into desired products.

to be much more tractable than attempting to integrate all the necessary functions into a single host. Selecting a base species/strain for each specialist member of the consortium requires careful consideration of organism physiology and engineering requirements. A primary criterion for the saccharolytic specialist is efficient cellulase production and overall rapid cellulose degradation with minimal nutrition requirements. Choice of the fermentation specialist depends largely on the desired final product. For instance, most of previous work aimed to produce ethanol and hence used species/strains capable of ethanol fermentation, which included both native species and recombinant strains.

The notion of using mixed microbial cultures to convert biomass carbohydrates into fuels or other products is not new and has come in and out of vogue in bioprocessing research several times over the past few decades. Thus the general approach of using two-member saccharolytic/fermentation (S/F) mixed cultures for direct conversion of biomass carbohydrates has been tried a number of times; a number of representative examples are given in Table 1.

Most of these studies focused on the production of ethanol, of which the physicochemical properties are suboptimal for motor fuel use. More recently, several works have reported the production of n-butanol or isobutanol from cellulose or starch.[21–24] Product tiers and yields, however, are considerably lower than those achieved for ethanol production. Another

Engineering Synthetic Microbial Consortia* 371

Table 1: Representative examples of two-member saccharolytic/fermentation (S/F) consortia for producing biofuels from carbohydrates

Product	Saccharolytic sp.	Fermentation sp.	Substrate	Yield[a] (g/g)	Titer (g/L)	Notes and Source
Ethanol	Trichoderma reesei RUTC30	Saccharomyces cerevisiae Y1	Glucose and solka floc (cellulose)	0.27	40	Rich media with added glucose[16]
	Fusarium oxysporum F3	S. cerevisiae 2541	Sweet sorghum (glucose, sucrose, lignocellulose)	0.46	49	17
	Kluyveromyces fragilis LOCK 0027	Zymomonas mobilis 3881	Jerusalem artichoke (inulin)	0.48	99	18
	Clostridium thermocellum M1570	Thermoanaerobacterium saccharolyticum ALK2	Avicel (cellulose)	0.41	38.1	Rich media[8]
	Acremonium cellulolyticus C-1	S. cerevisiae ATCC 4126	Solka floc (cellulose)	0.18	46.3	Rich media[19]
	Clostridium phytofermentans	S. cerevisiae cdt-1	α-cellulose	0.31	22	20
n-Butanol	Bacillus subtilis WD161	Clostridium butylicum TISTR 1032	Cassava starch	0.17	6.7	Rich media[21]
	Clostridium thermocellum ATCC 27,405	Clostridium saccharoperbutylacetonicum N1-4	Avicel (cellulose)	0.2	7.9	Rich media[22]
	Clostridium cellulovorans 743B	Clostridium beijerinckii NCIMB 8052	Corn cobs	0.12	8.3	23
Isobutanol	T. reesei RUTC30	Escherichia coli NV3	Avicel (cellulose) and corn stalks	0.25	1.88	Minimal/no nutrient supplementation[24]
Methyl halide	Actinotalea fermentans	S. cerevisiae	Poplar	n/a	0.14	25

[a]Yields given as g product/g estimated total carbohydrates consumed.

limitation is related to culture conditions and methods that are neither practical nor economical for industrial biofuel or chemical production. In particular, many studies use rich media containing costly supplements (e.g., yeast extract, peptone, amino acids, etc.), and/or use model substrates instead of real biomass. Furthermore, some of these works use sequential inoculation schemes in which the saccharolytic specialist is cultured on the feedstock for a period before inoculating with the fermentation specialist. Such strategies are advantageous when the saccharolytic and fermentation specialists have different physiological and environmental preferences and work well in bench-scale systems. However, sequential inoculation poses substantial barriers for process scale-up. To achieve the much more desirable simultaneous inoculation scheme, one needs to take into account compatibility between the two specialists in the overall consortium design. Physiological compatibility is a key determinant— the consortium members need to grow well and perform their specialized tasks in the same environment and therefore must be compatible in terms of preferences for pH, environmental redox state, temperature, nutrients, etc. The specialist organisms must also be compatible from an ecological standpoint (e.g., neutral or symbiotic interactions with minimal antagonism). Additionally, both consortium members should be tolerant to anticipated environmental stresses (including inhibitors present in the feedstock and accumulation of toxic biofuel products). More recent works have started to use these criteria in consortium design and engineering; notable examples include a *C. phytofermentans–S. cerevisiae* consortium for cellulosic ethanol production[20] and a *T. reesei–E. coli* consortium for cellulosic isobutanol production.[24]

Other Synthetic Microbial Consortia for CBP

In addition to constructing consortia consisting of cellulolytic and fermentative specialists, another emerging approach in developing consortia for biofuel production entails engineering specialist strains to secrete different synergistically acting cellulases and/or hemicellulases. Such an approach was pursued by Arai et al.,[26] Tsai et al.,[27] Goyal et al.,[28] and Bokinsky et al.[29] Tsai et al. engineered *S. cerevisiae* strains to secrete three different cellulases and a scaffold protein, which were then combined to form a synthetic consortium (Figure 4(a)).[27] The secreted cellulases assembled extracellularly on the scaffold protein, forming a synergistic cellulosome complex.[27] Through the production of extracellular cellulosomes, the consortium was able to convert phosphoric acid swollen cellulose (PASC) to ethanol at yield of up to 0.475 g/g (93% theoretical) and titer of 1.87 g/L, using an optimized strain ratio (SC:AT:CB:BF) of 7:2:4:2.[27] In contrast, a monoculture using a similar enzyme system achieved much lower ethanol yields and titers, possibly due to the burden of expressing and secreting multiple heterologous proteins in a single host.[27] As another example, Bokinsky et al. engineered *E. coli* strains to produce cellulases or xyalanases which could then be co-cultured to accomplish simultaneous conversion of both hemicellulose and cellulose[29] (Figure 4(b)). By integrating heterologous pathways for butanol, pinene, or fatty acid ethyl

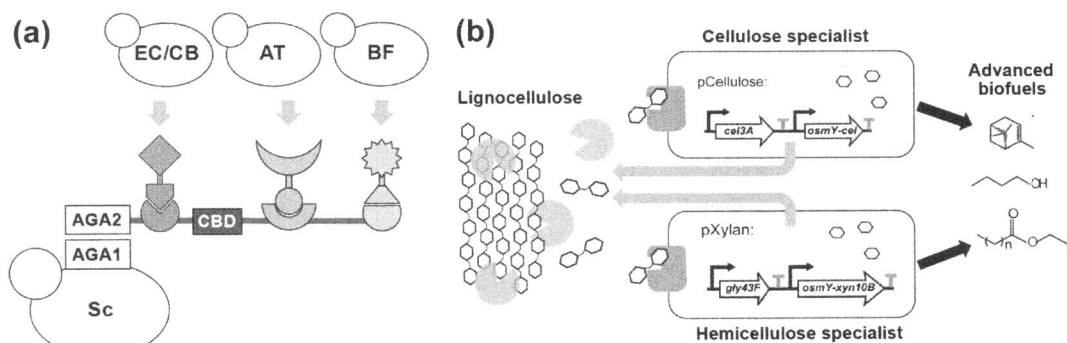

Figure 4
Cellulase producing consortia for CBP. (a) An *Saccharomyces cerevisiae* consortium for assembly of synthetic extracellular cellulosome complexes. SC cells surface display scaffold proteins and EC/CB, AT, and BF cells secrete cellulases. *(Adapted from Ref. 27.)* (b) An *E. coli* consortium for simultaneous conversion of cellulose and hemicellulose into pinene, n-butanol, or fatty acid ethyl esters (FAEE). Cells harbor plasmids containing oligosaccharide transporters (cel3A or gly43F) and secretable hydrolases (osmY-cel or osmY-xyn10B) for either cellulose (pCellulose) or xylan (pXylan). *(Adapted from Ref. 29.).*

esters (FAEE), consortia of xylanase/cellulase producing *E. coli* strains were able to directly convert ionic liquid (IL) pretreated switchgrass to these respective biofuels.[29] While the titers and yields achieved in this study are very low, it serves as an important proof of concept of using microbial consortia for CBP production of advanced biofuels.

Engineering hexose and pentose specialized microbes, each exclusively metabolizing it's respective carbon source, is another emerging trend in engineering microbial consortia for production of biofuels or other chemicals from lignocellulosic substrates (Figure 5). Co-cultures of hexose and pentose specialists would be expected to use both types of sugar simultaneously (in contrast to sequential utilization, observed in most natural species), thus improving the conversion rate of mixed sugars derived from lignocellulosic biomass. This approach was reported by Trinh et al.,[30] Eiteman et al.,[31] Eiteman et al.,[32] and Xia et al.[33] and was also pursued in our group independently during the same period,[34] all using *E. coli* as a base species. Eiteman et al. demonstrated that co-cultures of hexose and pentose specialized *E. coli* gave higher rates of fermentation and more complete utilization of glucose/xylose mixtures compared to fermentation by a single diauxic *E. coli* strain.[32] Furthermore, the hexose/pentose specialist co-culture was able to adapt to fluctuations in feed composition.[32]

Emerging Methods for Designing and Regulating Synthetic Microbial Consortia

Microbial consortia with stable population compositions and function would be highly advantageous for industrial bioprocessing and could enable lower-cost process

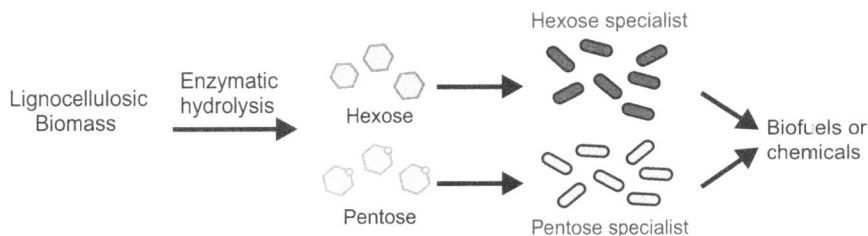

Figure 5

General scheme of consortia for co-fermentation of pentose and hexose sugars. (*Adapted from additional reference: Kerner, Alissa R. (2013) Design, Construction, and Application of Synthetic Microbial Consortia. PhD dissertation.*).

configurations. For instance, stable microbial consortia could be used in repeated batch fermentations (in which a fraction of the previous batch is used as inoculum for new batches) or in continuous processes; operating costs of these configurations are generally lower compared to standard batch fermentation. In contrast, microbial consortia with unstable population compositions could at best only be used in standard batch processes, and each batch would require preparation of individual inoculum cultures. It is widely acknowledged that stability, robustness, and control of population composition remain key challenges in engineering synthetic consortia.[15,35–37] Addressing these challenges requires additional consideration of ecological interactions within engineered consortia. In this section, we discuss emerging methods for population regulation in synthetic consortia.

Synthetic Cell–Cell Signaling

Several natural signaling systems have been co-opted for use in engineered genetic circuits, thus enabling synthetic intercellular communication. Some noteworthy examples include a synthetic predator-prey system (AHL signaling),[38] multicellular Boolean logic gates (*S. cerevisiae* α factor signaling),[39] and an oscillatory circuit synchronized across an entire cell population (AHL and H_2O_2 signaling).[40] Synthetic intercellular communication has even been extended to interspecies and interkingdom signaling; for example, Weber et al. used acetaldehyde-based signaling between and among mammalian, bacteria, yeast, and plant cells to engineer several demonstration synthetic ecosystems.[41] Synthetic cell–cell signaling represents a valuable tool for engineering synthetic microbial consortia. To date, synthetic intercellular communication has been used to construct canonical ecological and logic systems for proof-of-concept and fundamental study, but there are few reports of using synthetic signaling in consortia for biotechnology applications, despite much potential. One key challenge is that synthetic genetic circuits are subject to mutational inactivation, which often occurs in relatively few generations of growth depending on the size and host burden of

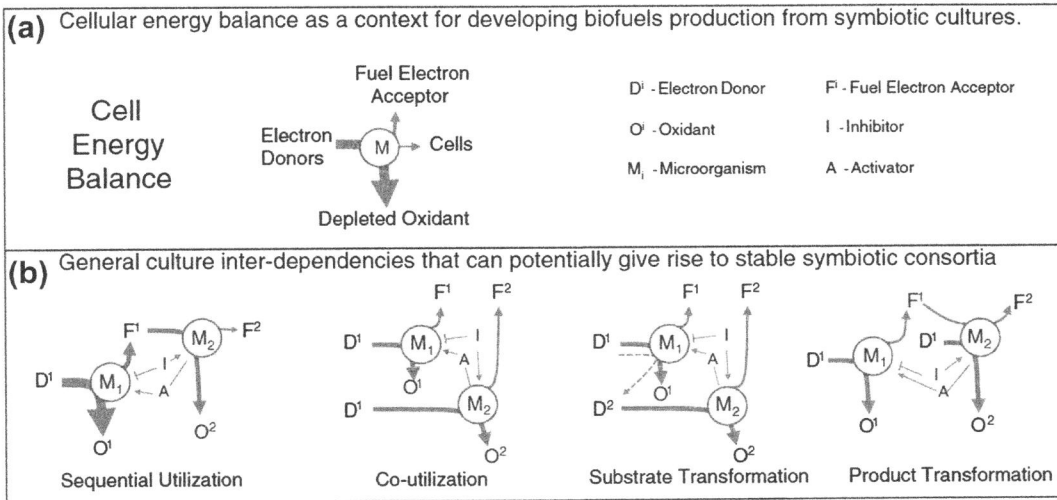

Figure 6

Symbiotic interaction schemes for designing biofuel producing consortia. (a) Energy balance for single biofuel producing microbe. (b) Interdependencies that could give rise to stable consortia (dual cultures shown for simplicity). Inhibitors and activators are generalized representations of various types of molecules such as quorum sensing (QS) signals and exchanged metabolites. *(Reprinted with permission from Ref. 36. Copyright 2012 Springer.).*

the circuit.[42] As a result, consortia with programmed intercellular communication are frequently highly unstable and often cannot be cultivated outside of microfluidic devices, because larger cell populations increase the probability of competitive loss-of-function mutants.[37,43] These stability issues need to be addressed before consortia with synthetic intercellular communication can be developed for bioprocessing applications.

Synthetic Ecologies

As a complement to synthetic genetic circuits and signaling, it is possible to engineer synthetic microbial consortia in which populations interact via ecologically stable motifs. By incorporating designs with ecological and evolutionary stability, such systems may stably persist over long time scales. Furthermore, in principle it should be possible to tune population compositions by modulating ecological interaction parameters (e.g., genetically or through environmental manipulation). One broad means of achieving ecological stability is to engineer mutualistic interactions between consortia members, wherein the different consortia members are interdependent on one another. There are a number of symbiotic interaction topologies that could be used to design synthetic consortia. Some examples relevant to CBP applications include sequential substrate utilization, co-utilization of substrate, substrate transformation, and product transformation (Figure 6).[36]

In sequential substrate utilization, one species metabolizes substrate to waste products that serve as substrate for a second species; in product transformation, the second species co-utilizes both the primary substrate and waste products (Figure 6(b)).[36] Sequential utilization and product transformation interactions can be mutualistic if the waste products produced by the first species are toxic, because the first species provides substrate for the second, while the second species aids the community by removing toxins. The product transformation approach was demonstrated by Bayer et al. in engineering a consortium for production of methyl halides from cellulose; the cellulolytic microbe *Actinotalea fermentans* ferments cellulose to toxic waste products ethanol and acetate, which are in turn converted to methyl halides by an engineered *S. cerevisiae* strain that also presumably co-utilizes cellulose hydrolysis products (Figure 7(a)).[25] Bernstein et al. used the sequential utilization scheme in engineering a two-member synthetic *E. coli* consortium, in which a primary producer metabolizes glucose to acetate and other waste products, while a glucose-negative *E. coli* strain consumes them (Figure 7(b)).[44] The consortium has higher biomass productivity compared to monocultures of either member strain, illustrating the mutualistic nature of the system.[44] While the product transformation and sequential substrate utilization motifs used in these two examples are expected to be stable, neither of these studies examined population dynamics over long time scales.

In addition to the interaction topologies shown in Figure 6, cross-feeding, in which species are obligately interdependent on one another for nutrients or growth substrates, represents another important class of ecologically stable interactions; substrate transformation (Figure 6(b)) is a specific instance of this general topology. As an example, Shou et al. engineered synthetic cross-feeding symbiosis between two *S. cerevisiae* strains.[46] In another noteworthy example, Kerner et al. engineered a tunable symbiotic system in which two auxotrophic *E. coli* strains cross-feed each other (Figure 7(c)).[45] The equilibrium population of this system depends on the growth requirements and export rates for the two cross-fed amino acids; by modulating amino acid export rates, Kerner et al. were able to tune the co-culture composition.[45] This study demonstrates that ecological parameters can be adjusted to tune the population composition of a synthetic consortium.

The synthetic consortia discussed above are all based on cooperative interactions between different species or strains. Other stable interaction topologies are also possible. For instance, game theory suggests that cooperators and cheaters can stably coexist. This type of co-existence was observed in a synthetic *S. cerevisiae* ecosystem growing on a disaccharide sucrose.[47] One *S. cerevisiae* strain produces periplasmic invertase for hydrolyzing sucrose into glucose/fructose that can support growth of the whole population and thus serves as a cooperator; a second *S. cerevisiae* strain feeds on the hydrolysis products without bearing the cost of making invertase, fulfilling the role of a cheater (Figure 8(a)). However, periplasmic localization of invertase leads to increased glucose/fructose concentrations near cooperator cells, thus affording them privileged access (Figure 8(a)). Gore et al. demonstrated in this study that cooperator-cheater co-existence is possible if (1) cooperators have privileged

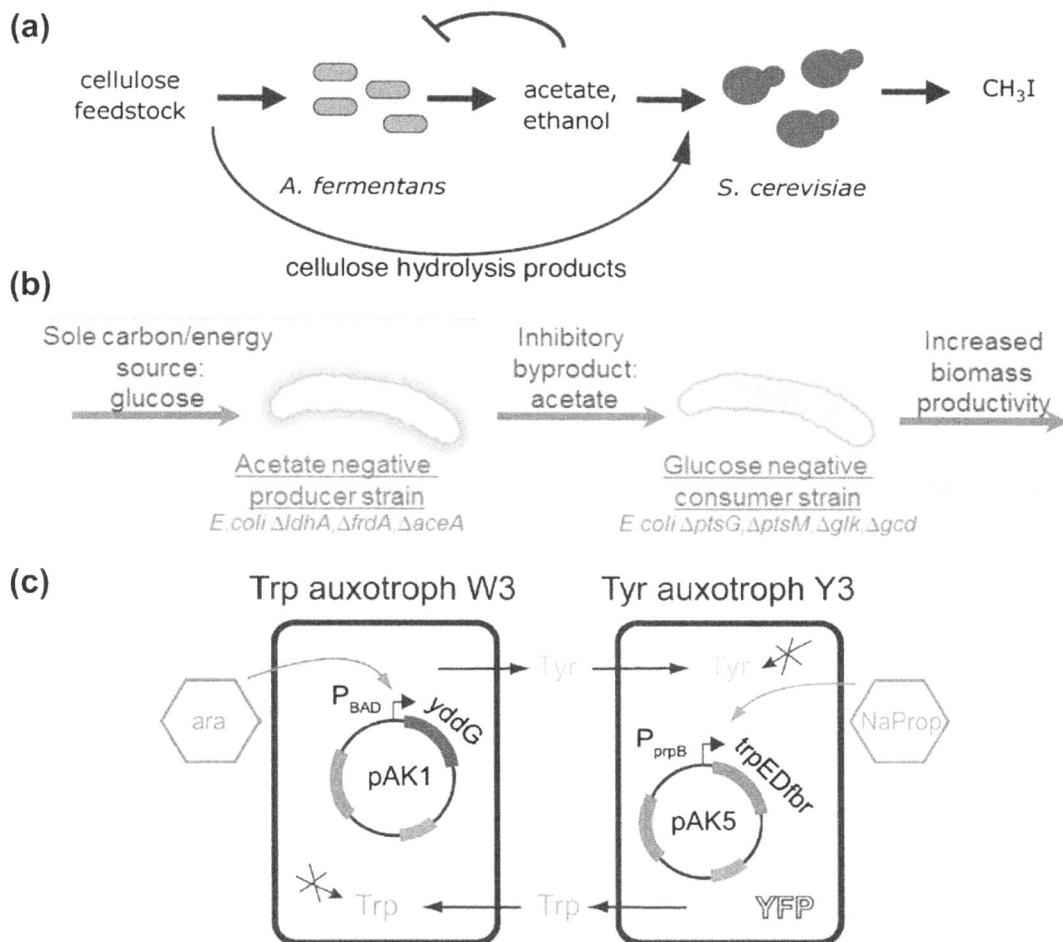

(a)

(b)

(c)

Figure 7

Synthetic consortia based on mutualistic interactions. (a) Product transformation system for conversion of cellulose to methyl halides. *(Reprinted with permission from Ref. 25. Copyright 2009 American Chemical Society.)* (b) Two-member synthetic *E. coli* consortium for sequential degradation of glucose. *(Reprinted from the additional reference below under a Creative Commons Attribution License. Bernstein, H. C. S. D. Paulson, et al. (2012). "Microbial Consortia Engineering for Cellular Factories: in vitro to in silico systems." Computational and Structural Biotechnology Journal 3(4): e201210017 (Figure 2B, pg 3))* (c) Tunable symbiosis between two *E. coli* strains auxotrophic for tryptophan and tyrosine. Equilibrium population composition depends on the growth requirements and export rates for each amino acid; by controlling amino acid export rates (via transcriptional regulation of plasmid-based exporters or pathways), population composition can be tuned. *(Reprinted from Ref. 45 under a Creative Commons Attribution License.)*

access to hydrolysis products, and (2) fitness benefits are a concave function of the substrate concentration.[47] Inspired by this work, Minty et al. exploited the cooperator–cheater dynamics in a synthetic consortium consisting of a saccharolytic fungus *T. reesei* and a genetically engineered *E. coli* for CBP production of isobutanol.[24] In this system, cellulase synthesis and

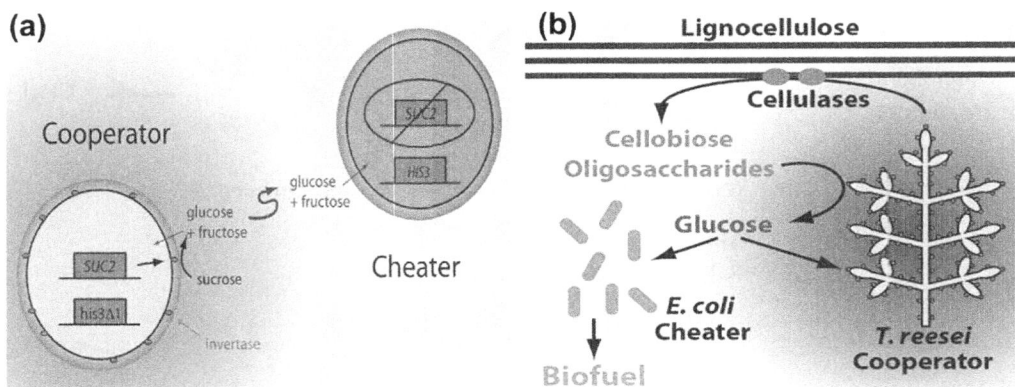

Figure 8

Cooperater-cheater co-existence. (a) A synthetic *Saccharomyces cerevisiae* consortium feeding on sucrose. *(Reprinted with permission from Ref. 47. Copyright 2009 Nature Publishing Group.)* (b) A synthetic *Trichoderma reesei–Escherichia coli* consortium converting cellulose to biofuel. *(Adapted from Ref. 24.)*.

secretion by *T. reesei* is cooperative because it is metabolically expensive, and *T. reesei* does not have exclusive access to hydrolysis products; whereas *E. coli* behaves as a cheater by using hydrolysis products without bearing the burden of cellulose production (Figure 8(b)). The two conditions for coexistence are satisfied by concave growth kinetics (i.e., Monod kinetics) and increased glucose concentration at the *T. reesei* cell surface due to hydrolysis of soluble oligosaccharides by cell wall-localized β-glucosidases (Figure 8(b)). It was shown that cooperator–cheater dynamics in this consortium led to stable population equilibria; in addition, it also provided a mechanism for tuning the consortium composition.[24] This work illustrated the power and potential of using ecology theories in the engineering of CBP consortia. We expect this approach to be further pursued by the research community, and a more diverse range of promising ecological designs for synthetic microbial consortia will emerge.

Elucidation for Engineering

Beyond proof-of-concept design and construction, synthetic microbial consortia require extensive optimization to reach optimal performance. At this stage, understanding the physiology of individual members and the intrinsic ecological interactions is crucial. It is widely observed that the behavior of an organism depends on the environmental context. In the case of synthetic consortia, it has been reported that the same species/strain can exhibit remarkably different phenotypes in the presence of other species in an "artificially" assembled consortium, compared to those in a monoculture under ideal conditions. For instance, Minty et al. observed that a genetically engineered *E. coli* strain capable of high-titer and high-yield production of isobutanol from glucose delivered suboptimal performance when it was teamed up with fungus *T. reesei* in a medium that contained cellulose as the sole carbon source, in part due to loss of plasmids used to overexpress native and heterologous pathways.[24] This example highlights the importance of engineering microbial strains under conditions representative of co-cultures. To facilitate this process, it is important to elucidate the physiological and metabolic state of specialist members in the consortium. Of various tools that can prove valuable for such investigations, omics approaches, as detailed in Chapter 13, can be particularly useful. As a remarkable example, Chignell et al. conducted quantitative proteomics on an artificial co-culture consisting of Gram-negative *Pseudomonas putida* KT2440 and Gram-positive *Bacillus atrophaeus* ATCC 9372; it was found that out of more than 1100 proteins that could be identified across pure and co-culture samples, several tens to more than 100 proteins were upregulated or downregulated in co-cultures compared to pure cultures for each species.[48] This recent study illustrated the power of omics approaches as well as the complexity of molecular–level interactions in synthetic microbial consortia. Differential expression patterns identified from such investigations can be used to generate hypothesis regarding the interactions of consortium members, which will ultimately contribute to the rational optimization of synthetic consortium for desired performance criteria.

Concluding Remarks

The past several years have seen accelerating progress in the engineering of synthetic microbial consortia for CBP of lignocellulosic biomass into fuels and chemicals. These works have demonstrated the potential of an alternative strategy for biofuels and microbial engineering and provide valuable foundations for future research. To enable real-world large-scale applications, however, this technology will require substantial further development to reach high efficiency and robustness. A fast growing suite of tools and approaches, particularly for understanding and exploiting the complex interactions and dynamics in microbial consortia, are emerging to achieve this goal. Ultimately, we envision that synthetic microbial consortia will become increasingly important in industrial bioprocessing and other biotechnology applications; it is likely that this strategy will lead to the most cost-effective route for biosynthesis of many different fuels and chemicals from lignocellulosic biomass.

References

1. Voet D, Voet JG. *Biochemistry*, vol. 1. Hoboken: John Wiley & Sons; 2004. 591.
2. Gelfand I, Sahajpal R, Zhang X, Izaurralde RC, Gross KL, Robertson GP. Sustainable bioenergy production from marginal lands in the US Midwest. *Nature* 2013;**493**:514–7.
3. Perlack RD, Wright LL, Turhollow AF, Graham RL, Stokes BJ, Erbach DC. *DTIC document* 2005.
4. Lynd LR, Weimer PJ, van Zyl WH, Pretorius IS. Microbial cellulose utilization: fundamentals and biotechnology. *Microbiol Mol Biol Rev* 2002;**66**:506–77. table of contents.
5. Lynd LR, van Zyl WH, McBride JE, Laser M. Consolidated bioprocessing of cellulosic biomass: an update. *Curr Opin Biotechnol* 2005;**16**:577–83.
6. Dellomonaco C, Fava F, Gonzalez R. The path to next generation biofuels: successes and challenges in the era of synthetic biology. *Microb Cell Fact* 2010;**9**:3.

7. Olson DG, McBride JE, Shaw AJ, Lynd LR. Recent progress in consolidated bioprocessing. *Curr Opin Biotechnol* 2012;**23**:396–405.

8. Argyros DA, Tripathi SA, Barrett TF, Rogers SR, Feinberg LF, Olson DG, et al. High ethanol titers from cellulose by using metabolically engineered thermophilic, anaerobic microbes. *Appl Environ Microbiol* 2011;**77**:8288–94.

9. Higashide W, Li Y, Yang Y, Liao JC. Metabolic engineering of *Clostridium cellulolyticum* for production of isobutanol from cellulose. *Appl Environ Microbiol* 2011;**77**:2727–33.

10. Desvaux M. *Clostridium cellulolyticum:* model organism of mesophilic cellulolytic clostridia. *FEMS Microbiol Rev* 2005;**29**:741–64.

11. Fan LH, Zhang ZJ, Yu XY, Xue YX, Tan TW. Self-surface assembly of cellulosomes with two miniscaffoldins on *Saccharomyces cerevisiae* for cellulosic ethanol production. *Proc Natl Acad Sci USA* 2012;**109**:13260–5.

12. Vasan PT, Piriya PS, Prabhu DI, Vennison SJ. Cellulosic ethanol production by *Zymomonas mobilis* harboring an endoglucanase gene from *Enterobacter cloacae*. *Bioresour Technol* 2011;**102**:2585–9.

13. Fierobe H-P, Mingardon F, Chanal A. Engineering cellulase activity into *Clostridium acetobutylicum*. *Methods Enzymol* 2012;**510**:301.

14. Mee MT, Wang HH. Engineering ecosystems and synthetic ecologies. *Mol Biosyst* 2012;**8**:2470–83.

15. Brenner K, You L, Arnold FH. Engineering microbial consortia: a new frontier in synthetic biology. *Trends Biotechnol* 2008;**26**:483–9.

16. Hahn-Hägerdal B, Häggström M. Production of ethanol from cellulose, Solka Floc BW 200, in a fedbatch mixed culture of *Trichoderma reesei*, C30, and *Saccharomyces cerevisiae*. *Appl Microbiol Biotechnol* 1985;**22**:187–9.

17. Mamma D, Koullas D, Fountoukidis G, Kekos D, Macris B, Koukios E. Bioethanol from sweet sorghum: simultaneous saccharification and fermentation of carbohydrates by a mixed microbial culture. *Process Biochem* 1996;**31**:377–81.

18. Szambelan K, Nowak J, Czarnecki Z. Use of *Zymomonas mobilis* and *Saccharomyces cerevisiae* mixed with *Kluyveromyces fragilis* for improved ethanol production from Jerusalem artichoke tubers. *Biotechnol Lett* 2004;**26**:845–8.

19. Park EY, Naruse K, Kato T. One-pot bioethanol production from cellulose by co-culture of *Acremonium cellulolyticus* and *Saccharomyces cerevisiae*. *Biotechnol Biofuels* 2012;**5**:64.

20. Zuroff TR, Xiques SB, Curtis WR. Consortia-mediated bioprocessing of cellulose to ethanol with a symbiotic *Clostridium phytofermentans/yeast* co-culture. *Biotechnol Biofuels* 2013;**6**:59.

21. Tran HTM, Cheirsilp B, Hodgson B, Umsakul K. Potential use of *Bacillus subtilis* in a co-culture with *Clostridium butylicum* for acetone' butanol' and ethanol production from cassava starch. *Biochem Eng J* 2010;**48**:260–7.

22. Nakayama S, Kiyoshi K, Kadokura T, Nakazato A. Butanol production from crystalline cellulose by cocultured *Clostridium thermocellum* and *Clostridium saccharoperbutylacetonicum* N1-4. *Appl Environ Microbiol* 2011;**77**:6470–5.

23. Wen Z, Wu M, Lin Y, Yang L, Lin J, Cen P. Artificial symbiosis for acetone-butanol-ethanol (ABE) fermentation from alkali extracted deshelled corn cobs by co-culture of *Clostridium beijerinckii* and *Clostridium cellulovorans*. *Microb Cell Fact* 2014;**13**:92.

24. Minty JJ, Singer ME, Scholz SA, Bae CH, Ahn JH, Foster CE, et al. Design and characterization of synthetic fungal-bacterial consortia for direct production of isobutanol from cellulosic biomass. *Proc Natl Acad Sci USA* 2013;**110**:14592–7.

25. Bayer TS, Widmaier DM, Temme K, Mirsky EA, Santi DV, Voigt CA. Synthesis of methyl halides from biomass using engineered microbes. *J Am Chem Soc* 2009;**131**:6508–15.

26. Arai T, Matsuoka S, Cho HY, Yukawa H, Inui M, Wong SL, et al. Synthesis of *Clostridium cellulovorans* minicellulosomes by intercellular complementation. *Proc Natl Acad Sci USA* 2007;**104**:1456–60.

27. Tsai S-L, Goyal G, Chen W. Surface display of a functional minicellulosome by intracellular complementation using a synthetic yeast consortium and its application to cellulose hydrolysis and ethanol production. *Appl Environ Microbiol* 2010;**76**:7514–20.

28. Goyal G, Tsai SL, Madan B, DaSilva NA, Chen W. Simultaneous cell growth and ethanol production from cellulose by an engineered yeast consortium displaying a functional mini-cellulosome. *Microb Cell Fact* 2011;**10**:89.
29. Bokinsky G, Peralta-Yahya PP, George A, Holmes BM, Steen EJ, Dietrich J, et al. Synthesis of three advanced biofuels from ionic liquid-pretreated switchgrass using engineered *Escherichia coli. Proc Natl Acad Sci USA* 2011;**108**:19949–54.
30. Trinh CT, Unrean P, Srienc F. Minimal *Escherichia coli* cell for the most efficient production of ethanol from hexoses and pentoses. *Appl Environ Microbiol* 2008;**74**:3634–43.
31. Eiteman MA, Lee SA, Altman E. A co-fermentation strategy to consume sugar mixtures effectively. *J Biol Eng* 2008;**2**:3.
32. Eiteman MA, Lee SA, Altman R, Altman E. A substrate-selective co-fermentation strategy with *Escherichia coli* produces lactate by simultaneously consuming xylose and glucose. *Biotechnol Bioeng* 2009;**102**:822–7.
33. Xia T, Eiteman MA, Altman E. Simultaneous utilization of glucose, xylose and arabinose in the presence of acetate by a consortium of *Escherichia coli* strains. *Microb Cell Fact* 2012;**11**:77.
34. Kerner A, Minty JJ, Kistler S, Singer ME, Faulkner I, Balan V, et al. A synthetic *E. coli* consortium for efficient conversion of lignocellulose-derived hexose and pentose sugars to isobutanol, submitted for publication.
35. Goldman RP, Brown SP. Making sense of microbial consortia using ecology and evolution. *Trends Biotechnol* 2009;**27**:3–4.
36. Zuroff TR, Curtis WR. Developing symbiotic consortia for lignocellulosic biofuel production. *Appl Microbiol Biotechnol* 2012;**93**:1423–35.
37. Shong J, Jimenez Diaz MR, Collins CH. Towards synthetic microbial consortia for bioprocessing. *Curr Opin Biotechnol* 2012;**23**:798–802.
38. Balagaddé FK, Song H, Ozaki J, Collins CH, Barnet M, Arnold FH, et al. A synthetic *Escherichia coli* predator-prey ecosystem. *Mol Syst Biol* 2008;**4**:187.
39. Regot S, Macia J, Conde N, Furukawa K, Kjellén J, Peeters T, et al. Distributed biological computation with multicellular engineered networks. *Nature* 2010;**469**:207–11.
40. Prindle A, Samayoa P, Razinkov I, Danino T, Tsimring LS, Hasty J. A sensing array of radically coupled genetic/biopixels. *Nature* 2011;**481**:39–44.
41. Weber W, Daoud-El Baba M, Fussenegger M. Synthetic ecosystems based on airborne inter-and intrakingdom communication. *Proc Natl Acad Sci USA* 2007;**104**:10435–40.
42. Arkin AP, Fletcher DA. Fast, cheap and somewhat in control. *Genome Biol* 2006;**7**:114.
43. Balagaddé FK, You L, Hansen CL, Arnold FH, Quake SR. Long-term monitoring of bacteria undergoing programmed population control in a microchemostat. *Science* 2005;**309**:137–40.
44. Bernstein HC, Paulson SD, Carlson RP. Synthetic *Escherichia coli* consortia engineered for syntrophy demonstrate enhanced biomass productivity. *J Biotechnol* 2012;**157**:159–66.
45. Kerner A, Park J, Williams A, Lin XN. A programmable *Escherichia coli* consortium via tunable symbiosis. *PLoS One* 2012;**7**:e34032.
46. Shou W, Ram S, Vilar JM. Synthetic cooperation in engineered yeast populations. *Proc Natl Acad Sci USA* 2007;**104**:1877–82.
47. Gore J, Youk H, van Oudenaarden A. Snowdrift game dynamics and facultative cheating in yeast. *Nature* 2009;**459**:253–6.
48. Chignell J, Reardon KF, Park S. Label-free, strain-resolved, shotgun proteomics of a defined bacterial co-culture. In: *2013 AIChE annual meeting, November 8, 2013, San Francisco, CA*. 2013.

CHAPTER 19

A Route from Biomass to Hydrocarbons via Depolymerization and Decarboxylation of Microbially Produced Polyhydroxybutyrate

Heidi Pilath[1], Ashutosh Mittal[2], Luc Moens[1], Todd B. Vinzant[2], Wei Wang[2], David K. Johnson[2]

[1]*National Bioenergy Center, National Renewable Energy Laboratory (NREL), Golden, CO, USA;*
[2]*Biosciences Center, National Renewable Energy Laboratory (NREL), Golden, CO, USA*

Introduction

To convert biomass components into hydrocarbon fuels, it is obvious that there are two main transformations that need to occur, deoxygenation and carbon chain extension. As shown in Figure 1 the heating values of components that can be made from biomass have much lower heating values than gasoline, jet, and diesel fuel components, and this is, of course, because of the higher oxygen contents of the biomass components. In addition, the majority of the biomass components have six carbons or fewer in their structures because they are made from the hexoses and pentoses present in the cellulose and hemicellulose in biomass, and so have fewer carbons than the hydrocarbon components in gasoline, jet, and diesel fuels. The potential routes for decreasing the oxygen content of biomass intermediates include dehydration, hydrodeoxygenation, and decarboxylation. Figure 2 compares the H/C ratios and O/C ratios for bioproducts to those of hydrocarbon components in gasoline, jet, and diesel fuels. The orange arrows in Figure 2 indicate the direction in which these ratios would change for the three deoxygenation reactions. Conversion of glucose into polyhydroxybutyrate (PHB) and then propene is a pathway that uses the decarboxylation route to produce a hydrocarbon. Because the hydrocarbon that is formed is an alkene, it can be oligomerized into hydrocarbons with chain lengths that are suitable for blending into any of the major transportation fuels. Patents[2,3] already exist for converting propene into trimers and tetramers that would make suitable gasoline and jet fuel components, and with further research, it should be possible to make hexamers that would be suitable for diesel fuel.

The conversion of polyhydroxyalkanoates (PHA) to alkenes appears promising because there are several microorganisms that incorporate high levels of PHA (up to 80% of dry cell

Direct Microbial Conversion of Biomass to Advanced Biofuels. http://dx.doi.org/10.1016/B978-0-444-59592-8.00019-1

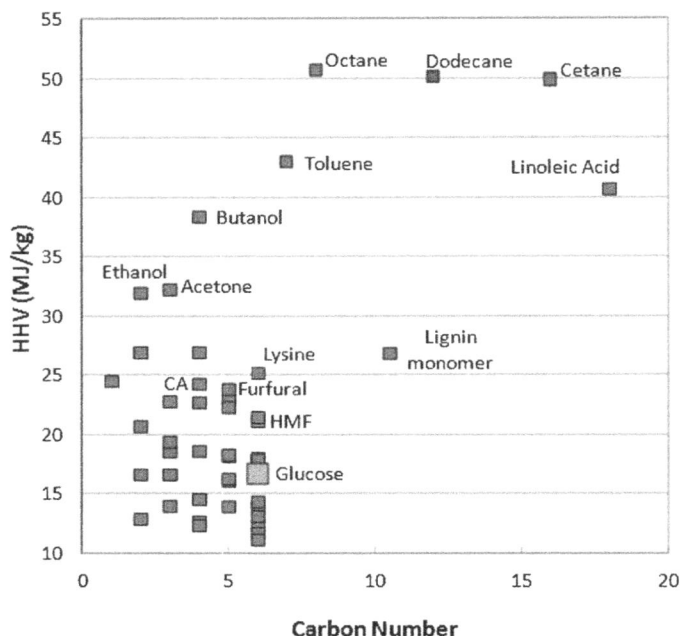

Figure 1
Heating value and carbon number of glucose (green square (Light grey in print version)) and biomass components (blue squares (dark grey in print version)) compared to gasoline, jet, and diesel fuel components (red squares (black in print version)). The biomass components were mostly those described in the report on the top value-added chemicals from Biomass Volume I.[1]

mass) as a form of energy storage molecule to be metabolized when other common energy sources are not available. PHAs are synthesized by more than 140 different genera of Eubacteria and a few members of the Halobactericeae family of Archaea[4,5] and can have alkyl chains ranging from 3 to 16 carbons long. These microorganisms synthesize PHAs and deposit these polyesters as cytoplasmic granules. These PHA-producing microorganisms have been found in a range of environments, including soil, marine, sewage, microbial mats, and groundwater aquifers.[6,7] Among the aliphatic polyesters, PHAs are one of the most attractive due to their excellent thermoplastic-like physical properties and inherent biodegradability. This ability makes them a viable green substitute for current fossil fuel-based thermoplastics (e.g., polypropylene) that are recalcitrant to any type of microbial action and can persist in the landscape for decades. Rehm et al. showed that the composition of PHA depends to a large extent on PHA synthases, the carbon source, as well as the metabolic routes.[8]

As potential feedstocks for fuels, PHAs have the advantage over sugars of having a lower degree of oxygenation. Our research is aimed at determining the most efficient routes for converting PHA into liquid fuels. To that end, we are working on the development of

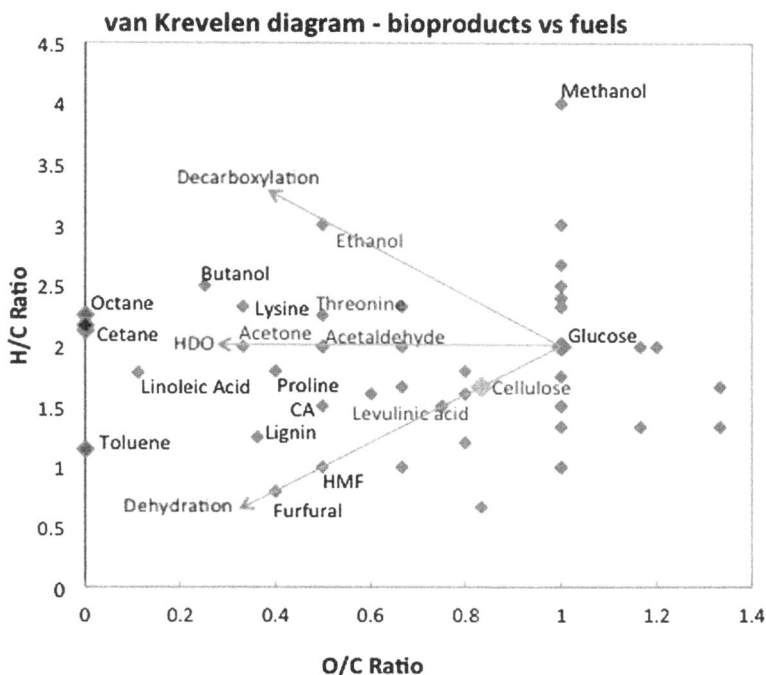

Figure 2

van Krevelen diagram of glucose and cellulose (green squares (light grey in print version)), bioproducts (blue squares (black in print version)) and hydrocarbon fuel components (red squares (dark grey in print version)). Orange arrows indicate the direction of change in H/C and O/C ratios for products of glucose that undergo decarboxylation, hydrodeoxygenation (HDO), and dehydration reactions. The biomass components were mostly those described in the report on the top value-added chemicals from Biomass Volume I.[1]

chemical transformation routes that efficiently convert PHA into fuel products. The majority of our work has focused on the PHA polyhydroxybutyrate (PHB), which is a short-chain length (scl)-PHA consisting of a C_4 repeat unit including a methyl group side chain. It is one of only a few commercially available PHAs.

Thermal breakdown of PHA proceeds via an intermediate carboxylic acid, which can then be decarboxylated to an alkene. Literature in the field of thermochemical conversion of biopolyesters into smaller fragments is sparse.[9–11] Literature on fatty acid decarboxylation relevant to biofuels is also limited. Heterogeneous catalysts, platinum and palladium on activated carbon (Pt/C and Pd/C), have been used in water at near critical conditions to decarboxylate lipid fatty acids in good yields (>90%).[12,13] Watanabe et al. demonstrated that metal oxide catalysts (CeO_2, Y_2O_3, ZrO_2) enhance the decarboxylation of the fatty acids to alkenes, whereas alkali hydroxides (NaOH and KOH) enhanced the decarboxylation of some fatty acids to alkanes.[14] Fischer et al. demonstrated the direct production of propylene, a basic industrial building block, via the hydrothermal decarboxylation/reforming of PHB without the need for an external hydrogen source with yields approaching 50% (mol/mol) of propene.[15] However, all these (hydrothermal) decarboxylation reactions were conducted in water at temperatures up to 400 °C (subcritical to supercritical conditions) and pressures approaching 4000 psi (>25 MPa). It has been our goal to demonstrate the production of propene without the use of water, which would eliminate the high pressures observed in the above work, so that scale-up could become a realistic possibility.

The overall reaction pathway for converting glucose to hydrocarbons via PHB and CA is shown below as Scheme 1:

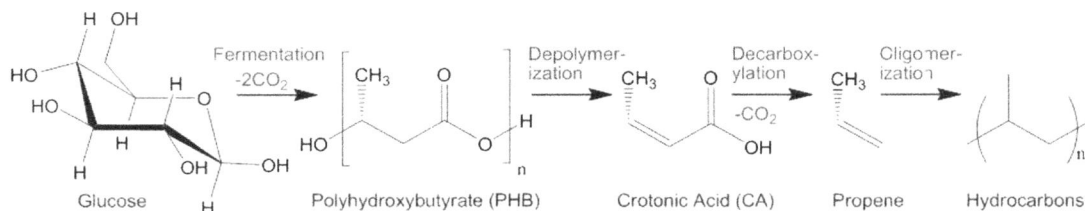

Scheme 1
Proposed route for transforming sugars into PHB and then into a hydrocarbon fuel component.

Experimental Section
Chemicals and Catalysts

CA was purchased from Sigma–Aldrich with a purity of 98%. Poly(3-hydroxybutyric acid) was also purchased from Sigma–Aldrich (no purity claim).

Stainless Steel Tube Reactor

Reactions were performed using a stainless steel tube reactor with a measured internal volume of 74 mL that was heated in a fluidized sand bath. Two shutoff valves were connected to the top of the reactor, one attached to a pressure gauge that allowed measurement of the postreaction pressure, and the other connected directly to sampling bags (SKC-West Inc., Houston, TX) to collect the produced gases. A thermocouple was inserted through the top of the reactor and permitted measurement of the temperature within the reactor. The reactant was added to an open glass tube, which was inserted into the reactor. The steel reactor was sealed and then leak tested using an inert gas to verify a good seal prior to the reaction. Any air was removed from the reactor by pressurization and depressurization with the inert gas. The reactor was submerged into the fluidized sand bath for a predetermined time and temperature. There was no agitation of the mixture during the reaction. At the end of each experiment, the reactor was rapidly quenched in an ice water bath until it reached room temperature. At this point, the pressure in the reactor and temperature were noted, and then the generated gas was collected in a gas sampling bag for subsequent GC analysis. Any product remaining after the reaction was washed from the reactor with methanol, including anything coating the inside of the reactor and the outside of the glass tube. The wash was evaporated to dryness with nitrogen to get the weight of the fraction and to prepare it for analysis.

Gas Analysis

Gas samples were analyzed by gas chromatography using an Agilent 7890A GC system equipped with a GS-Q column (J&W, length 30 m, ID 0.53 mm). The injection temperature was 250 °C. Helium served as the carrier gas. The temperature of the column oven was increased at a rate of 20 °C/min from a temperature of 35 °C (held initially for 3 min) to 135 °C. The concentrations of propene and CO_2 were quantified by use of an external standard (mixture of 1% propene and 1% CO_2 in N_2; Air Liquide). Various volumes of the mixture were injected into the GC to establish a six-point calibration curve. A secondary standard (Air Liquide, 3% propene and 3% CO_2 in N_2) was used to verify the initial calibration. The mole fraction of each gaseous product was determined from the gas concentrations, and the volume of gas injected on the GC. The molar yield of each gas was calculated from this mole fraction multiplied by the total moles of gas produced divided by the initial moles of CA used in each reaction.

PHB Analysis

The PHB content of bacterial cells was determined by a quantitative method that uses HPLC analysis to measure the quantity of CA formed by acid-catalyzed depolymerization of PHB.[16] Briefly, PHB-containing dried bacterial cells (15–50 mg) were digested in 96% H_2SO_4 (1 mL) at 90 °C for 1 h. The reaction vials were then cooled on ice, after which, ice-cold 0.01 N H_2SO_4 (4 mL) was added followed by rapid mixing. The samples were further diluted 20- to 150-fold with 0.01 N H_2SO_4 before analysis by HPLC.

The concentration of CA was measured at 210 nm using an HPLC equipped with a photodiode array detector (Agilent 1100, Agilent Technologies, Palo Alto, Calif.). A Rezex RFQ Fast Acids column (100×7.8 mm, 8 µm particle size, Phenomenex, Torrance, Calif.) and Cation H+ guard column (BioRad Laboratories, Hercules, Calif.) operated at 85 °C were used to separate the CA present in the reaction solutions. The eluant was 0.01 N H_2SO_4 at a flow rate of 1.0 mL/min. Samples and CA standards were filtered through 0.45 µm pore size nylon membrane syringe filters (Pall Corp., East Hills, NY) prior to injection onto the column. The HPLC was controlled and data was analyzed using Agilent ChemStation software (Rev.B.03.02).

The amount of unreacted PHB remaining in the methanol reactor wash could be measured by the same method. Unreacted CA could be quantified by direct analysis using the HPLC method.

Microbial Production of PHB

Strains

Cupriavidus necator DSM 428, DSM 542 and DSM 541 were purchased from the European DSMZ Culture Collection. The strains were maintained in 40% glycerol at −80 °C.

Cultivation of strains in shake flasks

For the seeding culture, the strains were first grown in 10 mL of Lysogeny Broth (LB) in 100 mL baffled flasks, cultured at 160 rpm and 28 °C. After 1 day, a 10% inoculum from the first seeding culture was transferred to the second LB seeding culture with a volume of 100 mL and cultured under the same conditions as above. After 1 day, a 10% inoculum was added to 1.0 L of PHB fermentation media in a 2.8 L flask and incubated in a shaker at 28 °C and 160 rpm for 5 days. PHB fermentation media were composed of different carbon sources fructose, glucose, or fructose plus glycerol, at 10, 20, or 40 g/L with 1.0 g/L NH_4Cl and 0.5 g/L yeast extract. Two milliliters aliquots of broth were taken daily for cell dry weight (CDW) measurement and PHB analysis.

Fermentation in 5-L bioreactor

Cupriavidus necator DSM 428 PHB production was carried out in a 5-L BIOFLO 3000 (New Brunswick Scientific, Edison, NJ) batch bioreactor. A seed culture was inoculated from a single colony into 50 mL of culture media (pH 7.0), incubated at 28 °C and 200 RPM, and then transferred after 24 h of incubation into 400 mL of fresh media. The secondary seed culture was subsequently transferred into the NBS BIOFLO 3000 bioreactor (4.0 L), where the fermentation was controlled at 28 °C, 30% OD, 200 RPM, and pH 7.0. Thirty-five milliliters samples were taken at regular time intervals over a period of 120 h for analysis of optical density (OD600), offline pH, CDW, and PHB content. Optical density was measured in the linear range of a Cary 4000 spectrophotometer, and the CDW was determined by freeze drying the cell pellet from each time point. These data were used to determine bacterial growth rate, and grams PHB yield/gram cell dry weight.

Results and Discussion

Thermal Decarboxylation of Crotonic Acid

Initially our goal was to find conditions that would give high molar yields of propene from the thermal treatment of CA. Decarboxylation of CA was found to occur in high yield (65–80%) by thermal treatment at 400 °C for 15 min without catalyst or solvent. Similar yields of decarboxylation products were also observed in 10 min at 450 °C. At 350 °C, propene and CO_2 yields increased from 20 to 60% as reaction time increased from 15 to 60 min (Figure 3). At 300 °C, CA decarboxylation products were not observed until 60 min reaction time (yields <10%), and none was observed at 250 °C.

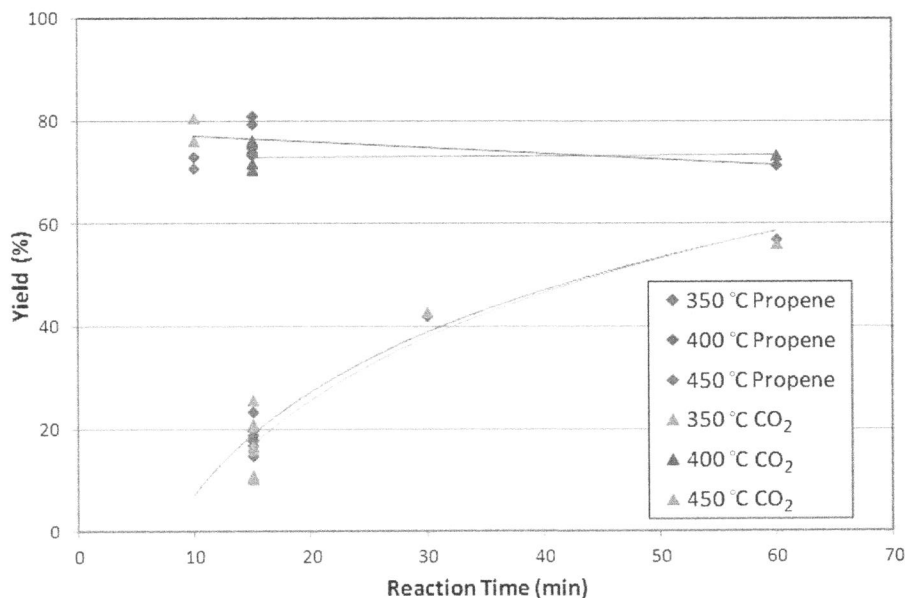

Figure 3
Effect of time and temperature on formation of CA decarboxylation products.

Thermal Depolymerization and Decarboxylation of Commercial PHB

Earlier results had shown that PHB could be thermally depolymerized at 280 °C producing CA monomer, and linear oligomers (dimer and trimer) having crotonyl end-groups. No higher oligomers were noted. This showed that depolymerization of PHB occurs below 300 °C without the use of a catalyst.

Our next goal was to demonstrate that the depolymerization and decarboxylation steps could be combined so that PHB could be simply converted to propene and CO_2 in a single step. The experiments starting with PHB proceeded similarly to those using CA. As can be seen from the results in Table 1, similar yields of propene and CO_2 were obtained from PHB as had been obtained from CA. These reactions were performed as before without the use of a catalyst or solvent, just by thermal treatment. Much of our research has focused on the decarboxylation step because this is believed to be the rate limiting step, but these results show that the depolymerization and decarboxylation steps can be combined so that propene can be produced directly from PHB polymer.

Microbial Production of PHB

PHAs are synthesized by microorganisms to serve as intracellular storage molecules for carbon resulting from the limitation of a nutrient, such as nitrogen, phosphate, etc.[17] In some

Table 1: Thermal treatment of CA and PHB (15 min reaction time; at least five replicate reactions)

Substrate	Temperature (°C)	Pressure (psig)[a]	Propene Yield (molar%)	CO₂ Yield (molar%)
CA	350	23 ± 3	19 ± 3	15 ± 7
CA	400	76 ± 2	74 ± 6	74 ± 3
PHB	350	24 ± 1	22 ± 3	20 ± 2
PHB	400	71 ± 7	70 ± 7	78 ± 6

[a]Autogenic pressure measured at 25 °C after thermal treatment.

bacterial strains, PHA can account for up to 80% of cell mass.[18] PHAs are synthesized by many types of living organisms mainly bacteria, including *Cupriavidus necator* (formerly known as *Ralstonia eutropha*), *Bacillus* spp., *Pseudomonas* spp., *Rhizobium* spp., etc.[19,20] Accordingly, the carbon sources metabolized by these strains also vary and include fructose, glucose, sucrose, fatty acids, glycerol, molasses, etc. Among the strains, *C. necator* has been most extensively studied and seems to be the most effective organism for producing PHB on an industrial scale.[21] To test the concept that whole cell mass containing PHB could be directly converted to propene, we chose to use the strain, *C. necator* DSM 428 and its derivatives, which have high growth rates and can accumulate high PHB contents, up to 70 to 80%.

Initially *Cupriavidus necator* DSM 428 was used to study cell growth and PHB production in shake flasks. By doing so, a general picture of the fermentation of this strain was obtained. *C. necator* DSM 428 metabolized fructose well; however, it was found to only take up glucose slowly. To use biomass-derived sugars, another strain *C. necator* DSM 542, a mutant of DSM 428, was introduced to try to get high cell density and PHB content in a shake flask using glucose as substrate. In addition, to check if cell components other than PHB could have some background contribution to propene and CO₂ production, another mutant of *C. necator* DSM 428, i.e., *C. necator* DSM 541, which is a PHB-synthesis negative strain, was also cultured and the cell mass was collected to be used as a control in thermal conversion reactions.

Production of PHB in C. necator *DSM 428 using a 5-L bioreactor*

Initially production of PHB from *C. necator* DSM 428 was studied in shake flasks using fructose (40 g/L) as the carbon source. It was found that PHB could be successfully grown on fructose, yielding cell mass at about 1.5 g/L with PHB content of ~65% (dry basis) after 96 h.

To achieve a higher cell density and collect additional cell mass with high PHB content for use in thermal conversion reactions, the fermentation was scaled up in a 5-L bioreactor. Cell growth and PHB accumulation were determined and are shown in Figure 4.

Although this fermentation was not optimized, it is clear that *C. necator* DSM 428 produces PHB from fructose with a high PHB content (up to 75% of the total cell mass) making it a realistic candidate for further study. A key element will be to produce PHB on process relevant carbohydrate streams such as glucose and xylose.

5-L Fermentor Results

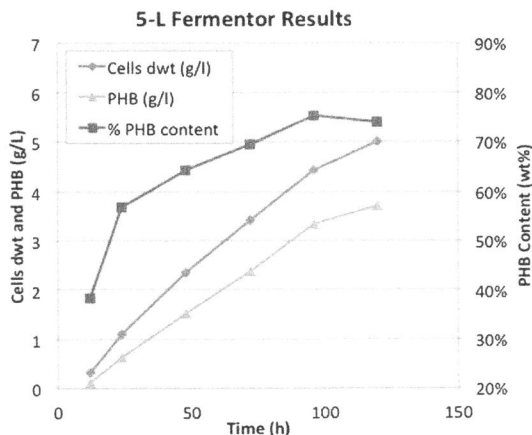

Figure 4

Production of PHB in *Cupriavidus necator* DSM 428 using fructose as the carbon source in a 5.0-L fermentor.

Table 2: Growth of *Cupriavidus necator* DSM 428 and DSM 542 on fructose and glucose at 48 h

	Cell Growth (OD$_{600}$)	
Strains	**On Fructose**	**On Glucose**
C. necator DSM 428	5.02	0.54
C. necator DSM 542	4.46	2.69

Fermentation of C. necator *DSM 541 bacterial control*

As PHB is only produced when there is an excess of carbon source *C. necator* DSM 541, was grown with solutions containing only 1% and 2% fructose solutions. Under these conditions, cell mass accumulated with zero PHB storage, giving the desired background control for the PHB thermal conversion reactions.

Fermentation of C. necator *DSM 542 using glucose as substrate*

Cupriavidus necator DSM 542 is a mutant of *C. necator* DSM 428, engineered to use glucose. However, the uptake of glucose by this mutant was obviously slower than fructose uptake, as the 48 h growth test indicated (see Table 2). Since our goal is to use biomass-derived sugars to produce PHB, our focus will be on *C. necator* DSM 542 and other glucose-using strains, such as *Alcaligenes eutrophus* NCIMB 11599. It has been reported that this glucose-using strain can produce high concentrations of PHB (121 g/L), with a maximum PHB content reaching 76% of dry cell weight. The productivity was reported as 2.42 g/L h and a yield of 0.3 g PHB/g of glucose (63% yield on a stoichiometric basis).[22]

Table 3: Thermal treatment of PHB-containing *Cupriavidus necator* (15 min reaction time)

Substrate	Cell Wt (g)	PHB Content (%)	Temperature (°C)	Pressure (psig)	Propene Yield (molar%)	CO_2 Yield (molar%)[a]
Negative Control—Cells with Zero PHB						
Cupriavidus 541 (shake flask)	0.50	0	400	7	0	0.71 µg/0.5 g
PHB-Containing Cells						
Cupriavidus 428 (shake flask)	1.0	56	350	17	14	52
Cupriavidus 428 (shake flask)	0.64	38	400	21	61	133
Cupriavidus 428 (5-L fermentor)	0.50	65	400	28	69	118

[a]Yield based on PHB content of cell mass, except for the negative control experiment, in which the yield was based on cell weight because the control did not contain PHB.

Thermal Depolymerization and Decarboxylation of PHB Containing Bacterial Cells

For a process converting PHB to propene to have any chance of being economically viable, it is essential that the process work not only on isolated PHB, but also on the PHB in whole cells. Isolation of PHB from the bacterial cell mass prior to the thermal conversion reaction would likely make the process too expensive for commercial production of propene. As can be seen in Table 3, significant yields of propene were obtained by thermal treatment of *C. necator* cells containing PHB. Thermal experiments on *C. necator* 428 containing various amounts of PHB resulted in the production of propene and CO_2. At 350 °C, yields of 14% propene and 52% CO_2 were obtained in 15 min. At 400 °C, *C. necator* 428 gave propene yields of 60 to 70% and CO_2 yields more than 100%, based on the PHB content of the cells. The propene yields were only slightly lower than were obtained with pure CA or PHB, whereas the CO_2 yields were much higher, probably due to the conversion of other cell components. A negative control was treated thermally at the same conditions, and this produced a significant amount of CO_2 but no propene, confirming that all propene generated comes from PHB and that other cell components do not breakdown to give propene, but will breakdown to give CO_2.

Conclusions

It has been demonstrated that CA, the depolymerization product from PHB, can be effectively decarboxylated to propene at molar yields of about 70%, using a relatively simple thermal treatment at 400 °C for 15 min. This has been accomplished by direct treatment of solid CA without use of water or another solvent, and without use of a catalyst. Pressures (about 180 psi at the reaction temperature) generated by this process are just those of the gases, propene and CO_2, produced by the process. Furthermore, we have demonstrated that PHB depolymerization

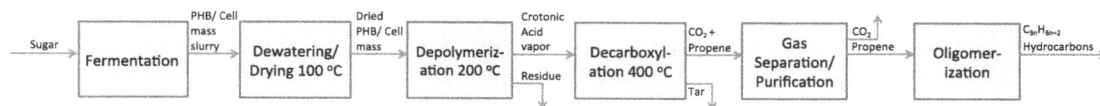

Figure 5
Proposed process flow diagram for converting sugars into hydrocarbons via PHB.

to CA can be combined with CA decarboxylation so that propene is produced in similar yields by direct thermal treatment of pure PHB at 400 °C. Furthermore *C. necator* containing PHB has been produced and used to test the direct conversion of PHB in bacterial cells. The yields (60–70 molar %) of propene from whole cells were slightly lower than from CA or pure PHB, but still acceptable. Importantly, propene was obtained from whole cells demonstrating that an expensive process step to isolate PHB from the cells was unnecessary.

Figure 5 shows a proposed process flow diagram for the production of hydrocarbons from sugars via a route that first involves the conversion of sugars into PHB that is then depolymerized and decarboxylated to propene. A critical task for the further development of this process will be the technoeconomic evaluation of the proposed process, to determine if it is feasible to produce hydrocarbons at a reasonable cost by this route and to determine the critical technical targets that need to be achieved to meet the desired cost goal.

Acknowledgment

We are grateful to the US Department of Energy, EERE Bioenergy Technologies Office (BETO), for the continued support of our research.

References

1. Aden A, Bozell J, Holladay J, White J, Manheim A. In: Werpy T, Peterson G, editors. 2004.
2. US Patent 6649802. 2003.
3. US Patent 7649123. 2010.
4. Steinbuchel A, Fuchtenbusch B. Bacterial and other biological systems for polyester production. *Trends Biotechnol* 1998;**16**:419–27.
5. Rehm BHA. Biogenesis of microbial polyhydroxyalkanoate granules: a platform technology for the production of tailor-made bioparticles. *Curr Issues Mol Biol* 2007;**9**:41–62.
6. Snell KD, Peoples OP. PHA bioplastic: a value-added coproduct for biomass biorefineries. *Biofuels Bioprod Biorefin* 2009;**3**:456–67.
7. Villanueva L, Del Campo J, Guerrero R. Diversity and physiology of polyhydroxyalkanoate-producing and -degrading strains in microbial mats. *FEMS Microbiol Ecol* 2010;**74**:42–54.
8. Rehm BHA, Mitsky TA, Steinbuchel A. Role of fatty acid de novo biosynthesis in polyhydroxyalkanoic acid (PHA) and rhamnolipid synthesis by pseudomonads: establishment of the transacylase (PhaG)-mediated pathway for PHA biosynthesis in *Escherichia coli. Appl Environ Microbiol* 2001;**67**:3102–9.
9. Ariffin H, Nishida H, Hassan MA, Shirai Y. Chemical recycling of polyhydroxyalkanoates as a method towards sustainable development. *Biotechnol J* 2010;**5**:484–92.
10. Aoyagi Y, Yamashita K, Doi Y. Thermal degradation of poly (R)-3-hydroxybutyrate, poly epsilon-caprolactone, and poly (S)-lactide. *Polym Degrad Stabil* 2002;**76**:53–9.

11. Li SD, He JD, Yu PH, Cheung MK. Thermal degradation of poly(3-hydroxybutyrate) and poly(3-hydroxybu-tyrate-co-3-hydroxyvalerate) as studied by TG, TG-FTIR, and Py-GC/MS. *J Appl Polym Sci* 2003;**89**:1530–6.

12. Fu J, Lu X, Savage PE. Catalytic hydrothermal deoxygenation of palmitic acid. *Energy Environ Sci* 2010;**3**:311–7.

13. Fu J, Lu X, Savage PE. Hydrothermal decarboxylation and hydrogenation of fatty acids over Pt/C. *Chemsus-chem* 2011;**4**:481–6.

14. Watanabe M, Iida T, Inomata H. Decomposition of a long chain saturated fatty acid with some additives in hot compressed water. *Energy Convers Manag* 2006;**47**:3344–50.

15. Fischer CR, Peterson AA, Tester JW. Production of C(3) hydrocarbons from biomass via hydrothermal carboxylate reforming. *Ind Eng Chem Res* 2011;**50**:4420–4.

16. Karr DB, Waters JK, Emerich DW. Analysis of poly-β-hydroxybutyrate in *Rhizobium japonicum* bacteroids by ion-exclusion high-pressure liquid chromatography and UV detection. *Appl Environ Microbiol* 1983;**46**(6):1339–44.

17. Anderson AJ, Dawes EA. Occurrence, metabolism, metabolic role, and industrial uses of bacterial polyhy-droxyalkanoates. *Microbiol Rev* 1990;**54**:450–72.

18. Khanna S, Srivastava AK. A simple structured mathematical model for biopolymer (PHB) production. *Biotechnol Progr* 2005;**21**:830–8.

19. Chen GQ. A microbial polyhydroxyalkanoates (PHA) based bio-and materials industry. *Chem Soc Rev* 2009;**38**:2434–46.

20. Verlinden RAJ, Hill DJ, Kenward M, Williams CD, Radecka I. Bacterial synthesis of biodegradable polyhy-droxyalkanoates. *J Appl Microbiol* 2007;**102**:1437–49.

21. Du G, Chen J, Yu J, Lun S. Continuous production of poly-3-hydroxybutyrate by *Ralstonia eutropha* in a two-stage culture system. *J Biotechnol* 2001;**88**:59–65.

22. Kim BS, Lee SC, Lee SY, Chang HN, Chang YK, Woo SI. Production of poly (3-hydroxybutyric acid) by fed-batch culture of *Alcaligenes eutrophus* with glucose concentration control. *Biotechnol Bioeng* 1994;**43**:892–8.

Index

Lightning Source UK Ltd.
Milton Keynes UK
UKOW07n2127280615

254192UK00002B/36/P